Integrating Educational Technology into Teaching

M.D. Roblyer
Florida A&M University

Jack Edwards
The Webster School

Mary Anne Havriluk
Distance Learning Associates

Merrill,
an imprint of Prentice Hall
Upper Saddle River, New Jersey   Columbus, Ohio

Library of Congress Cataloging-in-Publication Data
Roblyer, M. D.
    Integrating educational technology into teaching / M.D. Roblyer,
Jack Edwards, Mary Anne Havriluk.
        p.   cm.
    Includes bibliographical references and index.
    ISBN 0-02-402608-5
    1. Educational technology—United States.   2. Computer-assisted
instruction—United States.   3. Curriculum planning—United States.
I. Edwards, Jack.   II. Havriluk, Mary Anne.   III. Title.
LB1028.3.R595   1997
371.3'078—dc20   96-14921

Cover art: Diana Ong/Superstock
Editor: Debra A. Stollenwerk
Project Management, Text Design, and Composition: Elm Street Publishing Services, Inc.
Design Coordinator: Julia Zonneveld Van Hook
Cover Designer: Raymond Hummons
Production Manager: Laura Messerly

This book was set in Times Roman by Elm Street Publishing Services, Inc., and was
printed and bound by The Banta Company. The cover was printed by Phoenix Color Corp.

© 1997 by Prentice-Hall, Inc.
Simon & Schuster/A Viacom Company
Upper Saddle River, New Jersey 07458

Printed in the United States of America

10 9 8 7 6

ISBN: 0-02-402608-5

Prentice-Hall International (UK) Limited, *London*
Prentice-Hall Of Australia Pty. Limited, *Sydney*
Prentice-Hall of Canada, Inc., *Toronto*
Prentice-Hall Hispanoamericana, S. A., *Mexico*
Prentice-Hall of India Private Limited, *New Delhi*
Prentice-Hall of Japan, Inc., *Tokyo*
Simon & Schuster Asia Pte. Ltd., *Singapore*
Editora Prentice-Hall do Brasil, Ltda., *Rio de Janeiro*

To Bill Wiencke for making this book possible and to Paige Wiencke for her sweet distractions

To Robert W. Edwards, in memory of Mary E. Edwards, and to Jordan M. Burke

To Rod and Neely Havriluk

Preface

... of the computer, we ask more. We ask not just about where we stand in nature, but about where we stand in the world of artifact. We search for a link between who we are and what we might create, between who we are and what, through our intimacy with our own creations, we might become.
*Sherry Turkle, **The Second Self: Computers and the Human Spirit** (1983)*

The beginning of the information age may not have seemed much like a renaissance. By today's standards, the first computers were more like something from the ice age, as huge and slow-moving as glaciers, and in their refrigerated rooms, just as cold. Things have heated up and accelerated considerably since then. Already we have been through several generations of computers, a microcomputer revolution, and the birth of an electronic superhighway. It is a new age, a time of exciting discoveries and unexpected challenges. Those of us who remember the ice age sometimes feel as though we awakened suddenly to find the glaciers changed to rockets screaming at us at full throttle, racing around us, driving us. Life has become both as simple as we hoped and more complicated than we could have imagined.

Everything has changed so quickly, in fact, in education and in society at large, that it is often difficult to determine just what is happening and what response is required of us. We teachers stand before technology as we would a mirror. What we see is determined largely by what we are and what we consider important. We may see technology as a creature of speed, precision, and efficiency, able each day to squeeze more tasks into an hour of time than we could the day before. We may see an explorer and discoverer, able to lead others to see the images we see, to pursue the new worlds we envision. We may see something else, something novel and original, capable of making possible the unimagined and undreamt-of—an agent of reform, change, and progress.

The purpose of this book is to illustrate that technology products can help teachers to be all these things: efficiency expert, explorer, creator—the teacher as renaissance person. The authors of this text believe that the key to releasing this potential is to find practical, effective ways to integrate technology resources and technology-based methods into everyday practices, both inside and outside the classroom. Our approach to accomplishing this rests on three premises:

- **Integration methods should be based both in learning theory and teaching practice.** There seems no shortage of innovative ideas in the field of instructional technology. New and interesting methods come forward about as often as new and improved gadgets. We feel that those who would build on the knowledge of the past should know why they do what they do, as well as how to do it. Thus, we have linked various technology-based integration strategies to well-researched theories of learning, and we have illustrated them with examples of successful practices based on these theories.

- **Integration should match specific teaching and learning needs.** We feel technology has the power to improve teaching and learning, but it also can make a teacher's life more complicated. Therefore, we feel each resource should be examined for its unique qualities and its potential benefits for teachers and students. Teachers should not use a tool simply because it is new and available. Each integration strategy should be matched to a recognized need. We do not oppose experimentation, but we do advocate informed use.

• **Old integration strategies are not necessarily bad; new strategies are not necessarily good.** As technology products change and evolve at lightning speed, there is a decided tendency toward throwing out older teaching methods with the older machines. Sometimes this is a good idea; sometimes it would be a shame. Each of the integration strategies recommended in this book is based on methods with proven usefulness to teachers and students. Some of the strategies are based on directed methods that have been used for some time; some strategies are based on the newer, constructivist learning models. Each is recommended on the basis of its usefulness, rather than its age.

Some may say that this book leans too far toward the traditional technology uses of the past, focusing too much on efficiency and not enough on reform. Others may say it is too radical and recommends too much change too quickly. But, the authors feel that, like technology itself, this book acts only as a mirror. If readers look carefully at the descriptions and examples, they will perceive what they need to become; they will know what they need to do. As with each revolution and renaissance, what remains is the hard work.

Who Will Find This Book Helpful

This book is designed to help teach both theoretical and practical characteristics of technology integration strategies. It should be useful in several different types of education settings:

• **As a primary instructional material.** It should benefit instructional technology courses for preservice teachers and workshops and graduate courses for inservice teachers.

• **As a supplemental instructional material.** It should support research and content-area methods courses.

• **As a reference.** It should provide topical information in K–12 school libraries/media centers and university college of education library/media centers.

Chapter 1 establishes the authors' perspective on educational technology as a process rather than a set of devices. However, it also explains that the primary focus of the book is on a subset of educational technology: computer-related and information technologies and processes.

Organization of the Text

This text is organized into four sections—one of background and three of resources and applications:

Part I: Introduction and Background on Technology in Education. Einstein is said to have observed that "Everything should be made as simple as possible, but not more so." Using technology well becomes simpler when one understands the foundations upon which integration strategies are based—but that is no small task in itself. This section provides background on technology's role in education, reviews planning issues to be addressed prior to and during integration, and describes learning theories and teaching/learning models related to technology integration.

Part II: Using Instructional Software—Principles and Strategies. To paraphrase a popular jingle, "Software—it ain't just CAI anymore." This section describes more than 40 types of instructional software products ranging from drill and practice to integrated learning systems, from word processing to groupware. Each product description covers unique qualities, potential benefits, and sample integration strategies.

Part III: Using Technology Media and New Instructional Tools. Some of the most compelling technology resources are, with few exceptions, among the most complicated to learn and to implement. This section deals with these powerful and complex resources in four sections: the optical technologies, hypermedia authoring and production tools, "linking to learn" options ranging from e-mail to the Internet, and some of the technologies considered most futuristic based on their current and potential uses in education—artificial intelligence (AI), personal digital assistants (PDAs), and virtual reality (VR). As with Part II, example lesson plans or activities are given for each recommended integration strategy.

Part IV: Integrating Technology into the Curriculum. These five chapters describe and give examples of technology resources and integration strategies for five different areas: language arts and foreign languages, math and science, social sciences, the arts, and exceptional student education. While these chapters separate the areas into topics, the chapters themselves recognize and incorporate the current trends toward thematic, interdisciplinary instruction. Many of the examples cross discipline boundaries and serve to illustrate how the concepts of several content areas can be merged into a single lesson or learning activity—and how technology can support the process.

Special Features

Each chapter has the following features to help both the instructor and the student:

- **A list of descriptive topics and objectives.** This list appears at the beginning of each chapter.
- **Illustrative screens.** Figures show screen displays from software, media, and networks whenever possible.
- **Summary tables of important information.** These compilations aid recall and analysis.
- **Sample, teacher-designed lesson plans.** All from published sources, these materials match integration strategies.
- **Exercises.** End-of-chapter questions require both recall and application of concepts.
- **A list of sample resources.** References for further reading and a valuable resource list end each chapter.

Acknowledgments

The first thing that we would like to acknowledge is that this was a lot more work than we planned! When we began this task, each of us had about 20 years of experience in teaching and educational technology; we felt that we knew what we wanted to say and what would-be technology teachers needed to hear. We agreed on a short turnaround from drafts to final manuscript.

But as we started to write, the field changed around us; also, as we read and researched and worked together, we changed as well. We came to see concepts that we hadn't seen before, impediments we hadn't realized existed, and relationships among ideas that surprised and amazed us. We wrote and rewrote, conceptualized and reconceptualized.

The constructivists would say we generated our knowledge through authentic tasks. We agree, but we also had direction from some good teachers. We would like to acknowledge the contributions of many researchers and educators who shared their ideas in writing and in conversations. Jeanne Barker of the Maclay School and Bill Castine of the FAMU College of Education were especially patient and helpful in reading and commenting on early drafts. The students in FAMU Technology Applications in Education class, who used drafts of this text in their learning activities, were always most open and ready with their criticisms and suggestions.

Reviewers gave insightful and practical critiques and advice, all of which helped us clarify our prose and sharpen our focus: Morris I. Beers, SUNY College at Brockport; Donna Baumbach, University of Central Florida; Michelle Churma, Ashland University; Carol Dwyer, Pennsylvania State University; Mark Charles Fissel, Ball State University; Janet R. Handler, Mount Mercy College; Marianne Handler, National-Louis University; Mark A. Horney, University of Oregon; Scott D. Johnson, University of Illinois; Burga Jung, Texas Tech University; Gregory Sales, University of Minnesota; Janice R. Sandiford, Florida International University; and Robert Tennyson, University of Minnesota.

Colleagues like Donna Baumbach of the University of Central Florida, Eileen Pracek of the Brevard County School District, Janeen Clinton of Palm Beach County School District, Richard A. Smith of the Houston Independent School District, Sandra Hall and all the technology-using teachers at the Oak Ridge Elementary School, Harry Buerkle of the

ARC in Leon County, Norish Adams, Virginia Lawrence, and the faculty of the FAMU DRS, Debi Barett-Hayes, Dr. Martinez, and Connie Lane at the FSU Developmental Research School; Marilyn Comet of the Jostens Learning Corporation, and Anita Best of ISTE responded graciously and quickly to our requests for articles, sources, and advice. There is a saying in exceptional student education that "people don't care how much you know until they know how much you care." We have come to feel that the field of instructional technology and, indeed, education itself is fortunate to have such knowledgeable and caring professionals.

We must give special mention to Dr. Robert Gagne and Dr. Jerome Bruner, who were kind enough to send us updated photos. Also, we would like to acknowledge the assistance of many people whom we have never met but who took time from busy schedules to send a photo or give permission for a diagram. Among these are Yvonne Cekel, Transparent Language; Vanessa Dennen, Victor Maxx; Steve Goodman, Laureate Learning Systems; Gus Jackson, Pinnacle Micro; Michael Hageloh and Brigette Brooks, Apple Computer; Michele Hernandez and Jack Roberts, Scholastic; Billie Kingsley, Vanderbilt University; Cary Lafferty, Philips Consumer Electronics; Eve Nolan, National Geographic; Barbara Sistak, Don Johnson, Inc.; and Andre Rossi, Terrapin Software.

The enormous contributions of the Merrill editorial staff are impossible to measure. Editor Debbie Stollenwerk gave us the opportunity to share our ideas with others and supported us with wise counsel and endless encouragement. With skill and professionalism, support, editorial, and production team members (Penny Burleson, Carol Sykes, and Patty Kelly from Prentice Hall; and Phyllis Crittenden, Cathy Ferguson, Peter Kilander and Sue Langguth at Elm Street Publishing Services) made our ideas and words take on a form that readers would find useful.

Finally, we would like to thank our families for taking second place for so many weekends and holidays while we labored to give birth to this "child." Our own children learned in a most constructivist way what writing a book means in terms of time, work, and sacrifice.

M.D. Roblyer would like to give special recognition to Bill and Paige Wiencke for their enduring love and patience, and to Becky and Erin Kelley for their exemplary (and fast) library work. Jack Edwards would like to recognize the special support given by his father, Robert W. Edwards, his son, Jordan M. Burke, and colleagues Pat Andrews, Mary Lou Beverly, and Roger Coffee; and Mary Anne Havriluk would like to acknowledge her husband, Rod Havriluk, and daughter, Neely. Also, we would like to remember and acknowledge the contributions of those who are with us now only in memory: S.L. Roblyer, Raymond and Marjorie Wiencke, and Mary E. Edwards.

Finally, we would like to recognize the contributions of all the educators who have worked so long and so hard to develop the potential of technology in education. It's been an amazing evolution from the time of the hulking UNIVACs to the first 16K TRS-80 microcomputers to today's 1 gigabyte laptops. But the real wonder has never been the machines; it is the dedication, creativity, and tenacity of the educators who come into class every day with fresh energy and new ideas to make them work for students.

M.D. Roblyer
Tallahassee, Florida

Jack Edwards
St. Augustine, Florida

Mary Anne Havriluk
Tallahassee, Florida

Brief Contents

Contents

Part I

Introduction and Background on Integrating Technology in Education

The chapters in this part will help teachers learn:

1. How technology for education has evolved from its beginnings to its present day resources and applications

2. Issues and concerns that become important when implementing technology resources in schools and classrooms

3. How learning theories influence integration strategies

Introduction

"We Want to Be Ready ..."

About 15 years ago, around the time when microcomputers were first beginning to appear with some regularity in K-12 classrooms, one of the authors visited a middle school to see how two teachers were using some new purchases: two Apple II computers and instructional software, primarily very simple games and drills in mathematics. As these teachers demonstrated some of the programs and their classroom applications, they had to cope with a variety of technical problems.

Some of the software was designed for an earlier version of the Apple operating system, and each disk required a format adjustment each time it was used. Other programs would "hang" every now and then when students entered something the programmers hadn't anticipated; users would either have to adjust the code or restart the programs from the keyboard. The computer needed a small device to allow text to appear in both upper case and lower case on the screen, but this seemed to work with only some programs. In spite of these and other annoying problems, the teachers were very excited about their computer resources and talked with enthusiasm about their hopes, plans, and expectations.

"You guys are obviously doing a great job with your computers," the visitor said, "and I don't mean to seem negative about them. But this sure seems like an awful lot of trouble for what you get out of it. What motivates you to keep investing all the time you do?" Their answer was both instructive and prophetic. They said, "We know the time is coming when computers will be in all classrooms. Software will be better and equipment will be easier to use. When this time comes, we want to be among those prepared to use computers in teaching; we want to be ready."

As people today look at what technology is doing—and what it promises to do—in classrooms across the country, they see that those middle school teachers were right: What is happening now is worth the preparation. Computers and other technology resources have improved in capabilities and user-friendliness to educators. Some of the most innovative and promising practices in education today involve technology, and the promise of even more exciting capabilities foreshadow even greater benefits for teachers.

This book will present some of the most powerful and capable educational technology resources available today. It will also demonstrate how teachers can take advantage of this power and capability. Despite advances, being ready still requires some investment of time. This introductory section discusses the knowledge and skills that teachers need to prepare themselves to apply technology, especially computer technology, effectively in classrooms.

What Do Teachers Need to Be Ready for Technology?

In a field with such a wide range of powerful and complex tools, experts cannot help but disagree about what teachers need to know and even where they should begin. Not long ago, many experts advised teachers who wanted to become capable computer users learn to write programs in computer languages such as FORTRAN and BASIC. To become what was commonly called "computer literate," many assumed that teachers needed to know enough about the technical workings of computer systems to develop instructions for computers to follow.

Few people today believe that teachers need this much technical skill, but textbooks still provide wide varieties of information for beginning technology users. The background information in this section is based on the following steps that beginning technology users need to take:

- **Develop a philosophy.** Teachers must observe where current resources and types of applications fit in the history of the field. Then they must begin developing personal perspectives on the current and future role of technology in education—and in their own classrooms.

- **Purchase products.** Teachers must become informed, knowledgeable consumers of computer products and select wisely among available alternatives.

- **Identify problems.** Teachers must be able to troubleshoot computer systems they use frequently in order to discriminate between problems they can correct and those that will require outside help.

- **Speak the language.** Sufficient understanding of the terms and concepts related to technology allows users to exchange information with other teachers and experts and to ask and answer questions to expand their knowledge.

- **See where technology fits in education.** In perhaps the most important—and the most difficult—challenge, teachers must identify specific school activities where technology can help to improve existing conditions or to create important educational opportunities that did not exist without it. As part of this process, teachers decide what they need to make these changes occur. This process of determining where and how technology fits is known among users of educational technology as *integration*.

Required Background for Teachers

In this part, three chapters provide the information and skills that will help teachers accomplish their goals.

Chapter 1: Educational Technology— Evolution in Progress

Computer technology has nearly a 50-year history in education, and other kinds of technology have been in use for much longer. Obviously, classroom technology resources have changed dramatically over time. But a broad perspective of the field helps to illuminate many of today's concepts, terms, and activities. Chapter 1 will describe the history of computer resources and related applications of educational technology in order to show how they have evolved—and are still evolving—into the tools described later in this book. The chapter also provides a general overview of technology resources in education today and where computer technology fits into this picture.

Chapter 2: Planning and Implementation Issues for Effective Technology Integration

Educators must resolve many complex issues in order to apply technology solutions to educational problems. They must address many concerns before and during implementation to ensure that technology will have the desired effects on students and schools. These concerns range from funding to selection and placement of technology resources. Chapter 2 acts as a planning guide. It discusses each of these issues and recommends useful and practical steps that teachers may take to deal with each one.

Chapter 3: Learning Theories and Integration Strategies

The last chapter in this part provides an important link between learning and technology. It emphasizes the need to reach beyond the "nuts and bolts" of how technology resources work. Successful integration requires a connection between how people learn and how teachers employ technology to assist and enhance this learning. Chapter 3 begins with an overview of learning theories and related research findings, and it introduces two different perspectives on how to integrate technology into teaching and learning activities. These two perspectives are known as *directed* and *constructivist models*. Finally, Chapter 3 develops some specific integration strategies based on each of these models.

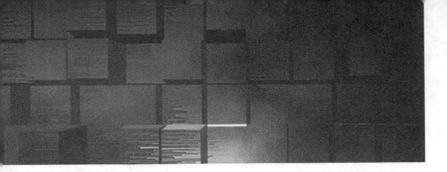

Chapter 1

Educational Technology— Evolution in Progress

This chapter will cover the following topics:

- Definitions and descriptions of the categories of technology in education

- A description of computer systems and how they work

- A brief history of computing technology in education

- An overview of configurations for computer-based systems used in education and their applications

Chapter Objectives

1. Given definitions of *educational technology* in common use in education, identify some historical background that led to the development of the definitions.

2. Identify the component parts of a computer system based on their functions: input, processing, output, I/O, internal (primary) memory or storage, external (secondary) memory or storage, and communications.

3. State which of these functions a given device performs.

4. Identify a computer program as systems software or applications software.

5. Describe the contributions of several people and projects to the development of current educational technology practices.

6. Classify educational technology products as stand-alone personal computers, networks, interactive video/multimedia applications, or virtual reality.

"...this author does not see any quick, revolutionary changes in educational technology in the remainder of this century. The more dramatic changes are likely to occur in the next century, but any changes in educational technology are likely to be evolutionary rather than revolutionary."
Paul Saettler, from *The Evolution of American Educational Technology* (1990).

When a classroom teacher clicks a mouse to activate a demonstration of a math principle or uses a bar-code reader to call up an image from a videodisc, that teacher is using some of the latest and best of what is commonly called *technology in education* or *educational technology*. Educational technology itself is not new at all, however, and it is by no means limited only to the use of equipment, let alone electronic equipment. Modern tools and techniques are simply the latest developments in a field that some believe is as old as education itself.

In his excellent, comprehensive historical description, *The Evolution of American Educational Technology,* Paul Saettler (1990) begins by pointing out that "Educational technology ... can be traced back to the time when tribal priests systematized bodies of knowledge, and early cultures invented pictographs or sign writing to record and transmit information.... It is clear that educational technology is essentially the product of a great historical stream consisting of trial and error, long practice and imitation, and sporadic manifestations of unusual individual creativity and persuasion" (p. 4).

This chapter explores the link between the early applications of educational technology and those of today and tomorrow. This exploration includes some historical and technical background. Many readers will grow impatient when they encounter these paragraphs of description and explanation and exhort the authors to "get on with the important stuff." This impatience is understandable and quite common in a field where the real excitement for teachers (as well as students) lies in hands-on exploration of the newest gadgets and techniques. We included this background, and we encourage you to read it, for three reasons:

• **Looking back before going ahead.** This information shows where the field is headed by demonstrating where it began. It points out the current status in the evolution of the technology of education along with changes in goals and methods over time. It provides a foundation on which to build ever more successful and useful structures to respond to the challenges of modern education.

• **Learning from past mistakes.** This background also helps those just embarking on their first applications of educational technology to make best use of their learning time by avoiding mistakes that others have made and by choosing directions that experience has shown to be promising.

• **Developing a "big picture."** Finally, this background helps new learners to develop mental pictures of the field, what Ausubel (1968) might call *cognitive frameworks* through which to view all applications—past, present, and future.

Technology in Education: Concepts and Definitions

Origins and Definitions of Key Terms

Teachers will probably see references to the terms *educational technology* and *instructional technology* in any professional journals they pick up. Perhaps no other topics are the focus of so much new development in so many content areas. It may seem easy to identify definitions for terms in such pervasive use, but in fact, no single, acceptable definition dominates the field. Even professional organizations provide conflicting definitions based on the perspectives and concerns of their members. Paul Saettler, a recognized authority on the history of the field, notes uncertainty even about the origins of the terms. The earliest reference he can confirm for the term *educational technology* was in an interview with W. W. Charters in 1948; the earliest known reference he finds for *instructional technology* was in a 1963 foreword written by James Finn for a technology development project sponsored by the National Education Association.

For many of today's teachers, any mention of technology in education immediately brings to mind the use of some device or a set of equipment, particularly computer equipment. Muffoletto (1994) says that "technology is commonly thought of in terms of gadgets, instruments, machines, and devices ... most (educators) will defer to technology as computers" (p. 25). Only about a decade ago, a history of technology in education since 1920 placed the emphasis on radio and television, with computers as an afterthought (Cuban, 1986).

In one sense, these views are correct, since any "definition" of state-of-the-art instruction usually mentions the most recently developed tools. But Saettler (1990) urges those seeking a precise use of the term to remember that "the historical function of educational technology is a *process* rather than a product. No matter how sophisticated the media of instruction may become, a precise distinction must always be made between the process of developing a technology of education and the use of certain products or media within a particular technology of instruction" (p. 4). Therefore, in the view of most writers, researchers, and practitioners in the field, any useful definition of *educational technology* must focus on the process of applying tools for educational purposes, as well as the tools and materials that are used. As Muffoletto (1994) puts it, "Technology ... is not a collection of machines and devices, but a way of acting" (p. 25).

In education, the combination of process and product merges instructional procedures with instructional tools. For the processes, or instructional procedures, guidance in the application of tools comes from learning theories based on the sciences of human behavior. The most modern of these tools, the electronic and computer-based ones, seem to present the most difficulty for teachers learning how to use them and integrate them into teaching.

Therefore, this textbook focuses primarily on the convergence of modern educational theory and practice with state-of-the-art electronic technology.

However, educators may want to become familiar with other perspectives on technology in education. Discussions of educational technology in the literature emerge from perspectives very different from the one in this book, and teachers want to know how these views arose and in what contexts they may be useful and valid.

Technology in Education as Media and Audiovisual Communications

The earliest view of educational technology, and one that continues today, emphasized technology as media. This view grew out of what Saettler (1990) calls the *audiovisual movement:* ways of delivering information that could be used as alternatives to lectures and books. This view, developed in the 1930s primarily by instructors in higher education, held that media such as slides and films delivered information in more concrete, and therefore more effective, ways. Later, this perspective developed into a field called *audiovisual communications,* the "branch of educational theory and practice concerned primarily with the design and use of messages which control the learning process" (Saettler, 1990, p. 9). However, the view of technology as media continued to dominate some areas of education and the communications industry. Saettler reports that as late as 1986, the National Task Force on Educational Technology used a definition that equated educational technology with media, treating computers simply as media.

A professional organization, the Association for Educational Communications and Technology (AECT), tends to represent this view of technology as media and communications systems. Originally a department of the National Education Association (NEA) that focused on audiovisual instruction, the AECT was until very recently concerned primarily with devices that carry messages and the applications of these devices in instructional situations. After a reorganization in 1988, it broadened its mission to include other concerns such as instructional uses of telecommunications and computer/information systems. However, three of its nine divisions still focus on the concerns of media educators, and many of its state affiliates still refer to themselves as media associations.

Technology in Education as Instructional Systems

The instructional design or instructional systems movement took shape in the 1960s and 1970s, adding another dimension to the media-and-communications view of technology in education. Systems approaches to solving educational problems originated in military and industrial training and migrated to K-12 schools by way of university research and development. These approaches were based on the belief that both human and nonhuman resources (teachers and media) could be parts of a system for addressing an instructional need. From this viewpoint, educational tech-

nology was seen not just as a system for communicating instructional information, but as a systematic approach to designing, developing, and delivering instruction matched to carefully identified needs (Heinich, Molenda, and Russell, 1996, p. 16). Resources for delivering instruction were identified only after detailed analysis of learning tasks and objectives and the kinds of instructional strategies required to teach them.

From the 1960s through the 1980s, applications of systems approaches to instruction were influenced and shaped by learning theories from educational psychology. Behaviorist theories held sway initially, and cognitive theories gained influence later. Views of instructional systems in the 1990s were also influenced by popular learning theories, but these theories criticized systems approaches as too rigid to foster some kinds of learning, particularly higher-order ones. Thus, the current view of educational technology as instructional systems seems to be in the process of changing once again. (See Chapter 4 for more detailed information on two approaches to educational technology as instructional systems and how each influences methods of integration.)

Just as the AECT had its origins in the media systems view of educational technology, the International Society for Performance Improvement (ISPI) grew out of the view of educational technology as a systems approach to instruction. Originally named the National Society for Programmed Instruction, it adopted its current name in 1974. As its original name indicates, ISPI is still concerned primarily with creating and validating instructional systems.

Technology in Education as Vocational Training Tools

Yet another popular view of technology in education has developed from the perspective of technology as tools used in business and industry. Generally referred to as *technology education,* this view originated with industry trainers and vocational educators, and it reflects their need for technology to enhance training in specific job skills. This perspective is based on two premises. First, it holds that one important function of school learning is to prepare students for the world of work. Therefore, students need to learn about and use technology in school that they will encounter after graduation. For example, technology educators believe that every student should learn word processing, since this skill will help them to perform in many jobs or professions. Second, technology educators believe that vocational training can be a practical means of teaching all content areas such as math, science, and language. Since computer skills are integral to carrying out both of these aims, many technology education activities focus on them. However, technology education includes other topics such as robotics, manufacturing systems, and computer-assisted design (CAD) systems.

The organization that espouses this view is the International Technology Education Association (ITEA), formerly the American Industrial Arts Association. This organization has helped shape a major paradigm shift in vocational training in K-12 schools. Most schools are currently in the process of changing from industrial arts curricula centered in wood shops to technology education courses taught in labs equipped with high-technology resources such as computer-assisted design (CAD) stations and robotics systems.

Technology in Education as Computers and Computer-based Systems

A final view of educational technology originated in the last half of this century to accompany the technology applications made possible by the development of a new electronic device: the computer. When computers came into use in the 1950s, business, industry, and military trainers, as well as educators in K-12 and higher education immediately recognized their potential as instructional tools. Many of these trainers and educators predicted that computer technology would quickly transform education and become the most important component of educational technology. Although instructional applications of computers did not produce the anticipated overnight success, they did inspire the development of another branch of educational technology. From the time that computers began to be used in classrooms in the 1960s until about 1990, this perspective was known as *educational computing,* and it encompassed both instructional and support applications of computers.

Originally, educational computing applications were strongly influenced by the input of technical personnel such as programmers and systems analysts. But by the 1970s, many of the same educators who were involved with media, audiovisual communications, and instructional systems were directing the course of research and development in educational computing. By the 1990s, these educators began to see computers as part of a combination of technology resources, including media, instructional systems, and computer-based support systems. At that point, educational computing became known as *educational technology.*

The organization that represents this view of technology in education is the International Society for Technology in Education (ISTE). This organization is the product of a merger between two computer-oriented groups: the International Council for Computers in Education (ICCE) and the International Association for Computers in Education (IACE). IACE was known for most of its existence from 1960 until 1986 by the name Association for Educational Data Systems or AEDS (Lidke, 1992). The title of one of ISTE's major publications, *The Computing Teacher,* reflected the original computer orientation of the organization. In 1995, it was renamed as *Learning and Leading with Technology.*

Technology in Education in This Textbook

Each of these perspectives on technology in education has made significant contributions to the current body of knowledge about processes and tools that address educa-

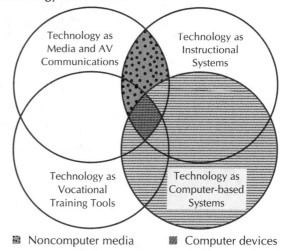

Figure 1.1 Various Approaches to Technology in Education

Technology as Media and AV Communications

Technology as Instructional Systems

Technology as Vocational Training Tools

Technology as Computer-based Systems

▩ Noncomputer media   ▨ Computer devices

tional needs. But, as Saettler points out, none of the individual paradigms that attempt to describe educational technology can satisfactorily characterize what is happening with technology in education today and what will happen in the future. Furthermore, all of the organizations described earlier appear to be engaged in a struggle to claim the high-profile term *educational technology.* Each seems determined to assign a definition based on the perspective and concerns of its member. Each wants to be identified with the future of educational technology and each wants to shape its directions. However, these often-conflicting views of the role of technology in education confuse newcomers to the field and inhibit their understandings of the purposes and issues involved; the resources and issues differ depending on whose descriptions a teacher hears and which set of publications he or she reads.

An emphasis on computer systems. This textbook looks at technology as a combination of media, instructional systems, and computer-based support systems. (See Figure 1.1.) However, this text emphasizes a subset of all available technology resources, focusing primarily on computers and their roles in instructional systems. There are three reasons for this focus:

1. Computers as media are more complex and more capable than other media such as films or overheads, and they require more technical knowledge to operate.

2. Computer systems are currently moving toward subsuming all other media within their own resources. For example, CD-ROMs and videodiscs now store films and slides. Presentation software can generate overhead transparencies.

3. The complexity of computer-based systems has traditionally complicated the efforts of educators to integrate various forms of software and computer-driven media into other classroom activities. They can see much more easily (some would say even intuitively) how to integrate individual media such as films or overheads.

Figure 1.2   The Basic Components of a Computer System

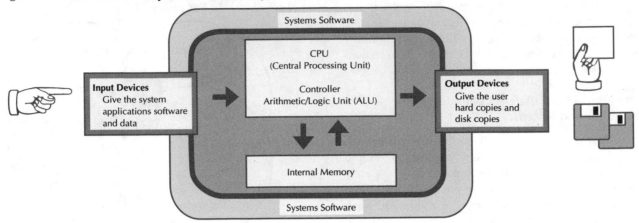

Thus, in this textbook, the phrase "integrating educational technology" refers to the process of determining which electronic tools and which methods for implementing them are appropriate for given classroom situations and problems.

Background on Computer-based Educational Technology

Most newcomers to educational technology define the term *computer* by referring to the equipment with which they have had some experience. A definition based on limited experience may or may not be precise enough to help them select or use other equipment effectively. This section will provide two kinds of definitions. First, it will present a general definition based on essential computer features and characteristics and a history of their development. Then it will define computers and related technology in education based on systems currently in use in schools and classrooms.

What Is a Computer?

What most people tend to consider one item—a computer—is really several different components that function together as a system. Therefore, what is commonly called a computer is more accurately called a *computer system*. In order to qualify as a computer system, the elements shown in Figure 1.2 must be present:

• **Hardware.** This classification includes all of the devices or equipment in the computer system. Required hardware includes input devices (to get user requests into the system), the central processing unit or CPU (which carries out user requests), output devices (which display the results of the CPU's action), and internal memory (which stores the instructions to the computer system). Every computer system must have a CPU, memory, and at least one input device and one output device. Optional hardware includes external storage devices such as disk drives and communications devices such as modems. All devices aside from the CPU and internal memory hardware are also referred to by the term peripheral devices, or simply *peripherals*. Table 1.1 summarizes

		Input/Output (I/O)	CPU and Internal	
Input Only	**Output Only**	**and Secondary Storage**	**Memory**	**Other**
Keyboard	Screen (TV, CRT, LED)	Floppy disk drive	ALU	Modem
Mouse	Printer	Internal or external hard disk drives	Controller	
Joystick	Speech synthesizer	Removable disk drive (Bernoulli drive, etc.)	RAM	
Game paddles	Speaker	Magneto-optical disc drive	ROM	
Scanner	Projection panel	Fax machine		
Bar-code reader	Graphics plotter			
Graphics tablet/pad				
Videodisc player				
CD-ROM player				
Touch panel/light pen				

Table 1.1   Hardware Components in a Computer System

Table 1.2 Software in a Computer System	
Systems Software (Types of Operating Systems)	**Applications Software for Education**
Unix	**Tool software:**
MS-DOS	Word processing
Apple DOS, ProDOS	Spreadsheets
Macintosh OS	Database management
Windows 95	Integrated packages (combinations of WP, SS, DB, etc.)
	Graphics software
	Communications software
	Other tools: test generators, CMI, etc.
	Programming software:
	BASIC
	Logo
	FORTRAN
	COBOL, SNOBOL
	C
	Pascal
	PL/1
	Courseware:
	Tutorial
	Drill (and practice)
	Simulation
	Instructional game
	Problem solving

the hardware elements commonly seen in educational computer systems.

• **Software.** Sets of instructions that let users communicate with the hardware and tell it what to accomplish are known as *programs* or *software*. (The term *software program* is actually a redundant phrase.) Programs are usually written in computer languages such as BASIC or FORTRAN, but they may also be prepared in machine language. Hardware needs two kinds of instructions, which define two general kinds of software. Systems software tells the computer how to do basic operations such as storing a program in memory or sending something to be printed. Applications software represents programs written to answer specific user requests such as word processing or performing a statistical test on some data. Table 1.2 lists some examples of common software elements in educational computer systems.

Computer media. A third category of computer resource is computer media. These are not really separate components in a computer system, but neither are they hardware or software. Rather they are the means by which users store software to make it more portable and easier to use. Diskettes (5 1/4-inch and 3 1/2-inch sizes) along with videodiscs and CD-ROMs are common computer media. Software is like music stored on tape or CD. A tape or CD is not actually

music, only a medium for storing it. The hardware is a tape or CD player, just as a disk drive/CD drive is part of a computer system's hardware.

A computer system may be defined as *a combination of electronic components (hardware) with which humans can communicate through sets of coded instructions called* software. Every computer system, from a handheld personal digital assistant (PDA) to a giant supercomputer, has these same basic elements: hardware and software. Actual combinations of hardware and software components vary considerably according to the power of computer systems and the purposes for which they were designed. The following description of basic hardware and software principles provides a foundation for dealing with computer systems that teachers are likely to encounter in schools.

Hardware: Providing the Electronic Capability

The hardware or electronic devices in a computer system perform an essential function. They give people a complex and capable interface between themselves and the potential power of electricity. The computer derives its name from the fact that it is part of an electronic system that performs rapid series of calculations or computations. Computer users

Input Devices

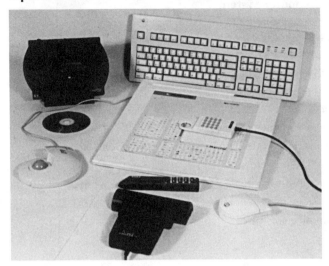

Output Devices

do not see these computations as they take place. Instead, users know only that they give instructions, and the systems do what they tell it to do and show the results. The activity of giving the system instructions is the input function; the computer's activity of doing what the user tells it to do is the processing function; the activity of showing the results is the output function. In addition to these readily apparent functions, invisible storage functions go on during all other activities. The system stores or "remembers" items of information it needs in order to carry out instructions. Finally, special communications functions give systems the capability of linking with other systems to exchange information. Hardware (equipment) provides the electronic capabilities required to accomplish the five functions: input, output, storage, processing, and communications.

Input functions. Some devices serve only as channels to let users give instructions and information to the computer system. The most common input devices that carry out this function are:

- **Keyboard.** Since most users need to type letters and numbers as input to their systems, nearly every computer system has a keyboard. The photo above illustrates many different styles. Most keyboards are based on the QWERTY format familiar from typewriter keyboards, but some keyboards, such as those designed especially for young children, use an ABCDE format.

- **Mouse.** Mice are becoming as common as keyboards and, like keyboards, they come in many styles. Some computer mice actually look a good bit like real mice! The mouse allows a user to select items from the screen and move quickly and easily to various screen locations.

- **Joystick and game paddle.** Whenever the computer system is being used for a game of skill, such as Ping-Pong or a video game system such as Nintendo®, the input device is usually a joystick, a set of game paddles, or both. These are used rather than a keyboard or mouse because they are designed to allow the user to manipulate an object on the screen rather than to type in letters and numbers.

- **Bar-code reader.** Since bar codes are becoming commonly used to identify the locations of items such as pictures or film clips stored on disk, the input device used to give this information to the computer is a bar code reader. These are usually attached to VCRs or videodisc players, but they be used as remote controls with one of these devices.

- **Scanner.** There are several different kinds of scanners, but all are designed to "read" printed information and transfer it into the computer system for storage and processing. Some scanners read bar codes or UPC symbols that identify specific items such as books or groceries. Some scanners digitize pictures or text from a page and read it into a file that can be held by a computer's storage device. Still other scanners read pencil marks or "bubbles" on paper and translate the marks into data for test answers or application form responses. Scanners use different technology, but they all carry out input functions in a computer system.

- **Touch screen and light pen.** Most computer screens are considered output devices, but some let users make choices or enter responses by touching the screen with a finger or a device called a *light pen.* Since these *touch screens* get information into the system, they are classified as input devices.

Other input devices include voice recognition units, microphones, optical character readers, and graphics tablets. Any device that lets a user send instructions or information *into* a system is classified as an input device.

Output functions. Some devices display the results of what the computer does. The most common of these devices are:

- **Printers.** Three kinds of devices in common use produce paper copies (also known as *hard copies)* of printed information. An inkjet printer sprays a controlled flow of ink onto a page to form characters or graphics, but it works so quickly and quietly that the actual spraying is difficult to see with the naked eye. Dot matrix printers strike patterns of dots through inked ribbons to form letters and other symbols. The dots are produced by the impact of small pins hitting against a typewriter-like ribbon place characters or graphics on paper, making dot matrix printers sound like very fast typewriters. The fastest and most expensive type of printer in current use is the laser printer, which uses a method very similar to that of a photocopier to place an image on paper. Several processes, including one with a laser, place an inky substance called *toner* on a page to form the desired images. The laser printer produces the highest-quality output of all three printers and is rapidly becoming the most popular. (Two other kinds of printers, daisy wheel printers and thermal printers, are still in use but are rapidly being replaced by the three described here.) Printers can also be classified by the way they receive data from a computer. Serial printers receive information one information unit (bit) at a time, while parallel printers receive a group of information units (bits) at a time. Serial printers are usually slower than parallel printers.

• **Monitors.** The video output for a computer system is displayed with monitors or cathode ray tube (CRT) screens. While these look very similar to television sets, they are designed to have higher resolutions than most TVs. (This may soon change with the introduction of high-definition television.) Monitors display images with individual dots of light called *pixels,* short for *picture elements.* More pixels give a monitor higher definition (and usually, a higher price). While most monitors employ cathode ray tube technology, some use liquid crystal displays or LCDs.

Other output devices include plotters and speech synthesizers. Any device that makes it possible to determine what the computer does with user requests is classified as an output device.

Input and output functions. Some devices, for example disk drives, can perform two different functions, although they can handle only one at a time. These devices are capable of both delivering input to the computer for processing and storing information once it is processed. Although the latter function is usually storage rather than output, they are generally referred to as *input/output* or *I/O devices.* Disk drives are described in more detail later in this chapter under the heading External Memory (Secondary Storage) Functions.

Processing functions. The central processing unit (CPU) performs the actual work of the computer system— processing user instructions. This is the part of the system that one can most accurately call the *computer.* In today's computer systems, the CPUs are series of electronic circuits arranged and stored on silicon chips. In microcomputers, these chips are housed on a component called a *motherboard.* (See accompanying photo.) Although the naked eye cannot see separate units, the CPU consists of two parts:

• **The control unit.** This part of the CPU directs the activities of the whole system. It resembles an airport's air traffic controller or the manager of a busy courier service. The control unit directs all parts of the computer system to work together to accomplish the tasks that the user gives it to do. Like an air traffic controller or service manager, the control unit completes no functions itself that are considered the actual work of the system; its job is to direct the work of the other members of the system.

• **Arithmetic/Logic Unit (ALU).** All operations in a computer system are based on computations. The basic arithmetic operations in a computer of adding, subtracting, multiplying, and dividing are actually all based on addition. But the com-

Example Motherboard

Source: Courtesy of Apple Computer, Inc.

puter's operations also include sorting and comparing bits of information. All of these computation operations—arithmetic, comparing, and sorting—are done by the ALU.

A computer's most complex activities are really just combinations of simple steps arranged in sequence. For any task, the CPU follows the same general sequence of functions over and over again:

Step 1 The controller gets an instruction from the system's internal memory (discussed in the next section) and puts it in a temporary storage location called a *register.* (These instructions got into internal memory in the first place when the controller directed an input or storage device to place them there.)

Step 2 The controller directs the ALU to do necessary computations. The result is stored in another register.

Step 3 The controller gets the result from the register and stores it once again in internal memory.

The CPU may complete hundreds of iterations of this three-step sequence to accomplish a basic task such as drawing a line on a screen. Although these are discrete steps, the CPU can perform millions of them in the blink of an eye. The speed at which it completes these operations is

INSET 1.1

Technical Information about Bits and Bytes: Computer Communications, Memory, and Storage Concepts

How the "Language of 1s and 0s" Got Started

The basic "language" of computers is electricity. All that a computer or any machine understands is whether or not its circuits are carrying electrical current. The first computer designers hit upon a coding system to allow communication with a machine by translating human language into the language of electricity. Here's how it works.

They decided that a single circuit in the computer could represent a basic unit of communication. The circuits could represent numbers, and the numbers could represent letters and other symbols. Most children have played at one time or another with writing secret messages in coding systems where numbers represent letters. But what coding system made sense to use with machines?

It has been said that human beings use the decimal numbering system because they have ten fingers, the first counting devices. If we all had been born with only eight fingers, we might prefer the octal numbering system instead of the decimal system. But if we all had *two fingers*, we might favor the numbering system that seems most logical for machines: the binary number system, which has only 0s and 1s. Computer designers reasoned that if they sent a pulse of electricity through a circuit, it could represent the number 1. A circuit with no pulse could represent a 0. Circuits would act like light switches that could be turned on or off.

In this way, every single circuit or switch could be a unit of memory or storage that could hold a binary digit, either a 1 or a 0. The term *binary digit* was shortened to the word *bit*. By combining banks of bits, designers could generate enough combinations of on/off switches or bits to represent all of the letters in the alphabet plus all the number symbols from 0 through 9 and some other useful symbols such as the math operators =, −, /, and *. It was decided that bits would be arranged in groups of eight and that each group would be called a *byte*. Each byte would represent a character, such as a letter. For example:

		Eight Switches (Bits) in On/Off Positions				**Binary Number**	**Letter Represented**
1 byte of memory	=	8 bits	=	off off on on off off off on	=	0 0 1 1 0 0 0 1	=	A

Two standard coding systems, the American Standard Code for Information Interchange (ASCII) and Extended Binary Coded Decimal Interchange Code (EBCDIC), were designed around this coding system. The ASCII coding scheme has come into more common use than EBCDIC throughout the world. Some example ASCII codes include:

ASCII Code	What It Represents
00110000	The number 0
00110001	The number 1
00110010	The number 2
etc.	
01000001	The letter A
01000010	The letter B
01000011	The letter C
etc.	
00001101	The symbol +
00111111	The symbol ?
00100001	The symbol !
etc.	

A Child's Coding System

1   0

Continued

The 1s and 0s Coding Scheme for Computer Communications
Early computers were programmed only with binary digits or machine language. Using principles of electronics that were based on logical operators, these on/off circuits could drive mechanical operations such as running disk drives and electrical operations such as displaying points of light on screens. But programming only with 1s and 0s is time-consuming and requires in-depth technical knowledge of computer electronics. As with any language, programmers can combine basic units of written communication (e.g., letters) to develop words and sentences. When groups of bytes were put together to represent more human-sounding words like *RUN* and *GO TO,* higher-level computer languages were born. These languages allowed people with less technical knowledge to write instructions for computers. The languages were designed to translate the instructions automatically back into machine language so the machine can understand them.

Bytes as Units of Computer Storage
The concept of bytes as memory units is used in various ways in computers today:

- **Bytes for memory size.** In today's microcomputers, users measure internal memory in terms of how many thousands of characters a machine has to store programs and data. Since each byte stands for a character, they evaluate RAM in terms of kilobytes or K (thousands of bytes), megabytes or M (millions of bytes), and gigabytes or G (billions of bytes). More memory makes a computer more expensive. A computer with 512K of memory (512,000 bytes) is less expensive than one with 1M (1 million bytes). Disks are also described in terms of storage in bytes. Microdisks can usually hold from 720K to 1.44M (see Chapter 3.)

- **Bytes (bits) for processor size.** The type of CPU in a microcomputer is based on the number of bytes it can process at one time. The first microcomputers had 8-bit processors; the more recently developed ones have 32-bit processors. The size of the processing unit also relates to how much memory a microcomputer can have; thus, 32-bit machines are more powerful than those with earlier processors.

---

measured in mips or millions of instructions per second, and it has much to do with how fast the task gets done.

Internal memory (primary storage) functions. A limited amount of space is arranged inside the computer for storage of instructions. On a microcomputer, this memory is all on chips on the motherboard. The size of this space is measured in units called *bytes.* The fundamental concept of a byte as a storage unit indicates how computer memory stores bits of information, and it also underlies how a computer functions as an electronic device. (See Inset 1.1: Technical Information about Bits and Bytes.) Teachers need to be familiar with two kinds of internal memory:

- **Random access memory (RAM).** This kind of memory serves as temporary storage for user requests, in the form of applications program commands, and the data that programs use. The control unit may place a program and data at any location (i.e., randomly) in RAM. RAM is also called *volatile memory;* it depends on electricity to hold information in the circuits. When electricity goes off, RAM is erased.

- **Read only memory (ROM).** ROM is a kind of memory designed to hold instructions permanently within the computer. "Read only" indicates that the user cannot place information in ROM circuits; the manufacturer stores only systems software there in building the computer. The computer system uses this software each time it starts up.

The power of a microcomputer is based, in part, on the amount of its RAM in bytes. Users may describe RAM in any of the following ways:

- **Kilobytes or K.** For example, 512K means that the computer has approximately 512,000 locations for storing characters of information (e. g., letters, numbers, symbols).

- **Megabytes or M.** For example, 5M indicates 5 million bytes of storage space.

- **Gigabytes or G.** For example, 1G indicates 1 billion bytes of storage space.

External memory (secondary storage) functions. Since the internal memory capacity of a computer system is limited, designers have devised ways of extending the amount of information that it can store or "remember." External storage devices carry out this function. Disk drives and, less frequently, tape drives do this job for microcomputers. The first disk drives served large, mainframe computers. They looked like large, shiny disks housed in groups within floor-standing cases. Today, teachers are most likely to use the disk drives that come with microcomputers. While the 5 1/4-inch diskettes used with floppy disk drives are still in use, they are rapidly being replaced by 3 1/2-inch microdisks and the microdisk drives that run them. Hard disk drives, usually shortened to *hard disks,* are also in common use. A hard disk is usually housed permanently within a computer's case, but some kinds (e.g., Bernouilli drives) come in removable, transportable cases. When a hard disk is housed in the same case that holds the CPU and memory chips, people may become confused about whether it represents internal or external storage. It is definitely an external memory device, since it lies outside the motherboard. Computer systems still use tape drives, but less frequently and usually as backup devices for hard disks.

An increasingly common and quite useful form of external storage is an optical disc drive. The most common optical disc storage employs a CD-ROM drive. The name stands for compact disc-read only memory, because users cannot store information there. However, this limitation is relaxing as new types of CD-ROM storage allow users to write to them as well as read from them. Innovators are

External Storage Device

5.25" disk (diskette, floppy disk)

5.25 disks on DOS systems:
Double-sided, double density (DS, DD) store 360K
Double-sided, high density (DS, HD) store 1.2 MB

3.5" disk (diskette, microdisk)

3.5 disks on DOS systems:
Double-sided, double density (DS, DD) store 720 K
Double-sided, high density (DS, HD) store 1.4 MB

3.5 disks on Mac systems:
Double-sided, double density can store 800K
Double-sided, high density can store 1.44 MB

CD-ROM (Compact Disc-Read Only Memory)

Store approx. 600 MB

constantly introducing new types of removable disk storage that can hold even more information than CD-ROMs. New types of removable disk storage include hard disk cartridges (designed for use on cartridge hard drives such as Syquest and Zip drives), Bernoulli cartridges, and rewriteable magneto-optical discs. Although currently less common than diskettes and CD-ROMs, these more recent arrivals are typical of the trend toward storing more and more information on smaller and smaller media.

External memory or storage devices require media such as disks, tapes, or CD-ROMs to store data. External devices store information on disks magnetically rather than in off/on electrical circuits that combine to form bytes. Still, disk or tape storage is also measured in bytes. The disk drive or CD-ROM drive accomplishes the actual act of storing or transferring information, but the storage itself takes place on the medium (floppy disks, microdisks, hard disks, or CD-ROM discs).

Disk drives as I/O devices or external storage devices. Different references refer to disk drives as one or the other, but the most accurate answer is that they are both. They function primarily as storage devices, but in order to use this secondary storage, they must also act as input/output devices.

Types of disk resources. The three most common types of removable disks (ones that can be inserted and taken out of disk drives) are 5 1/4-inch (or floppy) disks, 3 1/2-inch disks (or microdisks), and CD-ROM. (See accompanying photo.) More recent models of Macintosh and MS-DOS computer systems use mostly 3 1/2-inch disks and CD-ROMs. Disks are sometimes called *diskettes,* and 3 1/2-inch disks also are sometimes referred to as *floppy disks,* but the term *tapes* is a common but incorrect usage. Removable disks come in various storage formats. Double-density (DD) microdisks can store from 720K to 800K of data, while high-density microdisks have from 1.2 to about 1.4M of storage capacity. Disk drives on some older Macintosh models cannot read high-density disks, but all can read double-density disks.

Single density, double density, high density. The density of a disk refers to the format with which information is stored on the disk, and this determines how much infor-

mation can be placed there. Single-density disks can store the least, high-density disks the most. Most disks nowadays are either double-density or high-density. High-density disks, usually labeled *HD,* can be used only in high-density drives.

Single-sided, double-sided. Only one side of the first disks could store programs or data. Most disks today are 3 1/2-inch, double-sided disks. Information can be stored on and read from both sides. The other commonly used disk, the hard disk, cannot be removed in the same way as floppy disks or microdisks. It usually resides inside the computer system case in a sealed case of its own; the whole drive is removed only for repair or replacement. However, some types of hard disks are removable devices in themselves that can be connected to the computer with cables. These are often used to transfer large, multidisk programs from one computer to another. Hard disks range in size from about 20M (20 million bytes) up to as much as a gigabyte or more (1G or 1 billion bytes) of storage.

Communications functions. An optional, but increasingly desirable, capability in a computer system is the ability to communicate with other systems. To do this, a system needs a special device to change (modulate) the information from the digital format produced by computers into an audible format that can be sent across telephone lines. When the information reaches the receiving computer, that machine changes it back (demodulates it) into the digital format. The device that accomplishes this dual function is called a *modem,* a term derived from combining the words *modulator-demodulator.* Modems vary in price according to the

Example Modems

speeds at which they send and receive information. At one time, users rated speed as *baud rates,* but they currently measure it as bits (of information) per second or bps. Modem speeds range from 2,400 bps to 28,800 bps (28.8 bps), but this capability is increasing rapidly. Some kinds of modems can also send and receive faxed information.

Software: Communicating with the Hardware

Although hardware performs important tasks in the computer system, it can do nothing without user instructions. These instructions must be given to the computer system in a very structured sequence of steps called a *program.* Programs must be written in languages specially designed for communicating with computers. Although teachers do not need to become programmers, they should be aware of the kinds of programming languages, the types of programs used with computers, and how programs are developed and stored.

Programming languages. Just as humans speak and understand a variety of languages (e.g., Spanish, English, French, Swahili), several kinds of computer languages also serve various purposes. The first kinds of languages to be developed were low-level languages that were designed as interfaces between human beings and the "electrical language" of machines. Later, higher-level languages were designed that required less technical expertise, making them easier for nontechnical people (like teachers) to use.

- **Low level languages.** Original computer languages include machine language and assembly language or assembler. Machine language is based on the 1s and 0s that represent on/off circuits in a computer's memory. It is programmed into the machine when it is built. Assembler is considered one step up from the binary digits of machine language and consists of symbolic commands such as *ADD* and *LOAD*. Each assembler command represents several steps written in machine language. Although assembler is the lowest-level language that can be used to program today's computers, most people find it still too technical and time-consuming to use; they prefer high-level languages.

- **High-level languages.** These languages are actually programs themselves that are written in machine language. High-level languages consist of commands that resemble human words such as *RUN* and *GO TO*. Among the dozens of high-level languages, the most common choices for educators are: BASIC (Beginners All-purpose Symbolic Instruction Code), Pascal (named after an early pioneer in

programming concepts, Blaise Pascal), and Logo. Other languages educators may see include: C, FORTRAN, COBOL, and JAVA (used with Internet resources).

High-level languages are sometimes classified as compiled or interpreted, depending on the way in which they translate commands into machine language. Interpreted languages (such as BASIC and Logo) use programs called *interpreters* to translate commands (called *source code*) one-by-one into machine code. If the interpreter finds an error, it stops interpreting and alerts the programmer. Compiled languages (such as Pascal and FORTRAN) use programs called *compiler* to do the translation. Compilers translate entire programs into machine language, which is then called *object code*. Programmers discover errors in language syntax only after an entire program is compiled. Programmers must also identify and correct any errors in logic after a program is compiled.

Types of software. Every computer system requires two different types of software: systems software and applications software.

- **Systems software.** Software that makes the computer system perform its most basic operations, such as starting up and reading disks, is called *systems software*. The manufacturer includes some parts of the systems software in the computer's ROM when it is built; other parts are read into RAM from a disk or from a ROM chip when the computer starts up. Educators need to be most concerned with the latter kind of systems software: the operating system. This software is an important choice because it acts as a visible interface between the machine and the user. The operating system determines both what the user sees on the screen when the computer is turned on and the steps the user follows to do things like selecting options from menus. Since the CPU receives some of its instructions from the operating system each time the computer is turned on, the startup process came to be called *bootstrapping* the system (after the phrase "pulling oneself up by one's bootstraps"). The term later became shortened to *booting* the system. Starting up the system by turning on the power switch is sometimes called a *cold boot,* while using keys on the keyboard to re-initiate the installation of the operating system is sometimes called a *warm boot.*

Early microcomputer systems used either CP/M (Control Program for Microprocessors), Apple DOS (Apple Disk Operating System), or ProDOS. The UNIX operating system, originally designed by Bell Laboratories for use on larger computers such as the DEC VAX, was later modified for use on some microcomputer systems. Later, two other operating systems came to dominate the field: MS-DOS (Microsoft Disk Operating System) and Macintosh Operating System.

MS-DOS was succeeded by Windows, a program that serves as a graphic user interface (GUI pronounced "gooey") between the MS-DOS operating system and the user. (Windows essentially made the screen on a machine running MS-DOS look like a Macintosh screen.) Newer versions of Windows (e.g., Windows 95) are actually operating systems themselves. In the past, a user's choice of equipment brand determined which operating system he or she could use. MS-DOS ran only on IBM and IBM-compatible machines, and the Macintosh Disk Operating System ran only on Macintoshes. Newer versions of each brand can run both operating systems.

Figure 1.3   Flowcharting Symbols and Example Flowchart

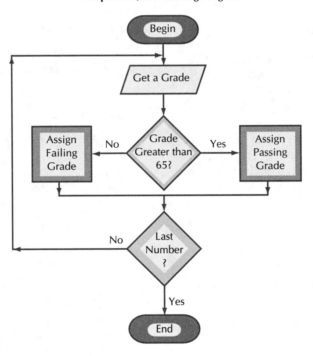

• **Applications software.** Programs written to do tasks that humans want to do, such as word processing or drawing pictures, is called *applications software,* or simply *programs.* Instructional programs (courseware) such as drill and practice, tutorial, simulation, and problem solving programs are applications software, as are in instructional games. Other applications software includes tools such as word processing, databases, and spreadsheets. Applications are written in higher-level languages such as BASIC or Pascal, or combinations of such languages.

Teachers buy most of their programs, but they could write their own software for some purposes. Usually, a programmer writes a program in a version of a language that is specific to an operating system (e.g., MS-DOS BASIC). This means that applications programs written for one type of machine (IBM) cannot run on another (Macintosh). But recent changes in hardware manufacturing have made it possible for software designed for one operating system to run on another. This very

recent improvement represents a major breakthrough for user convenience.

Software design. Although many people refer to the process of writing program commands in a computer language as "*programming,* writing the code is actually only one step in the total process. Program design is a team effort by system designers, content area specialists, and programmers. The following steps are usually considered essential to developing good programs:

Step 1   Analyzing the problem. Most programmers agree that this is the most difficult step in programming. The programmer has to identify the user's problem exactly before attempting to design a solution. If the problem is relatively simple, such as drawing a circle on the screen, a programmer may quickly break it into its component requirements. To write a new word processing program, however, problem analysis may take weeks or months.

Step 2 Developing the algorithm. The next step in programming is mapping out a problem solution or algorithm. One common method of showing problem solutions is with the aid of flowcharting symbols. (See Figure 1.3.) Flowcharts illustrate a proposed solution graphically, like a picture, so programmers can more easily follow its steps or flow.

Step 3 Coding. At this step, the program is translated into a series of commands in a computer language. Every coded program must meet two requirements: It must be in correct syntax (that is, it must follow the structure and spelling rules of the language), and it must follow a logical, step-by-step order. Many programming languages number commands, and the computer carries them out in order by their numbers.

Step 4 Testing. Once the program is drafted, it is run on the computer. If the program works exactly as expected and meets the design expectations of the users, the program development process is complete (an unusual occurrence). If not, the design team goes to Step 5.

Step 5 Revising and Debugging. In the final step, programmers revise the program to make it more efficient and capable and to locate errors. If the program does something unexpected and fails to meet the user's need, the design team must locate errors in syntax or logic that are causing the problem. This process is sometimes referred to as *debugging,* a term originated by an early computer designer named Grace Hopper when one of the computers she worked with in the 1950s broke down when a hapless moth entered its circuitry. "Bugs," or program errors, are corrected as they are found.

The summary of the program design process given here is necessarily brief and simplistic. In reality, a design effort (especially for a large system) can be a complex and ongoing process. The design team for a large, complicated program could continue the process of improving it and debugging it indefinitely. The description given here tells mainly the types of activities involved in program design.

Types of Computer Systems

Today's computer systems are often classified by size. A computer's size category usually depends on the amount of internal memory and the number of people who can use the system at once.

• **Personal computers.** These single-user computers come in several different types. Personal digital assistants (PDAs) are handheld, pen-operated devices that may prove of great practical use to teachers. (See Chapter 10 for more detail on PDAs.) Laptop computers are portable enough to rest on the user's lap. They often have liquid crystal display (LCD) monitors instead of CRTs, since LCD screens are flatter and more portable. Microcomputers are desktop systems that vary in size and price according to their amounts of RAM and hard disk storage.

• **Minicomputers and mainframes.** Up until about the mid-1980s, it was possible to differentiate between minicomputers and mainframes. Both could support many users and peripherals at the same time, but minicomputers usually featured less internal storage than mainframes, and they could support fewer users and peripherals. Minicomputers usually served single locations (e.g., a lab) or met individual needs (e.g., programming classes in a university). Mainframes were physically larger and more powerful, and they could support many kinds of applications at once (e.g., all of the computing/data processing needs of a university).

However, as computer components became smaller and systems became more powerful, the lines between these two kinds of systems began to blur. Today, the term *minicomputer* is still used, but an exact definition varies from company to company. Most manufacturers of systems that they call *minicomputers* sell them along with software for some single, designated purpose such as running a certain network or supporting the student advisement and registration activities at a university.

• **Supercomputers.** Again, no agreed-upon criteria make a computer a supercomputer. The term is usually reserved for extremely fast computer systems designed to perform complex mathematical operations such as those involved in predicting weather patterns.

A Brief History of Educational Computing Activities and Resources

Developments in computer technology have shaped the history of educational technology at least in part. The development of integrated circuits made computers both smaller and more accessible to teachers and students. Thus, the introduction of microcomputers was a major turning point in the history of the field. Much of the information in the following summary was gleaned from the history of computer uses in education given in Roblyer's "Education" entry in the *Macmillan Encyclopedia of Computing* (Roblyer, 1992). Niemiec and Walberg (1989) provide another source of historical perspective on this field. This history is told in two periods: before and after the introduction of microcomputers. (See Table 1.3.)

The era before microcomputers. Although many of today's technology-oriented teachers have been using computer systems only since microcomputers came into common use, a thriving educational computing culture predated that development by 20 years. The first documented instructional use of a computer was in 1950 with a computer-driven flight simulator used to train pilots at MIT. The first use with school children was in 1959, when an IBM 650 computer helped to teach binary arithmetic to New York City elementary school students. The next 20 years were a time of intense development and research with mainframe-based computer systems in schools, colleges, and universities, as well as in industry and military training.

This activity peaked in the early and mid-1970s with federal government funds supporting many large-scale projects. The professional literature of education during this time reflected growing excitement and interest in computer-based instruction, also known as *computer-assisted instruction* or

Table 1.3   Milestones and Trends in Educational Computing

The Era before Microcomputers

1950	Flight simulator used to train pilots at MIT.
1959	First use with school children: IBM 650.
1966	Dedicated mainframe instructional computing system (IBM 1500) offered.
	First authoring system (Coursewriter).
1967	CCC offers minicomputer-based instructional system (DEC PDP/1).
	Mitre Corporation offers the TICCIT instructional computing/television system.
1970s	CDC offers the PLATO instructional system.

The Microcomputer Era and Beyond

1977	The first microcomputers enter schools.
1980	Seymour Papert writes about Logo in *Mindstorms;* he starts the Logo movement.
1980s	MECC offers microcomputer software; publishers begin developing courseware.
	MicroSIFT, EPIE, and others offer courseware evaluations.
	The computer literacy movement emerges then wanes after 1987.
1990s	ILSs and other networked systems are used: multimedia use increases.

Widespread Use of the Internet

CAI. The list below reviews some of the key projects, activities, and developments during the period before microcomputers. While these implementations used earlier technologies, each has had an impact on current computer uses in education.

• **IBM 1500 system in universities and schools.** Following sporadic instructional uses of computers throughout the 1960s, IBM offered the first computer system dedicated solely to instruction and research on learning around 1966. The IBM 1500 system was the first multimedia learning station. In addition to a CRT screen, it had earphones, a microphone, an audiotape player, and a slide projector. The system was first placed at Stanford University, which used it to develop tutorial software to teach reading and mathematics. This instruction was offered to schools through terminals linked to the mainframe located in Palo Alto. Another significant development at this time was the first authoring system, a very-high-level language called *Coursewriter,* which was designed exclusively for writing CAI lessons on the IBM 1500. These lessons also began to be called *courseware,* or *instructional software.* Until 1975, when IBM discontinued support, some 25 other universities and school districts had purchased IBM 1500 systems, and they developed and offered instruction to schools via terminals in the same way that Stanford had done.

• **Stanford University and the CCC System.** Another instructional development project at Stanford University relied

on one of the first minicomputers: Digital Equipment Corporation's PDP-1. A dramatically simpler hardware configuration than the IBM 1500, the PDP-1 was used to develop and offer very basic kinds of instruction such as drill and practice activities in reading and mathematics (Niemiec and Walberg, 1989). In 1967, Stanford University Professor Patrick Suppes led an extensive research and development effort to create some of the first computer-assisted instruction systems. He was also the first president of the Computer Curriculum Corporation (CCC), a company that installed complete hardware and software systems in schools across the country. These early instructional computing efforts earned Suppes the honorary title "grandfather of computer-assisted instruction."

• **Control Data Corporation (CDC) and the PLATO System.** A third line of development occurred at about the same time as the IBM 1500 and the PDP/1 projects. In the early 1970s, CDC in conjunction with Dr. Don Bitzer at the University of Illinois developed an instructional system called Programmed Logic for Automatic Teaching Operations (PLATO). A PLATO terminal consisted of a plasma screen (argon/neon gas contained between two glass plates with wire grids running through them) and a specially designed keyboard. It also had an authoring system originally called *Tutor.* CDC developed tutorial lessons and complete courses rather than just drill and practice lessons. CDC's president, Dr. William Norris, had an almost messianic belief that PLATO would revolutionize classroom practice (Norris, 1977); he channeled significant funding and personnel into development of PLATO over the period from 1965 to 1980. From about 1972 to 1980, the development and marketing efforts of IBM, CCC, and CDC dominated the educational computing field.

• **Brigham Young University and the TICCIT System.** Another parallel line of development took place at Brigham Young University under the leadership of Dr. Victor Bunderson and Dr. Dexter Fletcher. They added color television to a computer learning station and developed Time-shared Interactive Computer-Controlled Information Television (TICCIT). This system, marketed by the Mitre Corporation, never enjoyed the same financial success as Suppes's CCC system (Niemiec and Walberg, 1989).

• **Computerized instruction management systems.** Other organizations focused on projects to apply the recordkeeping capabilities of computer systems to support mastery learning models. Two of these computer-managed instruction (CMI) systems were the Program for Learning in Accordance with Needs (PLAN) developed at the American Institutes for Research and the Individually Prescribed Instruction (IPI) system at the University of Pittsburgh.

• **University time-sharing systems.** During the 1960s and 1970s, as CAI and CMI development was taking place, a thriving computer culture was also developing at universities around the country. Some 22 universities had acquired mainframes and were using them to teach programming. Faculty and students also used these systems to develop programs and utilities, sharing them among members of the academic community. The first meeting of the National Education Computing Conference (NECC), now the largest educational technology conference in the country, occurred in Iowa City in 1979. This provided an opportunity for people in higher education to share computer resources and activities.

While interest in instructional applications grew, educational organizations also worked to computerize more and more of their administrative activities (e. g., student and staff records, attendance, report cards). Due in part to the expense and technical complexity of mainframe computer systems, control of both instructional and administrative computer hardware and applications resided in school district offices rather than with schools or individual teachers. Data processing specialists administered most of these systems.

The "microcomputer revolution" in education. As Roblyer (1992) points out, the intense interest in computer use for instruction from 1960 to 1975 sprang in large part from the belief that computers could revolutionize classrooms in the same way that they were changing business offices in post–World War II America. Funding agencies and computer companies alike seemed to anticipate computers taking over much of the teacher's role in delivering information. By the end of the 1970s, it was coming clear that this kind of revolution was neither feasible nor desirable.

The first microcomputers came into schools in 1977, and instructional computing development shifted its focus rapidly from mainframes to desktop microcomputer systems. But the introduction of these locally controlled resources also transformed the computer's role in education. Computer resources and their instructional applications were no longer managed by large companies or school district offices. Classroom teachers could decide what they wanted to do with computers. Even some administrative applications began to migrate to school-based computers, much to the dismay of personnel in district data processing centers. However, microcomputers made school-based management even more feasible.

Just as key projects shaped the course of development before microcomputers, some key events have occurred in the years since the first microcomputers entered schools. These developments are noteworthy in the impact they have had on educational technology and on education itself.

- **The Minnesota Educational Computing Consortium (MECC) and the software publishing movement.** Before the advent of microcomputers, courseware came primarily from hardware manufacturers such as IBM and CDC, software systems companies like CCC, and university development projects. As microcomputers gained popularity almost overnight, a new software market for education rapidly emerged driven primarily by teachers. The first organization to take advantage of this market was the nonprofit Minnesota Educational Computing Consortium (MECC), which received its original funding from the National Science Foundation and developed much of its instructional software on mainframes. In the 1980s, MECC began transferring these programs to microcomputers such as the Apple, for a time becoming the largest single provider of courseware. MECC later became a profit-oriented organization, retaining its well-known acronym but changing its name to the Minnesota Educational Computing *Corporation.*

  Major software publishing companies quickly jumped into the courseware development market, and a plethora of small

companies, many of them cottage industries, were also organized. Some mainframe companies such as CDC and IBM also tried to compete in this market. CDC first introduced its own microcomputers. When it finally converted many PLATO programs to microcomputer platforms such as Apple, it signaled an end to the dominance of mainframes.

- **MicroSIFT, EPIE, and other courseware evaluation efforts.** Teachers quickly learned that simply putting lessons on a microcomputer did not guarantee their quality or usefulness for students, and many organizations sprang up to offer expertise in evaluating courseware. Those with the highest profiles were a project out of the Northwest Regional Educational Laboratory called *MicroSIFT* and the Educational Products Information Exchange (EPIE). For a short time, EPIE was part of the consumer advocacy group that produces *Consumer Reports.*

  Many professional organizations, magazines, and journals also began to evaluate courseware. In fact, so many reviews were being produced that still other organizations were founded to compile and summarize reviews! Most of these latter groups eventually went out of business as courseware evaluation became less essential. Some educators attribute this waning interest in expert evaluation to the growing tendency of school districts to depend on their own committees to select courseware. Also, purchasers were channeling more funds to purchases of ILS-type systems rather than individual software packages.

- **Courseware authoring activities.** As teachers began to clamor for more input into the design of courseware, some companies saw another potential market for tools to let educators develop their own courseware. These so-called *authoring systems* were the predecessors of modern tools such as HyperCard and Linkway. Some authoring systems were more like high-level languages (e.g., PILOT and SuperPILOT), while others functioned primarily as prompting systems that allowed developers to choose from menus of options (GENIS, PASS). For a time, teacher-developed software became popular, but interest waned as teachers realized how much time, expertise, and work they would have to invest to develop courseware that would prove more useful than what they could buy.

- **The computer literacy movement.** A parallel movement to the growth of instruction in computer use sought to teach students *about* computers. This activity came to be called *computer literacy,* a term thought to have been coined by educational computing pioneer Dr. Arthur Luehrmann (Roblyer, 1992). While Luehrmann originally believed that computer literacy was equivalent to programming skills and use of tools such as word processing, it later came to be associated with a variety of skills. Many expressed a popular fear in the 1980s that students who were not "computer literate" would be left behind academically, further widening the gap between the advantaged and disadvantaged (Molnar, 1978). By 1985, computer literacy skills began to appear in required curricula around the country. By the 1990s, most educators became dissatisfied with the vague concept of computer literacy, which allowed almost unlimited definitions and could not be linked to any specific set of skills. Some school systems still require instruction in versions of computer literacy skills, but others feel that students develop computer skills simply by using computer resources in schools and elsewhere.

- **Logo and the problem-solving movement.** Logo had as profound an influence on instructional computing as any other single computer product. For the period from 1980 until about 1987, Logo-based products, activities, and research dominated the field. Logo was developed and promoted as a programming language for young children by an MIT mathematics professor named Seymour Papert (Cuban, 1986). Through Papert's prolific writings and speeches, it also became a challenge to traditional educational methods and to the computer uses that had supported them (e.g., drill and practice and tutorial uses). (See Chapter 4 for more information on Logo.) Papert based his philosophy of computer use on his interpretation of the work of his mentor, developmental theorist Jean Piaget.

In his popular 1980 book *Mindstorms,* Papert proposed that learning should occur primarily through child-directed exploration rather than teacher-directed instruction and that Logo-based projects could be the basis for such exploration. Several versions of the Logo language were developed, and derivative products were marketed (e.g., Logowriter, LegoLogo). Logo assumed many of the characteristics of a craze; Logo clubs, user groups, and T-shirts all filled schools. Although research showed that the applications Papert proposed could be useful in some contexts, by 1985 educators began to worry that "Logo promised more than it has delivered" (Papert, 1986, p. 46) and interest waned. Logo and Logo-based products are still in use, but the primary contribution of the Logo movement may have come from its example of how technology could be used to restructure educational methods.

• **The emergence of integrated learning systems (ILSs) and other networked systems.** One of the most recent major developments in instructional computing returns in some ways to the kinds of hardware-and-software packages first developed and marketed by Dr. Suppes in the 1970s. In fact, some ILSs employ curicula derived from these early systems. ILSs came about because both school districts and software companies realized that one of the most common applications of microcomputers, and one with strong validation by research, was instruction and practice in basic skills. Educators and software producers began to see that networked computers with courseware residing on a central computer could provide this instruction more cost-effectively than a system using disks on standalone microcomputers could do. An ILS offered the advantages of ability to track and report data on student progress, varied courseware types in one location, and a means of getting software to students quickly.

ILSs designed for instruction in basic skills were popular from about 1988 to 1991. But when curriculum trends moved toward methods that were less structured and teacher-directed, companies began to respond to this need. The most recent networked systems offer arrays of software tools, lessons, and media that teachers may integrate in many ways. These systems resemble the original ILSs technically, but they differ in curricular approaches. Some articles refer to these systems as "multimedia learning systems," "integrated technology systems," or "open learning systems" (Hill, 1993, p. 29), but no universally accepted alternative name for these kinds of products has come about. At this time, many educators still refer to any networked instructional delivery system as an ILS. By any name, all networked systems that operate from a central server mark a significant movement away from standalone computer systems under the control of individual teachers and back toward more centralized control of instructional computing resources.

What Have We Learned from the Past?

Learning about history is an interesting, but useless activity unless people apply this learning to future actions. What have teachers learned from some 50 years of applying technology to educational problems that can improve classrooms now? Educators are encouraged to draw their own conclusions from these and other descriptions they might read. However, the authors would like to emphasize three guidelines for future efforts at integrating technology:

• **Direct technology resources to specific problems and needs.** Many parents and educators want technology tools in the classroom because they feel that technical skills will help prepare students for the work place. But Collis (1988) and others emphasize the weakness of "employability" and the need for "computer/technology literacy" as rationales for investments in instructional technology. Computer resources are becoming increasingly easier to use and less technical, and technology changes so quickly that tasks or materials may alter substantially before a student leaves school to begin work. Some computer-based solutions are not the best instructional choices. Instead of an "employability" rationale, teachers must match the capabilities of specific technology resources and methods to areas that display obvious needs for improvement, e.g., reading, writing, and mathematics skills; research and information-gathering; and problem solving and analysis. Planning must always begin with the question: "What needs do my students and I have that these resources can help meet?"

• **Anticipate and plan for change.** The history of technology in education has shown that resources and accepted methods of applying them will change, sometimes quickly and dramatically. This places a special burden on already overworked teachers to continue learning new resources and changing their teaching methods. Gone are the days, if, indeed, they ever existed, when a teacher could rely on the same handouts, homework, or lecture notes from year to year .

Educators may not be able to predict the future of educational technology, but they know that it will be different from the present. This means that they must anticipate and accept the inevitability of change and the need for continual investment of their time.

• **Separate fad from fact.** It is also clear that technology in education is an area especially prone to what Roblyer (1990) called "the glitz factor." With so little emphasis on finding out what actually works, any "technological guru" who gives a glib rationale for new methods can lead a new movement in education. Sometimes, the mere presence of a new-and-improved technology suffices. It doesn't have to do anything; it just has to look good and promise to improve everything. Then, when dramatic improvements fail to appear, educators move on to the next fad. This approach fails to solve real problems, and it draws attention away from the effort to find legitimate solutions. Worse, sometimes teachers also throw out methods that had some potential if they only had realistic expectations. The past has shown that teachers must be careful and savvy consumers of technological innovation, looking to what has worked in the past to guide their decisions and measure their expectations. Of course, educational practice tends to move in cycles, and "new" methods are often old methods in new dressing. In short, teachers must be as informed and analytical as they want their students to become.

Current Types of Educational Technology Systems

To keep informed about how technology can improve education, one must remain aware of the equipment configurations that are available and their current uses. Availability is a key condition because schools cannot deploy all possible technology resources. Some are either too expensive or their places in the current structure and activities of schools and classrooms remain unclear. Some of the most powerful new technologies are in very limited use in K-12 classrooms for these reasons (National Education Association [NEA], 1993; U. S. Congress, 1995). Also, as the history of educational technology to date demonstrates, the field is most often driven by a combination of educational trends and priorities, economic factors, and the marketing efforts of individuals and companies. Schools do not carefully plan purchases and match technology resources to identified educational needs as often as one might hope; frequently, educators respond to what is most readily available and affordable.

This section describes four types of hardware/software configurations in use or coming into use in schools and classrooms. Most computing resources in education today are variations or combinations of these configurations. The final section of the chapter presents an overview of the current educational uses of these configurations. These sections serve both as an introduction and advance organizer for the more detailed descriptions of applications in the coming chapters. These overviews can also help educators become more aware of the options available to them and encourage careful selection of resources to help them better match appropriate technology resources to their needs.

Standalone Personal Computers

Although they have come into common use only in the last 15 years, individual microcomputers and other standalone devices are currently the most common kind of technology hardware in use in schools. These are usually configured in one of the following ways:

- **Portable units.** Units such as PDAs are carried around like calculators. Laptop computers are usually purchased for their portability, but sometimes they are networked. (See the next section.) In these cases, laptops are purchased mainly for their smaller size. Some schools stretch scarce hardware resources by placing microcomputers and peripherals on carts that teachers can check out from a central location such as the library/media center and take to classrooms for temporary use.
- **Workstations.** Sometimes two or three microcomputers or laptops are placed together with a shared printer and other resources such as a videodisc player to form one of several classroom learning stations. Another type of learning station

(a)   (b)

(c)

(d)

Source: Photo (a) Courtesy of Apple Computer, Inc., photographer Frank Pryor; (b) and (c) Courtesy of Apple Computer, Inc., photographer John Greenleigh; (d) Courtesy of Harry and Jennifer Buerkle.

may consist of a single microcomputer combined with other peripherals and media. Both of these configurations are called *workstations.*

- **Labs.** Microcomputer laboratories or labs, usually consisting of 10 to 30 units with shared printers and other peripherals, were more popular in the early years of microcomputer use but they are still in widespread use. Schools often set up single-purpose labs for tasks like programming or word processing, but they may also place labs in common areas for general, multipurpose use by groups of students or teachers.

Networked Stations: From LANs to MANs

The principle of linking standalone stations in a lab or school, sometimes called *connectivity,* is becoming increasingly popular in educational computing. Many schools view networks as essential steps toward preparing for future technology needs (Petrusco and Humes, 1994). A network centralizes resources and cuts down on handling of individual disks. It also helps to get software to individuals or classes comparatively easily and cheaply, since it can be downloaded or sent to individual computers from a central computer or file server. As Kee (1994) puts it, "networks are for sharing" (p. 6), and this sharing was designed to make computing activities more efficient and cost-effective.

Schools can implement several kinds of networks, and their design and maintainance can pose a technically com-

Figure 1.4   Network Types

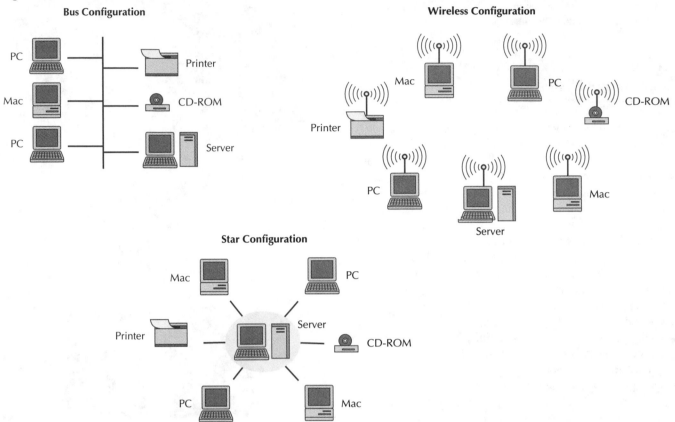

plex challenge for teachers. Those who face the task of designing networks for their organizations can refer to Carlitz and Lenz (1995) for some standards to assure the "interoperability, reliability, and maintainability of a school district's network" (p. 71). Hazari (1995) also gives a case-study example of the tasks and costs involved in designing a school local area network, or LAN, as well as a glossary of networking terms. While schools usually use LANs, they can also tap into wide area networks (WANs) and metropolitan area networks (MANs) that span greater distances. They can also establish network-to-network connections. Kee (1994) provides an easy-to-understand, illustrated introduction to the intricacies of all these kinds of networking.

LANs are groups of computers connected by cables or wireless methods. Many, but not all, networks are also functionally connected by network operating systems that run on separate computers or file servers.

Types of LANs. LANs can be classified by their physical designs and by how they function.

- **Physical design.** LANs are classified according to physical configuration, or "topology," as star, bus, or wireless networks (see Figure 1.4). At this writing, wireless LANs are very expensive and much less common than those connected by cables.

- **Function (transmission method or protocol).** LANs can be classified as token ring or Ethernet networks according to

how they transmit data to individual stations. Every token ring network uses a star topology, while the Ethernet method can be used with star, ring, or wireless networks.

- **Token ring method.** A token ring network allows only one station at a time to send and receive information. In such a network, a signal or "token" is continually sent around the system, and the station that receives the token can then receive or send data. This assures that only one data transmission is being sent at any given time. Kee (1994) describes several kinds of token-passing networks: ARCnet, IBM Token Ring, and Fiber Distributed Data Interface (FDDI).

- **Ethernet method.** The transmission method or protocol in most widespread use today is Ethernet. Unlike token-passing networks, an Ethernet network can send several data transmissions to different destinations at any given time. However, a station can receive only one data transmission at any given time; the network accepts and delivers transmissions on a first-come, first-served basis.

Other protocols (e.g., SPX, IPX, TCP/IP, and network pipes) fulfill special data transmission functions within networks.

Three kinds of cable can connect LAN stations to each other and to the server: coaxial, twisted pair, or fiber optic cable. Hazari (1995) points out that "a majority of network problems occur as a result of incorrect choice of cable type or improper connector installation" (p. 83).

- **Coaxial cable.** This cable is designed to transfer signals with minimal interference. At the cable's center is a conducting wire (usually copper) surrounded by an insulating jacket, then a foil shield, a braided shield, or both. Common uses of coaxial cable are in ARCnet networks and cable TV.

- **Twisted pair cable.** Unshielded twisted pair (UTP) cable is an inexpensive way of handling ARCnet, Ethernet, and token ring connections. This kind of cable has a center conductor such as copper covered by an outside layer of insulation.

- **Fiber optic cable.** This kind of cable is the most recent development in network connections. Designed to allow networks to send large amounts of information at high speeds (Kee, 1994), it consists of glass strands surrounded by a plastic coating and a plastic material called *kevlar* (also used in bulletproof vests). The cable also has an outer jacket made of PVC or Teflon. Fiber optic cable is a multipurpose conductor of data in computer networks as well as sound and video signals.

Types of networked systems in education. A network configuration forms the basis for several kinds of systems in education. Sometimes a network connects individual microcomputers in a lab or computers and/or workstations in classrooms throughout a school. Integrated learning systems (ILSs) and their derivative systems are also made possible by networks. (See more about ILSs in Chapter 4.)

Some network configurations are set up especially to receive instructional information broadcast from distant locations. The equipment in the stations of such a network can be more complex than that in other kinds of stations. This varies according to whether the network carries one-way or two-way transmissions, whether it handles audio-only or audio-plus-video signals. Equipment also varies to suit other kinds of media used in conjunction with the broadcasts. Each distance-learning station must have some kind of connection with the broadcast site (e.g., satellite link or microwave dish) and some devices for receiving and translating the audio and video signals. Chapter 9 on distance learning and telecommunications gives further information on long-distance networks that make possible applications such as videoconferencing and telecomputing.

Interactive Video and Multimedia Stations

Teachers use video technologies such as videodiscs and CD-ROMs by themselves or in combination with computers in configurations called *interactive video* or *multimedia stations*. Interactive videodisc stations are often classified according to their level of interactivity as Level I, II, III, and IV. Level III and IV setups make possible some types of multimedia applications. The computer displays text and graphics information, and it runs a program called a *driver* that allows the user to select and view parts of the videodisc. Some of the more recently developed computer systems are multimedia stations in themselves, since they include CD-ROM players and audio devices. Video images, text, and graphics all are

Example VR Wind Tunnel

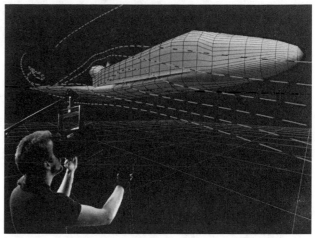

Source: Courtesy of NASA Ames Research Center.

displayed on computer monitors. (See Chapter 7 for more information.)

Virtual Reality Systems

The newest types of technology configurations, and those most rarely found in education, allow a person to feel like a part of a computer simulation with the ability to interact with it. The basic elements virtual reality technology usually include a computer with software that creates simulated environments, headgear consisting of videogoggles and headphones that allows the user to feel the experience of being physically inside an environment, and a DataGlove to provide a realistic feeling of moving around and manipulating things within that environment. (See Chapter 10 for more information.)

Current Applications of Educational Technology Systems

These kinds of hardware/software systems support hundreds of kinds of activities related to education. Some of them touch students directly; some benefit students only by helping teachers prepare for teaching. All the applications discussed in this textbook fulfill one or more of the following roles: instructional applications, productivity applications, or student tools.

- **Instructional technology applications.** These technology resources work directly with students to help them learn some information or skills through demonstration, examples, or explanation. Examples of such uses and resources are:

  - **Instructional software.** This includes drill and practice applications, tutorials, simulations, problem-solving programs, and combinations of these types (discussed further in Chapter 4).

  - **Interactive video-based materials.** This includes information summaries such as *Martin Luther King, Jr.* from ABC

News, problem-solving scenarios such as Optical Data's *Jasper Woodbury Problem Solving Series,* and interactive tutorials (discussed further in Chapters 7 and 8).

- **Courses through distance learning.** These are discussed in Chapter 9.

• **Productivity applications for teachers.** Educators apply these technology resources to help them complete teaching preparations (e.g., developing materials), followup, and/or recordkeeping faster, easier, or more efficiently. Examples of such uses and resources include those which help with the following tasks (discussed further in Chapters 5 and 6):

- **Prepare print instructional materials.** This includes word processing and desktop publishing systems along with test, worksheet, certificate, and puzzle generators.

- **Keep records and analyze data.** This includes database, spreadsheet, and gradebook programs along with statistical packages and computer-managed instruction (CMI) tools.

- **Prepare and make instructional and informative presentations.** This includes graphics tools, multimedia/hypermedia authoring programs, presentation software, digitizing programs, video systems, and bar-code generators.

- **Organize time and materials.** This includes lesson planning tools, schedule/calendar makers, and time management tools.

• **Tools for students.** These technology resources help students complete tasks faster, easier, or more efficiently. Many tools used by students are the same as those used by teachers to enhance productivity. The most popular applications are:

- **Writing assignments in all subject areas.** Students use word processors and desktop publishing systems for these tasks.

- **Helping with research.** Students refer to electronic encyclopedias, databases, and online systems.

- **Assisting with learning tasks in various content areas.** CAD systems, music editors, and probeware serve this function.

- **Developing products and presentations.** For this purpose, students use graphics tools, multimedia/hypermedia authoring programs, presentation software, digitizing programs, and video systems.

Sometimes a given technology product can be applied in many ways. For example, word processing software can demonstrate language concepts (instruction), prepare handouts for students (teacher productivity), or allow students to produce their own written work (student tool). Some products, on the other hand, may have only one type of use. For example, a tutorial program is intended to present instruction in a specific skill area. These general categories help to define the field of technology in education rather than to rigidly classify particular applications. Integration strategies described in each chapter of this textbook are based on instructional or productivity applications for teachers and/or students.

This book is based on the belief that knowing *why* teachers use a given technology resource is as important as knowing *how* to use it. Thus, each integration strategy is discussed in terms of the reason a product is being used, the context in which the applications should occur, and some specific procedures or criteria for using it effectively.

Exercises

Exercise 1a. For each of the following definitions of technology in education, tell which group of people in education was responsible for *originating* it and which professional organization currently represents this view.

1.1 Educational technology means the process of applying audiovisual (AV) and communications equipment as media to meet instructional needs.

1.2 Educational technology means using systematic processes to select methods and materials to solve instructional problems.

1.3 Educational technology means "technology education," or teaching students technology skills that will prepare them to obtain and hold jobs.

1.4 Educational technology means using computers and their related devices as instructional delivery systems and as tools to prepare for and support instruction.

Exercise 1b. Tell which aspect of educational technology this book emphasizes and why.

Exercise 2a. Identify each of the following materials as hardware (H), software (S), or media (M):

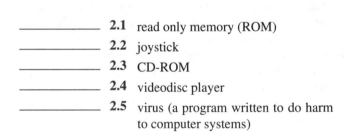

_____ **2.1** read only memory (ROM)
_____ **2.2** joystick
_____ **2.3** CD-ROM
_____ **2.4** videodisc player
_____ **2.5** virus (a program written to do harm to computer systems)

Exercise 2b. Locate or produce a diagram or photograph that shows devices in a computer system that perform the following functions:

• Input functions
• Output functions
• Processing functions
• Internal memory (primary storage) functions
• Secondary storage functions
• I/O or input/output functions
• Communications functions

Label correctly each component with its function and its name. (EXAMPLE - Output device: printer).

Exercise 3. Identify the kind of function(s) each of the following devices fulfills: input (I), output (O), input/output (I/O), internal memory (IM), external memory (EM), or communications (C):

_____ **3.1** hard disk drive

_____ **3.2** CD-ROM player

_____ **3.3** bar-code reader

_____ **3.4** RAM

_____ **3.5** graphics tablet

Exercise 4. Tell whether each of the following kinds of programs is systems software (SS) or applications software (AS):

_____ **4.1** An electronic gradebook that lets teachers store and calculate grades

_____ **4.2** Part of an operating system that tells the CPU to send text on the screen to the printer

_____ **4.3** Optional software one may buy for a microcomputer that checks disks for viruses and helps retrieve lost files

_____ **4.4** Spreadsheet program that lets teachers maintain budgets for clubs or grants

_____ **4.5** Program stored in ROM that checks memory each time the system starts up

Exercise 5. Tell what each of the following people or projects contributed to the development of current educational technology practices:

5.1 What was the IBM 1500? Why is this particular system significant in the history of educational computer systems?

5.2 Who was Patrick Suppes and what did he contribute to the development of the educational technology field?

5.3 For what contribution is Dr. Arthur Luehrmann best known?

5.4 For what contribution is Dr. Seymour Papert best known?

5.5 What does the acronym *MECC* stand for now? What did MECC *first* stand for? What was the major contribution of this organization to the development of instructional computing?

Exercise 6. Identify each of the following products as a type or *combination of types* of computer system configurations (perhaps after reading pages 21–24 on possible configurations):

_____ **6.1** TYCHO 2OOO: A handheld device on which a teacher can use a stylus to enter handwritten notes. The notes can then be transferred into digital information and stored in a computer word processing file.

_____ **6.2** Knowledge Navigator: A hardware and software system that lets students search for text, audio, and video items and then show or play the items they select.

_____ **6.3** A Jostens integrated learning system (ILS): Several microcomputers linked together by cables and served by a central computer. The central computer can download individual programs (e.g., a copy of *Compton's Encyclopedia)* to any connected microcomputer.

_____ **6.4** Acceleration 2000 workstation: A Macintosh microcomputer, monitor, mouse, and keyboard plus a laser disc player, scanner, fax machine, modem, and printer.

_____ **6.5** A learning system consisting of computer and software, headgear that a person wears, and a "steering wheel" that allows the person to learn how to drive a car in simulated driving conditions.

References

Ausubel, D. (1968). *Educational psychology: A cognitive view.* New York: Holt, Rinehart and Winston.

Carlitz, R., and Lenz, M. (1995). Standards for school networking. *T. H. E. Journal, 22* (9), 71–74.

Collis, B. (1988). *Computers, curriculum, and whole-class instruction: Issues and ideas.* Belmont, CA: Wadsworth.

Cuban, I. (1986). *Teachers and machines: The classroom use of technology since the 1920s.* New York: Teachers College Press.

Hazari, S. (1995). Multi-protocol LAN design and implementation: A case study. *T. H. E. Journal, 22* (9), 80–86.

Heinich, R., Molenda, M., Russell, J., and Smaldino, S. (1996). *Instructional media and technologies for learning.* Englewood Cliffs, NJ: Merrill, an imprint of Prentice Hall.

Hill, M. (1993). Chapter 1 revisited: Technology's second chance. *Electronic Learning, 12* (1), 27–32.

Kee, E. (1994). *Networking illustrated.* Indianapolis, IN: Que Corporation.

Lidke, D. (1992). History of computers. In G. Bitter (Ed.), *Macmillan encyclopedia of computers.* New York: Macmillan.

Molnar, A. (1978). The next great crisis in American education: Computer literacy. *T. H. E. Journal, 5* (4), 35–38.

Muffoletto, R. (1994).Technology and restructuring education: Constructing a context. *Educational Technology, 34* (2), 24–28.

National Education Association Communications Survey. (1993). *Report of the findings.* Princeton, NJ: Princeton Survey Research Associates.

Niemiec, R. and Walberg, R. (1989). From teaching machines to microcomputers: Some milestones in the history of computer-based instruction. *Journal of Research on Computing in Education, 21* (3), 263–276.

Norris, W. (1977). Via technology to a new era in education. *Phi Delta Kappan, 58* (6), 451–459.

Papert, S. (1980). *Mindstorms—Children, computers, and powerful ideas.* New York: Basic Books.

Papert, S. (1986). Different visions of Logo. *Classroom Computer Learning, 7,* 46–49.

Petrusco, S. and Humes, V. (1994). Hybrid fiber/copper LAN meets school's 25-year networking requirements. *T. H. E. Journal, 21* (10), 86–90.

Roblyer, M. (1990). The glitz factor. *Educational Technology, 30* (10), 34–36.

Roblyer, M. (1992). Computers in education. In G. Bitter (Ed.), *Macmillan encyclopedia of computers.* New York: Macmillan.

Saettler, P. (1990). *The evolution of American educational technology.* Englewood, CO: Libraries Unlimited.

Spencer, D. (1982). *Exploring the world of computers.* Ormond Beach, FL: Camelot.

Spencer, D. (1992). Charles Babbage. In G. Bitter (Ed.), *Macmillan encyclopedia of computers.* New York: Macmillan.

U. S. Congress, Office of Technology Assessment (1995). *Teachers and technology: Making the Connection.* OTA-EHR-616. Washington, DC: U. S. Government Printing Office.

Chapter 2

Planning and Implementation for Effective Technology Integration

This chapter will cover the following topics:

- Justification for technology purchases by relating them to potential improvements in teaching and learning practices

- Technology's role in restructuring education in the future

- Long-range and short-range plans for acquiring and using technology resources

- Problems with and recommendations for funding technology purchases

- How to put necessary resources in place: hardware, software, physical space, support personnel, and teacher training

- Effective procedures for maintenance, security, and virus protection

- Ongoing ethical, legal, and equity issues in the implementation of technology

- Effective technology use with constant changes in resources

Chapter Objectives

1. Identify some reasons that would and would not help to justify technology purchases.

2. Describe essential elements of a plan for obtaining and using technology resources.

3. Recommend an appropriate reaction for a teacher to deal with:

    - Funding for technology
    - Using technology to help restructure and improve school and classroom practices
    - Choosing hardware and/or software resources
    - Setting up physical spaces for computer resources
    - Selecting and/or training personnel
    - Maintenance and security of computer resources (including virus protection)
    - Equity
    - Legal and ethical concerns
    - Dealing with constant changes in technology

"Let the ideas speak for themselves," more than one scientist told me, "and never mind the people involved." Alas, it isn't quite that simple.
Paula McCorduck, from *Machines Who Think* (1979)

The literature on educational technology is full of glowing promises of dramatic and meaningful improvements to classroom activities and outcomes. But the mere presence of technology is not an automatic guarantee for improved education. In spite of its potential power, educational technology has had some well-documented, high-profile failures (Ferrell, 1986; Morehouse, Hoaglund, and Schmidt, 1987; The revolution that fizzled, 1991). Success with any technology is rarely serendipitous. Certain clear factors profoundly affect whether technology helps education take a leap forward or a pratfall.

What conditions determine the influence of technology? Chapter 2 answers this question by describing three areas that have profound impacts on how well a school can integrate technology into the curriculum: preparation tasks, obtaining appropriate resources, and implementation issues.

Preparing for Technology Integration

Why Use Technology? Developing a Sound Rationale

Many educators, parents, and students already believe that technology should be an integral part of K-12 education. To them, the reasons seem so obvious that everyone should recognize them. This "common sense rationale" for using technology is based on two major points:

* **Technology is everywhere.** A widely accepted belief holds that technology already plays a high-profile role in the educational system and that schools and classrooms cannot deliver high-quality education without using technology-based methods. People tend to believe that since technology tools play important roles in other areas of society, education should also reflect this growing trend.

  Technology is certainly a part of the landscape of society. There is no place one can go, no job one can choose to avoid it. Many people conclude, then, that technology logically should also play a major role in educating children. Many also observe many of the country's most successful educators employing technology in key ways.

* **Technology has been shown to be effective.** Since computers and other technology resources have been in widespread use in education for many years, people assume that a substantial body of research shows the effectiveness of computer-based methods as compared to other methods, at least for certain kinds of learning needs. However, extensive research with computer-based methods supports only a general conclusion that technology has made a difference—sometimes.

Both of these commonly held beliefs have some validity, and both provide rationales for using technology—at least as far as they go. But both also tend to be too general to show specifically how to use technology in education. That requires answers to some practical questions:

* Should technology take over most or all of the teacher's role? If not, how should it fit in with what teachers already do?
* Should schools rely on computers at all levels, for all students, or for all topics? If not, which levels, students, or topics suit computer-based methods?
* Does some reliable information suggest specific benefits of using technology in certain ways?

To justify the expensive and time-consuming task of integrating technology into education, teachers must identify specific contributions that technology can and should make to improvements in an education system. Funding agencies, for example, can reasonably ask why a school should choose a technology-based resource or method over another path to reach its desired goals. As Roblyer (1993) noted, "Answering the question, 'Why use technology in education?' seems not only necessary, but fundamental to all our efforts with technology. It is important ... for assuring that ... technology is used to shape the kind of future we want for education and for society itself" (p. 13).

A specific rationale for choosing technology will guide goals for technology use and help identify the skills and resources needed to accomplish these goals. Before looking at some aspects of this rationale, it seems important to take a careful look at the source from which many educators draw evidence of technology's present and potential benefits: educational research.

Problems with research-based justifications for educational technology. Although technology (especially computers) has been in use in education since the 1950s (Roblyer, 1992), research results to date have not made a strong case for its impact on teaching and learning. In general, the number and quality of studies on educational impact have been disappointing (Roblyer, Castine, and King, 1988). Researchers have finished too few studies to make definitive statements about current and projected benefits, and their results have frequently contradicted each other. For many years, supporters of technology found hopeful, if tentative, evidence in applications of statistical methods such as meta-analysis to summarize results across studies comparing computer-based and traditional methods. These methods later fell under criticism, however (Thompson, Simonson, and Hargrave, 1992).

Clark (1983, 1985, 1991, 1994) has been the most vocal critic of "computer-based effectiveness" research. After empirical and statistical analyses of reviews of research in this area, he concluded that most such studies suffered from "confounding variables." They attempted to show a greater impact on achievement of one method over the other without controlling for other factors such as instructional methods, curriculum contents, or novelty. These differences could either increase or decrease achievement. Clark (1985) exhorted educators to "avoid rationalizing computer purchases by referencing the achievement gains" (p. 259) in such studies. Kozma (1991, 1994) responded to these challenges by proposing that research should look at technology in a different way: not as a medium to deliver information but in the context of "the learner actively collaborating with the medium to construct knowledge" (p. 179). Thompson, Simonson, and Hargrave (1992) also questioned the usefulness of past research that focused on computers as delivery systems; both they and Kozma proposed new models in the continuing search for a research base that can justify the expense and logistical difficulties of using technology in education.

At this point in the evolution of the field, with the scarcity of agreed-upon guidelines for the benefits of technology resources for education, it seems best to follow Clark's advice to refrain from using past reviews of research to justify investments in technology. However, several promising lines of research are under way, and meanwhile we can point to several aspects of technology use that have improved educational practice and probably will continue to do so.

Some trends in technology use have found theoretical support in basic research on learning and cognition; others are so new that researchers have not yet designed adequate methods to measure their impact. Still other applications do not lend themselves to behavioral research, but their practical value has been validated by several years of use in schools. Some of these trends may provide the most powerful—and durable—evidence of technology's benefits to education. Table 2.1 lists some arguments that could form a rationale for continuing or expanding the use of technology in education.

Justifying technology use: The case for motivation. Motivating students to learn, to enjoy learning, and to want to learn more has assumed more importance in recent years than it ever had before. Recognizing strong correlations between dropping out of school and undesirable outcomes such as criminal activity, society has identified the drive to keep students in school as an urgent national priority. A growing belief holds that technology has an important role to play in achieving this goal. Kozma and Croninger (1992) described several ways in which technology might help to address the cognitive, motivational, and social needs of so-called "at-risk" students; Bialo and Sivin (1989) listed several software packages that were either designed or adapted to appeal to these kinds of students. Technology-based methods have successfully promoted several kinds of motivational strategies that may be used individually or in combination:

• **Gaining learner attention.** In 1965, renowned learning theorist Robert Gagne proposed a need to gain the attention of

---

Table 2.1   Elements That May Serve as a Rationale for Using Technology in Education

1. **Motivation**
   - Gaining learner attention
   - Engaging the learner through production work
   - Increasing perceptions of control

2. **Unique instructional capabilities**
   - Linking learners to information sources
   - Helping learners visualize problems and solutions
   - Tracking learner progress
   - Linking learners to learning tools

3. **Support for new instructional approaches**
   - Cooperative learning
   - Shared intelligence
   - Problem solving and higher-level skills

4. **Increased teacher productivity**
   - Freeing time to work with students by helping with production and recordkeeping tasks
   - Providing more accurate information more quickly
   - Allowing teachers to produce better-looking, more "student-friendly" materials more quickly

---

the learner as a critical first "event" in providing optimal conditions for instruction of any kind. Although other aspects of instruction must direct this attention toward meaningful learning, teachers widely recognize that the visual and interactive features of many technology resources does, indeed, effectively help to focus students' attention and encourage them to spend more time on learning tasks (Pask-McCartney, 1989; Summers, 1990–91). Substantial empirical evidence indicates that teachers frequently and beneficially capitalize on the novelty and television-like attraction of computers and multimedia to achieve the essential instructional goal of capturing and holding students' attention.

• **Engaging the learner through production work.** In one highly successful way to make learning more meaningful to students, teachers often try to engage them in creating their own technology-based products. This strategy has been used effectively with word processing (Tibbs, 1989; Franklin, 1991), hypermedia (Volker, 1992; LaRoue, 1990), computer-generated art (Buchholz, 1991), and telecommunications (Taylor, 1989; Marcus, 1995). Reports of such uses reveal that students like the activities because they promote creativity, self-expression, and feelings of self-efficacy and because they result in professional-looking products students can view with pride.

• **Increasing perceptions of control.** Many successful users of technology-based materials say that students find strong motivation in the feeling that they are in control of their own learning (Arnone & Grabowski, 1991; Relan, 1992). Learner control seems to have especially important implications for at-risk students and others who have experienced academic failure. When students perceive themselves as in control of their learning—either through setting the pace of movement through a drill or tutorial or by creating their own computer-generated products with Logo or word processing software—it often seems to result in "intrinsic motivation." That is, students become caught up in and motivated by the awareness that they are learning. This finding, which has been reported from the earliest uses of computer-based materials, continues to be one of the most potentially powerful reasons for using technology resources as motivational aids. Exceptions to this notion of learner control is when learning paths become very complex (e.g., with hypertext environments and interactive videodisc applications). In these cases, learners with weak learning skills seem to profit most when teachers supply some structure to the activities (Kozma, 1991, 1994; McNeil and Nelson, 1991).

Justifying technology use: Unique instructional capabilities. Another extremely powerful case for using technology resources is that some technological media can facilitate unique learning environments or contribute unique features to make more traditional learning environments more powerful and effective.

* **Linking learners to information sources.** Hypertext systems are computer-based products that provide readers with links between information from a variety of sources. A student can select a keyword from a screen and get options to see several other sources with other information on the same topic. These, in turn, can lead to other, related sources and topics, forming an endless chain of information to peruse. Kozma (1991, 1994) reports that, while little research has focused on hypertext to date, encouraging preliminary findings suggest that a hypertext learning environment "both calls on and develops skills in addition to those used with standard text" (1991, p. 203) and "helps the reader build links among texts ... and construct meaning based on these relationships" (1991, p. 204). Computers handle the logistics of this complex activity and, though it remains a complicated process, they make it more feasible for classroom activities.

* **Helping learners to visualize problems and solutions.** Kozma (1991) also reports that interactive visual media such as videodisc applications seem to have unique capabilities for instruction in topics that involve social situations or problem solving. He notes that these media provide powerful visual means of "representing social situations and tasks such as interpersonal problem solving, foreign language learning, or moral decision-making" (p. 200). The growing number of videodisc products designed for these kinds of topics (e.g., the *AIDS* videodisc from ABC News, Computer Curriculum Corporation's *SuccessMaker,* and *A Right to Die? The Case of Max Cowart* [Covey, 1989]) confirms that designers and educators are beginning to recognize and exploit these unique and powerful qualities.

* **Tracking learner progress.** Integrated learning systems (ILSs) and subsequent products based on them have capitalized on the computer's unique ability to capture, analyze, and present data on students' performance during learning (*Electronic Learning,* 1990, 1992; *Educational Technology,*

1992). This ability for data gathering and reporting is central to all efforts to design efficient and meaningful instructional paths tailored to individual students' learning needs.

A teacher attempting to teach a set of skills to a large group of students needs accurate and up-to-date information on what each student is and is not learning. The teacher needs this information in a format that can be quickly reviewed and analyzed. A well-designed computer-based system for data collection (sometimes called a *computer managed instruction* or CMI system) offers a unique capacity to provide this essential information. In addition, new technology products such as pen-activated devices allow teachers and researchers alike to keep moment-to-moment records of their observations of students. These important records can later be analyzed for indications of appropriate learning experiences. Progress is also being made toward affordable expert systems that can provide instruction, analyze students' errors and learning styles as they go through instruction, and provide feedback tailored to unique learning needs.

* **Linking learners to learning tools.** The ability to link learners at distant sites with each other and with widely varied online resources has long been recognized for its unique potential to support instruction and enhance learning (Clark, Kurshan, and Yoder, 1989; Kurshan, 1990; Roblyer, 1991; U. S. News and World Report, 1993; Marcus, 1995). These capabilities include getting access to information not available through local sources, developing research and study skills that will benefit students in all future learning, and providing multicultural activities without leaving the classroom. Some unique affective benefits have also been observed, including increased multicultural awareness as students of different cultures interact online (Roblyer, 1991) and enhanced communication skills when students correspond with each other (Cohen and Riel, 1989).

Justifying technology use: Support for new instructional approaches. The educational system is struggling to revamp its instructional goals and methods in preparation for the complex demands of life in the technology-driven 21st century (SCANS Report, 1992). Educators are beginning to look at technology resources to help make these new directions at once feasible and motivational to students. Several new instructional initiatives can benefit from applications of technology:

* **Cooperative learning.** There is a growing realization in American society that its traditional cultural emphasis on individualism as opposed to group activities will not promote success in the complex problem solving that lies ahead. This has led to an increase in emphasis on small-group instructional activities that involve cooperative learning. Technology-based activities that lend themselves to cooperative, small-group work include development of hypermedia products and Logo programs, development of special-purpose databases, research projects using online and offline databases, and research projects using videodiscs and multimedia.

* **Shared intelligence.** In a concept related to cooperative learning, educators are exploring the potential for intelligence to function not simply as an individual capability, but also as a product of individuals and tools, each of which contributes

to desired goals. Technology resources such as those described above make possible this "shared intelligence" or "distributed intelligence." According to some theorists, the capabilities afforded by new technologies make the concept of intelligence as something that resides in people's heads too restrictive. "Intellectual partnership with computers suggests the possibility that resources enable and shape activity and do not reside in one or another agent but are genuinely distributed between persons, situations, and tools" (Polin, 1992, p. 7). Therefore, some educators hypothesize that the most important role for technology might be to change the goals of education themselves, as well as the measures of educational success.

- **Problem solving and higher-order skills.** While basic communications and mathematics skills are still recognized as essential, educators are also increasingly aware that they must emphasize the learning of specific information less than learning to solve problems and think critically about complex issues. In addition, curriculum is beginning to reflect the belief that students need not master basic skills before going on to higher-level skills. The engaging qualities of technology resources such as videodiscs, multimedia, and telecommunications allow teachers to set complex, long-term goals that call for basic skills, thus motivating students to learn the lower-level skills they need at the same time they acquire higher-level skills.

Justifying technology use: Increased teacher productivity. An important but often-overlooked reason for using technology resources is to help teachers cope with their growing paperwork load. Teachers and organizations alike have recognized that if they spend less time on recordkeeping and preparing teaching materials, they can spend more time analyzing student needs and having direct contact with students (Adams, 1985; Minnesota State DOE, 1989; George Mason University, 1989). Teachers can become more productive through training in technology-based methods and quick access to accurate information that can help them meet individual needs. Many technology resources can help teachers increase their productivity in these ways: word processing, spreadsheet, database, gradebook, graphics, desktop publishing, instructional management, and test generator programs, along with online communications between teachers (e.g., e-mail) and other online services (e.g., Prodigy).

Technology's Role in Restructuring Education: Dilemmas and Directions

Still another part of the rationale for integrating technology into education comes from its widely perceived role in school reform and restructuring. Many educators are convinced that technology is essential to the curriculum reform and school restructuring that is needed to improve the educational system (Bruder, Buchsbaum, Hill, and Orlando, 1992; Hill, 1993). The proper role for computers and related technology in education has stimulated continued and often

intense debate for some years. Although computers captured the imaginations of educational innovators early in the 1960s, no commonly held vision has ever emerged to show how technology would enhance the educational process. Even now, with an apparently growing dissatisfaction with traditional teaching and learning systems and a consensus on the need to change or restructure American education, considerable disagreement persists over the part that technology will play in the restructured system.

Replacing teacher functions versus changing teacher roles. In the early days of educational technology, when resources were available only through centrally controlled, mainframe computer systems, some foresaw technology eventually replacing the teacher as the primary instructional delivery system (Norris, 1977). However, the advent of standalone microcomputers placed the power of technology directly in the hands of teachers, and the image of technology shifted from replacing teachers to supplementing and enhancing teacher-based instruction. Today, as mounting criticism assails the educational system as expensive, inefficient, and outdated, technology is again proposed as an alternative to delivering instruction primarily through teachers (Reigeluth and Garfinkle, 1992).

This proposal asserts that technology-based delivery systems will achieve better results by standardizing instructional methods and decreasing personnel costs (Smith, 1991; Reigeluth and Garfinkle, 1992; U. S. News and World Report, 1993). Some critics advocate technology-based systems as replacements for the traditional roles of both schools and teachers (Perelman, 1993). The opposing view seems to anticipate that teachers and schools must remain an important part of the instructional process, but that technology tools will empower them to teach better and use their time more productively. As calls for curricular reform increase, however, it is apparent that far-reaching changes in traditional teacher roles will be a part of the total restructuring package.

Enhancing existing methods versus changing the nature of education. Even if one discounts the option of eliminating or decreasing the role of teachers, considerable debate remains over the related question of just how technology will change those teachers' roles. As Neuman (1991) observed, depending on how technologies are implemented, they can either help restructure a school's fundamental operations and educational goals or support existing structures. She points out that integrated learning systems (ILSs), for example, are designed to fit in with both the goals and operations of the existing school organization. However, other kinds of resources such as local area networks can add flexibility to a school's curriculum and schedule. This flexibility facilitates long-term, open-ended student projects, the essence of a restructured curriculum.

Papert (1980) was an early critic of traditional approaches to teaching and learning that emphasize isolated skills. He advocated a less structured environment that would let students use computers to learn to think and solve problems. His vision of Logo "microworlds" as a basis for this kind of teaching received widespread attention in the late 1980s, but it later gave way to a broader view of learner-directed methods that has become known as *constructivism* (Bagley and Hunter, 1992; Strommen and Lincoln, 1992). This framework calls for assigning tasks that emphasize learners' creativity and allow them to construct or build their own knowledge rather than giving them knowledge to absorb. A separate but related view would restructure learning around "whole language" or interdisciplinary student projects that emphasize cooperative work and collaborative teaching (Butzin, 1991; David, 1991). Proponents of approaches like these view technology as a way to facilitate fundamental changes to learning methods. Technology resources allow easy access to information and help the teacher cope with the complexities of managing individual and small-group work in the classroom (Ahearn, 1991).

Preparing for an uncertain role. The educational system clearly is responding to recent criticisms of its productivity by making profound changes in its goals and methods. Technology will certainly play a key role in the new system. However, the nature of that role remains uncertain, since it will depend upon the paradigm or combination of paradigms that are eventually adopted. As Sheingold (1991) emphasized, "… it is not the features of the technology alone, but rather the ways in which those features are used in human environments that shape its impact" (p. 18). The "ways in which those features are used" (i.e., integration strategies) are still being decided. Meanwhile, teachers face the difficult task of preparing appropriately for a future that is still in the process of being shaped. The set of skills and integration strategies needed to use technology effectively could differ radically depending on which restructuring direction a school or district takes.

Predictions on technology's role in restructuring education. Literature on technology's role in restructuring yields some common principles (Ahearn, 1991; Norris and Reigeluth, 1991; Foley, 1993; U.S. News and World Report, 1993; Muffoletto, 1994; Luterbach and Reigeluth, 1994; Chesley, 1994; Reigeluth and Garfinkle, 1994; Jostens Learning Corporation, 1995). The following recurring themes seem to be perceived as central to all efforts at building a more effective system of education:

- **Teachers will retain a key role.** Although teacher roles will undergo radical changes, few consider replacing teachers with technology-based delivery systems as a viable option. Even where teachers are not available or in short supply (e.g., in rural schools and highly technical subject areas), the technology strategy of choice seems to be networking or distance learning to optimize the power of available teachers. Technology resources will also help teachers to shift their emphasis from delivering information to facilitating learning.

- **Interdisciplinary approaches will flourish.** Curriculum will change from a disjointed collection of isolated skills training to integrated activities that incorporate many disciplines and call for teacher collaboration. The theme-based projects described in Part IV of this book illustrate how technology resources can both focus and facilitate these cross-disciplinary activities.

- **Research and problem-solving skills will gain attention.** Pure constructivist principles may prove difficult to implement under conditions of current constraints and resource limitations, but educational goals are already undergoing two kinds of shifts. First, an increasing emphasis on general-purpose study and research skills seeks to help learners in any content area. Use of databases, online information services, and hypermedia systems will promote success in this new direction of studies. Second, the emphasis is shifting from learning isolated skills and information within each content area to learning how to solve problems specific to each area. Again, the engaging qualities of technology resources such as videodiscs, multimedia, and telecommunications help teachers to focus students on such complex goals that call for underlying basic skills.

- **Assessment methods will change to reflect the new curriculum.** New calls for "authentic assessment" methods mirror the need to make both instruction and evaluation of progress more relevant to student needs. Assessment of performance is shifting from paper-and-pencil tests to performance-based methods and student portfolios. Technology-based production tasks can serve both as means of accomplishing this assessment goal and ways to track acquisition of underlying skills.

A Technology Planning Guide

Although no one is ever sure exactly what the future will bring, teachers know that they can strongly influence events in schools. Setting appropriate goals and developing sound plans for reaching them are such common-sense prerequisites for success in any endeavor that someone might assume that any technology project would follow a well-conceived plan. Sadly, this is not always the case.

Recent surveys indicate that schools and districts often purchase technology resources without first adopting technology usage plans (Dyrli and Kinnaman, 1994a). Lack of planning does not guarantee failure of an educational technology project any more than planning assures success. Still, technology experts and technology-oriented educators generally agree that developing and maintaining a school-level and/or a district-level plan increases significantly the likelihood of receiving the full benefits of technology's potential for improving teaching, learning, and productivity.

A technology plan helps a school or district make sure that its investment in technology pays expected dividends. However, the process of planning itself requires an investment of time and resources. Technology planners can spend a substantial amount of time researching various products and services, meeting to discuss options and make decisions, documenting their findings, and communicating them to others. Agreement may not come easily on issues such as which brands of computers and software to adopt and who gets computers first. In fact, these issues can spark ongoing, heated debate among faculty and staff. Anyone who undertakes this task must recognize that technology planning is worth the time and effort it requires. Several statements summarize the rationale for this preliminary investment:

- **Planning saves time and money.** A technology plan helps to prevent purchases and activities that do not move the organization toward its goals. For example, if preset criteria guide equipment and software purchases, it is less likely that someone will buy products in a casual or uninformed way. Also, thorough, basic research on products and services ahead of time by a central committee avoids wasteful duplication of efforts later.

- **Planning helps achieve goals.** As Robert Mager (1984) once said, "If you're not sure where you're going, you're liable to end up someplace else" (p. *v*). Without a clear idea of what a technology initiative should accomplish, it is difficult to know whether or not technology is achieving its goals and, if not, how to make changes. Technology plans require educators to set goals, periodically evaluate their progress toward achieving them, and revise them based on concrete evidence.

- **Planning builds motivation.** Any effort to take advantage of technology's benefits must overcome a major problem of convincing people in the school or district that these resources justify the effort to integrate them. Planning for technology forces participation by key people from each group in the organization. As they review resources and set goals for technology use, they become acquainted with the potential benefits; they are also more likely to begin using technology resources that they have helped to select. Finally, participants in the planning process are more likely to become advocates for technology, working to convince other members of their groups to use resources that become available.

In sum, even the smallest school or district can find an abundance of good reasons to develop and adopt its own technology plan. Indeed, it hardly makes sense to use technology *without* completing the planning process as an essential first step.

Planning strategies and steps. Before planning can begin, the planners must be identified. Most reports of first-hand experience with planning for technology (Apple Computer Company, 1991; Association for Media and Technology, 1991; See, 1992; Bruder, 1993; Dyrli and Kinnaman, 1994a; Wall, 1994; Brody, 1995) recommend assigning the task to a technology planning committee made up of both educators and technology experts, as well as representatives from all groups in the school or district. As Dyrli and Kinnaman (1994a) and Brody (1995) point out, such committees are most effective when appointed by top-level administrators who give them authority to implement what they recommend. (See Figure 2.1 for sample planning screens.)

Several good sources document the steps that a planning committee should follow to develop a sound technology plan. In 1991, the Apple Computer Company developed a planning guide entitled *Teaching, Learning, and Technology—A*

Planning Guide. This recently updated multimedia package describes these steps in detail and gives examples in both written and video formats. Dyrli and Kinnaman (1994a) also describe a good sequence of planning steps, and Brody (1995) gives a well-prepared summary of planning steps and guidelines. A recommended sequence common to these and other sources includes six steps:

Step 1   Create a "merged vision." As a critical first step, planners should envision potential applications of technology. As part of this process, they should identify a clear statement of the organization's mission and philosophy in order to articulate a role for technology. For example, a school's central goal may emphasize ethnic and cultural diversity. Technology planners should then emphasize applications that will promote and reflect this priority. Dyrli and Kinnaman (1994a) advocate collecting and analyzing all available materials that document the organization's mission, curricular goals and objectives, and educational guidelines. With this kind of information in hand, the committee can begin to research technology resources and activities with the aim of merging the educational vision of the school or district with a vision of the benefits of technology to promote organizational goals and priorities.

Step 2   Assess the current status. In the next step, technology planners review the organization's current uses of technology. This usually requires a survey instrument to collect data on current resources and activities. The members of the planning committee may also want to visit classrooms and labs to observe technology uses first-hand and talk to those involved. Whenever possible, the committee should present data in visual ways such as charts and graphs so anyone can easily see who is doing what with which technology resources.

Step 3   Set goals. Dyrli and Kinnaman (1994a) call this activity "developing a guiding framework" (p. 53). At this stage, planners specify concrete goals that direct the organization's later actions. These principles should address instructional, administrative, and teacher productivity uses as specifically and in as much detail as possible. For example, a school may specify a goal that by a certain date, all teachers will keep their grades on an electronic gradebook program and that all students in secondary grades will make at least one presentation via presentation software or a multimedia system. To keep these performance aims practical and feasible, the committee will probably want to review other, previously developed plans that talk to a variety of experts and technology-oriented educators who have successfully adopted technology resources. Apple (1995) also recommends careful review of and reflection on potential goals, leading to revisions that produce final statements.

Step 4   Develop activities. After developing technology goals, the committee must outline specific activities that will take the organization from where it is to where it wants to be. This part of the plan specifies needed purchases and training and a time frame for accomplishing them. The Apple Computer Company (1995) model calls for several events at this step: identifying human resources, developing a time line, developing a budget and identifying funding sources, and deciding how to evaluate implementation. It also recommends developing a presentation package to communicate the plan to everyone involved.

Figure 2.1   Sample Planning Program from Apple Computer Company

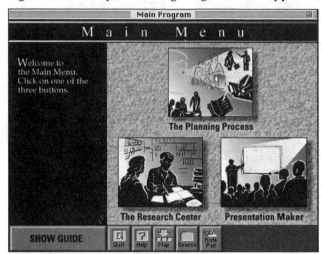

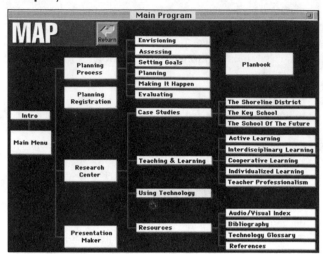

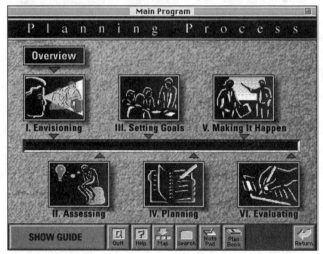

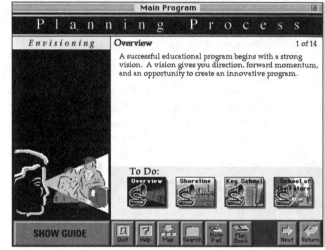

Source: Courtesy of Apple Computer, Inc.,(1995) *Teaching, Learning, and Technology: A Planning Guide.*

Step 5   Implement the plan. To make sure that a plan leads to actions, planners begin by obtaining the approval and endorsement of key decision makers. They may present their findings to a school board, principal, and/or PTO board. Once the plan is approved, several individuals and groups will play key roles in its implementation. The planning committee will continue to supply guidance and direction. A district technology coordinator can also help to oversee all the activities.

Step 6   Evaluate and revise the plan. Implementation is not really the end of planning; in fact, technology planning should never really end. Technology changes so quickly and dramatically that periodic review and revision of any plan is an absolute necessity. Activities should be monitored continuously and adjusted as necessary to assure accomplishment of the overall goal: to use technology to improve education and promote the organization's educational agenda.

Characteristics of good planning. Apple (1991), See (1992), Dyrli and Kinnaman (1994a), and Wall (1994) offer good advice to assure effective completion of all phases of technology planning. There are several important points to consider:

• **Planning should continue at both district and school levels.** Some decisions are best made at the district level, and some issues are best left for each school to consider; but plans each level should coordinate with other plans. Dyrli and Kinnaman (1994a) recommend that each school designate a technology "liaison/coordinator" to act as the school's representative in a districtwide planning committee.

• **Involve teachers and other personnel at all levels.** To obtain widespread support for a plan, the planning team should include parents, community leaders, school and district administrators, and teachers. Involving teachers is especially important. Any technology plan must show where and how technology resources will fit into instructional plans for all grade levels and content areas. Just as curriculum plans require input from teachers, technology plans depend on direct guidance from those who will implement them.

• **Budget yearly amounts for technology purchases.** Technology changes too rapidly for schools to expect one-time purchases of equipment or software to suffice. A technology plan should allow for yearly upgrades and additions to keep resources current and useful.

- **Make funding incremental.** Few schools' yearly budgets allow the purchase of all needed resources or teacher training. A plan should identify a specific amount to spend each year and a priority list of activities to fund over the life of the plan.

- **Emphasize teacher training.** Knowledgeable people are as important to a technology plan as up-to-date technology resources. Successful technology programs hinge on well-trained, motivated teachers. A technology plan should acknowledge and address this need with appropriate training activities. See (1992) recommends close coordination between technology training plans and staff development plans.

- **Apply technology to needs and integrate curriculum.** To paraphrase the old adage, "If technology is the answer, what's the question?" Effective planning focuses on the correct questions. For example, planners should ask "What are our current unmet needs and how can technology address them?" Too many skip this question and jump to "How can we use this equipment and software?" It is difficult to identify needs since the emergence of new technology has a way of changing them! Many educators didn't realize that they needed faster communications until the fax machine, e-mail, and cellular telephones became available.

  Curriculum integration should also focus on "unmet needs." Technology should become an integral part of new methods to make education more efficient, exciting, and successful. Planners should ask, "What are we teaching now that we can teach better with technology?" and "What can we teach with technology that we couldn't teach before but that should be taught?"

- **Keep current and build in flexibility.** Both technology and users' opinions about how to implement it change daily. As textbooks of only 2 or 3 years ago show, leading-edge technology solutions can become out-of-date soon after their development as more capable resources emerge and new research and information clarify what works best. To keep up with these changes, educators must constantly read and attend conferences, workshops, and meetings—a full-time job in itself! Each school's and district's technology plan should address how it will obtain and use technology resources over a 3-year to 5-year period (New York State School Boards Association, 1989; Mageau, 1990; Orlando, 1993). But any technology plan should be designed to incorporate new information and changing priorities through yearly reviews and revisions (See, 1992).

- **Planning essentials and mistakes.** See (1992) and Palazzo (1995) cite some critical attributes and criteria for successful technology plans. These include: planning committees made up of parents, teachers, administrators, and business leaders; provisions for on-site technical support; access to hardware and software; long-term staff development and in service training; assessment of present technology status and future needs; and ongoing assessment and evaluation methods. On the other hand, Wall (1994) and Dyrli and Kinnaman (1994a) note some common pitfalls to avoid:

- Failing to link the organization's education goals to its technology planning goals

- Preoccupation with overly detailed recordkeeping or surveys that obscure or overlooks the "big picture" of technology use

- Making plans too general (e.g., stating goals too vaguely) or too specific (e.g., requiring purchases of certain hardware that will become obsolete over time)

- Making massive investments in untried, first-generation technology

Example technology plans. The Apple Computer Company multimedia package (1995), Dyrli and Kinnaman (1994a), Van Dam (1994), and Palazzo (1995) offer good examples of plans that have already been developed. Apple demonstrates planning and implementation activities of four example schools. Dyrli and Kinnaman's article cites sample plans from the National Center for Technology Planning (NCTP) at Mississippi State University. They note that these plans can be obtained either by ftp (file transfer protocol) via the Internet at RA.MSSTATE.EDU in the directory /PUB/ARCHIVES/NCTP or by mail for the costs of copying and mailing. Van Dam (1994) gives a very down-to-earth description of one school's experience in renovating its facility to accommodate and promote the use of new technologies. Palazzo (1995) describes five "great technology plans" that won a planning contest sponsored by a magazine.

Obtaining the Right Material and Personnel Resources

Funding for Technology Resources: Problems and Recommendations

In a field known for its lack of consensus, it is remarkable that there is the general agreement that adequate funding can mean the difference between the success or failure of even the best technology plans (November and Huntley, 1988; Bullough and Beatty, 1991). Formal studies of obstacles to technology integration have reached the same conclusion (Bailey, 1990; Mahmood and Hirt, 1992). The most important issues in educational technology reflect those in the education system itself, and both place funding at the top of the list. Funding issues may be defined by three critical questions:

1. What do schools need to improve the present situation?
2. What kind of investments will it take?
3. Where and how will schools get the funds?

The first question is the most difficult to answer. Educators invest time and money in technology because they believe that it will help to improve their ability to teach and students' ability to learn. Teachers devote great effort in locating resources to accomplish these aims. Once a school system—or an individual teacher—decides what to do, a wealth of guidelines and advice suggest resources that will meet the identified need and how to find money to buy them. However, several problems can complicate the identification of resources and the search for funding.

The high price of keeping up with technology. Besides the high initial cost, the primary problem with investing in technology is the changing pattern of technology usage along with revisions in the associated definition of "adequate resources." When microcomputers first entered schools in the late 1970s, educators strived to get enough microcomputers to lower their computer-to-student ratios and enough drill, tutorial, and simulation software packages matched to all slots in the "curriculum matrix" (all content areas and all grade levels). Schools that invested heavily in early microcomputers were often surprised not only at how quickly their equipment became out-of-date, but also at its incompatibility with newer models. Within a relatively short period of time, a completely new generation of more capable and "friendly" equipment became available.

In addition, the philosophy of the benefits of technology for teaching and learning was evolving rapidly. The problem of providing adequate teacher training, always a difficult and expensive need, became even more difficult without agreed-upon directions for how best to integrate technology into instruction. Maintenance and security for existing resources also became important cost issues. In the 1980s and 1990s, new directions in technology use replaced the emphasis on microcomputers with the trend toward multimedia and integrated learning systems. Schools now face a dual challenge that seems likely to remain the only constant amid changing educational technology: how to acquire technology resources adequate for today's needs while keeping an eye on emerging trends in the field that could affect future purchases and training.

The lack of hard evidence on effectiveness. Traditionally, the keepers of the purse strings for any area want concrete evidence that a planned investment is likely to bring about the desired improvement. As Roblyer, Castine, and King (1988) observed, "society currently has some very specific measures for the effectiveness of its educational system: student achievement, student attitudes (toward learning), dropout rate, and learning time" (p. 12). Unfortunately, advocates of technology often have trouble isolating clear evidence about technology's impact on any of these criteria. As budget consciousness increases the expense of keeping up with technological change rises, it is difficult to propose expensive technology programs or resources with benefits that are hypothetical rather than proven.

Special problems for rural schools. The usual problems with funding for technology seem magnified for schools in rural communities (Clauss and Witwer, 1989; Clark, 1990; Inman-Freitas, 1991; Freitas, 1992; Holland, 1995). As Freitas points out, financial problems in rural school districts often stem from state funding formulas that tend to favor larger, more urban districts. This frequently results in less state funding for areas that need it most (Clark, 1990). Since nearly a quarter of funding for educational technology comes from state revenues (Mernit, 1993), this is a sig-

nificant problem for a large segment of the country's educational community. For many schools this will have a clear impact on potential funding for technology.

Recommended funding strategies. More positive trends seem likely, however, because most people are becoming aware of the increasingly pervasive influence of technology throughout society, and this influence cannot avoid education. Investments in technology are at an all-time high in education because educators and parents alike recognize its critical role in current and planned efforts to make a foundering educational system more efficient and more responsive to the needs of today's students (Branson, 1988, Dede, 1992). Current uses of technology based on past experience help to define and shape this future role. This accompanies a growing awareness among legislators and funding agencies that technology in education will require major investments—both initially and continually (Clark, 1990; Rose, 1992). Several tactics can help educators who need funding for technology resources to identify the most promising technology-based activities and maximize their chances for finding financial support for their plans.

Partnerships and other funding sources. Business and industry partners have become part of a major strategy for funding education in general in recent years (McCarthy, 1993). Many companies have come to share a special interest in funding technology in education, and other potential funding sources abound. Several recent publications have documented these sources and how schools can tap them (*Technology and Learning,* 1992; *Electronic Learning,* 1993). These journals' special issues, which also include advice on grant writing and fund raising, provide invaluable assistance in locating and obtaining support for technology.

Stretching scarce resources. In the best American tradition of American frugality and economy, educators have created many ways of making do with their current technology resources (Smith, 1992; Finkel, 1993). Some strategies for optimizing resources emphasize:

- Requiring competitive bids for large items or frequently used supplies
- Upgrading current software whenever possible
- Recycling whenever possible (e.g., re-inking printer cartridges)
- Using older equipment to meet lower-profile, noninstructional needs
- Sharing resources among groups whenever feasible

Choosing the Right Software for Your Needs

The first question that technology planners must answer asks, "What do you want technology to help you do?" Software and hardware selections hinge on the answer to this question. In the past, planners have followed the maxim, "Software drives hardware." That is, they first identify needed software to accomplish their goals, then they get hardware that will run the software. This remains generally true, but sometimes hardware and software come packaged together, as in an integrated learning system (ILS). However they choose software, educators should consider several factors in this decision:

- **Quality.** Although technical problems (e.g., program "breaks") do not plague today's software as they did in early applications, programs can still have little flaws and idiosyncrasies that are discovered only through extensive use. Teachers should become fairly familiar with a software package's features and characteristics before they buy it. For the best strategy in selecting good quality software that will meet one's needs, Bunson (1988) advocates seeking out software reviews and recommendations, either from informal sources (e.g., people you know who have used a particular package) or published ones. Teachers and students should then pilot test the software before buying it. Suppliers or publishers of more expensive software should also offer support and supply documentation on program operation.

- **Number (and type) of copies.** In the early years of microcomputers in education, software companies stuck to a fairly rigid policy: Users had to buy one full-price copy of a software package for each microcomputer on which the software would be used. Backup disks were rare. Now most companies offer flexible pricing structures in addition to supplying backup copies. For example, they often offer "lab packs," or price discounts for buying five, ten, or more copies at a time. They may sell site licenses that allow one copy on each of many computers in a lab or building. Finally, a publisher may offer a networked version that legally can be placed on a central computer and downloaded to individual stations. Software purchasers must consider how many copies they will need and the options available for the software titles they want.

- **Source of best prices.** Sometimes a software publisher is not the best (i.e., cheapest) source for obtaining a program. Publishers frequently offer their software through dealers who may set substantially lower prices than the publisher itself. Companies such as Educational Resources and Educational Software Institute are examples of dealers that offer hundreds of discounted software titles.

- **Match with curriculum and students.** The market seldom includes only one software package that meets a specific need. To choose between two packages that seem to do the same thing, buyers must match software carefully to the level of their students and the specific needs of their curriculum. For example, two software packages may advertise support for acquisition of pre-algebra skills, but one may be designed for secondary and adult students while the other suits advanced middle school students; the first might take a tutorial approach, while the second is based on a problem-solving strategy. A decision between these packages should focus on a careful match between educational goals and program features.

Choosing the Right Hardware for Your Needs

Planners should carefully consider several factors before buying a single computer or peripheral. Each of these factors directly affects the extent and effectiveness of technology-based activities in the curriculum.

Type of hardware platforms. The first, and by far the most important, influence on hardware selection is the needs of desired software. Schools should not, however, buy out-of-date hardware simply because the organization already owns older software. This choice would set up a Catch-22 cycle that would keep its systems always out-of-date. But software compatibility certainly does affect hardware decisions.

Schools have grappled with another major issue: whether to buy Apple/Macintosh or MS-DOS machines. However, as more machines are developed to run both Mac and MS-DOS operating systems, this problem may be in the process of being resolved. At this writing, even the Macintosh Power PC computers designed to run both Mac and MS-DOS software will actually run only about 90 to 95 percent of the total available software. (Problems occur primarily with software that has many graphics.) Teachers can hope that these incompatibility issues will soon be moot.

Those still considering purchasing one platform or the other should consider several criteria:

- **Ease of use.** The Macintosh operating systems have generally been considered slightly "friendlier" than the MS-DOS environment; that is, Macs are designed to be easier for less technically skilled people to use. Windows (both the program and the operating system) has made this "desktop difference" less noticeable, but some computer users still perceive a substantial edge in favor of the Macintosh. Generally though, people seem most comfortable with the system they began using and were trained on first, regardless of its characteristics.

- **Price.** MS-DOS machines usually carry lower prices than Macintoshes with comparable features such as memory and hard drive size. Prices vary considerably, however, and they always depend on what kind of deal is available to the school, district, or state.

- **Support.** The old warning your parents gave you about sticking to the well-known brands is often good advice with computers. If you buy a Macintosh or an IBM microcomputer, you can count on help with repairs and maintenance. If a school purchases a third firm's duplicate or "clone" of a Macintosh or MS-DOS machine, staff should be sure that someone can repair and maintain the equipment and fulfill its warranty.

- **Intended use.** For equipment to be used by students, educational goals strongly influence the selection of a computer brand and type. Although Macintosh computers are currently popular in education, Apple equipment represents only about 15 percent of the total microcomputer market. To buy computers for vocational training, for example, educators may want to go with MS-DOS/Windows brands that will prepare their students for likely conditions in the workplace. Ideally, students should learn to use both platforms, since they cannot anticipate what they will encounter in their future jobs.

How many computers? A school determines the number of computers to buy based on how it will configure and use the equipment. If each classroom will have a three-computer workstation, the school will probably need more computers than if it is placing computers in a centralized learning lab or media center. The number of computers per lab will also depend on how many people the lab must serve and how often.

Other kinds of equipment. Computers are not the only technology that should be considered an essential purchase. As multimedia and communications applications become increasingly visible in classroom learning activities, workstations and labs must include hardware such as modems, videodisc players, CD-ROM players, and large-screen projection devices. Printers are also required purchases, and laser printers are desirable if students and/or teachers will be doing desktop publishing.

Individual computer characteristics. Any computer has three important characteristics: size of RAM, size of the hard drive, and peripherals (I/O hardware and other devices such as modems). RAM is the memory available when the machine is running software, and the hard drive is the internal device where most users store programs and files. Most newer machines are now sold with built-in CD-ROM players. Of course, all of these characteristics affect the computer's speed.

The problem of hardware incompatibility. As was mentioned previously in this chapter (in the section headed Funding for Technology Resources), incompatibility problems add to the expense of technology investments. Educators must guard against two kinds of incapability: differences between brands and those between older and the upgraded versions of a single brand. Both can pose obstacles to educators with limited funds.

Incompatibility between older and newer versions of a single company's equipment quickly became a major problem in the early days of microcomputers. Some indications suggest, however, that computer companies are becoming more sensitive to the needs of their customers, and they are attempting to maintain some degree of compatibility when they upgrade their products. Since they ultimately aim to get consumers to buy newer (i.e., better) hardware, this problem is never fully resolved.

Some companies that emphasize the education market have recognized that making their equipment and software incompatible with other products inhibits teachers when they try to integrate technology effectively. Manufacturers have attempted to address these problems by providing hardware "fixes" such as circuit boards or cards that may be added to one microcomputer to emulate another brand and run its software. However, these short-term solutions never prove to be completely satisfactory. Meanwhile, teachers cannot afford to wait until industry responds to demands for solutions to these problems.

Strategies for dealing with incompatibility. Some larger school districts and schools minimize compatibility problems by purchasing and upgrading most or all of their technology resources at the same time. These larger entities frequently use their purchasing power to get computer and software companies to help them find solutions to incompatibilities between versions and brands. This preferred strategy requires a great deal of top-down management authority, careful planning, and arduous negotiations with computer companies.

However, most schools either cannot or do not operate this way. They always encounter incompatibility problems because they never have the funds they need to replace older equipment and software with up-to-date versions. The most common school technology inventory consists of a variety of resources acquired at different times for different purposes. Such a school should try to make the most effective possible use of the resources it has while allocating a part of its yearly budget for upgrades and new technology. Several strategies help educators to make the best use of available resources while dealing with incompatible brands and versions:

- **Grouping similar resources.** Most schools that cope effectively with incompatibility tend to group like brands together or assign them similar purposes. For example, all of the Apple II machines may serve programming classes while Macintoshes reside in science labs, and the IBMs serve teacher workstations.

- **Using older equipment for simpler purposes.** Many schools continue to use their oldest pieces of working equipment by relegating them to lower-order functions that the outdated computers can still do effectively. For example, one school district placed all of its Atari machines in elementary classrooms for use with instructional games. Other schools use older equipment for preparing flyers and other simple but often-used graphic products (Jordahl, 1995).

- **Upgrading or selling older equipment.** Jordahl (1995) provides some helpful criteria to guide the decision to upgrade older equipment. Some schools hold used equipment sales; others delegate this task to companies that specialize in resale. Proceeds from equipment sales can fund purchases of new resources.

In all of these strategies, schools do not attempt to make different brands compatible or older equipment compatible with newer acquisitions. They simply isolate incompatible systems from each other by having each perform different functions.

Setting Up Physical Facilities

Schools have developed several common arrangements for technology equipment. Table 2.2 details the benefits and limitations of each. Each school is likely to need several of these configurations, but which it will select depends on practical factors such as how much funding is available and how many students it serves. As Milone (1989) observes, the kinds of instruction that a school needs and wants to emphasize also influences these choices. Labs, for example, are usually considered more useful for providing group instruction, and they are more common at secondary levels; individual workstations seem better

	Table 2.2 Types of Technology Facilities and Their Uses		
	Benefits/Possibilities	**Limitations/Problems**	**Common Uses**
Laboratories	Centralized resources are easier to maintain and keep secure; software can be networked and shared.	Need permanent staff to supervise and maintain resources. Students must leave their classrooms	See below
Special-purpose labs	Permanent setups group resources specific to the needs of certain content areas or types of students.	Usually exclude other groups. Isolate resources	Programming courses; word processing classes of students in mathematics, science, etc.; teacher work labs; vocational courses (CAD, robotics); Chapter I students; multimedia production courses and activities
General-use computer labs open to all school groups	Accommodate varied uses by different groups	Difficult to schedule specific uses. Usually available to only one class at a time	Student productivity tasks (preparation of reports, assignments); class demonstrations; followup work
Library/media center labs	Same as general-use labs, but permanent staff are already present. Ready access to all materials to promote integration of computer and noncomputer resources	Same as general use labs Staff will need special training. Classes cannot do production or group work that may bother other users of the library/media center	Same as general-use labs
Mobile workstations	Stretch resources by sharing them among many users	Moving equipment increases breakage and other maintenance problems. Sometimes difficult to get through doors or up stairs	Demonstrations
Mobile PCs (laptops, PDAs)	On-demand access	Portability increases security problems	Individual student or teacher production tasks; teachers' assessment tasks
Classroom workstations	Easily accessible to teachers and students	No immediate assistance available to teachers. Only a few students can use at one time	Tutoring and drills; demonstrations; production tasks for cooperative learning groups; e-mail between other teachers
Standalone classroom computers	Easily accessible to teachers and students	Same as classroom workstations	Tutoring and drills; whole-class demonstrations; pairs/small workgroups

suited to small-group, classroom work, and they appear more often at lower grades.

Ideally, however, a school would have access to both classroom and lab resources. Each classroom should have a workstation capable of performing the full gamut of technology-based instructional and productivity activities from word processing to multimedia applications. This station should act as a learning station to support either individual or small-group work. In addition to classroom resources, every school with an enrollment of 1,000 students or more should also have at least one general-purpose lab with at least 15 to 20 stations to serve the productivity needs of students and teachers. Generally, larger schools have more special-purpose labs.

Designing technology resources for the classroom. Dyrli and Kinnaman (1994b) describe how today's classrooms should "target for technology." They advise schools to plan to supply four computers per classroom, network and telecommunications access, CD-ROM and laserdisc players, and display capability for both computers and large-screen projection. Although every school may not be able

Example Computer Lab

to attain these ideal conditions (at least not right away), each school should identify the facilities that it wants in its technology plan and set up a priority list that will help it work toward achieving them.

Designing a microcomputer lab. Bunson (1988) gives a rather complete list of concerns to address when setting up a microcomputer lab in a media center. These include:

- **Environmental factors.** A lab's layout must provide spatial arrangements for equipment and traffic flow; furniture; power outlets, uninterrupted power sources, and backup power; antistatic mats and sprays; and proper temperature, lighting, and acoustics.

- **Equipment (resource) acquisition.** Software and hardware needs govern design criteria.

- **Administration.** A lab's design must set policies for copyright enforcement; equipment distribution, control, and access; staff responsibilities and training; budgeting for hardware, software, personnel, supplies, and maintenance; and public relations.

Manczuk (1994) updated this list with some additional factors to address. These include equity and access issues to assure that special populations (e.g., physically handicapped users) can benefit from the center and selection of an automated system to allow the center to maintain and locate resources easily. Security measures and safety features (e.g., preventing electrical shocks) are also major concerns in lab design and placement. Apple Computer Company (1995) has developed a helpful guide that addresses all these important factors. Wilson (1991) also adds design concerns specific to elementary schools, which need to "scale down" workstations for smaller students.

Redesigning school facilities. Van Dam (1994) is among a growing number of educators who urge schools to provide facilities that allow teachers "access to information via

voice, video, and computer data, anytime, anyplace" (p. 56). For many schools, this requires complete redesigns—sometimes referred to as *retrofit*—of their facilities, including new wiring and power supplies. Van Dam describes how her school went about this effort. Such dramatic change is an expensive undertaking, but some organizations consider it so important to the future of technology integration that they have decided to allocate special funds to support these redesign or retrofit activities (Macon, 1992).

Training Teachers

Observers generally agree that properly trained teachers make the difference between success or failure of an integration effort (Sheingold, 1991; Munday, Windham, and Stamper, 1991; Dyrli and Kinnaman, 1994b; Siegel, 1995). Recent studies have settled on the kinds of areas in which teachers should be trained. The National Council for Accreditation (NCATE), the agency responsible for accrediting colleges of education, enlisted the help of the International Society for Technology in Education (ISTE) to develop standards for teaching about technology in education. Todd (1993) and Dyrli and Kinnaman (1994b) summarized fundamental technology goals that ISTE recommended for every teacher:

- Operate a computer system to use software successfully.

- Evaluate and use computers and other technologies to support instruction.

- Explore, evaluate, and use technology-based applications, communications, presentations, and decision making.

- Apply current instructional principles and research and appropriate assessment practices to the use of computers and related technologies.

- Demonstrate knowledge of uses of computers for problem solving, data collection, information management, communications, presentations, and decision making.

- Develop student learning activities that integrate computers and technology for a variety of student grouping strategies and for diverse student populations.

- Evaluate, select, and integrate computer/technology-based instruction in the curriculum in a subject area and/or grade level.

- Demonstrate knowledge of uses of multimedia, hypermedia, and telecommunications tools to support instruction.

- Demonstrate skills in using productivity tools for professional and personal use, including word processing, database management, spreadsheet software, and print/graphic utilities.

- Demonstrate knowledge of equity, ethical, legal, and human issues of computing and technology use as they relate to society, and model appropriate behavior.

- Identify resources to keep current in applications of computing and related technologies in education.

- Use technology to access information to enhance personal and professional productivity.

- Apply computers and related technologies to facilitate emerging roles of learners and educators.

All widespread recognition of the importance of teacher training has accompanied the recent concurrence on the list of required skills. Still, Sheingold (1991) pinpoints a fundamental stumbling block that will complicate teacher training for some time to come: "Teachers will have to confront squarely the difficult problem of creating a school environment that is fundamentally different from the one they themselves experienced" (p. 23). Using technology doesn't stop with computer-based grades or assigning students to use word processing to produce traditional book reports. Instead, technology confronts teachers with both new possibilities and imperatives for radical changes in teaching behaviors. Collins (1991) describes how these new teaching/learning environments differ from those of the past by citing eight trends identified from observations of schools that have begun using technology. He notes the following shifts in classroom behaviors:

- From whole-class to small-group instruction
- From lecture and recitation to coaching
- From working with better students to working with weaker ones
- Toward more engaged students
- From test-based assessment to that based on products, progress, and effort
- From competitive to cooperative social structures
- From all students learning the same things to different students learning different things
- From primarily verbal learning to an integration of visual and verbal thinking

Since most preservice and inservice teachers experienced educational environments far different from the one Collins describes, their technology training must provide first-hand experience with these new methods. Effective training must model the desired environment as it teaches about the new technologies. Brooks and Kopp (1989) and Roblyer (1994) describe ways of modeling technology by using it in the regular activities of teacher education programs; these same methods could also improve inservice training. Suggestions for teacher trainers include:

- Using cooperative learning activities, telecommunications-based projects, and other nontraditional/nonlecture methods to carry out training
- Using presentation software to teach groups and requiring its use for learner presentations to classes and other groups
- Requiring use of technology products (e. g., software and videodiscs) in trainees' research projects or demonstrations for other courses or training workshops
- Requiring learners to do research for class projects using online, CD-ROM, or disc-based databases (e.g., on ERIC via LUIS)
- Having each learner develop and maintain a personal database of recommended teaching resources that includes technology products and projects

Experts also generally agree that technology training requires an ongoing school or district program rather than a one-shot, learn-it-now-or-else session. This new learning introduces too many new concepts and too much information for a teacher to absorb at one time, however long the course. Finally, effective training requires "just in time" exposure to new ideas. Resources should be in place so that teachers can apply what they learn immediately after the training experience.

Protecting Your Investment: Maintenance and Security Issues

With all their power and capabilities, computers and related technologies are simply machines. They are subject to the same mundane and frustrating problems as any equipment; that is, they can break down, malfunction or be damaged, or stolen. As microcomputers came into schools in greater numbers in the 1980s, these problems became increasingly important—and expensive. Schools found that the initial cost of equipment was only a fraction of the funds required to keep it available and useful to teachers. They have found no easy answers to maintenance and security issues, and these subjects represent an important aspect of planning for technology use. This section describes some ongoing maintenance and security concerns that will continue to powerfully affect teachers' ability to integrate technology.

Technology Labs and Workstations: House Rules and Procedures

Most labs adopt rules intended to extend the lives of the resources they buy and make sure that the labs fulfill the purposes for which they were designed. Teachers will find that most of these same rules should apply to classroom workstations. Lab rules and regulations should be posted prominently and should apply to everyone who uses the lab, from the principal to the teacher aides:

- No eating, drinking, or smoking should be allowed near equipment.
- Lab resources should be reserved for instructional purposes (e.g., no one should play noninstructional games).
- Only authorized lab personnel should check out lab resources.
- Group work should be encouraged, but lab users should show respect for others by maintaining appropriate noise levels.
- Schedules for use should be strictly observed.
- Problems with equipment should be reported promptly to designated personnel.

Gray offers a dozen "gems" for managing a microcomputer lab effectively. Although written in 1988 for use in higher education, these guidelines apply equally well to labs in any educational organization, and they are as useful now as when they were written. They include:

1. Conduct a needs assessment.
2. Improve staff communication.
3. Use written operational guidelines.
4. Be cost-conscious.
5. Use wish lists.
6. Inspire student assistance.
7. Manage time effectively.
8. Provide staff development.
9. Keep accurate utilization records.
10. Perform frequent evaluations.
11. Practice hands-on management.
12. Stay abreast of new developments.

Maintenance Needs and Options

Each teacher who uses technology needs training in simple troubleshooting procedures (e.g., how to confirm that the printer is plugged in and the "online button" lit, what to do if a computer says disk is "unreadable"). Educators should not be expected to address more complicated diagnostic and maintenance problems, though. Nothing is more frustrating than depending on a piece of equipment to complete an important student project only to discover it is broken or functioning peculiarly. A technology plan must make some provision ahead of time to expediently replace and repair equipment designated for classroom use.

Schools can minimize technology repair problems if users follow good usage rules and do preventive maintenance procedures (e.g., regularly cleaning disk drives). Even under the best circumstances, however, computers and other equipment will break or suffer damage. A school with more resources can expect to need a larger repair budget. Schools and districts have tried to deal with these problems in many ways. Whole businesses have sprung up to provide maintenance for microcomputers. Educational organizations usually choose one of the following maintenance options:

- **Maintenance contracts.** Like health insurance for machines, these contracts guarantee that equipment will be repaired if and when it breaks. Equipment owners pay per-machine annual fees to outside suppliers that provide this service.

- **In-house maintenance office.** Some educational organizations are large enough to hire special personnel and set up internal offices to service their equipment. Brody (1995) offers some tips on how to set up an effective in-house maintenance program.

- **Built-in maintenance.** Some kinds of equipment, notably integrated learning systems (ILSs), cover maintenance costs as part of their purchase or lease prices.

- **Repair and maintenance budget.** Still other organizations choose to pay for repair and replacement of equipment as needed by allocating portions of their operating budgets for this purpose.

Each of these methods has its problems and limitations, and debate continues over which method or combination of methods is most cost-effective for an organization of a given size with a given number of computers and peripherals.

Security Requirements

Microcomputers and peripherals such as disk drives and printers can be very portable. Indeed, they often seem to walk away on their own! Security is a separate, but equally important, equipment maintenance issue. Loss of equipment from vandalism and theft is a common problem in schools. Again, several options are available to deal with this problem:

- **Monitoring and alarm systems.** Some schools install security systems for their entire facilities or for areas that house technology equipment (Brody, 1995). As with home security systems, these systems typically monitor door or window openings, noises, and/or movement within protected areas. If any problem is detected, the system automatically sets off an alarm and notifies the monitoring office which, in turn, calls police and/or other prearranged contacts.

- **Security cabinets.** Specially-designed cabinets are available that enclose whole microcomputer stations, allowing teachers to close and lock them when not in use.

- **Lock-down systems.** A variety of other methods can make equipment less easy to move. These include devices that attach computers to tables, and wires that tie equipment to furniture or floors.

As with maintenance strategies, each method of protecting equipment from loss is less than perfect, and each involves considerable expense. Depending on the problems encountered at a specific site and the methods selected for dealing with them, equipment maintenance and security arrangements can easily take up a significant portion of the technology budget. But no school should leave security to chance. Everyone should start with the assumption that unprotected equipment *will* be stolen. Although security can be a significant technology-related expense, it is usually cheaper than replacing stolen or vandalized equipment.

Viruses: Causes, Prevention, and Cures

Computer viruses are programs written specifically to cause damage or do mischief to other programs or to information (Hansen and Koltes, 1992). Like real viruses, these programs can pass to other programs they contact. Unlike viruses that affect humans, however, computer viruses can be passed only by connecting one computer to another via telecommunications or by inserting a disk containing the virus into a computer. Some viruses are carried into a computer system on "Trojan horses," or attractive programs ostensibly designed for another, productive purpose but also carry instructions that get around protection codes (Lee, 1992). Some viruses are "worms," or programs designed specifically to run within (at the same time as) other programs; others are "logic bombs" that carry out destructive activities at certain dates or times. Many different strains of viruses plague computer systems, and more are being generated all the time. Hansen and Koltes (1992) hypothesize that most viruses are written out of curiosity or as intellectual challenges. Less often, they seem to have been produced as destructive forms of political or personal protest or revenge. However, Mungo and Clough (1992) warn that this latter kind of activity may be on the increase.

The impact of a virus can take many forms. Some viruses eat through data stored in a computer. Others replicate copies of themselves in computer memory and destroy files. Still others print mischievous messages or cause unusual screen displays. No matter what their purposes, viruses have the general effect of tying up computer resources, frustrating users, and wasting valuable time. Even after a virus has been detected and removed from hard drives, it can return if users do not diligently examine their floppy disks as they insert them into the "cleaned" computer.

Since computer viruses are currently as widespread and as communicable as the common cold, and they can interfere with planned activities nearly as much, teachers and schools must take precautions against contracting these electronic diseases. Dormady (1991) recommends a four-point program of activities to minimize the impact of viruses:

- **Establish good practices.** Scan systems and disks regularly for infections and foreign, suspicious software. Always back up important data or files.

- **Enforce safety policies.** Do not allow users to run illegal copies of software on your computers. Allow only authorized programs to be placed on hard drives.

- **Use virus detection programs.** Consider low-cost virus detection and removal (i.e., disinfectant) programs as required purchases for labs and individual stations.

- **Educate users.** Train all personnel who store information on disks in how to prevent, detect, and remove viruses and how to prevent their spread among computers.

Other Ongoing Implementation Issues

In order for technology to make a real difference in education, teachers must recognize the limitations and problems that surround its use and be prepared to deal with them. Only then can they identify and help shape a realistic, appropriate role for technology resources. Three of the issues that educators must address continually are multicultural and equity issues, ethical/legal issues, and the challenge of keeping up with change.

Equity Issues

As Molnar pointed out in his landmark 1978 article ("The Next Great Crisis in American Education: Computer Literacy"), the power of technology is a two-edged sword, especially for education. While it presents obvious potential for changing education and empowering teachers and students, technology also has potential for further dividing members of our society along socioeconomic, ethnic, and cultural lines and for widening the gender gap. Teachers will lead the struggle to make sure technology uses promote, rather than conflict with, the goals of a democratic society.

Economic inequity. Some evidence supports Molnar's prediction that students with initial educational advantages will get more access to technology resources than those who could use the extra help. Demographic studies by Becker (1985, 1986a, and 1986b) confirmed the predictable correlation between school districts' socioeconomic levels and their levels of microcomputer resources. As Lockard, Abrams, and Many (1994) pointed out, this discrepancy is to be expected since "Computers only call further attention to the fact that schools in the U.S. are anything but equal. Inequities affect everything from basic supplies such as paper and pencils to library resources and even the quality of teachers" (p. 411). They observe that students from wealthier families are also far more likely to have computers and other technology resources at home than those from poorer families. All of these conditions are well-documented (Sanders and Stone, 1986; Neuman, 1991).

Evidence of the educational and/or economic crises that Molnar predicted has been more difficult to obtain, however. Widespread recognition of the need for computer literacy as conceived in 1982 never really emerged. (Indeed, no definition of *computer literacy* was ever established!) However, it seems logical that students who have more access to computers will also have better, more efficient learning tools at their disposal, and this access seems likely to become increasingly important as technological learning tools increase in power. Use of technology tools may also logically correlate to students' ability to enter mathematics, science, and technical areas such as engineering. Despite the lack of concrete evidence of an economic impact, the possibility seems very real that poorer students could be hampered in their learning (and, therefore, earning) potential by their unequal access to technology tools.

Ethnic inequity. The same problems with differences in technology access between economic subgroups may also apply to ethnic groups. Merrill, Hammons, Tolman, Christensen, Vincent, and Reynolds (1992) observe that "minority students are more likely to be in poorer school districts where computer equipment is limited" (p. 296). Engler (1992) reported that recent surveys show "clear racial/ethnic differences in computer competence favoring white students over Black and Hispanic students" and "the greatest competence was found among those who had computers at home and were studying computers at school" (p. 360). Engler also reported that the number of course prerequisites for computer programming studies (usually in mathematics classes) often increase as the percentage of minority enrollment increases. While this may be an unintended inequity, minority students clearly do not have the same access to these courses as their white peers have.

Data show that minorities are underrepresented in the fields of mathematics, science, and engineering. Though less access to technology in K-12 schools may not have caused this problem, it clearly has the potential to make it worse and prolong its effects.

Multicultural issues. Although ethnicity and culture overlap, cultural issues are more pervasive, include more factors, and have even more potential to affect the quality of students' educational opportunities. Dozier-Henry (1995) and others point out that technology has important but limited usefulness in promoting multicultural education goals. Roblyer, Dozier-Henry, and Burnette (1996, in press) cover these issues thoroughly. They describe several current uses of technology related to multicultural education, including telecommunications activities to promote communications between people of different cultures; applications that address the special language, visual, and experience needs of ESL and ESOL students; and multimedia applications with examples that enhance understanding of cultures. These authors also describe four problems related to technology use in this important area. Two of these issues (access and equity issues and biases in selection and applications of technology) are addressed in other parts of this section. Two others are summarized here:

- **Technology's built-in bias.** Western culture tends to view science and its offspring, technology, as beneficial for everyone and nearly everything. "The Western idea of progress is 'the more technology, the better!' (However) … the reverence with which technology is held in the U.S. may be in direct contradiction to the perceptions of cultures that are heavily relationship oriented" (Roblyer, Dozier-Henry, and Burnette, 1996, in press). These authors also note a growing "counter-computer culture" in the United States that is based on social, psychological, and even religious grounds.

- **Multicultural goals that technology cannot achieve.** As Roblyer, Dozier-Henry, and Burnette (1996) emphasize, technology can help to achieve some goals of multicultural education, namely making students aware of cultures other than their own, creating an interest in interacting with people of other cultures, and teaching about common attributes of all cultures despite their many differences. But these authors also say that "The next steps are more difficult, since they require accepting, learning from, and appreciating people" of other cultures. Schools must build upon the relatively superficial activities of "tele-pals" and learning about various foods and holidays in other cultures. In this deeper and more meaningful study, technology may have a limited role.

The gender gap. Research has thoroughly documented the fact that girls tend to use computers, often by choice, less than boys (Bohlin, 1993; Sanders, 1993). This unequal proportion extends to vocational areas where computers are more frequently used: again, mathematics, science, and technical areas such as engineering and computer science (Sanders and Stone, 1986; Fredman, 1990; Holmes, 1991; Nelson and Watson, 1991; Engler, 1992; Fear-Fenn and Kapostacy, 1992). A variety of reasons have been proposed for this disparity. Children may be reacting to stereotypes on television and in publications where men appear as primary users of computers. Depictions of women's use of computers tend to involve clerical tasks. The association of technology with machines, mathematics, and science—all stereotypically male areas—makes girls think of computers as "unfeminine."

Gender bias may spring up in software that features competitive activities preferred more by males than females and an emphasis on violent video games that appeal more to boys. Finally, many blame subtle and overt classroom practices for making girls think that computers are not intended for them. These range from a lack of female teacher role models to teachers' assumptions that girls are simply not as interested in computer work. Whatever combination of factors is involved, females clearly are being excluded in large numbers from using the power of technology and jobs that require technology skills.

Equity for special populations. Warning that technology and equity are not "inevitable partners," Neuman (1991) points out a variety of factors that can inhibit equitable access to technology for several kinds of groups. These groups include rural, handicapped, and "differently abled" students. The special problems of rural schools in supplying their students with adequate technology resources have already been discussed in the section headed Funding for Technology Resources. Thurston (1990) and Holland (1995) observe that rural schools have more severe equity problems of all types (i.e., economic, gender, and ethnic equity) than their urban counterparts.

A variety of adaptive devices have been designed to allow handicapped students to take advantage of the power of technology and to enhance personal freedom. Fredman (1991) says that "using the computer unlocks their potential. It is an enabling tool—allowing them to function as other students function without the barriers that their handicaps impose" (p. 47). However, Neuman (1991) and Engler (1992) report potential inequities in funding for these devices, and computer resources are often housed in locations that are not wheelchair accessible.

A more subtle kind of technology inequity has been observed with handicapped, lower-ability, and learning-disabled students (Fredman, 1991; Engler, 1992; Dozier-Henry, 1995). Frequently, these students' uses of computers have been limited to low-level ones such as remedial drill and practice applications. The more powerful, higher-level applications such as hypermedia and Logo production work is often directed toward higher-ability students. This finding is especially disconcerting since many of these subgroups are also at-risk students who might profit from the motivation stimulated by higher-level uses.

Recommendations to address equity issues. Several kinds of successful strategies have been documented in recent years as educators have become more aware of equity problems and their implications for society. Engler (1992) recommends several strategies that can help state and district leaders and policy makers to assure more equitable access to technology for all students. These include:

- **Accountability measures.** Monitor and document disproportionate participation to increase awareness of the problem.
- **Incentives and priority funding.** Tie state grant funds and entitlements to districts' efforts to address to needs of underrepresented students.
- **Innovative programs.** Develop and support new initiatives aimed at improving student access to technology.
- **Enrichment programs.** Supply funding for computer contests, summer camps with technology themes, and similar activities directed toward offering special opportunities to disadvantaged students.
- **Recognition.** State-level and district-level awards and publications could feature successful equity-related technology programs.
- **Business and community partnerships.** Place business-sponsored magnet schools and other programs in neighborhoods where students have usually had lower access to technology resources.
- **Staff development.** Educate teachers, parents, and school personnel to increase their expectations and support of girls, minorities, and other groups.
- **Student recruitment.** Set entrance requirements for special and gifted programs to include alternatives to test scores (e.g., portfolios) to encompass a wider variety of student groups, and actively seek participation of these students.

Engler (1992), Fredman (1990), and Sanders and Stone (1986) also recommend many classroom-level strategies by which teachers can encourage greater participation by girls, minorities, and disadvantaged students. Rosenthal and Demetrulias (1988) supply helpful guidelines to help detect gender bias as teachers select software for classroom use. Another major strategy recommended by Engler (1992), Salehi (1990), and others uses networking, either through statewide systems or local projects. Networking can attack the problems of inequitable access and disparate use on many fronts. It can provide many resources to rural schools that they may lack on-site. Communicating electronically can also allow minority students and teachers to exchange information with others of similar or different ethnic backgrounds in the United States and other countries. This opportunity can create a powerful motivation to learn about and use higher-level technology skills and, at the same time, build multicultural awareness.

Educators must pioneer other strategies to ensure that physically handicapped students get the technological devices they need to achieve learning opportunities more equal to those of their nondisabled counterparts. This begins with an increased level of awareness of two kinds. First, the students themselves and their families must learn about assistive devices and the funding sources they can use to purchase them. Second, when decision makers plan their resources and facilities, they must recognize the obstacles to technology access for physically disabled students and understand how to bring about that access. In many instances, technology-oriented teachers will be the ones to keep students, families, and administrators apprised of these important issues.

Recommendations to address multicultural issues. As one of their more difficult challenges, teachers who want to use technology must remain sensitive to cultural differences among their students, some of whom may be very deeply opposed to using machines (even very attractive ones). Teachers must consider very carefully the implications of ignoring or trying to change these cultural views in order to accomplish technology goals. Also, those who would address the goals of multicultural education must recognize how technology can and cannot contribute. They will have to make special efforts to supplement the technology-based activities made possible by telecommunications and multimedia tools with what Roblyer, Dozier-Henry, and Burnette (1996, in press) call "deeper study and interpersonal experiences with members of the groups being studied." Finally, Miller-Lachman (1994) provides a checklist of criteria for selecting software free of built-in cultural biases:

1. What is the purpose of presenting other cultures? (p. 26) (Is it integral to the program or an irrelevant add-on?)

2. Do people of color and their cultures receive as much attention as people of European descent? (p. 27)

3. How accurate is the presentation of various nationalities? (p. 27) (For example, portraying ancient Egyptians as having pink-colored skin would not be accurate.)

4. Are the language and terms appropriate? (p. 27) (For example, referring to people from any culture as "savages" is pejorative.)

5. Do the illustrations or sounds distort or ridicule members of other cultures? (p. 28)

6. Does the program present a true picture of the culture's diversity and complexity? (p. 28)

7. Who are the characters and what roles do they play? (p. 28) (Are the "good guys" always white, but the villains always include persons of color?)

8. From whose perspective is the story presented? (p. 29) Is the viewpoint of the software always the settlers and never the native Americans?

9. Does the documentation allow the instructors to go beyond the program itself? (p. 29)

10.   Should some simulations not be played? What kinds of actions are the players asked to undertake so that they may succeed, e.g., engaging in treachery and theft in order to capture an Aztec treasure? (p. 29)

Ethical and Legal Issues

In many ways, technology users represent the society in a microcosm. The culture, language, and problems of the larger society emerge among technology users, as well, and their activities reflect many of the rules of conduct and values of society in general. The same array of problems arise when people try to work outside those values and rules. Applications of technology in education create two major kinds of ethical and legal issues that educators should be prepared to address. They should know both the causes and the implications of both problems.

Copyright infringements. Software packages are very much like books. Companies put up development money to produce them and then sell copies in the hope of recouping their initial investments and earning profits. Like book publishers, the companies protect their products against illegal copying under U.S. copyright law. When microcomputer software became an industry, the problem of illegal copying of disks, called *software piracy,* became widespread. Forester (1990) reported on large-scale illegal copying operations in some foreign countries that produce thousands of copies of best-selling programs and sell them for as little as $10 each. Illegal copying has also become common among individuals, especially in education where teachers usually need multiple copies (e.g., for lab uses) but cannot afford per-copy prices. Many school personnel either are not aware of laws protecting software copyrights or do not feel the same compunction about copying software that they do about making illegal copies of books or films. Many educators have not clearly understood when copying is illegal and when they are permitted to make copies (Becker, 1992). Even when teachers clearly grasp these issues, their students may make illegal copies, and schools are legally responsible for these infractions (Becker, 1992).

Software publishers initially responded to illegal copying by placing protection codes within the software on each disk. These quickly proved ineffective, as many computer enthusiasts set about breaking these codes as an entertaining challenge. Subsequently, software producers omitted such codes, put stern copyright warnings on their products, and began to prosecute offenders.

Illegal access. Another ethical problem has received increasing notoriety in the media in recent years: computer users gaining illegal access to computerized information. These problems are often classified as either "computer crime" or "hacking," although the definitions tend to overlap. In the usual image of computer crime, individuals gain illegal access to computerized records for illicit purposes, e.g., transferring bank funds into their own accounts or gathering confidential information from which they can profit. Computer crime also includes software piracy (see Figure 2.2) and acts of mischief such as viruses and destruction of information. Profit is often the motive for computer crimes, but former employees have destroyed computer information as acts of revenge, and students have accessed school or district computers to change grades.

Hackers are people who are so captivated by the power and intricacies of computer systems that they adopt computer activities as a hobby. Hacking is not illegal in itself, but when this fascination turns toward exploring ways to invade privately held information, it becomes a crime. Hacking can become an especially serious problem in education, since students just learning about the computer can easily cross the line between harmless exploration and illegal access.

Recommendations to address ethical and legal issues. Educators' general response to these problems should take two forms. First, they must keep their students and others informed of rules and expectations for ethical and legal computer use. Second, they must adhere to strict rules of conduct themselves. This is not always easy to do but educators must remember that by modeling ethical behavior with computers, they impart in their students principles that are just as important as skills in computer use—probably more important. Additional suggestions can help teachers deal with specific ethical issues:

- **Stopping illegal copying.** One noted authority on copyright issues for educational media has documented many pertinent copyright problems, laws, and punishments, how the problems come about, and how to prevent them (Becker, 1992). The Software Publishers Association (1994) has also developed a summary of guidelines for software copying and a video entitled *Don't Copy that Floppy,* both of which are available on request. Technology-oriented teachers should accept responsibility for obtaining and using these materials to keep themselves and others informed on this important issue. As Becker points out, educational organizations would be well-advised to protect themselves against copyright infringement suits by stating and publicizing a policy regarding software copying, requiring teacher and staff training on the topic, and maintaining hard drive and network programs that discourage users from making illegal copies. Schools should also consider options for providing adequate numbers of copies for their users (e.g., by purchasing site licenses, lab packs, or networkable versions).

- **Restricting illegal access.** Although computer crime poses a greater threat in business and industry settings than in education, schools and districts that maintain computer files on students and staff must take steps to restrict illegal access. Teachers of programming and computer applications should guard against giving future white-collar criminals the tools they need. Teachers should be sure to cover the topics of computer crime and ethical behavior and help students to understand the implications of illegal access.

Figure 2.2   Software Publishers Association (SPA) Copying Guidelines

No, it's not okay to copy your colleague's software. Software is protected by federal copyright law, which says that you can't make such additional copies without the permission of the copyright holder. By protecting the investment of computer software companies in software development, the copyright law serves the cause of promoting broad public availability of new, creative, and innovative products. These companies devote large portions of their earnings to the creation of new software products and they deserve a fair return on their investment. The creative teams who develop the software — programmers, writers, graphic artists and others — also deserve fair compensation for their efforts. Without the protection given by our copyright laws, they would be unable to produce the valuable programs that have become so important in our daily lives: educational software that teaches us much needed skills; business software that allows us to save time, effort and money; and entertainment and personal productivity software that enhances leisure time.

That makes sense, but what do I get out of purchasing my own software?

When you purchase authorized copies of software programs, you receive user guides and tutorials, quick reference cards, the opportunity to purchase upgrades, and technical support from the software publishers. For most software programs, you can read about user benefits in the registration brochure or upgrade flyer in the product box.

What exactly does the law say about copying software?

The law says that anyone who purchases a copy of software has the right to load that copy onto a single computer and to make another copy "for archival purposes only." It is illegal to use that software on more than one computer or to make or distribute copies of that software for any other purpose unless specific permission has been obtained from the copyright owner. If you pirate software, you may face not only a civil suit for damages and other relief, but criminal liability as well, including fines and jail terms of up to one year.

So I'm never allowed to copy software for any other reason?

That's correct. Other than copying the software you purchase onto a single computer and making another copy "for archival purposes only," the copyright law prohibits you from making additional copies of the software for any other reason unless you obtain the permission of the software company.

At my company, we pass disks around all the time. We all assume that this must be okay since it was the company that purchased the software in the first place.

Many employees don't realize that corporations are bound by the copyright laws, just like everyone else. Such conduct exposes the company (and possibly the persons involved) to liability for copyright infringement. Consequently, more and more corporations concerned about their liability have written policies against such "softlifting". Employees may face disciplinary action if they make extra copies of the company's software for use at home or on additional computers within the office. A good rule to remember is that there must be one authorized copy of a software product for every computer upon which it is run.

Do the same rules apply to bulletin boards and user groups? I always thought that the reason they got together was to share software.

Yes. Bulletin boards and user groups are bound by the copyright law just as individuals and corporations. However, to the extent they offer shareware or public domain software, this is a perfectly acceptable practice. Similarly, some software companies offer bulletin boards and user groups special demonstration versions of their products, which in some instances may be copied. In any event, it is the responsibility of the bulletin board operator or user group to respect copyright law and to ensure that it is not used as a vehicle for unauthorized copying or distribution.

What about schools and professional training organizations?

The same copyright responsibilities that apply to individuals and corporations apply to schools and professional training organizations. No one is exempt from the copyright law.

I'll bet most of the people who copy software don't even know that they're breaking the law.

Because the software industry is relatively new, and because copying software is so easy, many people are either unaware of the laws governing software use or choose to ignore them. It is the responsibility of each and every software user to understand and adhere to copyright law. Ignorance of the law is no excuse. If you are part of an organization, see what you can do to initiate a policy statement that everyone respects. Also, suggest that your management consider conducting a software audit. Finally, as an individual, help spread the word that the users should be "software legal."

Source: Software Publishers Association. "Is it okay to copy my colleague's software?" Used by permission of SPA.

Keeping Up—When Change Is the Only Constant

Most experts acknowledge that technology involvement can pose an intimidating challenge under the best of circumstances (Dyrli and Kinnaman, 1994b). Many teachers feel threatened by this challenge, for one reason, because it represents a journey into the unknown. "Technology-induced feelings of vulnerability can arise" (p. 20). Technology's well-recognized pattern of rapid change complicates this problem; just when you get used to one machine or software option, it changes and you have to learn another one. Some educators hesitate to buy any one kind of computer because they fear it will quickly become outdated (Jordahl, 1995).

There are no easy answers to these problems. Some teachers will have more trouble than others dealing with this rapid rate of change, perhaps because some people feel challenged and energized by new situations, while others strongly prefer familiar things. For planning purposes, however, both kinds of people may benefit from a recognition that some changes are inevitable and predictable and that many changes will be good ones. Everyone should anticipate some predictable changes:

- **Interfaces will get "friendlier."** As computer systems change, they are also getting increasingly easy to use. The invention of the on-screen "desktop" was a major leap forward in ease-of-use, and it or something like it will probably be around for a long time. This means that skills in using a desktop will probably transfer to whatever microcomputer one uses in the future. Devices such as personal digital assistants (PDAs) and input devices that allow voice recognition will also become more prevalent. Interfaces, the means by which users communicate with computers, are destined to change, but these changes should make them easier to learn.

- **More software will be on CD-ROM.** Media for storing programs and files are getting more durable and reliable. CD-ROMs represent the latest development in this trend. Whenever possible, teachers should get microcomputers equipped with CD-ROM drives and software in CD-ROM versions.
- **A computer will be out of date when you buy it.** There is no such thing anymore as a state-of-the-art microcomputer. Technical developments are happening so rapidly that microcomputer buyers should follow practices similar to those for car purchases. Look for the features you need and those that will probably stay around a while, and be aware that a computer is out-of-date when you "drive it off the lot."

Dyrli and Kinnaman (1994b) seem to give teachers the best advice: "... embrace, (do) not fear, technological advance ... [T]he earlier you get in the game, the better your position will be for taking advantage of what is to come" (p. 48). For many teachers, the bad news is that change is inevitable; the good news is that the changes are usually for the better!

Exercises

Exercises for Objective 1: Rationales for justifying use of technology

1.1   Tell whether or not each of the following statements is a good rationale for justifying technology purchases. On a separate sheet, explain why or why not.

_____ 1.   Extensive research results show clearly that computer-based instruction is highly effective in raising test scores for low-achieving students.

_____ 2.   Kids love computers and will spend more time on computer-based activities than on paper-based ones.

_____ 3.   Every teacher needs daily access to a computer to produce word processing documents like handouts, reports, and other paperwork.

_____ 4.   Videodisc-based resources would be a good way to teach students about sex education and the human body.

_____ 5.   Every school should have an Internet connection because everyone will need to know how to use e-mail in the future.

1.2   Identify a specific idea for a technology-based activity that would require your school to purchase some resources (e.g., multimedia hardware and software, telecommunications access, computer, and software). Write a convincing justification for these purchases by explaining the unique benefits you expect from your activity.

Exercises for Objective 2: Identify a school in your local area that does not currently have a technology implementation plan. Describe how this school should go about developing its plan. Specify the following:

_____ 1.   Who should serve on the planning committee? (Be sure to represent all necessary groups.)

_____ 2.   What items should they plan to address?

_____ 3.   What steps should the committee follow to develop the plan?

If it is appropriate to do so, have your class present your planning proposal to the school principal and get his/her reactions.

Exercise for Objective 3: What would you do or say in each of the following situations? (Put your descriptions on a separate piece of paper.)

1.   You are a sixth grade teacher in an inner-city school. You would like to do some multimedia production work with your low-motivated, lower-achieving students. However, the Chapter I teacher in your school says that these students should use computers strictly for drill and practice in remedial skills because it is important that low-achieving stu-

dents get a good grounding in basic skills before attempting any activities that require higher-order thinking and problem solving.

2. In addition to your teaching duties, you have been placed in charge of the school's one general-use computer lab. One of the teachers in your school has just purchased an expensive science software package for use in the 15-computer lab. When you ask how she is going to get the software on all the machines, she replies that she will simply transfer the software from the disks to the hard drives on each computer. When her students are finished using it, she will erase it from the hard drives.

3. As the new principal of a rural high school, you realize that most students in the school's Advanced Placement computer science courses and its "Komputer Klub for Kids" are white males. The teacher for the courses (who is also the club's sponsor) says it has always been that way and is not likely to change.

4. You are one of several teachers who worked on a successful grant to get multimedia equipment, software, videodiscs, and an Internet connection for your students. The school has suffered break-ins before, and you are concerned about the security of these new resources. You talk to the principal about your concerns, but the principal says the school probably cannot afford any security measures.

5. You are a teacher of Grade 11–12 social studies. You are excited because your school has just received six computers, some printers, and a videodisc player. You have plans to begin using these resources with your classes. The principal says that space is tight in the school and he is considering placing these resources in a hallway outside the library/media center. He says that way they would be both accessible and highly visible to everyone.

References

Preparing for Technology Integration

Ferrell, B. (1986). Evaluating the impact of CAI on mathematics learning: Computer immersion project. *Journal of Educational Computing Research, 2* (3), 327–336.

Morehouse, D., Hoaglund, M., and Schmidt, R. (1987). *Technology demonstration project final report.* Menomonie, WI: Quality Evaluation and Development.

The revolution that fizzled. (1991, May 20). *Time,* p. 48.

Effectiveness Research

Bialo, E. and Sivin, J. (1989). Computers and at-risk youth: Software and hardware that can help. *Classroom Computer Learning, 9* (5), 48–55.

Clark, R. (1983). Reconsidering research on learning from media. *Review of Educational Research, 53* (4), 445–459.

Clark, R. (1985). Evidence for confounding in computer-based instruction studies: Analyzing the meta-analyses. *Educational Communications and Technology Journal, 33* (4), 249–262.

Clark, R. (1991). When researchers swim upstream: Reflections on an unpopular argument about learning from media. *Educational Technology, 31* (2), 34–40.

Clark, R.E. (1994). Media will never influence learning. *Educational Technology Research and Development, 42* (2), 21–29.

Clark, C., Kurshan, B., and Yoder, S. (1989). *Telecommunications in the classroom.* Eugene, OR: International Society for Technology in Education.

Kozma, R. (1991). Learning with media. *Review of Educational Research, 61* (2), 179–211.

Kozma, R. and Chroninger, R. (1992). Technology and the fate of at-risk students. *Education and Urban Society, 24* (4), 440–453.

Kozma, R. (1994). Will media influence learning? Reframing the debate. *Educational Technology Research and Development, 42* (2), 5–17.

Roblyer, M., Castine, W., and King, F.J. (1988). *Assessing the impact of computer-based instruction: A review of recent research.* New York: Haworth Press.

Roblyer, M. (1992). Computers in education. In G. Bitter (Ed.), *Macmillan encyclopedia of computers.* New York: Macmillan.

Thompson, A., Simonson, M., and Hargrave, C. (1992). *Educational technology: A review of the research.* Washington, DC: Association for Educational Communications and Technology.

Rationales for Using Technology

Adams, R. (1985). Why I moved my micro to the teacher's desk. *Instructor, 94* (5), 56–58, 62.

Arnone, M. and Grabowski, B. (1991). Effects of variations in learner control on children's curiosity and learning from interactive video. Proceedings of Selected Research Presentations at the Annual Convention of the AECT (ERIC Document Reproduction No. ED 334 972).

Buchholz, W. (1991). A learning activity for at-risk ninth through twelfth grade students in creating a computer-generated children's storybook design. Master's thesis, New York Institute of Technology (ERIC Document Reproduction No. ED 345 695).

Clark, C., Kurshan, B., and Yoder, S. (1989). *Telecommunications in the classroom.* Eugene, OR: International Society for Technology in Education.

Cohen, M. and Riel, M. (1989). The effect of distant audiences on children's writing. *American Educational Research Journal, 26* (2), 143–159.

Covey, P. (1989). Project THEORIA: New media for values education. *Educational Technology, 29* (5), 31–32.

Educational Technology (1992), *32,* 9.

Electronic Learning (1990), *10,* 1.

Electronic Learning (1990), *11,* 1.

Franklin, S. (1991). Breathing life into reluctant writers: The Seattle Public Schools laptop project. *Writing Notebook, 8* (4), 40–42.

Gagne, R. (1965). *The conditions of learning.* New York: Holt, Rinehart and Winston.

George Mason University (1989). CBI project evaluation phase II: Data analysis results (ERIC Document Reproduction No. ED 325 090).

Kozma, R. (1991).

Kurshan, B. (1990). Educational telecommunications connections for the classroom. Part I. The *Computing Teacher, 17* (6), 30–35.

LaRoue, A. (1990). The M.A.P. shop: Integrating computers into the curriculum for at-risk students. *Florida Educational Computing Quarterly, 2* (4), 9–21.

Marcus, S. (1995). E-meliorating student writing. *Electronic Learning, 14* (4), 18–19.

McNeil, B. and Wilson, K. (1991). Meta-analysis of interactive video instruction: A 10-year review of achievement effects. *Journal of Computer-Based Instruction, 18* (1), 1–6.

Minnesota State Department of Education. (1989). Computer tools for teachers: A report (ERIC Document Reproduction No. ED 337 130).

Pask-McCartney, C. (1989). A discussion about motivation. Proceedings of Selected Research Presentations at the Annual Convention of the AECT (ERIC Document Reproduction No. ED 308 816).

Polin, L. (1992). Looking for love in all the wrong places? *The Computing Teacher, 20* (2), 6–7.

Relan, A. (1992). Motivational strategies in computer-based instruction: Some lessons from theories and models of motivation. Proceedings of Selected Research Presentations at the Annual Convention of the AECT (ERIC Document Reproduction No. ED 348 017).

Roblyer, M. (1991). Electronic hands across the ocean: The Florida-England connection. *The Computing Teacher, 19* (5), 16–19.

SCANS (Secretary's Commission on Achieving Necessary Skills) Report (1992). Washington, DC: U. S. Department of Labor.

Summers, J. (1990–91). Effect of interactivity upon student achievement, completion intervals, and affective perceptions. *Journal of Educational Technology Systems, 19* (1), 53–57.

Taylor, D. (1989). Communications technology for literacy work with isolated learners. *Journal of Reading, 32* (7), 634–639.

Tibbs, P. (1989). Video creation for junior high language arts. *Journal of Reading, 32* (6), 558–559.

Volker, R. (1992). Applications of constructivist theory to the use of hypermedia. Proceedings of Selected Research Presentations at the Annual Convention of the AECT (ERIC Document Reproduction No. ED 348 037).

Technology and Restructuring

Ahearn, E. (1991). Real restructuring through technology. *Perspectives 3,* 1, Council for Basic Education, Washington, DC (ERIC Document Reproduction Service No. ED 332 318).

Bagley, C. and Hunter, B. (1992). Restructuring, constructivism, and the future of classroom learning. *Education and Urban Society, 24* (4), 66–76.

Brodinsky, B. (1993). How "new" will the "new" Whittle American school be? *Phi Delta Kappan, 75,* 540–547.

Bruder, I., Buchsbaum, H., Hill, M., and Orlando, L. (1992). School reform: Why you need technology to get there. *Electronic Learning, 11* (8), 22–28.

Butzin, S. (1991). Project CHILD: Progress in restructuring schools to tap technology's potential. *The Computing Teacher, 19* (3), 45–47.

Chesley, G. (1994). The engineering of restructuring: What do we do and how do we do it? *NASSP Bulletin, 78* (565), 21–27.

Collins, A. (1991). The role of computer technology in restructuring schools. *Phi Delta Kappan, 73* (1), 28–36.

David, J. (1991). Restructuring and technology: Partners in change. *Phi Delta Kappan, 73* (1), 37–40, 78.

Foley, D. (1993). Restructuring with technology. *Principal, 72* (3), 22, 24–25.

Hill, M. (1993). Math reform: No technology, no chance. *Electronic Learning, 12* (7), 24–32.

Jostens Learning Corporation. (1995). *Educating Jessica's generation.* San Diego, CA: Author.

Luterbach, K. and Reigeluth, C. (1994). School's not out, yet. *Educational Technology, 34* (41), 47–54.

Neuman, (1991). Technology and equity. ERIC Digest. (ERIC Document Reproduction No. ED339 400).

Norris, W. (1977). Via technology to a new era in education. *Phi Delta Kappan, 58* (6), 451–453.

Norris, C. and Reigeluth, C. (1991). A national survey of systemic school restructuring experiences (ERIC Document Reproduction No. ED 335 001).

Papert, S. (1980). *Mindstorms—Children, computers, and powerful ideas.* New York: Basic Books.

Perelman, L. (1993). *School's out: Hyperlearning, the new technology, and the end of education.* New York: William Morrow.

Reigeluth, C. and Garfinkle, R. (1992). Envisioning a new system of education. *Educational Technology, 22* (11), 17–22.

Reigeluth, C. and Garfinkle, R. (1994). Systemic change in education (ERIC Document Reproduction No. ED 367 055).

Sheingold, K. (1991). Restructuring for learning with technology: The potential for synergy. *Phi Delta Kappan, 73* (1), 17–27.

Smith, R. (1991). Restructuring American education through technology: Three alternative scenarios. *ISTE Update, 3* (8), 1.

Strommen, E. and Lincoln, B. (1992). Constructivism, technology, and the future of classroom learning. *Education and Urban Society, 24* (4), 466–476.

Roblyer, M. (1993). From the editor's clipboard. *The Florida Technology in Education Quarterly, 5* (1–2), 2.

U. S. News and World Report (1993, January 11). The perfect school: Nine reforms to revolutionize American education.

Planning for Technology

Association for Media and Technology in Education (1991). K-12 educational technology planning: The state of the art. A selected bibliography. Hot topic No. 1. (ERIC Document Reproduction No. 346 819).

Apple Computer Company (1995). *Teaching, learning, and technology: A planning guide* (Multimedia package). Cupertino, CA: Author.

Brody, P. (1995). *Technology planning and management handbook.* Englewood, NJ: Educational Technology.

Bruder, I. (1993). Technology in the USA: An educational perspective. *Electronic Learning, 13* (2), 20–28.

Dyrli, O. and Kinnaman, D. (1994a). District wide technology planning: The key to long-term success. *Technology and Learning, 14* (7), 50–54.

Mager, R. (1984). *Preparing instructional objectives.* Belmont, CA: David S. Lake.

Mageau, T. (1990). ILS: Its new role in schools. *Electronic Learning, 10* (1), 22–24, 31–32.

New York State School Boards Association, (1989). Instructional technology: Policies and plans. A position paper (ERIC Document Reproduction Service No. ED314 870).

Palazzo, A. (1995). Great technology plans. *Electronic Learning, 14* (7), 31–39.

Roblyer, M. (1993). Why use technology in teaching: Making a case beyond research results. *The Florida Technology in Education Quarterly, 5* (4), 7–13.

See, J. (1992). Ten criteria for effective technology plans. *The Computing Teacher, 19* (8), 34–35.

Van Dam, J. (1994). Redesigning schools for 21st century technologies. *Technology and Learning, 14* (4), 54–58, 60–61.

Wall, T. (1994). A technology planning primer. *The American School Board Journal, 81* (3), 45–47.

Obtaining the Right Resources

Funding

Bailey, T. (1990). The superintendent's perception of the benefit of instructional technology in Virginia school divisions. Virginia State Department of Education, Richmond, VA (ERIC Document Reproduction Service No. ED 329 233)

Branson, R. (1988). Why the schools can't improve: The upper limit hypothesis. *Journal of Instructional Development, 10* (4), 15–26.

Bullough, R. and Beatty, L. (1991). *Classroom applications of microcomputers.* New York: Merrill/Macmillan.

Clark, C. (1990). Linking educational finance reform and educational technology in Texas. Paper presented at the Annual Meeting of the American Education Finance Association, Las Vegas (ERIC Document Reproduction Service No. ED 324 751).

Clauss, W. and Witwer, F. (1989). Establishing legislative support for the funding of technology in rural schools. *Rural Special Education Quarterly, 9* (4), 25–28.

Dede, C. (1992). The future of multimedia: Bridging to virtual worlds. *Educational Technology, 32* (5), 54–60.

Electronic Learning (1993) *12,* 5.

Finkel, L. (1993). Planning for obsolescence: Upgrading and replacing old computers. *Electronic Learning, 12* (7), 18–19.

Freitas, D. (1992). Managing smallness: Promising fiscal practices for rural school district administrators. ERIC Clearinghouse on Rural Education and Small Schools, Charleston, WV (ERIC Document Reproduction Service No. ED 348 205).

Holland, H. (1995). Needles in a haystack. *Electronic Learning, 14* (7), 26–28.

Inman-Freitas, D. (1991). Efficient financial management in rural schools: Common problems and solutions from the field. ERIC Clearinghouse on Rural Education and Small Schools, Charleston, WV (ERIC Document Reproduction Service No. ED 335 206).

Mahmood, M., and Hirt, S. (1992). Evaluating a technology integration causal model for the K-12 public school curriculum: A LISREL analysis (ERIC Document Reproduction Service No. ED 346 847).

November, A. and Huntley, M. (1988). Kids and computers '88: How one state's CUE organization changed the way politicians think about technology. *Classroom Computer Learning, 9* (2), 5–12.

November, A. (1990). Big dreams, no money. *Classroom Computer Learning, 10* (5), 14, 18–19.

Rose, A. (1992). Financing technology. *American School Board Journal, 178* (7), 17–19.

Smith, R. (1992). Teaching with technology: The classroom manager. Cost-conscious computing. *Instructor, 10* (6), 60–61.

Technology and Learning (1992). *12* (4), 36–47.

Webb, D. Everything I needed to know about fund raising for technology was learned the hard way. *ISTE Update, 6* (1), 6–7.

Hardware, Software, and Facilities

Apple Computer Company (1991). *Apple technology in support of learning: Creating and managing an academic computer lab.* Sunnyvale, CA: Apple Computer Co.

Bunson, S. (1988). Design and management of an IMC micro center. *Educational Technology, 28* (8), 29–36.

Dyrli, O. and Kinnaman, D. (1994b). Gaining access to technology: The first step in making a difference for your students. *Technology and Learning, 14* (4), 16–20, 48–50.

Finkel, L. (1993). Planning for obsolescence: Upgrading and replacing old computers. *Electronic Learning, 12* (7), 18–19.

Freitas, D. (1992). Managing smallness: Promising fiscal practices for rural school district administrators. ERIC Clearinghouse on Rural Education and Small Schools. (ERIC Document Reproduction No. ED 348 205).

Holland, H. (1995). Needles in a haystack. *Electronic Learning, 14* (7), 26–28.

Macon, C. (1992). The retrofit for technology project: A Florida initiative. *Florida Technology in Education Quarterly, 4* (4), 65–76.

Manckzuk, S. (1994). Planning for technology: A newcomer's guide. *Journal of Youth Services in Libraries, 7* (2), 199–206.

Milone, M. (1989). Classroom or lab: How to decide which is best. *Classroom Computer Learning, 10* (1), 34–43.

Muffoletto, R. (1994). Technology and restructuring education: Constructing a context. *Educational Technology, 34* (2), 24–28.

Neuman, D. (1991). Technology and equity. *ERIC Digest.* (ERIC Document Reproduction No. ED 339 400).

Van Dam, J. (1994). Redesigning schools for 21st century technologies. *Technology and Learning, 14* (4), 54–58, 60–61.

Wilson, J. (1991). Computer laboratory workstation dimensions: Scaling down for elementary school children. *Computers in the Schools, 8* (4), 41–48.

Support Personnel and Teacher Training

Brooks, D. and Kopp, T. (1989). Technology in teacher education. *Journal of Teacher Education, 40* (4), 2–8.

Dyrli, O. and Kinnaman, D. (1994b). Gaining access to technology: The first step in making a difference for your students. *Technology and Learning, 14* (4), 16–20, 48–50.

International Society for Technology in Education (ISTE). (1992). *Curriculum guidelines for accreditation of educational computing and technology programs.* Eugene, OR: ISTE.

Marker, G. and Ehman, L. (1989). Linking teachers to the world of technology. *Educational Technology, 29* (3), 26–30.

Marshall, G. (1993). Four issues confronting the design and delivery of staff development programs. *Journal of Computing in Teacher Education, 10* (1), 4–10.

Munday, R., Windham, R., and Stamper, J. (1991). Technology for learning: Are teachers being prepared? *Educational Technology, 31* (3), 29–32.

Roblyer, M. (1994). Creating technology-using teachers: A model for preservice technology training. Final report of the model preservice technology integration project. Tallahassee, FL: Florida A&M University.

Sheingold, K. (1991). Restructuring for learning with technology: The potential for synergy. *Phi Delta Kappan, 73* (1), 17–27.

Siegel, J. (1995). The state of teacher training. *Electronic Learning, 14* (8), 43–53.

Todd, N. (1993). A curriculum model for integrating technology in teacher education courses. *Journal of Computing in Teacher Education, 9* (3), 5–11.

Ongoing Implementation Issues

Security and Maintenance

Brody, P. (1995). *Technology planning and management handbook.* Englewood, NJ: Educational Technology.

Gray, B. (1988). Twelve 'gems' for managing a micro lab in higher education. *T.H.E. Journal, 16* (4), 70–72.

White, L. (1992) Beyond the educational structure: A conceptual framework for computer security. *Florida Technology in Education Quarterly, 4* (4), 77–83.

Equity Issues

Becker, H. (1985). The second national survey of instructional uses of school computers: A preliminary report (ERIC Document Reproduction Service No. ED 274 307).

Becker, H. (1986a). Instructional uses of school computers. Reports from the 1985 National Survey. Issue No. 1 (ERIC Document Reproduction Service No. ED 274 319).

Becker, H. (1986b). Instructional uses of school computers. Reports from the 1985 National Survey. Issue No. 3 (ERIC Document Reproduction Service No. ED 279 303).

Bohlin, R. (1993). Computers and gender difference: Achieving equity. *Computers in the Schools, 9* (2–3), 155–166.

Dozier-Henry, O. (1995). Technology and cultural diversity: The "uneasy alliance." *Florida Technology in Education Quarterly, 7* (2), 11–16.

Engler, P. (1992). Equity issues and computers. In G. Bitter (Ed.), *Macmillan encyclopedia of computers.* New York: Macmillan.

Fear-Fenn, M. and Kapostacy, K. (1992). Math + science + technology = Vocational preparation for girls: A difficult equation to balance. Columbus, OH: Ohio State University, Center for Sex Equity (ERIC Document Reproduction No. 341 863).

Fredman, A, (1990). *Yes, I can. Action projects to resolve equity issues in educational computing.* Eugene, OR: International Society for Technology in Education.

Holmes, N. (1991). The road less traveled by girls. *School Administrator, 48* (10, 11, 14), 16–20.

Lockard, J., Abrams, P., and Many, W. (1994). *Microcomputers for 21st century educators.* New York: HarperCollins.

McAdoo, M. (1994). Equity: Has technology bridged the gap? *Electronic Learning, 13* (7), 24–34.

Merrill, D., Hammons, K., Tolman, M., Christensen, L., Vincent, B., and Reynolds, P. (1992). *Computers in education.* Boston, MA: Allyn and Bacon.

Miller-Lachman, L. (1994). Bytes and bias: Eliminating cultural stereotypes from educational software. *School Library Journal, 40* (11), 26–30.

Molnar, A. (1978). The next great crisis in American education: Computer literacy. *AEDS Journal, 12* (1), 11–20.

Nelson, C. and Watson, J. (1991). The computer gender gap: Children's attitudes, performance, and socialization. *The Journal of Educational Technology Systems, 19* (4), 345–353.

Neuman, D. (1991). Beyond the chip: A model for fostering equity. *School Library Media Quarterly, 18* (3), 158–164.

Polin, L. (1992). Looking for love in all the wrong places? *The Computing Teacher, 20* (2), 6–7.

Roblyer, M., Dozier-Henry, O., and Burnette, A. (1996). Technology and multicultural education: The "uneasy alliance." *Educational Technology, 35* (3), 5–12.

Rosenthal, N. and Demetrulias, D. (1988). Assessing gender bias in computer software. *Computers in the Schools, 5* (1–2), 153–163.

Salehi, S. (1990). *Promoting equity through educational technology networks.* Baltimore, MD: Maryland State Department of Education (ERIC Document Reproduction Service No. ED 322 897).

Sanders, J. and Stone A. (1986). The neuter computer. Computers for girls and boys. New York: Neal Schuman Publishers.

Sanders, J. (1993). Closing the gender gap. *Executive Educator, 15* (9), 32–33.

Thurston, L. (1990). Girls, computers, and amber waves of grain: Computer equity programming for rural teachers. Paper presented at the Annual Conference of the National Women's Studies Association (ERIC Document Reproduction Service No. ED 319 660).

Ethical and Legal Issues

Becker, G. (1992). *Copyright: A guide to information and resources.* Gary H. Becker Consultants, 164 Lake Breeze Circle, Lake Mary, FL 32746-6038.

Dormady, D. (1991). Computer viruses: Suggestions on detection and prevention. *Florida Technology in Education Quarterly, 3* (4), 93–98.

Forester, T. (1990, March). Software theft and the problem of intellectual property rights. *Computers and Society,* 2–11.

Hansen, B. and Koltes, S. (1992). Viruses. In G. Bitter (Ed.), *Macmillan encyclopedia of computers.* New York: Macmillan.

Lee, J. (1992). Hacking. In G. Bitter (Ed.), *Macmillan encyclopedia of computers.* New York: Macmillan.

Mungo, P. and Clough, B. (1992). *Approaching zero: The incredible underworld of hackers, phreakers, virus writers, and keyboard criminals.* New York: Random House.

Software Publishers Association. (1994). Is it okay to copy my colleague's software? Washington, DC: SPA.

Keeping Up with Change

Dyrli, O. and Kinnaman, D. (1994b). Gaining access to technology: The first step in making a difference for your students. *Technology and Learning, 14 (*4), 18–20, 48–50.

Jordahl, G. (1995). Getting equipped and staying equipped, Part 1: Grappling with obsolescence. *Technology and Learning, 15* (6), 31–36.

Chapter 3

Learning Theories and Integration Models

This chapter will cover the following topics:

- Background on behavioral and cognitive learning theories

- How these learning theories contributed to current models of instruction

- Technology integration strategies based on each model of instruction

- An example of how these approaches are combined in a curriculum unit

Chapter Objectives

1. Define instructional terms and describe models associated with behaviorist, information processing, and other cognitive learning theories.

2. Identify the kind of instructional approach that served as the foundation for a given technology integration strategy.

3. Demonstrate how directed and constructivist approaches to thinking about teaching and learning have led to very different ways of integrating technology.

4. Describe and give examples of technology integration strategies that reflect each approach.

5. Show how both approaches can be—and must be—combined effectively in the same curriculum.

For education the crucial question is not whether skills are implicit theories ... but whether it facilitates learning to get students to think of skills in this way.
Hubert Dreyfus and Stuart Dreyfus, from "Putting Computers in Their Proper Place" in *The Computer in Education in Critical Perspective*

Introduction

Educators today debate the most appropriate instructional role for technology, particularly computer technology. Prior to about 1980, the question inspired much less disagreement. According to respected writers of the time (Taylor, 1980), the major disagreement divided people into three groups: those who advocated using computers primarily as tools (e.g., for word processing and numerical calculations), those who viewed them mainly as teaching aids or "tutors" (e.g., for drills, tutorials, and simulations), and those who believed the most powerful use was programming (the "tutee" use). But these groups would have generally agreed that each of these approaches had its place, and they could refer to a substantial and growing information base on classroom strategies for each use. These were simpler times, both for educational technology and for education itself, although few would have believed it then.

Changes Brought about by Technology

Subsequent years have witnessed two trends with unprecedented effects on the course of educational technology: (a) an increase in the number and types of technology resources available and (b) dramatic shifts in beliefs about the fundamental goals and strategies of education itself. These two trends have not developed in isolation; their roots are intertwined in the larger social and economic conditions that define and shape our modern world. In the past, educational goals reflected society's emphasis on "the need for basic skills" (e.g., reading, writing, and arithmetic) and a certain body of information that was considered essential for all citizens. Students were considered educated if they could demonstrate the abilities to read at a certain comprehension level; apply grammar, usage, and punctuation rules in written compositions; solve arithmetic problems that required addition, subtraction, multiplication, and division; and state certain series of facts (e.g., events leading up to the Civil War).

But as technology has become more capable and pervasive and as more types of technology resources have become available, everyday life has also become more complex and demanding. Many students just entering school today will eventually take jobs that do not even exist yet. More information is deemed important to learn than ever before, and the base of essential information is growing constantly. Many educators have come to believe that the world is changing too quickly to confine educational goals to specific information or skills; they believe that education should emphasize more general capabilities for "learning to learn" that will help future citizens to cope with inevitable changes. For example, instead of learning specific items of information, they want to emphasize training in ways of acquiring, sorting through, and using information. Knowing what questions to ask will be as important as, or more important than, giving the "right answers." In short, technology seems to have both increased the number of decisions that people must make and forced them to become more skilled decision makers.

Current Educational Goals and Methods: Two Views

As the goals of education begin to change to reflect new social and educational needs, teaching strategies also change, and so, consequently, do strategies for integrating technology into teaching and learning. Today, educators' definition of the appropriate role of technology depends on their perceptions of the goals of education itself and appropriate instructional methods to help students attain those goals.

Most educators seem to agree that changes are needed in education. But identifying exactly how new skills and methods will differ has become increasingly controversial. Some education experts argue that schools should teach almost none of the skills traditionally considered important. Others feel that many of these skills are still necessary but that schools should implement different kinds of instructional models to help foster them.

Growing disagreements among learning theorists have centered on which strategies will prove most effective in achieving today's educational goals. This controversy has served as a catalyst for two very different views on teaching and learning. One view, which we will call *directed instruction,* is grounded primarily in behaviorist learning theory and the information processing branch of the cognitive learning theories. The other view, which we will refer to as *constructivist,* evolved from other branches of thinking in cognitive learning theory. A few technology applications (e.g., drill and practice, tutorials) are associated only with directed instruction; most others (e.g., problem solving, multimedia applications, telecommunications) can enhance either directed instruction or constructivist environments, depending on how teachers integrate them into classroom instruction.

While acknowledging firmly held, widely differing opinions about the directions educators should be taking and the kinds of technology resources they should be selecting, this book sees meaningful roles for both directed instruction and constructivist strategies and the technology applications associated with them; both can help schools meet the many and varied requirements of learning. That belief guides the purposes of this chapter.

An Overview of Directed and Constructivist Instructional Methods

A Comparison of Terminologies and Models

Differences in terminologies. People with radically different views on an issue frequently use different terms to describe essentially the same things. The terminology guide in Table 3.1 can help prepare the reader for some of the differences between directed instruction models and constructivist models.

These differences in language signal some fundamental differences between the two models. How did these differences come about? It is important to recognize that both directed instruction and constructivist approaches attempt to identify what Gagne (1985) would call the "conditions of learning" or the "sets of circumstances that obtain when learning occurs" (p. 2). Both are based on the work of respected learning theorists and psychologists who have studied both the behavior of human beings as learning organisms and the behavior of students in schools and classrooms. The two approaches diverge, indeed they go in opposite directions, when they define *learning* and describe the conditions required to make learning happen and the kinds of problems that interfere most with learning. They disagree because they attend to different philosophies and learning theories and they take different perspectives on improving current educational practice. Yet some indications suggest that both kinds of strategies may prove useful to teachers in addressing commonly recognized instructional and educational problems.

Differences in philosophical foundations. The differences begin with underlying epistemologies (beliefs about the origins, nature, and limits of human knowledge). Constructivists and their opposites (also called *objectivists)* come from separate and very different epistemological "planets," although both planets nurture many different tribes or cultures (Molenda, 1991; Phillips, 1995). On the objectivist side, philosophers believe that knowledge has a separate, real existence of its own outside the human mind; advo-

cates of directed instruction believe that learning happens when this knowledge is transmitted to the learner. On the constructivist side, philosophers believe that humans construct all knowledge in their minds, so that learning happens when a learner constructs both mechanisms for learning and his or her own unique version of the knowledge, colored by background, experiences, and aptitudes (Willis, 1995). Two issues of *Educational Technology* (May 1991 and September 1991) do a good job of explicating these philosophical differences and the instructional approaches that sprang from them.

Merging the two approaches. As Molenda (1991) so sagely observed, an either–or stance seems to gain little for an educator. Rather, both sides need to find a way to merge the two approaches in a way that will benefit learners and teachers. They need to forge a link between the two planets so that students may travel freely from one to the other, depending on the characteristics of the topics at hand and each person's learning needs.

Bereiter (1990) is one theorist whose writings reflect initial support for directed instruction methods and a later shift toward constructivist principles. He suggests that much of what educators want students to achieve in school is sufficiently complex that none of the existing learning theories can account for how it is actually learned, let alone the conditions that should be arranged to facilitate learning. He says, "Let us recognize that (such learning) is genuinely problematic—chancy, susceptible to failure, in need of all the help it can get" (p. 604). He points out the futility of theory and research that attempts to identify relevant social, environmental, or individual influences on learning (e.g., prior experiences, types of reinforcement, learning styles) and quantify their comparative contribution to what he calls "difficult learning," that is, higher-order thinking and problem solving.

He observes that each of these contributing factors tends to interact with the others, thus changing their relative importance at any given time or for any given person. He quotes Cronbach's vivid metaphor: "Once we attend to interactions [between these relevant factors], we enter a

Table 3.1 Differences in Terminology in Directed Instruction and Constructivism		
	To Describe Directed Instruction	**To Describe Constructivism**
Advocates of directed instruction say	Teacher-directed	Discovery learning
	Systematic instruction	Unstructured learning
	Systems approaches	Self-directed learning
Advocates of constructivism say	Teacher-centered	Student-centered
	Knowledge transfer	Knowledge construction
	Transmission models	Generative learning models

hall of mirrors that extends to infinity" (Bereiter, 1990, p. 606). Practicing teachers could encounter endless variations of explanations about how people learn or fail to learn. Escaping from this hall of mirrors will require, Bereiter maintains, a more all-inclusive learning theory than those currently available. In light of Bereiter's observations, the debate between directed and constructivist proponents seems likely to inspire different methods primarily because they focus on different kinds of problems (or different aspects of the same problems) confronting teachers and students in today's schools. Like the blind men trying to describe the elephant, each focuses on a different part of the beast, and each is correct in limited observations. Needs addressed primarily by each model are described here and summarized in Table 3.2.

Instructional Problems Addressed by Directed Instruction Strategies

Although they are based primarily on early theories of learning, directed instruction methods target some very real, ongoing problems that originated many years ago. Still, schools and teachers can expect to encounter these

---

Table 3.2   Instructional Needs Met by Two Instructional Models

Needs Addressed by Directed Instruction

1. Individual pacing and remediation, especially when teacher time is limited

2. Making learning paths more efficient (e.g., faster), especially for instruction in skills that are prerequisite to higher-level skills

3. Performing time-consuming and labor-intensive tasks (e.g., skill practice), freeing teaching time for other, more complex student needs

4. Supplying self-instructional sequences, especially when human teachers are not available, teacher time for structured review is limited, and/or students are already highly motivated to learn skills

Needs Addressed by Constructivism

1. Making skills more relevant to students' backgrounds and experiences by anchoring learning tasks in meaningful, authentic (e.g., real-life), highly visual situations

2. Addressing motivation problems through interactive activities in which students must play active rather than passive roles

3. Teaching students how to work together to solve problems through group-based, cooperative learning activities

4. Emphasizing engaging, motivational activities that require higher-level skills and prerequisite lower-lever skills at the same time.

---

issues in the foreseeable future. Post–World War II America acted on a burgeoning awareness of the importance of education. More students were staying in school than ever before, and schools faced ever-increasing numbers of students with widely varying capabilities. More students than ever before aspired to college studies, which they viewed as a key to realizing "the American dream." Despite this dramatic increase in the numbers of students and the pressure it placed on school resources, schools were still required to certify that students had attained skills considered necessary for high school diplomas and entrance into higher education. At the same time, teachers had to meet the individual pacing and remedial needs of each student while assuring that all students were learning required skills. Individualization became both the goal and the terror of teachers in the 1960s.

By the 1970s, so-called *systems approaches* were being widely proposed as a way for teachers and others to design self-instructional packages for students to separate directed instruction from the need for the teacher to deliver it. Self-instruction was more efficient than trying to serve the pacing and content needs of each student, and it assured that instruction was replicable; that is, quality was uniform from presentation to presentation. However, systems approaches also were seen as a way to design more effective teacher-delivered presentations.

In the 1970s and 1980s, many educators began to recognize how technology resources such as computer software (e.g., drills and tutorials) could help them overcome some of the logistical obstacles to individualized instruction. Some courseware can help students get needed practice; other courseware can guide their learning of difficult concepts through step-by-step, self-paced teaching sequences; still other courseware can let students change the variables in given situations (e.g., population growth, stock market purchases) and see the effects of their decisions. All of these activities could free teacher time to work with students who needed personal help. Teachers were encouraged to design more systematic instruction and to insert computer-based materials as needed to carry out the sequences they designed.

In the 1990s and for the foreseeable future, teachers still face the problems of too many students, too many required skills to teach, and not enough time to deal with individual learning differences. Systematically designed self-instructional materials have frequently shown the capability of teaching many skills identified as important. They have proven especially useful for students who seem to profit from a structured learning environment. Depending on other important factors (e.g., their particular role in the teacher's instructional activities), various kinds of drills, tutorials, and other "older" kinds of packages have effectively supplemented and, more rarely, replaced teacher-led directed instruction. Studies comparing teacher-led versus computer-based instruction in certain skill areas and with certain kinds of students have frequently found that students can learn faster via computer-based learning systems

Of course, this is not always the case; the key requirements seem to include students' motivation to learn, a well-designed overall instruction routine, and an integral role for the technology resource in the plan.

Instructional Problems Addressed by Constructivist Strategies

In the late 1970s and 1980s, criticism of the educational system accelerated and a critical perspective of curriculum gained prominence. Many educators began to echo critics from years before that education pursued inappropriate and outdated goals. They felt that education should go beyond programs to learn isolated skills and memorize facts that were characteristic of much school curriculum. They called for more emphasis on the abilities to solve problems, find information, and think critically about information; in other words, critics called for more emphasis on learning how to learn instead of learning specific content.

They also decried the large number of required skills and traditional learning activities that seemed abstract and unrelated to any practical skills. Students could see little relevance between skills they learned in school and those they used in their daily lives. The emphasis on individualized learning also drew criticism because students did not develop the ability to work well together in groups, an important workplace competency for the 1990s and beyond. These were not new criticisms, but the increase in the school dropout rate and poor national performance in comparison to students in other countries gave the issues the status of a national crisis; the United States was "a nation at risk" (National Commission on Excellence in Education, 1983), and some changes had to be made.

New ideas from cognitive science propose the importance of "anchoring instruction" in activities that students find meaningful and "authentic" (e.g., related to real-life situations) in the context of their own experiences. Proponents of these theories say that students who learn skills in isolation from such real-life problem solving will not remember to apply this prerequisite information when they require it. They also believe that "passive learners," students who view learning as something that happens *to* them rather than something they generate, are more likely to be poorly motivated to learn. To answer all these needs, constructivists propose arranging instruction around problems that students find compelling and that require them to acquire and use skills and knowledge to formulate solutions. Constructivists call for more emphasis on engaging students in the process of learning than on finding a single correct answer.

Many newer technology applications such as Logo and hypermedia seem to provide ideal conditions for nurturing constructivist curriculum goals. They provide vivid visual support which helps students develop better mental models of problems to be solved. These visual media seem to help make up for many students' deficiencies in such prerequisites as reading skills; they help to involve and motivate students by using graphics and other devices that students find interesting and attractive. Also, they let students work together in cooperative groups to construct products. In short, they meet all of the requirements for fulfilling the constructivist prescription for improving learning environments and refocusing curriculum.

Where Does Technology Fit? What Do Teachers Need to Know?

Clearly, the instructional problems identified by both objectivists and constructivists are common to any school or classroom, regardless of level or type of students or content. Teachers will always use some directed instruction as the most efficient means of teaching students certain required skills; teachers will always need motivating, cooperative learning activities to make sure that students want to learn and that they can transfer what they learn to problems they encounter. Consequently, proficient technology-oriented teachers must learn to combine directed instruction and constructivist approaches. To implement each of these strategies, teachers will select technology resources and integration methods that are best suited to carrying them out. Together, these two ostensibly different views of reality may merge to form a new and powerful approach to solving some of the major problems of the educational system, each contributing an essential element of the new instructional formula.

Some practitioners believe that constructivism will eventually dominate certain overall educational goals and objectives (e.g., learning to apply scientific methods), while systematic approaches will assure specific prerequisite skills (e.g., grammar/usage skills and basic math skills). Tennyson (1990) has suggested, for example, that about 30 percent of learning time should be spent on what he terms "acquiring knowledge" (e.g., verbal information and procedural knowledge), while about 70 percent should be spent on the "employment of knowledge" (e.g., contextual skills, cognitive strategies, and creative processes).

Over the next decade, teachers will test Tennyson's and others' proposals for merging systematic and constructivist methods in classrooms across the country; educators will confront the task of identifying the best mix of approaches for each content area. The decade will also bring challenges to traditional views on curriculum organization (e.g., interdisciplinary courses versus single subject ones), as well as how schools can best help students to learn (e.g., direct teaching or transmission of knowledge versus providing resources and guiding learning). As they prepare to meet this challenge, teachers need to know how these methods came about, how each addresses classroom needs, and how each suggests that they integrate technology resources. Subsequent sections of this chapter will give more specific information on the origins and uses of each of the two approaches described here. Table 3.3 lays out descriptive characteristics of the directed instruction and constructivist models.

Table 3.3 Summary of Characteristics of the Two Instructional Models

Directed instructional models tend to:

1. Focus on teaching sequences of skills that begin with lower-level skills and build to higher-level skills

2. Clearly state skill objectives with test items matched to them

3. Stress more individualized work than group work

4. Emphasize traditional teaching and assessment methods: lectures, skill worksheets, activities and tests with specific expected responses

Constructivist learning models tend to:

1. Focus on learning through posing problems, exploring possible answers, and developing products and presentations

2. Pursue more global goals that specify general abilities such as problem solving and research skills

3. Stress more group work than individualized work

4. Emphasize alternative learning and assessment methods: exploration of open-ended questions and scenarios, doing research and developing products; assessment by student portfolios, performance checklists, and tests with open-ended questions; descriptive narratives written by teachers

Theoretical Foundations of Directed Instruction

Learning Theories Associated with Directed Instruction

Two different theories of learning contributed to the development of directed instruction:

- **Behavioral theories.** Behavioral theorists concentrated on immediately observable (thus, "behavioral") changes in performance (e.g., tests) as indicators of learning.

- **Information-processing theories.** These theories developed from a branch of cognitive psychology that focused on the memory and storage processes that make learning possible. They viewed the process of learning in human beings as similar to the way a computer processes information. Theorists in this area explored how a person receives (senses) information and stores it in memory, the structure of memory that allows the learning of something new to relate to and build on something learned previously, and how a learner retrieves information from short-term and long-term memory and applies it to new situations.

The early work of giants in behavioral psychology such as B. F. Skinner and Edward Thorndike preceded work by information-processing theorists such as Richard Atkinson and David Ausubel. Robert Gagne was a leader in building

upon both of these behavioral and cognitive theories to recommend approaches to instruction. Gagne also played a key role in an area of development referred to as *instructional systems* design or the systematic design of instruction. Others associated with research and development underlying these systems approaches include Leslie Briggs, Robert Glaser, Lee Cronbach, David Merrill, Charles Reigeluth, Michael Scriven, and Robert Tennyson.

The Contributions of Behavioral Theories

Considered the "grandfather of behaviorism," B. F. Skinner generated much of the experimental data that serves as the basis of behavioral learning theory. (See Inset 3.1, p. 60). He and other behavioral theorists were concerned mainly with observable indications of learning and what those observations could imply for teaching. They recognized that internal processes operated in the brain during learning, but they did not attempt to describe those processes. They reasoned that since no one could prove these processes with any available scientific procedures, researchers should concentrate instead on "cause-and-effect relationships" that could be established by observation.

Skinner and others viewed the teacher's job as modifying the behavior of students by setting up situations to reinforce students when they exhibit desired responses, teaching them to exhibit the same response in all such situations. Behaviorists viewed learning as a sequence of stimulus and response action in the learner. They reasoned that teachers could link together responses involving lower-level skills and create a learning "chain" to teach higher-level skills. For example, in order to learn how to solve word problems in mathematics, students would first have to learn a series of lower-order behaviors such as following simple directions and performing arithmetic operations. "Chaining" the lower-order skills would enable students to solve the higher-order problems. The teacher would determine all of the skills needed to lead up to the desired behavior and make sure students learned them all in a step-by-step manner.

These behavioral principles underlay two well-known trends in education: (1) behavior modification techniques in classroom management and (2) programmed instruction. Although current use of programmed instruction itself is limited, its principles form much of the basis of effective drill and practice and tutorial courseware.

The Contributions of Information-Processing Theories

Many educational psychologists found the emphasis on observable outcomes of learning unsatisfying. They did not agree with the behaviorists' view that stimulus-response learning alone could form the basis for building higher-level skills. As they focused on capabilities such as rule learning and problem solving, they became more concerned with the internal processes (those inside the brain) that went on during learning. With this knowledge, they hoped to

Inset 3.1

Skinner's Behaviorist Theories of Learning: Building on the S-R Connection

Skinner patterned his fundamental premises and experimental approaches after those of Edward Thorndike, who was "a pioneer in … efforts to understand the learning of animals by performing experiments rather than by collecting anecdotes about animal behavior" (Gagne, 1985, p. 6). Like Thorndike, Skinner's early work was with animals (Skinner, 1938). But some 30 years later, in a book called *The Technology of Learning* (Skinner, 1968), he used this and his subsequent work with human behavior to develop a detailed theory of how classroom instruction should reflect behavioral principles. Many of his principles led to development of effective classroom management techniques (Sultzer and Mayer, 1972).

Since internal processes (those inside the mind) involved in learning could not be seen directly, Skinner concentrated on cause-and-effect relationships that could be established by observation. The cause variables in his theory were:

- **Stimulus.** An event, combination of events, or relationship among events that affect a learner's senses

- **Reinforcement.** An event that increases the probability of an act that immediately preceded it

- **Contingencies of reinforcement.** Arranging situations for the learner in which reinforcement is made contingent upon a correct (i.e., desired) response

The "effect variables" in Skinner's theory are:

- **Respondents.** Reflex actions elicited by a given stimulus

- **Operants.** Responses without any obvious stimulus, which is, therefore, attributed to internal processes in the brain (e.g., learning)

For example, a teacher might have students practice the multiplication tables. Each flash card with a problem on it (e.g., $2 \times 2 =$) is a stimulus. Their answer is the response. The teacher may praise students intermittently (reinforcement). The students repeat this process over and over until neither the card or the praise is needed. Whenever they hear "2×2" they automatically give the required response.

To Skinner, teaching was a process of arranging contingencies of reinforcement effectively to bring about learning. Teachers and instructional materials are the stimuli; the skills that students demonstrate are the responses. Skinner believed that even such high-level capabilities as critical thinking and creativity could be taught in this way; it was simply a matter of establishing chains of behavior through principles of reinforcement. His recommendations resulted in the development of programmed instruction, which he viewed as the most efficient means available for learning skills. Educational psychologists such as Benjamin Bloom also used Skinnerian principles to develop methods that became known as *mastery learning*.

Source: Gagne, R. (1985). *The conditions of learning* (4th ed.). New York: Holt, Rinehart and Winston; Skinner, B. F. (1938). *The behavior of organisms*. New York: Appleton; Skinner, B. F. (1968). *The technology of teaching*. New York: Appleton; and Sulzer, B. and Mayer, R. (1972). *Behavior modification procedures for school personnel*. Hinsdale, IL: Dryden Press. Photo from Corbis-Bettman.

arrange appropriate instructional conditions to promote learning of these kinds of skills. (See Inset 3.2, page 61.)

Information-processing theorists based their work on a model of memory proposed by Richard Atkinson (Atkinson and Shiffrin, 1968). This model proposed processes and structures through which an individual receives and stores information. They derived their constructs from computer science and linguistic theories. They proposed that the human act of learning consists of three components analogous to the parts of a computer: input variables (information to be learned), a "processing component" consisting of executive controls (such as attention and expectancies) working in conjunction with short-term and long-term memory, and output variables (outward indications that a process has taken place).

One information-processing theorist, David Ausubel (1968), proposed that the way a learner receives and stores information can contribute to its usefulness, for example, by transferring current learning to learning of other skills. One of Ausubel's better-known recommendations was that teachers could take advantage of the "cognitive structures"

around which information is organized in the brain by using "advance organizers" to bridge the gap between what the learner already knows and the new information.

Characteristics of Directed Instruction

Teaching methods based primarily upon behaviorist and information-processing learning theories are usually associated with the more traditional, teacher-directed forms of instruction. Robert Gagne (see Inset 3.3, page 62) is considered a leader in developing instructional guidelines for directed instruction that combine the behavioral and information-processing learning theories. He asserted that teachers must accomplish at least three tasks to link these learning theories with teaching practices:

1. **Assure prerequisite skills.** Teachers must make sure that students have all the prerequisite skills they need to learn a given new skill. This may involve identifying component skills and the order in which they should be taught. Gagne referred to this group of skills as a *learning hierarchy*.

Inset 3.2
The Information-Processing Theorists: The Mind as Computer

Behaviorists focused only on external, directly observable indicators of human learning. Many people found this explanation insufficient to guide instruction. During the 1950s and 1960s, a group of researchers known as the *cognitive-learning theorists* began to hypothesize "explanations for learning that focus on the *internal* mental processes people use in their effort to make sense of the world" (Eggen and Kauchak, 1994, p. 305). The information-processing theorists were among the first and most influential of the cognitive-learning theorists. They hypothesized processes inside the brain that allow human beings to learn and remember. Although no single, cohesive information-processing theory of learning, summarizes the field, the work of the information-processing theorists is generally based on a model of memory and storage proposed by Atkinson and Shiffrin (1968). This model proposed that the brain contains certain structures that process information much like a computer does. This model of the mind as computer hypothesizes that the human brain has three kinds of memory or "stores":

- **Sensory registers.** The part of memory that receives all the information a person senses (sees, hears, feels, tastes, or smells)
- **Short-term memory (STM).** Also known as *working memory,* the part of memory where new information is held temporarily until it is either lost or placed into long-term memory
- **Long-term memory (LTM).** The part of memory which has an unlimited capacity and can hold information indefinitely

According to this model, learning occurs in the following way. First, information is sensed through receptors: eyes, ears, nose, mouth, and/or hands. This information is held in the sensory registers for a very short time (perhaps a second), after which it either enters STM or it is lost (forgotten). Many information-processing theorists believe that information can be sensed but lost before it gets to STM if the person is not paying attention to it. "Whatever people pay attention to moves into working memory" (Ormrod, 1995, p. 316), where it can stay for about 5 to 20 seconds (Ormrod, 1995). After this time, if information is not processed or practiced in a way that causes it to transfer to LTM, then it, too, is lost. Information-processing theorists believe that for new information to be transferred to LTM, it must be linked in some way to prior knowledge already in LTM. Once information does enter LTM, it is there essentially permanently, although some psychologists believe that even information stored in LTM can be lost if not used regularly (Ormrod, 1995).

The information-processing views on learning have become the basis for many common classroom practices. For example, teachers use a variety of methods to increase the likelihood that students will pay attention to new information. While presenting information, they give instructions that point out important points and characteristics in the new material and suggest methods of "encoding" or remembering them by linking them to information that students already know. Teachers also give students practice exercises to help assure the transfer of information from short-term to long-term memory.

Educational psychologists such as Gagne (see next inset) and Ausubel provided many instructional guidelines designed to enhance the processes of attention, encoding, and storage. Gagne proposed that teachers use a hierarchical "bottom-up approach," making sure that students learn lower-order skills first and building upon them. Ausubel, by contrast, recommended a "top-down" approach; he proposed that teachers provide "advance organizers" or overviews of the way information will be presented to help students develop mental frameworks on which to "hang" new information (Gage and Berliner, 1988). Information-processing theories have also guided the development of artificial intelligence (AI) applications, an attempt to develop computer software that can simulate the thinking and learning behaviors of humans.

A model of the human memory system

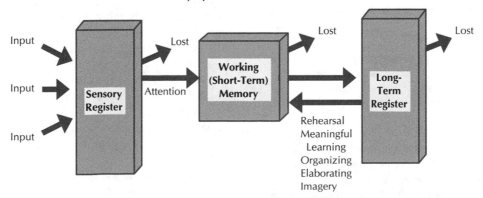

Source: Ormrod, J. (1995), *Educational Psychology: Principles and Applications.* Englewood Cliffs, NJ: Prentice Hall, p. 315.

Source: Atkinson, R. and Shiffrin, R. (1968). Human memory: A proposed system and its control processes. In K. Spence and J. Spence (Eds.). *The psychology of learning and motivation: Vol. 2.* New York: Academic Press; Eggen, P. and Kauchak, D. (1994). *Educational psychology: Classroom connections.* New York: Macmillan; Gage, N. and Berliner, D. (1988) *Educational psychology* (4th ed.). Boston: Houghton Mifflin; Gagne, R. (1985). *The conditions of learning* (4th ed.). New York; Holt, Rinehart and Winston; Klatzky, R. (1980). *Human memory: Structures and processes* (2nd ed.). San Francisco: Freeman; and Ormrod, J. (1995). *Educational psychology: Principles and applications.* Englewood Cliffs, NJ: Prentice-Hall.

Inset 3.3
Gagne's Principles: Providing Tools for Teachers

Gagne built on the work of the behavioral and information-processing theorists by translating principles from their learning theories into practical instructional strategies that teachers could employ with directed instruction. He is best-known for three of his contributions in this area: the events of instruction, the types of learning, and learning hierarchies.

Events of instruction. Gagne used the information-processing model of internal processes to derive a set of guidelines that teachers could follow to arrange optimal "conditions of learning." His set of nine "Events of Instruction" were perhaps the best-known of these guidelines (Gagne, Briggs, and Wager, 1988):

1. Gaining attention
2. Informing the learner of the objective
3. Stimulating recall of prerequisite learning
4. Presenting new material
5. Providing learning guidance
6. Eliciting performance
7. Providing feedback about correctness
8. Assessing performance
9. Enhancing retention and recall

Types of learning. Gagne identified several types of learning as behaviors students demonstrate after acquiring knowledge. These

differ according to the conditions necessary to foster them. He showed how the Events of Instruction would be carried out slightly differently from one type of learning to another (Gagne, Briggs, and Wager, 1988):

1. Intellectual skills
   - Problem solving
   - Higher-order rules
   - Defined concepts
   - Concrete concepts
   - Discriminations
2. Cognitive strategies
3. Verbal information
4. Motor skills
5. Attitudes

Learning hierarchies. To develop "intellectual skills," Gagne believed, requires learning that amounts to a building process. Lower-level skills provide a necessary foundation for higher-level ones. For example, in order to learn to work long division problems, students first would have to learn all the prerequisite math skills, beginning with number recognition, number facts, simple addition and subtraction, multiplication, and simple division. Therefore, to teach a given skill, a teacher must first identify its prerequisite skills and make sure the student possesses them. He called this list of building block skills a *learning hierarchy*.

Source: Gagne, R. (1985). *The conditions of learning* (4th ed.). New York: Holt, Rinehart and Winston; and Gagne, R., Briggs, L., and Wager, W. (1988). *Principles of instructional design*. New York: Holt, Rinehart and Winston. Photo courtesy of Robert Gagne.

2. **Supply instructional conditions.** Teachers must arrange for appropriate instructional conditions to support the internal processes involved in learning; that is, they must supply sequences of carefully structured presentations and activities that help students understand (process), remember (encode and store), and transfer (retrieve) information and skills.

3. **Determine the type of learning.** Finally, teachers must vary these conditions for each of several different kinds of learning. (The kinds of learning, along with brief descriptions of related instructional conditions, are shown in the inset.)

The behaviorist and information-processing theories have not only helped establish key concepts such as types of learning and instructional conditions required to bring about each type; they also laid the groundwork for more efficient methods of creating directed instruction. These methods, known as *systematic instructional design* or *systems approaches* (see Inset 3.4, page 63), incorporated information from learning theories into step-by-step proce-

dures for preparing instructional materials. Systematic methods came about largely in response to teachers' logistical problems in meeting individual needs of large numbers of learners. They were adopted more by military and industrial trainers, however, than by K-12 classroom teachers (Saettler, 1990; Wager, 1992).

Systems approaches contribute to courseware development primarily through the design of self-contained tutorial packages. However, when teachers plan their own directed instruction with technology, thinking about instruction as a system may help them develop guidelines to evaluate their own teaching effectiveness and the usefulness of their computer-based resources. For example, they may pose and answer the following kinds of criterion questions about the components of their instructional systems in order to evaluate and improve their plans and materials:

- **Instructional goals and objectives.** Am I teaching what I intended to teach? Do the goals and objectives of the courseware materials match my own?

Inset 3.4
Systems Approaches and the Design of Instruction: Managing the Complexity of Teaching

Saettler (1990) says that the development of "scientifically based instructional systems" precedes this century, but he also points out that modern instructional design models and methods have their roots in the collaborative work of Robert Gagne and Leslie Briggs. These notable educational psychologists developed a way to transfer "laboratory-based learning principles" gleaned from military and industrial training to create an efficient way of developing curriculum and instruction for schools. Gagne specialized in the use of instructional task analysis to identify required subskills and conditions of learning for them. Briggs's expertise was in systematic methods of designing training programs to save companies time and money in training their personnel. When they combined these two areas of expertise, the result was a set of step-by-step processes known as a *systems approach to instructional design* or *systematic instructional design* which came into common use in the 1970s and 1980s.

Although there are many versions of the systematic design process, Bradens (1996) describes the steps of this model in the following diagram. His article also gives a comprehensive discussion of the ongoing controversies surrounding the terms *instructional design* and *systems approaches.*

One component of a systematic instructional design process was the use of learning hierarchies to develop curriculum maps (Gagne, Briggs, & Wager, 1988, p. 24). According to Saettler (1990), "the 1960s produced most of the major components of the instructional design process." Names associated with this era include Robert Mager (instructional objectives), Glaser (criterion-referenced testing), and Cronbach and Scriven (formative and summative evaluation). Other major contributors to modern instructional design models include David Merrill (Component Display Theory) and Charles Reigeluth (Elaboration Theory).

Braden's 1996 Instructional Design Model

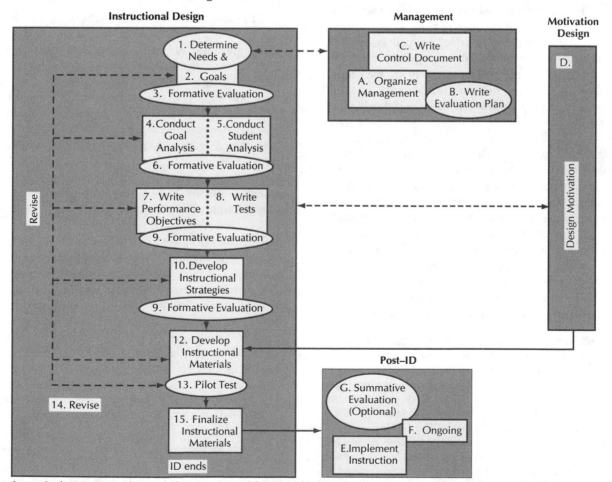

Source: Bradens, R. (1996). The case for linear instructional design and development: A commentary on models, challenges, and myths. *Educational Technology, 36* (2), 5–23.

Source: Gagne, R., Briggs, L., and Wager, W. (1988). *Principles of instructional design.* New York: Holt, Rinehart and Winston; and Saettler, P. (1990). *The evolution of American educational psychology.* Englewood, CO: Libraries Unlimited.

- **Instructional analysis (task analysis).** Do my students have all of the lower-level skills they need to learn successfully what I want to teach them? Does the courseware require skills that my students lack?

- **Tests and measures.** Do the tests I will use measure what I will teach? Do the items included in the courseware materials match my own measures?

- **Instructional strategies.** Are my instructional activities providing appropriate conditions (instructional events) for the kind of learning involved (e.g., supplying examples and explanation as well as gaining attention)? What part do chosen courseware resources play in the activities and why?

- **Evaluating and revising instruction.** Have I successfully presented the instruction I envisioned? How could I improve it to make it more effective (e.g., clearer, more motivating)? Has the courseware successfully played the part I envisioned for it? Do I need better strategies for using it? Do I need better courseware?

Directed Methods: Problems and Possibilities

The learning theories and instructional design approaches associated with directed instruction have profoundly affected American curriculum and classroom practices over the past five decades. Some would say that at least part of the impact has been negative. Programmed instruction, an early method based on behavioral principles and systematic methods, was found to teach skills successfully, but many users it found it quite boring. During the 1970s, the behavioral emphasis on observable outcomes resulted in strong attention to required performance objectives and individual skill testing in K-12 schools. These requirements were often very unpopular among teachers. In many cases, schools did not apply systematic methods according to their designs; for example, many school districts required specific performance objectives for all curricula but never linked the objectives to any instructional materials or tests.

In other cases, widespread "teaching to the tests" made curriculum very dry and apparently disconnected from any application outside the classroom. Constructivist approaches are, in part, a backlash against the perceived regimentation that arose from this emphasis. The greatest current criticisms of directed methods charge irrelevance to the needs of today's students. Critics frequently cite several problems:

- **Students cannot do problem solving.** Many parents and educators feel that traditional methods focus too narrowly on breaking topics into discrete skills and teaching them systematically. They blame this limitation for poor national test scores on more global skills of problem solving and reasoning (Cognition and Technology Group at Vanderbilt [CTGV], 1991b). The CTGV report says that, "The thinking activities that are of concern include the ability to write persuasive essays, engage in informal reasoning, explain how data relate to theory in scientific investigations, and formulate and solve moderately complex problems that require mathematical reasoning" (1991b, p. 34)

- **Students find directed instruction activities unmotivating and irrelevant.** Some critics of directed methods feel that teaching isolated skills also tends to isolate students from each other and from the authentic situations that they find motivating and relevant. This makes learning repetitive and predictable—what the CTGV call an "industrial assembly line" approach to transmitting knowledge. The CTGV repeatedly cited Corey's (1944) article entitled "Poor Scholar's Soliloquy," in which a student of obvious intelligence and capability describes how he performs poorly in school because he cannot relate to the tasks his teachers assign. The CTGV said that this 50-year-old article highlighted an old, ongoing problem: "... many students seem to learn effectively in the context of authentic, real-life activities yet have great difficulty learning in the decontextualized, arbitrary-task atmosphere of schools" (p. 9). They also indicate that students' lack of interest in school tasks leads directly to higher drop-out rates.

- **Students cannot work cooperatively.** Observers of economic trends in this country and throughout the world seem to feel that national economic survival depends, in large part, on how well workers work together to solve problems of mutual concern. Cooperative group work has rarely been emphasized in American schools, especially at secondary levels. Directed instruction seems geared toward individual learning, so it has been accused of isolating learners from each other and neglecting much-needed social skills.

In modern classrooms, teachers do not use programmed instruction to teach skills, nor do they design many individual lessons with specific objectives and tests for each one. Teachers use lesson plans primarily to communicate clearly to others (e.g., supervisors, substitutes) what will happen in the classroom, but lesson plans are not usually considered strict sequences to be followed exactly. A growing emphasis also highlights methods other than objective test items to determine what students have learned.

But teachers must still arrange conditions of learning, and they are still largely responsible for answering the question from directed instruction, "What behaviors will I look for in my students to show me they have learned what I expected them to learn?" Teachers can find powerful tools in traditional methods if they perceive that certain kinds of students need more structured learning than others do, or that certain required skills can best be learned through directed instruction. Although many educators shun behaviorism as an archaic and outmoded theory, a recent special issue of *Educational Technology* (October 1993) details some current uses of directed instruction based on behaviorist principles. Some examples include:

- Fluency practice in precision teaching of basic reading and math skills to young learners (Spence and Hively, 1993)

- Performance management contingencies to improve the study habits and achievement of college students (Mallott, 1993)

- Structured, teacher-directed techniques to teach problem solving and higher-order thinking skills to at-risk students (Carnine, 1993)

- Proposed application of behavioral techniques to teach the required behaviors leading to creativity (Epstein, 1993)

Theoretical Foundations of Constructivism

Molenda (1991) has said that "constructivism comes in different strengths … from weak to moderate to extreme" (p. 47). Phillips (1995) referred to constructivism as made up of many "sects, each of which harbors some distrust of its rivals" (p. 5). The differences among those who think of themselves as constructivists makes it very difficult to settle on a single definition for constructivism, but these differences may be explained by examining the variations in learning theories that underlie constructivist approaches.

Learning Theories Associated with Constructivism

Constructivist strategies are based on principles of learning that were derived from branches of cognitive science. This area focused specifically on students' motivation to learn and their ability to use what they learn outside the "school culture." Constructivist strategies attempt to account for and remedy perceived deficiencies in behaviorist and information-processing theories and the teaching methods based on them. Constructivists try to inspire students to see the relevance of what they learn, and to prevent what the CTGV (1990) call "inert knowledge," or students' failure to transfer what they already know to the learning of other skills that require the prior knowledge.

These theories are based on the ideas of revered educational philosophers such as John Dewey and renowned educational psychologists such as Lev Vygotsky, Jerome Bruner, and Jean Piaget. Later work by educational theorists such as Seymour Papert and educational psychologists and practitioners such as John Seely Brown, the Cognition and Technology Group at Vanderbilt, Rand Spiro, D. N. Perkins, Ann Brown, Joe Campione, Carl Bereiter, and Marlene Scardamalia expanded on these principles and developed specifications for translating these theories of cognition into teaching practices.

The Contributions of Early Cognitive Learning Theories

Educators credit theorists such as John Dewey, Lev Vygotsky, Jean Piaget, and Jerome Bruner with some of the fundamental premises of constructivist thinking. Dewey, is well-known, of course, for laying the theoretical groundwork for many characteristics of today's educational system. He was responsible for the progressive movement in American education, many principles of which are now being re-examined for possible applications in school restructuring efforts. Several of Dewey's ideas support constructivist models of teaching and learning. Among these ideas is the need to center instruction around activities that are relevant and meaningful to a student's own experience. Prawat (1993) recalled Dewey's label of "worse than useless" any instruction that did not center around problems "already stirring in the child's experience" (p. 6).

The work of renowned human development theorist Lev Vygotsky also contributed key support for constructivist approaches. His twin concepts of "scaffolding" and the "zone of proximal development" are important for constructivists. Vygotsky said that children differ from adults in their levels of "knowing" or understanding concepts. In a typical classroom, the student-novice represents one end of the continuum of understanding and the teacher-expert represents the other. The "zone of proximal development" is the gap between them. Schools can help children develop their level of understanding and bridge the gap between teacher-supervised work and independent work through what he called a "scaffolding" process. Scaffolding relies on assistance provided by expert problem solvers in an area (e.g., teachers) to novices (students). Prawat (1993) observed that, "Vygotsky emphasized the importance of social relations in all forms of complex mental activity" (p. 10). Constructivists feel that teachers can most effectively provide scaffolding, or assistance in acquiring new knowledge through supervised collaborative learning activities (see Inset 3.5, p. 66).

The internationally famous developmental psychologist Piaget is generally regarded as a major contributor of theoretical principles for constructivist thinking. While some educators feel that Piaget's ideas have been applied inappropriately, some of his basic premises seem related to constructivist approaches. Piaget felt that a child passes through a series of stages of cognitive development ranging from the sensorimotor stage characteristic of infants and very small children to the formal operations stage typical of teenagers (see Inset 3.6, p. 67).

According to Piaget, children's capabilities and characteristics at each of these stages arise from the way they are able to understand the world at those stages in their development. Piaget also felt that children progress from stage to stage through experiences in which they adapt to their environment and organize patterns of behavior based on what they learn. Sometimes they fit new experiences into their existing schemes or patterns of behavior, a process he called *assimilation;* sometimes they change their existing schemes to incorporate new experiences, which he called *accommodation.* These stages, he felt, occur naturally. Thus, he believed that much of what a child needs to learn cannot and should not be consciously taught. Rather, it should emerge as the natural by-product of experiences. There, Piaget advocated nonintervention, saying that "Everything one teaches a child prevents him from inventing or discovering" (Bringuier, 1980, p. 102).

Some of the principles associated with educational theorist Jerome Bruner seem to coincide with those of Vygotsky and Piaget, providing further theoretical support for constructivist theory. Like Piaget, Bruner believed children go through various stages of intellectual development (See Inset 3.7). But unlike Piaget, Bruner supported intervention. He was primarily concerned with making education more relevant to students' needs at each stage, and he believed that teachers could accomplish this by encouraging active participation in the learning process. Active participation, he felt,

Inset 3.5
The Contributions of Lev Vygotsky: Building a Scaffold to Learning

For a long time, the writings of Russian philosopher and educational psychologist Lev Semenovich Vygotsky seemed to have more influence on the development of educational theory and practice in America than in his own country. Davydov (1995) notes that Vygotsky's landmark book *Pedagogical Psychology*, though written in 1926, was not published in Russia until 1991. Davydov attributes this lack of attention to the nature of the Russian government up until the time of perestroika. "… Vygotsky's general ideas could not be used for such a long time in the education system of a totalitarian society—they simply contradict all of its principles" (p. 13). What were these educational concepts that were so threatening to a communistic state but found such a warm reception in a democracy?

Vygotsky felt that cognitive development was directly related to and based on social development (Gage and Berliner, 1988; Ormrod, 1995). What children learn and how they think are derived directly from the culture around them: "… children begin learning from the world around them, their social world, which is the source of all their concepts, ideas, facts, skills, and attitudes. … [O]ur personal psychological processes begin as social processes, patterned by our culture" (Gage and Berliner, 1988, p. 124). An adult perceives things much differently than a child does, but this difference decreases as children gradually translate their social views into personal, psychological ones.

Vygotsky referred to the difference between these two levels of cognitive functioning (adult/expert and child/novice) as the "zone of proximal development." He felt that teachers could provide good instruction by finding out where each child was in his or her development and building upon the child's experiences. He called this building process "scaffolding." Ormrod (1995) said that teachers promote students' cognitive development by presenting some classroom tasks that "they can complete only with assistance—that is, within each student's zone of proximal development" (p. 59). Gage and Berliner (1988) feel that problems occur when the teacher leaves too much for the child to do independently, thus slowing the child's intellectual growth. "The

mastery of more complex levels of functioning can proceed with the help of an adult … [I]t is the more knowledgeable person who provides the intellectual scaffolding for the child to climb. In the zone of proximal development, social knowledge—knowledge acquired through social interaction—becomes individual knowledge and individual knowledge grows and becomes more complex" (p. 126).

Davydov (1995) found five basic implications for education in Vygotsky's ideas (p. 13):

1. Education is intended to develop children's personalities.

2. The human personality is linked to its creative potential, and education should be designed to discover and develop this potential to its fullest in each individual.

3. Teaching and learning assume that students master their inner values through some personal activity.

4. Teachers direct and guide the individual activities of the students, but they do not force their will on them or dictate to them.

5. The most valuable methods for student learning are those that correspond to their individual developmental stages and needs; therefore, these methods cannot be uniform across students.

A review of these ideas clearly shows the influence of Vygotsky's work on constructivist thought. Constructivist concepts of basing instruction on each child's personal experiences and the need for learning through collaborative, social activities seem very much in tune with Vygotsky's premises. It is also apparent that Vygotsky's theories, with their emphasis on individual differences, personal creativity, and the influence of culture on learning, were discordant with the aims of the USSR, a government designed to "subjugate the education of young people to the interests of a militarized state that needed citizens only as devoted cogs" (Davydov, 1995, p. 12).

Source: Davydov, V. (1995). The influence of L. S. Vygotsky on education theory, research, and practice. *Educational Researcher, 24* (3), 12–21; Gage, N. and Berliner, D. (1988). *Educational psychology* (4th ed.). Boston: Houghton Mifflin; and Ormrod, J. (1995). *Educational psychology: Principles and applications.* Englewood Cliffs, NJ: Prentice-Hall.

was best achieved through providing discovery learning environments that would let children explore alternatives and recognize relationships between ideas (Bruner, 1973).

Constructivists agree that educational experiences should foster a child's progress through stages of development. They tend to perceive much of today's education as too structured and oriented too far toward activities that are inappropriate for children's current developmental levels. These experiences, they say, can actually slow students' progress by inhibiting their innate desire to make sense of their world at each stage of their cognitive development. Like Bruner, most constructivists call for instruc-

tional intervention, that is, for teachers to provide learning activities designed not only to match but even to accelerate movement through these stages. They also feel that education should provide children with more opportunities for cognitive growth through exploration, unstructured learning, and problem solving.

The Contributions of Research Based on Cognitive Principles

Several lines of research and development based on principles from cognitive science have profoundly affected educational practice, particularly instructional applications of

Inset 3.6
Piaget's Theories: Cognitive Development in Children

As Flavell (1985) observed, "Piaget's contributions to our knowledge of cognitive development have been nothing short of stupendous" (p. 4). His examination of how thinking and reasoning abilities develop in the human mind began with observations of his own children and developed into a career that spanned some 60 years. He referred to himself as a "genetic epistemologist," or a scientist who studies how knowledge begins and develops in individuals. Both believers in and critics of Piagetian principles agree that his work was complex, profound, sometimes misunderstood, and usually oversimplified. However, at least two features of this work are widely recognized as underlying all of Piaget's theories: his stages of cognitive development and his processes of cognitive functioning.

Piaget believed that all children go through four stages of cognitive development. While the ages at which they experience these stages vary somewhat, he felt that each developed higher reasoning abilities in the same sequence:

- **Sensorimotor stage (from birth to about 2 years).** Characteristics of children:

  - Explore the world around them through their senses and through motor activity. In the earliest stage, they cannot differentiate between themselves and their environments (e.g., if they cannot see something, it does not exist).

  - Begin to have some perception of cause and effect; develop the ability to follow something with their eyes.

- **Preoperational stage (from about age 2 to about age 7).** Characteristics of children:

  - Develop greater abilities to communicate through speech and to engage in symbolic activities such as drawing objects and playing by pretending and imagining

  - Develop numerical abilities such as the skill of assigning a number to each object in a group as it is counted

  - Increase their level of self-control and are able to delay gratification, but are still fairly egocentric

  - Unable to do what Piaget called *conservation tasks* (tasks that call for recognizing that a substance remains the same even though its appearance changes, e.g., shape is not related to quantity)

- **Concrete operational stage (from about age 7 to about age 11).** Characteristics of children:

  - Increase in abstract reasoning ability and ability to generalize from concrete experiences

  - Can do conservation tasks

- **Formal operations stage (from about age 12 to about age 15).** Characteristics of children:

  - Can form and test hypotheses, organize information, and reason scientifically

  - Can show results of abstract thinking in the form of symbolic materials (e.g., writing, drama)

Piaget believed that a child's development from one stage to another takes place through a gradual process of interacting with the environment. Children develop as they confront new and unfamiliar features of their environment that do not fit with their current views of the world. When this happens, he said, a "disequilibrium" occurs that the child seeks to resolve through one of two processes of adaptation. The child either fits the new experiences into his or her existing view of the world (a process called *assimilation*) or changes that schema or view of the world to incorporate the new experiences (a process called *accommodation*).

Ormrod (1995) summarizes Piaget's basic assumptions about children's cognitive development in the following way:

1. Children are active and motivated learners.

2. Their knowledge of the world becomes more integrated and organized over time.

3. Children learn through the processes of assimilation and accommodation.

4. Cognitive development depends on interaction with one's physical and social environment.

5. The processes of equilibration (resolving disequilibrium) helps to develop increasingly complex levels of thought.

6. Cognitive development can occur only after certain genetically controlled neurological changes occur.

7. Cognitive development occurs in four qualitatively different stages.

Recent research has raised questions about the ages at which children's abilities develop, and it is widely believed that age does not determine development alone (Ormrod, 1995). Educators do not always agree on the implications of Piaget's theories for classroom instruction. One frequently expressed instructional principle based on Piaget's stages is the need for concrete examples (e.g., pictures, manipulatives) and experiences when teaching abstract concepts to young children who may not yet have reached a formal operations stage. Piaget himself repeatedly expressed a lack of interest in how his work applied to school-based education, calling it "the American question." He pointed out that much learning occurs without any formal instruction, as a result of the child interacting with the environment.

Source: Flavell, (1985). *Cognitive development* (2nd ed.). Englewood Cliffs, NJ: Prentice-Hall; and Ormrod, J. (1995). *Educational psychology: Principles and applications.* Englewood Cliffs, NJ: Prentice-Hall. Photo from Corbis-Bettman.

Inset 3.7
The Contributions of Jerome Bruner: Learning as Discovery

Like Piaget, Jerome Bruner was interested in children's stages of cognitive development. Bruner described development in three stages (Gage and Berliner, 1988):

- **Enactive stage (from birth to about age 3).** Children perceive the environment solely through actions that they initiate. Objects are described and explained solely in terms of what a child can do with them. The child cannot tell how a bicycle works, but can show what to do with it. Showing and modeling have more learning value than telling for children at this stage.

- **Iconic stage (from about age 3 to about age 8).** Children can remember and use information through imagery (mental pictures or icons). Visual memory increases and children can imagine or think about actions without actually experiencing them. Decisions are still made on the basis of perceptions, rather than language.

- **Symbolic stage (from about age 8).** Children begin to use symbols (words or drawn pictures) to represent people, activities, and things. They have the ability to think and talk about things in abstract terms. They can also use and understand what Gagne would call "defined concepts." For example, they can discuss the concept of toys and identify various kinds of toys, rather than defining them only in terms of toys they have seen or handled. They can better understand mathematical principles and use symbolic idioms such as "Don't cry over spilt milk."

Bruner also identified six indicators or "benchmarks" that revealed cognitive growth or development (Owen, Froman, and Moscow, 1981, p. 49; Gage and Berliner, 1988, p. 121–122):

1. Responding to situations in varied ways, rather than always in the same way
2. Internalizing events into a "storage system" that corresponds to the environment
3. Increased capacity for language
4. Systematic interaction with a tutor (parent, teacher, or other role model)
5. Language as an instrument for ordering the environment
6. Increasing capacity to deal with multiple demands

Unlike Piaget, Bruner was very concerned about arrangements for school instruction that acknowledged and built upon the stages of cognitive development. The idea of discovery learning is largely attributed to him. Discovery learning is "an approach to instruction through which students interact with their environment—by exploring and manipulating objects, wrestling with questions and controversies, or performing experiments" (Ormrod, 1995, p. 442). Bruner felt that students were more likely to understand and remember concepts they had discovered in the course of their own exploration. However, research findings have yielded mixed results for discovery learning, and the relatively unstructured methods recommended by Bruner have not found widespread support (Eggen and Kauchak, 1994; Ormrod, 1995). Teachers have found that discovery learning is most successful when students have prerequisite knowledge and undergo some structured experiences.

Source: Eggen, P. and Kauchak, D. (1994). *Educational psychology: Classroom connections*. New York: Macmillan; Gage, N. and Berliner, D. (1988) *Educational psychology* (4th ed.). Boston: Houghton Mifflin; Goetze, E., Alexander, P., and Ash, M. (1992). *Educational psychology: A classroom perspective*. New York: Merrill; Bruner, J. (1966). *Toward a theory of instruction*. Boston: Little, Brown; Ormrod, J. (1995). *Educational psychology: Principles and applications*. Englewood Cliffs, NJ: Prentice-Hall; and Owen, S., Froman, R., and Moscow, H. (1981). *Educational psychology*. Boston: Little, Brown. Photo courtesy of Jerome Bruner.

technology. This section discusses some major contributors to this research.

Papert's "microworlds." Seymour Papert, a mathematician and pupil of Piaget, was one of the first vocal critics of using technology in the context of traditional instructional methods. In his 1980 book *Mindstorms,* he also became one of the first to raise the national consciousness about the potential role of technology in creating alternatives to what he perceived as inadequate and harmful educational methods. Papert credited Piaget with helping to shape his views on learning and school practices. Like Piaget, Papert characterized children as "builders of their own intellectual structures" (Papert, 1980, p. 7), and he asserted that these structures developed in a certain order. But as Papert claimed in his book, "I give more weight than [Piaget] does to the influence a particular culture provides in determining that order [of development]" (p. 20). Papert believed that, given the right resources and experiences, even very young children could accelerate their development and learn concepts involving formal operations. Bass (1985) observed that Papert's position "may have been controversial at first, but Piagetian theory has never been static and Papert's interventionist perspective merely presaged a new focus of the Genevan group. …In many ways, the development of Logo has paralleled the recent development of Piagetian theory" (p. 107). However, even Bass admitted that while the focus of Piaget's colleagues may have changed, Piaget himself was never "particularly concerned with what he called 'the American question' of how to influence [children's] development through planned learning environments" (p. 113).

Papert felt strongly that school instruction was frequently counterproductive to children's natural cognitive development. He echoed Piaget's beliefs that the most important learning was "learning without being taught," and that schools put too much emphasis on structured teaching. Papert had vivid memories of his delight at learning logical principles related to gears and how this delight motivated

Inset 3.8
Seymour Papert: Turtles and Beyond

One of Piaget's most famous American pupils, Seymour Papert, has profoundly influenced the field of educational technology. Papert began his career as a mathematician. After studying with Jean Piaget in Geneva from 1959 to 1964, however, Papert became impressed with Piaget's way of "looking at children as active builders of their own intellectual structures" (Papert, 1980, p. 19). Papert subsequently joined the Artificial Intelligence Laboratory at the Massachusetts Institute of Technology and began experimenting with Logo, a new programming language, and its use with young children. One of his colleagues was also working with children, teaching them to control a robot in the shape of a turtle. The MIT team decided to combine the two concepts, integrating an on-screen "turtle" into the Logo language. This addition provided the vital link that Papert felt would allow children to move more easily from the concrete operations of earlier stages of Piaget's hypothesis to more abstract (formal) ones. In 1980, Papert published his theories in a book entitled *Mindstorms: Children, Computers, and Powerful Ideas.* This book challenged then-current instructional goals and methods for both mathematics and educational technology, and it became the first widely recognized constructivist statement of educational practice with technology resources.

As Papert himself observed, "I make a slightly unorthodox interpretation of [Piaget's] theoretical position and a very unorthodox interpretation of the implications of his theory for education" (Papert, 1980, p. 217). Piaget himself was not concerned with instructional methods or curriculum matters, and he had no interest in trying to accelerate the stages of cognitive development. Papert, on the other hand, felt that children could advance in their

intellectual abilities more quickly with the right kind of environment and assistance. He described the requirements of such an environment in his 1980 book. Some key concepts included:

- **Logo and the microworlds concept.** Papert perceived Logo as a resource with ideal properties for encouraging learning. Since Logo is graphics-oriented, it allows children to see cause-and-effect relationships between the logic of programming commands and the pictures that result. This logical, cause-and-effect quality of Logo activities makes possible "microworlds," or self-contained environments where all actions are orderly and rule-governed. He called these microworlds "incubators for knowledge" where children could pose and test out hypotheses.

- **Discovery learning and "powerful ideas."** Although he never used the term *discovery learning,* Papert felt that children should be allowed to "teach themselves" with Logo. Reflecting the Piagetian concept of disequilibrium, he explained that "in a Logo environment, new ideas are often acquired as a means of satisfying a personal need to do something one could not do before" (Papert, 1980, p. 74). He felt that children need great flexibility to develop their own "powerful ideas" or insights about new concepts.

When research studies on educational applications of Logo failed to yield the improvements expected by educators, many of Papert's concepts became even more controversial. Papert criticized these research efforts as "technocentric," saying that they focused more on Logo itself than on the methods used with it.

Source: Papert, S. (1980). *Mindstorms: Children, computers, and powerful ideas.* New York: Basic Books. Photo courtesy of Scholastic, Inc.

---

him to learn more. Therefore, he also emphasized the role of the affective domain during children's development more than Piaget did. Papert believed that the purpose of education was to provide rich, motivational environments to foster cognitive growth, and he felt that computers could make possible such environments. "The internal intelligibility of computer worlds offers children the opportunity to carry out projects of greater complexity than is usually possible in the physical world. Many children imagine complex structures they might build with an erector set or fantasize about organizing their friends into complex enterprises. But when they try to realize such projects, they too soon run into [problems with] matter and people. Because computer programs can . . . be made to behave exactly as they are intended to, they can be combined more safely into complex systems" (Papert, 1980, p. 118). His development and uses of the Logo programming language grew naturally out of these theories. Logo offered what he called "microworlds," or self-contained, orderly environments that children could use as "incubators for knowledge" (p. 120). (See Inset 3.8.)

John Seely Brown and the importance of cognitive apprenticeships. Brown and his colleagues at the Institute for Research on Learning focused on the work of cognitive psychologists who built on Vygotsky's hypotheses (Brown, Collins, and Duguid, 1989). Brown et al. were especially concerned with the relationship between this work and a problem they observed throughout much school learning. They refer to this problem as *inert knowledge,* a term introduced in 1929 by Whitehead. Brown et al. find that many school practices reduce the likelihood that children will transfer the skills they learn to later problem solving that requires those skills. The researchers observe that "it is common for students to acquire algorithms, routines, and decontextualized definitions that they cannot use and that, therefore, lie inert" (p. 33). They say this is because skills are often taught in very abstract ways and in isolation from any actual, authentic application. For example, students often learn multiplication facts and procedures, but they fail to recognize applications of these skills to real-life problems they encounter or to word problems that require multiplication. Students often seem to feel that activities like multiplication are something you do in a school culture with no real utility or application, especially outside school.

What students learn, Brown et al. argue, should not be separated from *how* they learn it. Students must come to understand how to transfer knowledge, they say, by learning it at the same time as they apply it in meaningful ways. This can best be accomplished by providing "cognitive apprenticeships." These are activities that call for authentic problem solving, that is, problem solving in settings that are familiar and useful to the student. Such tasks require students to use knowledge in given content areas as "tools," much as an apprentice tailor would use tools such as a sewing machine and scissors. For example, the researchers suggest teaching multiplication in the context of coin problems "because in the community of fourth grade students, there is usually a strong, implicit, shared understanding of coins" (p. 38). Through these kinds of activities, multiplication becomes something that is useful and real beyond the school culture. They refer to this kind of teaching and learning as *situated cognition,* and like Vygotsky, they feel it can best be accomplished through collaborative (group) learning.

Vanderbilt's Cognition and Technology Group and the concept of anchored instruction. A group of researchers at Vanderbilt's Learning and Technology Center built on the concepts of situated cognition and collaborative learning introduced by Brown et al. as well as Vygotsky's concept of scaffolding. The Cognition and Technology Group at Vanderbilt (CTGV, 1993) also criticized many of today's educational practices, especially those for teaching mathematics, as ineffective and harmful. They described an alternative approach: *anchored instruction,* or teaching that is "situated in engaging, problem-rich environments that allow sustained exploration by students and teachers" (p. 65). They hypothesized that "self-generated information is better remembered than passively received information" (p. 68). Thus, they recommended anchoring instruction in situations where students not only create answers to problems, but also generate many aspects of the problem statements themselves. The researchers referred to this active involvement in problem solving as *generative learning,* and they pointed out that video-based technologies have unique qualities to support these kinds of problem-solving environments. (See Inset 3.9.)

Cognitive flexibility theory and radical constructivism. Rand Spiro and a group of researchers that included Feltovich, Jacobson, Coulson, Anderson, and Jehng developed a constructivist theory in reaction to a perceived failure of many current instructional approaches—including some constructivist ones! Spiro et al. say that current classroom methods are more suited to learning in well-structured knowledge domains, while much of what is students should learn lies in "ill-structured domains." For example, "… basic arithmetic is well-structured, while the process of

applying arithmetic in solving word problems drawn from real situations is more ill-structured" (Spiro et al., 1991, p. 26). For learning in these ill-structured domains, Spiro et al. say, students need a different way of thinking about learning. "The interpretation of constructivism that has dominated much of cognition and educational psychology for the past 20 years or so has frequently stressed the retrieval of organized packets of knowledge, or schemas, from memory. … We argue that … ill-structured knowledge domains often render [these] schemas inadequate" (p. 28).

Spiro et al. say that the "new constructivism" of their Cognitive Flexibility Theory is "doubly constructive." That is, it calls for students to generate not only solutions to new problems, but also the prior knowledge needed to solve the problems. This kind of constructivism seems to demand even less direct teaching and an even more unstructured exploration on the part of students than those of Brown et al. and CTVG. Perkins (1991) calls this the difference between "BIG (beyond the information given) instruction" and "WIG (without the information given) instruction" (p. 20). Because they call for an even greater departure from directed instruction methods than other constructivists, those who hold views similar to those of Spiro et al. are sometimes referred to as *radical constructivists.*

Characteristics of Constructivism

The image constructivism usually evokes is that of changing the traditional goals of education and making possible restructured, innovative teaching approaches. The methods emphasize students' ability to solve real-life, practical problems. In this model, learners construct knowledge themselves rather than simply receiving it from knowledgeable teachers. Students typically work in cooperative groups rather than individually; they tend to focus on projects that require solutions to problems rather than on instructional sequences that require learning of certain content skills. In contrast to directed instruction, where the teacher sets the goals and delivers most of the instruction, the job of the teacher in constructivist models is to arrange for required resources and act as a guide to students while they set their own goals and "teach themselves."

Putting the principles of cognitive theorists into practice in classrooms creates a variety of new problems and tasks for educators. Constructivism calls for teachers to rethink traditional views on both objectives and methods of instruction and to experiment with new ways of facilitating students' learning. For example, rather than teaching an isolated objective such as identifying animals by phylum and genus, teachers may try to get their students to carry out cooperative projects that investigate the behavior of animals in the local environment. The latter activity may be far more meaningful, but educators can rely on fewer guidelines for carrying it out. Also, many teachers are still bound by the constraints of required curricula, and they must ensure that their students accomplish existing district objectives as well as newer, more constructivist ones.

Inset 3.9
The Cognition and Technology Group at Vanderbilt (CTGV) : Tying Technology to Constructivism

A research team located at the Learning Technology Center at Vanderbilt University has helped establish some practical guidelines for integrating technology based on constructivist principles. This team, known as the Cognition and Technology Group at Vanderbilt (CTGV), proposed an instructional approach based on concepts introduced by Vygotsky; Whitehead; and Brown, Collins, and Duguid (1989). It has also developed several technology products modeling this approach that have achieved widespread use in American education.

Several related concepts provide the theoretical foundation for the CTGV team's approach:

- **Preventing inert knowledge.** The CTGV hypothesized without a direct relationship to children's personal experience often resulted in their acquiring what Whitehead referred to as *inert knowledge*. That is, students never actually applied the knowledge they had learned because they could not see its relationship to problems they encountered. Inert knowledge is "knowledge that can usually be recalled when people are

explicitly asked to do so, but is not used spontaneously in problem solving even though it is relevant" (CTGV, 1990).

- **The nature of situated cognition and the need for anchored instruction.** Brown, Collins, and Duguid (1989) suggested that teachers could prevent the problem of inert knowledge by situating learning in the context of what they called *authentic experiences* and practical apprenticeships— activities that learners considered important because they emulated the behavior of experts (e. g., adults) in the area. In this way, students see the link between school learning and real-life activities. The CTGV felt that teachers can meet the criteria for situated cognition by anchoring instruction in highly visual problem-solving environments. "Anchored instruction provides a way to recreate some of the advantages of apprenticeship training in formal educational settings involving groups of students" (CTGV, 1990, p. 2).

- **Building knowledge through generative activities.** Like Vygotsky, the CTGV believes that learning is most meaningful to students when it builds (e. g., scaffolds) on experiences they have already had. Students are also more likely to remember knowledge that they build or "generate" themselves, rather than that which they simply receive passively (CTGV, 1991).

The CTGV proposed that the best way of providing instruction that would meet all the required criteria was to present it as videodisc-based scenarios posing interesting (but difficult) problems for students to solve. The first of these technology-based products, the *Jasper Woodbury Problem Solving Series,* focused on mathematics problems. Another, *The Young Children's Literacy Series,* addressed reading and language skills (CTGV, 1993). Both of these products were designed to build on children's existing knowledge in a way that would emphasize knowledge transfer to real-life situations.

Source: Brown, J. S., Collins, A., and Duguid, P. (1989). Situated cognition and the culture of learning. *Educational Researcher 18* (1), 32–41; Cognition and Technology Group at Vanderbilt (1990). Anchored instruction and its relationship to situated cognition. *Educational Researcher 19* (6), 2–10; Cognition and Technology Group at Vanderbilt (1991). Technology and the design of generative learning environments. *Educational Technology 31* (5), 34–40; and Cognition and Technology Group at Vanderbilt (1993). The Jasper experiment: An exploration of issues in learning and instructional design. *Educational Technology Research and Development 40* (1), 65–80. Photo courtesy of Billie Kingsley, Vanderbilt University, photographer David Crenshaw.

Sometimes instructional activities based on constructivist models are more time-consuming, since they may call for teachers to organize and facilitate group work and to evaluate in authentic ways (e.g., by gauging performance on activities). By comparison, paper-and-pencil tests are both quicker to develop and easier to administer. Many commercially available instructional materials based on constructivist models are of recent design. Since these materials may have been in use only briefly, teachers often have limited information available on how to smooth classroom implementation or what problems to anticipate. This is especially true with activities involving newer technologies such as interactive video and multimedia. The knowledge base on these kinds of classroom activities is growing daily, but teachers are still in the process of adapting class-

room strategies and coping with the logistics of setting up and using such equipment and media.

It is also important to recognize potential contradictions in theorists' views on how teachers should carry out constructivist approaches. As Spiro, et al. (1991) observe, "there are many variations on what is meant by 'constructivist'" (p. 22). For example, Papert feels that learning activities should be fairly unstructured and open-ended, frequently with no goal in mind other than discovery of "powerful ideas." Spiro et al. also call for varied opportunities for exploration when learning in "ill-structured knowledge domains," but they seem to advocate at least some acquisition of specific skills and information. The guidelines set forth by CTGV are still more goal-oriented and call for students to generate solutions to specific problems

Teachers must analyze the needs of their students and decide which constructivist strategies seem most appropriate for meeting these needs.

Required characteristics of constructivist approaches. The work of researchers and theorists such as Papert, Brown et al., CTGV, and Perkins have contributed especially important guidelines on how to develop instructional activities according to constructivist models. Since these guidelines do not always agree, teachers cannot usually follow all of them within the same instructional activity. But each of the following principles is still considered characteristic of constructivist purposes and designs:

- **Problem-oriented activities.** Most constructivist models focus on students solving problems, either in a specific content area such as mathematics or using an interdisciplinary approach; for example, such a problem might require a combination of mathematics, science, and language arts skills. Jungck (1991) says that constructivist methods frequently combine problem posing, problem solving, and "persuasion of peers" (p. 155). Problems may be posed in terms of specific goals (e.g., how to develop an information package to help persuade classmates to stop littering the beach,) as "what if" questions (e.g., what would life be like on earth if we had half the gravity we now have?), or as open-ended questions (e.g., in light of what you know about the characters and the times in which they lived, what is the best ending for this story?) These kinds of problems are usually more complex than those associated with directed instruction, and they require students to devote more time and more diverse skills to solve them.

- **Visual formats and mental models.** CTGV is especially concerned that instructional activities help students build good "mental models" of problems to be solved. They feel that teachers can promote this work most effectively by posing problems in visual (as opposed to written) formats. These researchers say that "Visual formats allow students to develop their own pattern recognition skills," and they are "dynamic, rich, and spatial" (1990, p. 3). This degree of visual support is felt to be particularly important for low-achieving (e.g., at-risk) students who may have reading difficulties and for students with little expertise in the area in which the problems are posed.

- **"Rich" environments.** Many constructivist approaches seem to call for what Perkins (1991) terms "richer learning environments" (p. 19) in contrast to the "minimalist" classroom environment that usually relies primarily on the teacher, a textbook, and prepared materials like worksheets. Perkins observes that many constructivist models are facilitated by combinations of five kinds of resources: information banks (e.g., textbooks and electronic encyclopedias) to get access to required information; symbol pads (e.g., notebooks and laptop computers) to support learners' short-term memories; construction kits (e.g., Legos, Tinkertoys and Logo) to let learners manipulate and build; phenomenaria (e.g., a terrarium or computer simulation) to allow exploration; and task managers (e.g., teachers and electronic tutors such as CAI/CMI systems) to provide assistance and feedback as students complete tasks.

- **Cooperative or collaborative (group) learning.** Most constructivist approaches heavily emphasize work in groups rather than as individuals to solve problems. This arrangement achieves several aims that advocates of constructivism and directed instruction alike consider important. CTGV observes that gathering students in cooperative groups seems to be the best way to facilitate generative learning. Perkins (1991) points out that cooperative learning illustrates distributive intelligence at work. In a distributive definition of intelligence, accomplishment is not a function simply of individual capabilities but the product of individuals and tools, each of which contributes to the achievement of desired goals. Finally, cooperative learning seems an ideal environment for students to learn how to share responsibility and work together toward common goals, skills they will find very useful in a variety of settings outside school.

- **Learning through exploration.** All constructivist approaches call for some flexibility in achieving desired goals. Most stress exploration rather than merely "getting the right answer," and a high degree of what the advocates of directed instruction would call *discovery learning*. Constructivists differ among themselves, however, about how much assistance and guidance a teacher should offer. Only a few constructivists seem to feel that students should have complete freedom and unlimited time to discover the knowledge they need. As Perkins (1991) says, "Education given over entirely to WIG (without any given) instruction would prove grossly inefficient and ineffective, failing to pass on in straightforward ways the achievements of the past" (p. 20).

- **Authentic assessment methods.** When the goals and methods of education change in the ways described here, teachers also need new methods of evaluating student progress. Thus, constructivist learning environments exhibit more qualitative assessment strategies rather than quantitative ones. Some popular assessment methods center on student portfolios with examples of students' work and products they have developed (Bateson, 1994; Young, 1995); narratives written by teachers to describe each student's work habits and areas of strength and weakness, and performance-based assessments in combination with checklists of criteria for judging students' performance (Linn, 1994).

Constructivist Methods: Problems and Possibilities

Despite the current popularity of constructivism, its principles and practices have also stimulated a variety of criticisms. Two special issues of *Educational Technology* magazine (May and September 1991) provided a forum for describing and debating the merits of constructivist learning strategies. This discussion focused on the following issues:

- **How can one certify skill learning?** Reigeluth (1991) pointed out that, although constructivists deplore formal tests or "objective measurements," schools must sometimes certify that students have learned key skills. "It is not sufficient to know that a doctor was on a team of medical students that performed the operation successfully; you want to know if the doctor can do it without the team" (p. 35).

- **How much prior knowledge is needed?** Constructivist strategies often call for students to approach and solve complex problems. But both Tobias (1991) and Molenda (1991) point out that, regardless of their motivation, many students may lack the prerequisite abilities that would allow them to handle this kind of problem solving. Molenda uses a biblical metaphor to describe the problem: "He who can swim may bring up pearls from the depth of the sea; he who cannot swim will be drowned; therefore, only such persons as have had proper instruction should expose themselves to the risk" (p. 44).

- **Can students choose the most effective instruction?** Constructivist tasks often require students to learn how to teach themselves, to choose methods by which they will learn and solve problems. But Tobias quotes a study by Clark (1982) which indicated that students often learn the least from instructional methods they prefer most.

- **Which topics suit constructivist methods?** Many educators feel that constructivist methods serve some purposes more effectively than others. For example, constructivist activities frequently seek to teach the problem-solving methods used by experts in a content area (e.g., think like a historian), rather than to learn any specific content or skills (e.g., historical facts). Molenda points out that constructivists may be surprised to learn that this is not what parents and many educators have in mind at all. "Parents and school people [are] … much more interested in communicating our cultural heritage to the next generation. Facts are viewed as powerful ends in themselves" (Molenda, 1991, p. 45). Tobias notes that constructivists often favor depth of coverage on one topic over breadth of coverage on many topics. "Students taught the first term of American history from a constructivist perspective may have a very profound understanding of the injustices imposed by taxation without representation. … However, would they learn anything about the War of 1812, Shay's Rebellion, the Whiskey Rebellion, or the Monroe Doctrine?" (Tobias, 1991, p. 42).

- **Will skills transfer to practical situations?** Constructivists assume that problem solving taught in authentic situations in school will transfer more easily to problems that students must solve in real life. Yet Tobias (1991) found little evidence from related research to indicate that such transfer will occur.

- **What objective evidence demonstrates the effects of constructivist methods?** Tobias (1991) Also points out that many constructivist claims (e.g., transfer of school learning to real-life problem solving) seem appealing, but they have never been scientifically investigated. He decries "esoteric jargon which is unsubstantiated by research findings" (p. 42) and says that "Perhaps the time has come … to devote more attention to the conduct of research" (p. 42).

Despite these criticisms, interest in constructivist methods is on the rise. More interest than ever before is also focused on carrying out research to measure the impact of learning based on student problem solving and product development (CTGV, 1995). It seems likely that the next decade will witness some dramatic shifts in curriculum goals and methods that follow constructivist principles in large part.

In recent years, the movement in education to integrate technology into teaching has become closely identified with the restructuring movement. Many educators seem to believe that they cannot make curriculum reflect constructivist characteristics *without* technology. Constructivists offer this combination of problem-oriented activities, cooperative group work, tasks related to students' interests and

backgrounds, and highly visual formats provided by technology resources as components of a powerful antidote to some of the country's most pervasive and recalcitrant social and educational problems.

Technology Integration Strategies: Directed and Constructivist Approaches

Subsequent chapters in this book describe and give examples of integration strategies for various types of courseware materials and technology media. However, all of these strategies implement a group of general integration principles. Some draw on the unique characteristics of a technology resource to meet certain kinds of learning needs. Others take advantage of a resource's ability to substitute for materials lacking in a given school or classroom. Teachers may find themselves using many or all of the following strategies at the same time. However, it is important to recognize that each of the integration strategies described here addresses a specific instructional need. They are not employed to make students computer literate or because technology is the wave of the future or because students should use computers once in a while because it will be good for them. The authors advocate making a conscious effort to match technology resources to problems that educators cannot address in other, easier ways.

Integration Strategies Based on Directed Models

Integration to remedy identified weaknesses. In previous educational eras, curriculum was often made up largely of training in discrete skills introduced and practiced in isolation from each other and from applications to practical problems. In those days, technology resources were more likely directed toward whole classes to facilitate teacher-directed learning and to individualize the pace of practice. For example, a teacher might have given a classroom lecture and demonstration on one or more mathematics skills and then sent the whole class to the microcomputer lab to practice the skills.

Today, trends in school instruction clearly lead toward more motivating, interactive, cooperative learning activities in which the teacher is more a facilitator and manager of resources than a means of delivering information to passive receivers (Perkins, 1991; CTGV, 1991). In these kinds of learning environments, students complete activities and create products that call for high-level skills such as critical thinking and problem solving. Those who develop such activities recognize that these higher-level skills assume the presence of certain prerequisite, lower-level skills. But some developers expect highly motivated students to see the need for the lower-level skills and, if they do not already possess them, to acquire them through various indirect methods. Others feel that prerequisite skills should be taught first. The importance of prerequisite skills to the desired higher-level behaviors is, however, still universally acknowledged.

One premise of constructivist, whole-language approaches is that students will be motivated to learn prerequisite skills if they see their relevance when the need arises in the context of group or individual projects that they want to do. However, experienced teachers know that even the most motivated students do not always learn skills as expected. These failures occur for a variety of reasons, many of which are related to learners' internal capabilities and not all of which are thoroughly understood. Curriculum is currently moving toward allowing students to acquire given skills on more flexible schedules. But when the absence of prerequisite skills presents a barrier to higher-level learning, directed instruction is usually the most efficient way of providing them. For example, if a student does not learn to read when it is developmentally appropriate, research has shown great success in identifying and remedying specific weaknesses among the component skills. (Torgeson, Waters, Cohen, and Torgeson, 1988; Torgeson, 1993). Materials such as drills and tutorials have proven to be valuable resources that help teachers provide this kind of individualized instruction. Well-designed resources like these not only can give students very effective instruction, but they are also frequently more motivating and less threatening than teacher-delivered learning to students who are having difficulties.

Integration to promote fluency or automaticity of prerequisite skills. Some kinds of prerequisite skills benefit students more if they can apply the skills without conscious effort. Gagne (1982) and Bloom (1986) referred to this as *automaticity* of skills, and Hasselbring and Goin (1993) call it *fluency* or *proficiency*. Students need rapid recall and performance of a wide range of skills throughout the curriculum (e.g., simple math facts, grammar and usage rules, and spelling). Some students acquire this automaticity through repeated use of the skills in practical situations. Others acquire this automatic recall more efficiently through isolated practice. Drill and practice courseware provides an ideal means of getting practice tailored to individual skill needs and learning pace.

Integration to make learning efficient for highly motivated students. Current educational methods are sometimes criticized for failure to interest and motivate students because, it is said, activities and skills are irrelevant to students' needs, experiences, and interests. However, some students' motivation to learn seems to spring from internal rather than external sources. These internally motivated students do not seem to need explicit connections between specific skills and practical problems to which they apply. These students may be motivated more by desire to please parents or other authority figures or by long-range goals of going to particular colleges or pursuing given vocations. In addition, interest in a subject kindled originally by a cooperative class project may spur them to learn everything they can about the field.

These self-motivated students pursue any skills they believe are related to their topics or provide foundations for later concepts. For such learners, the most desirable method of learning is the most efficient one. Directed instruction for these students can frequently be supported by self-instructional tutorials and simulations, assuming the teacher can locate high-quality materials on the desired topics.

Integration to optimize scarce resources. Anyone associated with public schools will readily admit that current resources and personnel are not optimal. For example, very real problems result from schools having too many students and not enough teachers. Many of the courseware materials described in later chapters can help to make up for the lack of required resources in the school or classroom. These needs range from consumable supplies to qualified teachers. For example, drill and practice programs can replace worksheets, a good tutorial program can offer instruction in topics for which teachers are in short supply, and a simulation can let students repeat an experiment over and over again without using up chemicals or other materials.

Integration to remove logistical hurdles. Some technology tools offer no instructional sequence or tasks in themselves, but they help students complete learning tasks more efficiently. These tools support directed instruction by removing or reducing logistical hurdles to learning. For example, word processing programs have no features that actually teach students how to write, but they let students write and rewrite more quickly and without the physical labor of handwriting. Computer-assisted design (CAD) software does not teach students how to design a house, but it allows them to easily try out various designs and features to see what they look like before building models or real structures. A videodisc may contain only a series of pictures of sea life, but it can let a teacher illustrate concepts about sea creatures more quickly and easily than other alternatives would allow. Chapters in Part II will address tools that make learning more efficient and less laborious for students.

Integration Strategies Based on Constructivist Models

Integration to generate motivation to learn. Teachers who work with at-risk students often point to the need to capture students' interest and enthusiasm as a key to success and frequently as their most difficult challenge. Some educators assert that today's television-oriented students are increasingly likely to demand more motivational qualities in their instruction than students in previous generations expected. As an important part of their rationale, constructivists argue that instruction must address students' affective needs as well as their cognitive ones. They hypothesize that students will learn more if they find what they are learning to be interesting and relevant to their lives. Whenever a teacher feels the need for stronger student motivation, the highly visual and interactive qualities of videodisc and multimedia resources have been shown to be valuable (e.g., the *Adventures of Jasper Woodbury* problem solving videodiscs).

Integration to foster creativity. While creative work is not usually considered a primary goal of education, many educators and parents alike consider it highly desirable. Some may argue that a student can be educated without being creative, but few schools would like to graduate students who could not think or act creatively. Resources such as Logo, problem-solving courseware, and computer graphics tools require neither consumable supplies or any particular artistic or literary skill. They also allow students to revise "creative works" easily and as many times as they desire. These qualities have been shown to provide uniquely fertile, nonthreatening environments for fostering development of students' creativity.

Integration to facilitate self-analysis and metacognition. If students are conscious of the procedures they use to go about solving problems, perhaps they can more easily improve upon their strategies and become more effective problem solvers. Consequently, teachers often try to get students to analyze their procedures to increase their efficiency. Resources such as Logo, problem-solving courseware, and multimedia applications have often been considered ideal environments for constructivist activities that get students to think about how they think.

Integration to increase transfer of knowledge to problem solving. The CTGV team pointed out unique capabilities of certain technology resources to address the problem of inert knowledge. They observed that this problem often occurs when students learn skills in isolation from any applications to problems. When students later encounter problems that require the skills, they do not realize how the skills could be relevant. Problem-solving materials in highly visual videodisc-based formats can allow students to build rich mental models of problems to be solved (CTGV, 1991a). Students need not depend on reading skills, which may be deficient, to build these mental models. Thus, supporters hypothesize that teaching skills in these highly visual, problem-solving environments helps to assure that knowledge will transfer to higher-order skills. These technology-based methods seem especially desirable for teachers who work with students in areas where inert knowledge is frequently a problem (e.g., mathematics and science).

Integration to foster group cooperation. One skill area currently identified as an important focus for schools' efforts to restructure curriculum (U.S. Department of Labor, 1992) is the ability to work cooperatively in a group to solve problems and develop products. While schools can certainly teach cooperative work without technology resources, a growing body of evidence documents students' appreciation of cooperative work as both more motivating and easier to accomplish when it uses technology. For

example, descriptions of students developing their own multimedia products are becoming more common in the literature on teaching cooperative skills to at-risk students.

Combining Integration Strategies in Curriculum Planning

This chapter and subsequent chapters describe an established, well-documented base of knowledge on how to integrate technology resources effectively into directed instruction; more recently, integration strategies for constructivist models have become increasingly popular. This chapter has proposed that directed and constructivist models each address specific classroom needs and problems, and that both will continue to be useful.

But the authors of this text go one step further in this description; they propose that neither model in itself can meet the needs of all students in a classroom. Teachers need to merge directed and constructivist activities to form a new and more useful school curriculum. Even Gagne, who has led proponents of systematic, directed methods, recently proposed that effective, useful instruction sometimes calls for "integration of objectives" in the context of a complex, motivational learning activity that he referred to as an *enterprise* (Gagne and Merrill, 1990). His description of the nature of this enterprise sounds very much like the kinds of activities often proposed by constructivists. In fact, one of his three kinds of enterprises calls for a "discovering" schema.

At this time, however, teachers can find few practical guidelines for combining directed and constructivist approaches and integration strategies into a single curriculum. In fact, the concept of combining them at all is currently still in an exploratory stage. Although teachers across the country are, indeed, probably combining approaches in the course of classroom activities, formal descriptions of effective applications are rare and lacking in detail. Some recommended guidelines and an example will indicate how this curriculum development might occur based on discussions with teachers in the context of informal forums such as workshops and interviews.

Recommended Guidelines for Developing a Technology-Integrated Curriculum

- **Plan for a grading period (e.g., 6 to 9 weeks, a semester, or a school year).** Curriculum that includes constructivist activities requires a long-term view of skills development. Ideally, students will experience a combination of teacher-directed and self-directed work throughout their time in school. Curriculum planned for short, discrete lessons (e.g., 1 or 2 weeks) leaves little of the flexibility that both teachers and students need to accomplish curriculum goals.

- **Allow enough time.** When constructivist strategies are new to teacher and students, learning activities often take more time than the more familiar routines of directed instruction would require. Teachers who are just beginning to introduce these kinds of activities may find that it takes them a while to learn the management techniques they need to guide and facilitate learning. Students also need enough time to complete their tasks and follow up on newly discovered interests.

• **Match the assessment to the activity.** It is especially important to measure students' accomplishments in a way that suits the kinds of learning activities they have done. For example, a teacher should not guide students through a problem-solving activity based on Optical Data's *Adventures of Jasper Woodbury* and then give them a multiple choice test on how to solve mathematics word problems. Students know they must earn grades of some kind. As with all classroom learning, they should understand from the start the criteria by which the teacher will measure performances.

• **Be flexible.** Although planning is essential to successful technology integration, any plan should be flexible. If the teacher notices in the middle of a mathematics problem-solving project that students lack prerequisite skills (e.g., knowledge of formulas and procedures for calculating the area of a square), the lesson may have to stop to allow some direct teaching before it can proceed.

• **Don't be afraid to experiment.** Despite the demands on teachers to prepare their students well, they should have the same opportunity for exploration and risk taking that they give to students. Teachers will not develop new and more effective methods without making some mistakes along the way. Combining directed and constructivist curriculum requires a delicate balance of informal and formal situations, problem solving and drill and practice, generated and memorized knowledge. It is a difficult balance to strike—but it is worth the effort!

An Example Curriculum Scenario

The following section gives an example of a curriculum unit that combines directed and constructivist methods in a series of learning activities. This example incorporates a new multimedia resource for the social studies curriculum entitled *Vital Links: A Multimedia History of the United States*. The unit uses several of the integration strategies described in this chapter. Based on an actual situation, it represents a model of the kind of restructured curriculum that is emerging in classrooms across the country.

Some teachers may choose to spend more of their total class time on the constructivist activities described in this example; some may choose to spend less. The mix will be determined by the perceived needs of the students. The point is that both strategies serve important purposes, and they can be combined effectively in a single unit.

Vital Links: A Technology-Integration Social Studies Unit

Mr. Corley, a social studies teacher at Challenger Middle School, has typically thought of himself as a traditional teacher. He believes that his system of lectures and good questioning techniques, along with textbook-based reading

and writing assignments, has prepared his students to perform well at the next level. He regularly receives excellent evaluations from his supervisor, and he is popular with students and parents alike.

For the past 2 years, though, Corley has been going through a sort of professional metamorphosis. He has come to realize that he may need to do more than prepare his students for the next grade level; he now believes that he should do more to prepare them for the real world. With that in mind, he is about to embark on a month-long unit that will examine diversity in the United States between 1830 and 1880.

In developing this unit, Corley plans to draw heavily on a social studies resource that his school recently purchased. This program, *Vital Links: A Multimedia History of the United States*, includes a large collection of multimedia resources for both teacher and student use. He wants the unit to culminate with a student project, but he knows that he must provide enough background information to give them a solid grasp of the issues. The following outline provides an overview of the strategies and methods that Corley plans to employ.

Stage 1: Introducing the Concepts

• The unit begins with a lecture introducing the unit, accompanied by a bar-code-controlled presentation of slide and video segments from the *Vital Links* videodisc.

• Students view the *Vital Links* video, which reinforces the concepts presented in lecture. Corley tells students about information to look for in the video, and he encourages them to take notes as they view it.

• Students complete a written worksheet in class based on information covered in the lecture and video. They then review the worksheet in class.

• Students complete a homework assignment based on reading from a *Vital Links* handout and vocabulary words. They review homework in class.

• Students take a quiz based on the background information. They watch a tour of the *Vital Links* CD-ROM resource.

• Corley assigns independent readings to acquire more background information on cultural diversity.

• Smaller activities and teacher-guided discussion are interspersed through these major activities.

Stage 2: Investigating the Concepts and Content

• Corley introduces the scrapbook project. Students complete some of the background information activities that accompany *Vital Links*.

• Homework assignments require students to gather information on diversity in their community, write letters to important persons, and telecommunicate to other parts of the country or world.

Stage 3: Producing the Project

• Corley assigns the project: students will use the "Present" part of *Vital Links* to assemble a multimedia scrapbook of the era. Each group will cover a different region of the country—north, south, east, or west.

• Corley assigns groups. Roles within groups can include leader, writer, editor, researchers, sound editor, and picture editor.

• Students work in their groups to develop their projects. Corley helps to pull together and locate resources they can use. He also shows them how to use storyboards to structure their projects.

• For homework, students gather resources for their projects.

Stage 4: Presenting the Project

• As part of preparing the presentation, students invite visitors to view their presentations.

• Students present their scrapbook projects. Other students and visitors ask followup questions on the information presented. Individuals plan for further work to follow up their own interests and the lines of inquiry they have uncovered.

• Corley assesses the projects using a master and criteria included with program.

Exercises

1.1 Identify the learning or instructional theorist(s) associated with each of the following instructional terms, models, and processes. Choose your answers from this list: **Skinner, information-processing theorists, Ausubel, Gagne, systematic instructional designers, Piaget, Vygotsky, Bruner, Papert, Cognition and Technology Group at Vanderbilt.**

_____ **a.** Discovery learning

_____ **b.** Reinforcement contingencies

_____ **c.** Stages of cognitive development

_____ **d.** Logo and powerful ideas

_____ **e.** Zone of proximal development

_____ **f.** Performance objectives

_____ **g.** Events of instruction

_____ **h.** Anchored instruction

_____ **i.** Short-term memory

_____ **j.** Advance organizers

1.2 Identify each of the instructional methods and characteristics below as being associated with **directed instructional models** (based on behavioral/information-processing theories) and **constructivist learning models** (based on cognitive theories).

_____ **a.** Scaffolding

_____ **b.** Discovery learning

_____ **c.** Systematic instructional design

_____ **d.** Structured approaches

_____ **e.** Criterion-based testing

_____ **f.** Learning how to learn

_____ **g.** Authentic activities

_____ **h.** Automaticity

_____ **i.** Learning hierarchies

_____ **j.** Anchored instruction

1.3. Given a description of a technology integration strategy, identify the kind of instructional approach (**directed** or **constructivist**) that served as the foundation for it.

_____ **a.** Using Logo to accelerate the development of cognitive stages in young children

_____ **b.** Having small groups of students work together to develop their own multimedia presentations

_____ **c.** Letting students learn problem-solving skills by giving them videodisc scenarios and helping them solve problems presented in the scenarios

_____ **d.** Giving advanced students a structured computer tutorial on basic physics principles while the rest of the class works on grade-level science objectives

_____ **e.** Giving students drill and practice in homonym usage to make sure they can apply the knowledge with automaticity in their written compositions

References

Alessi, S. and Trollip, S. (1991). *Computer-based instruction: Methods and development.* Englewood Cliffs, NJ: Prentice-Hall.

Atkinson, R. and Shiffrin, R. (1968). Human memory: A proposed system and its control processes. In K. Spence and J. Spence (Eds.). *The psychology of learning and motivation: Vol. 2.* New York: Academic Press.

Ausubel, D. (1968). *Educational psychology: A cognitive view.* New York: Holt, Rinehart and Winston.

Barker, T., Torgeson, J., and Wagner, R. (1992). The role of orthographic processing skills in five different reading tests. *Reading Research Quarterly, 27* (4), 335–345.

Bass, J. (1985). The roots of Logo's educational theory: An analysis. *Computers in the Schools, 2* (2–3), 107–116.

Bateson, D. (1994). Psychometric and philosophic problems in "authentic" assessment: Performance tasks and portfolios. *Alberta Journal of Educational Research, 40* (2), 233–245.

Bereiter, C. (1990). Aspects of an educational learning theory. *Review of Educational Research, 60* (4), 603–624.

Bloom, B. (1986). Automaticity. *Educational Leadership, 43* (5), 70–77.

Branden, R. (1996). The case for linear instructional design and development: A commentary on models, challenges, and myths. *Educational Technology, 36* (2), 5–23.

Bringuier, J. (1980). *Conversations with Jean Piaget. Translated by Basia M. Gulati.* Chicago, IL: University of Chicago Press.

Brown, J. S., Collins, A., and Duguid, P. (1989). Situated cognition and the culture of learning. *Educational Researcher, 18* (1), 32–41.

Bruner, J. (1973). *The relevance of education.* New York, NY: W. W. Norton and Company.

Carnine, D. (1993). Effective teaching for higher cognitive functioning. *Educational Technology, 33* (10), 29–33.

Clark, R. E. (1982). Antagonism between achievement and enjoyment in ATI studies. *Educational Psychologist, 17* (2), 92–101.

Cognition and Technology Group at Vanderbilt (1990). Anchored instruction and its relationship to situated cognition. *Educational Researcher, 19* (6), 2–10.

Cognition and Technology Group at Vanderbilt (1991a, May). Integrated media: Toward a theoretical framework for utilizing their potential. Proceedings of the Multimedia Technology Seminar, Washington, DC.

Cognition and Technology Group at Vanderbilt (1991b). Technology and the design of generative learning environments. *Educational Technology, 31* (5), 34–40.

Cognition and Technology Group at Vanderbilt (1993). The Jasper experiment: An exploration of issues in learning and instructional design. *Educational Technology Research and Development, 40* (1), 65–80.

Cognition and Technology Group at Vanderbilt (1995). Looking at technology in context: A framework for understanding technology and education research. In D. C. Berliner (Ed.). *The handbook of educational psychology.* New York: Macmillan.

Corey, S. M. (1944). Poor scholar's soliloquy. *Childhood Education, 33,* 219–220.

Educational Technology (May 1991), *31* (5); and (September 1991), *31* (9). Special issues on constructivist versus directed approaches.

Educational Technology (October 1993), 32 (10). Special issue on current uses of behavioral theories.

Epstein, R. (1993). Generativity theory and education. *Educational Technology, 33* (10), 40–45.

Gagne, R. and Merrill, R. (1990). Integrative goals for instructional design. *Educational Technology Research and Development, 38* (1), 23–30.

Gagne, R. (1985). *The conditions of learning.* New York: Holt, Rinehart and Winston.

Gagne, R. (1982). Developments in learning psychology: Implications for instructional design. *Educational Technology, 22* (6), 11–15.

Gagne, R., Briggs, L., and Wager, W. (1988). *Principles of instructional design.* New York: Holt, Rinehart and Winston.

Gagne, R., Wager, W., and Rojas, A. (1981). Planning and authoring computer-assisted instruction lessons. *Educational Technology, 21* (9), 17–26.

Hasselbring, T. and Goin, L. (1993). Integrated technology and media. In Polloway and Patton (Eds.) *Strategies for teaching learners with special needs.* New York: Merrill Publishing Co.

Jungck, J. (1991). Constructivism, computer exploratoriums, and collaborative learning: Construction scientific knowledge. *Teaching Education, 3* (2), 151–170.

Linn, R. (1994). Performance assessment: Policy promises and technical measurement standards. *Educational Researcher, 23* (9), 4–14.

Mallott, R. (1993). The three-contingency model of performance management and support in higher education. *Educational Technology, 33* (10), 21–28.

Molenda, M. (1991). A philosophical critique on the claims of "constructivism." *Educational Technology, 31* (9), 44–48.

National Commission on Excellence in Education. (1983). *A nation at risk.* Washington, DC: U. S. Department of Education.

Papert, S. (1980). Mindstorms: *Children, computers, and powerful ideas.* New York: Basic Books.

Perkins, D. (1991). Technology meets constructivism: Do they make a marriage? *Educational Technology, 31* (5), 18–23.

Phillips, D. C. (1995). The good, the bad, and the ugly: The many faces of constructivism. *Educational Researcher, 24* (7), 5–12.

Prawat, R. (1993) The value of ideas: Problems versus possibilities in learning. *Educational Researcher, 22* (6), 5–16.

Reigeluth, C. (1991). Reflections on the implications of constructivism for educational technology. *Educational Technology, 33* (10), 34–37.

Saettler, P. (1990). *The evolution of American educational technology.* Englewood, CO: Libraries Unlimited.

Skinner, B. F. (1938). *The behavior of organisms.* New York: Appleton.

Skinner, B. F. (1968). *The technology of teaching.* New York: Appleton.

Spence, I. and Hively, W. (1993). What makes Chris practice? *Educational Technology, 35* (6), 5–23.

Spiro, R., Feltovich, P., Jacobson, M., and Coulson, R. (1991). Knowledge representation, content specification, and the development of skill in situation-specific knowledge assembly: Some constructivist issues as they relate to cognitive flexibility theory and hypertext. *Educational Technology, 31* (9), 22–25.

Sulzer, B. and Mayer, R. (1972). *Behavior modification procedures for school personnel.* Hinsdale, IL: Dryden Press.

Taylor, R. (1980). *The computer in the school: Tutor, tool, tutee.* New York: Teachers College Press.

Tennyson, R. (1990). Integrated instructional design theory: Advancements from cognitive science and instructional technology. *Educational Technology, 30* (7), 9–15.

Tobias, S. (1991). An eclectic examination of some issues in the constructivist-ISD controversy. *Educational Technology, 31* (9), 41–43.

Torgeson, J., Waters, M., Cohen, A., and Torgeson, J. (1988). Improving sight word recognition skills in learning disabled children: An evaluation of three computer program variations. *Learning Disabilities Quarterly, 11,* 125–133.

Torgeson, J. (1986). Using computers to help learning. Disabled children practice reading: A research-based perspective. *Learning Disabilities Focus, 1* (2), 72–81.

U. S. Department of Labor (1992). SCANS (The Secretary's Commission on Achieving Necessary Skills) report. Washington, DC: U.S. Government Printing Office.

Wager, W. (1992). Instructional systems fundamentals: Pressures to change. *Educational Technology, 33* (2), 8–12.

Willis, J. (1995). A recursive, reflective model based on constructivist-interpretist theory. *Educational Technology, 33* (10), 15–20.

Young, M. (1995). Assessment of situated learning using computer environments. *Journal of Science Education and Technology, 4* (1), 89–96.

Part II

Using Software Tutors and Tools: Principles and Strategies

The chapters in this part will help teachers learn:

1. To identify the various teaching and learning functions that instructional software can fulfill

2. The unique capabilities of each of the resources referred to as *technology tools*

3. To describe the various features of word processing, spreadsheet, and database tools

4. To match specific kinds of instructional software and technology tools to classroom needs

---

Introduction

As chapters in Part I have illustrated, the field of educational technology is characterized by controversy and change, and the dynamic nature of computer technology has only reinforced this characteristic. Lack of consensus about the terminology of instructional technology reflects this changing and evolving nature. As with the terms used to describe directed and constructivist methods (see Part I Overview), no agreement has ever emerged about the terms for various educational computing resources or how to categorize them. Until microcomputers entered schools, classroom computing resources were usually classified under three general headings: computer-assisted instruction (CAI), computer-managed instruction (CMI), and other. CAI usually referred mainly to drill and practice, tutorial, and simulation software; CMI encompassed testing, recordkeeping, and reporting software. During the 1980s, other authors began using many, more inclusive terms to refer to instructional uses of computers such as computer-based instruction (CBI), computer-based learning (CBL), and computer-assisted learning along with several similar derivatives.

In 1980, Taylor proposed a classification system for instructional technology that grouped computer resources according to the functions they fulfilled. This classification system consisted of three terms: tutor, tool, and tutee. The tutor functions included those in which the "computer [was] programmed by 'experts' in both programming and in a subject matter … and the student [was] then tutored by the computer" (p. 3). In its tool functions, the computer "had some useful capability programmed into it such as statistical analysis … or word processing" (p. 3). Tutee functions helped the student learn about logic processes or how computers worked by teaching those subjects, i. e., students programmed computers to perform various activities in languages like BASIC or FORTRAN.

Although many technology-oriented educators still observe these categories by broadening the definitions of *tutor* and *tools* (and dropping the tutee designation), the field has produced no neat classification system for technology resources. There are simply too many different resources and applications, and experts have found no feasible way to agree on terminology. This textbook recommends that teachers prepare themselves for this lack of consensus on terms and labels so they can recognize the functions of technology products under any name or in any medium. After teachers can recognize the functions of each resource, they will be ready to match its functions to instructional needs.

The three chapters in this section deal with technology resources that were developed primarily through programming in computer languages (or combination of languages) such as BASIC, Pascal, C, or Assembler. Each of these resources centers around or involves a computer program or software. Some resources (e. g., probeware or a music editor) combine software with hardware, such as a probe or a music synthesizer, in order to accomplish their functions. The resources in Part II are addressed in three chapters: instructional software or courseware; word processing, database, and spreadsheet tools; and other tools. Each of these chapters will describe resources and suggest integration strategies for them based on directed and constructivist models.

Chapter 4: Instructional Software

This text usually substitutes the term *courseware* for *instructional software*. Most people think of instructional software as performing a tutor function. However, it functions as a tool in integration strategies based on constructivist models. Software described in this chapter fulfills the following instructional roles:

- Tutorial
- Drill and practice
- Simulation
- Instructional game
- Problem solving

Chapter 4 also covers networked delivery systems that both provide these resources and keep track of students' usage of them. In directed models, these systems are usually termed *integrated learning systems* (ILSs). Those who implement constructivist models may also refer to ILSs, but they mean to indicate very different resources and uses of those resources. Finally, Chapter 4 describes and discusses integration strategies for still another distinct category of instructional software resources: a programming language known as Logo and the software products based on it. Logo was at one time the focus of a large movement in educational technology, and it helped spark interest in constructivist models. Though popularity of Logo has since waned, it remains a useful and potentially powerful type of instructional software.

What are Computer Tools?

Most people think of the word *tool* as an all-encompassing term describing any implement used to help accomplish an activity. Thus, classroom computer tools could include any technology resource that teachers use to accomplish the activities of teaching. But this definition is probably too broad to be helpful. A more limited and useful definition would identify computer *tools* as resources that help primarily with the more mundane, mechanical operations of teaching. Computer tools can help learners with many mechanical operations such as:

- Handwriting, when the focus of instruction is writing a story or a composition

- Arithmetic calculations, when the purpose of the lesson is solving algebra problems
- Organizing information, when the purpose of the lesson is showing how to classify animals according to common features
- Presenting information clearly and attractively, when the purpose of the presentation is showing the results of a research project

The remaining two chapters in Part II focus on technology tools as resources that can ease the logistics of the more mechanical activities involved in learning, so that both teachers and students may concentrate more on achieving learning objectives. Such tools contribute to teaching in the same way that tools such as electric saws and battery-operated screwdrivers contribute to designing and building a house as compared to using hand saws and manual screwdrivers. The more advanced tools make it easier to carry out the mechanics of building the house, but they can also profoundly affect the complexity of the designs that builders may attempt. In the same way, technology tools can not only make learning faster and easier, but they can also allow it to employ more complex, higher-level methods than would be possible without such tools.

Chapter 5: Three Types of Tools

Chapter 5 describes and gives integration strategies for three of the most commonly discussed computer programs under the category of classroom technology tools: word processing, spreadsheet, and data-base software. The popularity of these tools probably results from two facts. They were among the first tools to be developed and used in education, especially with microcomputers; thus, the literature documents more uses and classroom integration strategies for them than for most of the tools developed later. Also, these three functions are most commonly found as components of so-called *integrated software packages*. (These are different from the integrated learning systems or ILSs discussed in Chapter 4.) As parts of a single integrated software package, the products of these tools may be used separately or "cut and pasted" from one to the other. For example, a teacher may prepare an illustration of a budget on a spreadsheet program, copy it, and paste it into a word processed handout that describes the contents of budgets and how to prepare them. Although word processing, spreadsheet, and database programs are usually discussed and taught together as a group, each has a distinct set of characteristics and range of applications in education.

- Word processing software is arguably the most widely used of all the classroom technology tools. Whenever teachers are asked to list essential technology skills, they always cite word processing, usually near the top of the list. In addition to its versatility, this computer application substitutes for (and surpasses in capability) a technology with which most people are already familiar and comfortable: the typewriter.

- Spreadsheet software also substitutes for a familiar and widely accepted technology: the calculator. Spreadsheets are to numbers what word processing is to text. Many teachers rely on spreadsheet software primarily to keep grades; others prefer using specific grade keeping (or gradebook) software. Regardless of this preference, spreadsheets can also be versatile instructional tools to support learning in many curriculum areas.

- Database software came into common use in classrooms somewhat later than word processing, but it quickly became a very popular classroom tool. These programs correspond to no noncomputer tool, but they are usually likened to versatile file cabinets. Databases offer a way of organizing information to make it readily available to support various purposes. As research and study skills involved in locating and using information become more important in our modern, information-based society, schools place increasing emphasis on using databases in all curriculum areas.

Each of the chapters in Part II on technology tools describes and illustrates two kinds of applications: productivity and instructional applications. Both are necessary to classroom activities, but each fulfills a very different purpose to support and enhance teachers' activities.

- **Productivity uses.** These activities may have little to do with actual teaching (e. g., explaining or illustrating new concepts, or making information clearer or learning activities more enjoyable). These functions do help teachers use their time more efficiently, however, so that their subsequent interaction with students can be more meaningful. For example, teachers may record and calculate grades and track club budgets on a spreadsheet; they can then redirect the time they save on this activity to allow more contact with their students. Teachers may keep track of classroom instructional resources suitable for various instructional levels and purposes in a database; this can help them select materials targeted to the needs of each of their students. Productivity applications earn their name because they help make teachers more productive in accomplishing their professional tasks in the same way that technology tools can make office staff more productive for the companies in which they work.

- **Instructional uses.** Most of the same tools that improve teacher productivity can also enhance instruction. For example, a spreadsheet can help students keep track of data during an experiment or predict the effects on sums or averages of changing numbers in a column or row. Students can be taught to develop their own databases of information to help them learn how to organize and search for information. These tools are called *instructional applications* because they serve valuable purposes in direct learning activities, even though they may not deliver information in the same way a tutorial, drill, or simulation program does.

Chapter 6: Using Other Tools in Teaching and Learning

Chapter 6 is a catch-all category describing the features and applications of many more software tools that support teaching and learning tasks. These include resources to save teacher time (e.g., gradebooks and worksheet and test generators); products to enhance and support production and presentation tasks (software for printing and presentation graphics, drawing, CAD, and desktop publishing); tools for managing data (statistical packages and CMI software); and other support tools specific to the needs of certain disciplines (e.g., music editors and IEP generators).

Chapter 4

Using Instructional Software in Teaching and Learning

This chapter will cover the following topics:

- Definitions, issues, integration strategies, and example lesson activities based on a directed instructional model for:
  - Drill and practice functions
  - Tutorial functions

- Definitions, issues, integration strategies, and example lesson activities based on both directed and constructivist models for:
  - Simulation functions
  - Instructional game functions
  - Problem-solving functions

- Definitions, issues, integration strategies, and example lesson activities based on constructivist models for Logo and Logo-related products

- Characteristics and uses of integrated learning systems (ILSs) and other technology-oriented learning systems

- Criteria and methods for software selection

Chapter Objectives

1. For each description of a classroom need for instructional materials, identify one or more types of instructional technology functions that could meet the need.

2. Plan lesson activities that integrate technology resources using a directed learning strategy.

3. Plan lesson activities that integrate technology resources using a constructivist learning strategy.

"The fact that individuals bind themselves with strong emotional ties to machines ought not to be surprising. The instruments [we] use become ... extensions of [our] bodies."
Joseph Weizenbaum in *Computer Power and Human Reason* (1976, p. 9)

Introduction

What Is Instructional Software?

From the time that people began to recognize the potential power of the computer to do tasks quickly and systematically, they also began exploring and experimenting with its capability to emulate (and improve on) the functions of a human teacher. If computer programs could be written to do essentially anything, why could computers not be programmed to teach? Many educators and developers pursued this goal of the computer as teacher during the 1960s and 1970s. Some, like William Norris who developed Control Data's PLATO teaching systems, believed that computer-based education was the only logical alternative to education's "outdated, labor-intensive ways" (1977, p. 451). Norris believed that education could become more productive if computers were to take over much of the traditional role of teachers. Today, after about 30 years of development and experimentation, teachers hear less talk about computers taking their places, but programs are still available to perform various teaching functions. While these programs are not the alternatives to human teachers envisioned by Norris, they can enhance teaching and learning in many ways.

As was explained in Chapter 1, programs written in computer languages (like BASIC, Assembler, and C) can do tasks that humans want done; this is called *applications software,* or simply *programs*. Instructional software (or courseware) delivers all or part of a student's instruction on a given topic or assists with learning in some key way. Although software such as word processing, database, and spreadsheet programs also can enhance instructional activities, this textbook differentiates between such tools and instructional software. Software tools can serve a variety of purposes other than teaching; instructional software includes programs developed for the sole purpose of delivering instruction or supporting learning activities.

Problems in Identifying and Classifying Software Functions

Computer-assisted instruction (CAI) originated in the early days of educational technology as a name for instructional software, and the term is still in common use. However, some kinds of instructional software are designed with more constructivist purposes in mind and they do not actually deliver instruction per se, so many people consider the term CAI outdated and misleading. Teachers may hear instructional software called by names like *computer-based instruction* (CBI), *computer-based*

learning (CBL), and *computer-assisted learning* along with more generic terms such as *software learning tools*. Names for the types of instructional software also vary, but they are usually identified as:

- **Drill (or drill and practice) software.** Programs that allow learners to work problems or answer questions and get feedback on correctness

- **Tutorial software.** Programs that act like tutors by providing all the information and instructional activities that a learner needs to master a given topic (e.g., information summaries, explanation, practice routines, feedback, and assessment)

- **Simulation software.** Programs that model real or an imagined systems to show how those systems or similar ones work

- **Instructional games.** Programs designed to increase motivation by adding game rules to learning activities (usually either drills or simulations)

- **Problem-solving software.** (1) Programs that teach directly (through explanation and/or practice) the steps involved in solving problems, or (2) programs that help learners acquire problem-solving skills by giving them opportunities to solve problems

In the early days of their computer use, educators could relatively easily classify a software package as one of the above types. However, much of today's software complicates or defies this kind of classification. For one thing, developers have reached no consensus on the above terms or on the characteristics that define them. For example, some developers refer to a drill program that gives extensive feedback as a *tutorial*. Others refer to games as *drill programs*. Some packages contain several different activities, each of which serves a different purpose. For example, a program like Millie's Math House has a number of straight drill activities along with some problem-solving and game activities.

In light of these issues, educators who would use software for instruction should analyze all of the activities in a given package and classify each one according to its instructional function. For example, one may not be able to refer to an entire package as a tutorial or a drill, but it is possible and desirable to identify a particular activity according to whether it provides practice or opportunities for solving problems. As this chapter will show, each software function serves very different purposes during learning and, consequently, has its own integration strategies.

Drill and Practice Activities

Drill and Practice: Definition and Characteristics

Drill and practice activities provide exercises in which students work example items, usually one at a time, and receive feedback on their correctness. Programs vary considerably in the kind of feedback they provide in response to student input. They can range from a simple display like "OK" or "No, try again" to elaborate animated displays or

Figure 4.1 Example Software with Drill and Practice Activities: Millie's Math House

Source: Used by permission of Edmark Corporation

Figure 4.2 Example Software with Drill and Practice Activities: Math Rabbit

Source: Used by permission of the Learning Co.

verbal explanations. Some programs simply present the next item if the student answers correctly.

Types of drill and practice are sometimes distinguished by the sophistication with which the program tailors the practice session to student needs (Merrill and Salisbury, 1984). The most basic drill and practice function is often described as a *flashcard* activity. A student sees a set number of questions or problems on the screen and answers one at a time. Such a program may either select questions randomly or always show them in the same order. Whenever students answer, the program usually tells them whether their responses are correct and gives them either another chance to answer correctly or another question to answer. Examples of this type are shown in Figures 4.1 and 4.2.

A more sophisticated form of drill and practice moves students on to more advanced questions after they get a number of questions correct at some predetermined mastery level; it may also send them back to lower levels if they answer a certain number wrong. Some programs also automatically review questions that students get wrong before going on to other levels. Movement between levels is often transparent to students since the program may do it automatically without any indication. Sometimes, however, the program may congratulate students on good progress before proceeding to the next level, or it may allow them to choose their next activities.

In addition to meeting general criteria for good instructional courseware (see Inset 4.1), well-designed drill and practice programs should also meet the following criteria:

- **Control over the presentation rate.** Unless the questions are part of some timed review, students should have as much time as they wish to answer and examine the feedback before proceeding to later questions. If the program provides no specific feedback for correct answers, it is usually acceptable to present later questions without any further entries from students.

- **Appropriate feedback for correct answers.** While some courseware designers stress the importance of positive feedback for correct answers, not all programs provide it. If student's answers are timed, or if their session time is limited, they may

find it more motivating simply to move quickly to later questions. Positive feedback should not be so elaborate and time-consuming to display that it detracts from the lesson's purpose. No matter how attractive the display, students tend to tire of it after a while during a practice session and it ceases to motivate them.

- **Better reinforcement for correct answers.** Some programs inadvertently motivate students to get wrong answers. This happens when a program gives more exciting or interesting feedback for wrong answers than for correct ones. The most famous example of this design error occurs in an early version of a popular microcomputer-based math drill series. Each correct answer got a smiling face, but two or more wrong answers produced a full-screen, animated crying face that students found very amusing! Consequently, many students tried to answer incorrectly to see it. The company corrected this flaw, but this classic error still can be seen today in other programs.

Issues Related to Drill and Practice

Drill and practice courseware activities were among the earliest and most well-recognized instructional uses of computers, and they are still used extensively in schools. These activities have frequently been shown to allow the effective rehearsal (practice) that students need to transfer newly learned information into long-term memory (Merrill and Salisbury, 1984; Salisbury, 1990). However, drill and practice is also the most maligned of the courseware activities, sometimes informally referred to among its critics as "drill and kill." This derision results, in part, from perceived overuse. Many authors have criticized teachers for presenting drills for overly long periods or for teaching functions that drills are ill-suited to accomplish. For example, teachers may expose students to drill and practice courseware as a way of introducing new concepts rather than just practicing and reinforcing familiar ones.

Probably the most common reason for the rising criticism of drill and practice courseware is its identification as an easily targeted icon for what many people consider an outmoded approach to teaching. Critics claim that introducing isolated skills and directing students to practice them directly contradicts the trend toward restructured curriculum in which students learn and use skills in an integrated way within the context of their own projects that specifically require the skills.

Although curriculum increasingly emphasizes problem solving and higher-order skills, teachers still give students on-paper practice (e.g., worksheets or exercises) for many skills to help them learn and remember correct procedures. Many teachers seem to feel that such practice gives students more rapid recall and use of basic skills as prerequisites to more advanced concepts. They like students to have what Gagne (1982) and Bloom (1986) called *automaticity* or automatic recall of these lower-order skills to help them master higher-order ones faster and more easily. The usefulness of drill programs in providing this kind of practice has been well-documented, but the programs seem especially popular among teachers of students with learning disabilities (Hasselbring, 1988; Okolo, 1992; Higgins and Boone, 1993). The following examples cite basic skills that are prerequisite to higher-order skills:

- Automatic recall of multiplication facts is required for most higher-level mathematics ranging from long division to algebra.

- Keyboard proficiency is a prerequisite for assignments that require extensive typing.

- Graded compositions require rapid recall and application of correct sentence structure, spelling, and principles of grammar and usage (e.g., punctuation, subject-verb agreement).

- Many schools still require students to memorize historical facts (e.g., states and capitals).

Despite the increasing emphasis on problem solving and higher-order skills, it seems likely that some form of drill and practice courseware will probably be useful in many classrooms for some time to come. Such programs address needs for these and other required skills.

How to Use Drill and Practice in Teaching

Drill and practice programs may be used whenever teachers feel the need for on-paper exercises (e.g., worksheets or exercises). Drill courseware provides several acknowledged benefits as compared to paper exercises:

- **Immediate feedback.** When students practice skills on paper, they frequently do not know until much later whether or not they did their work correctly. To quote a common saying, "Practice does not make perfect; practice makes permanent." As they complete work incorrectly, students may actually be memorizing the wrong skills. Drill and practice courseware informs them immediately whether or not their responses are accurate, so they can make quick corrections. This helps them in two ways: "debugging" and retention. Students may think they know how to solve problems, but when they begin to practice, they may identify errors in their procedures. Practice helps them locate and get rid of the bugs in their thinking. Also, a certain amount of practice is usually necessary to place the skills in long-term memory for ready access later.

- **Motivation.** Sometimes teachers cannot get students to do the practice they need on paper. Perhaps a student has failed so much that the whole idea of practice in the area is abhorrent; another student may have poor handwriting skills or simply dislike writing. In these cases, computer-based practice may motivate students to do the practice they need. Computers don't get impatient or give disgusted looks when a student gives a wrong answer.

- **Saving teacher time.** Since teachers do not have to present or grade drill and practice, students may do this activity essentially on their own while the teacher addresses other student needs.

On some instructional occasions, even the most creative and innovative teacher may want to take advantage of the benefits of drill and practice courseware to have students practice using isolated skills.

- **In place of or to supplement worksheets and homework exercises.** Whenever students have difficulty with higher-order tasks ranging from reading and writing to mathematics, teachers may have to stop and identify specific prerequisite skills that these students lack and provide the instruction and practice they need to go forward. A similar response can promote rote learning and memorization. In these cases, "learning" is simply a rehearsal activity to make sure information becomes stored in long-term memory in a way that allows students to retrieve it easily. The benefits of drill and practice activities related to motivation, immediate feedback, and self-pacing can make it more productive for students to practice required skills via courseware rather than in worksheets or paper exercises.

- **In preparation for tests.** Despite the new emphasis on student portfolios and other authentic assessment measures, students can expect to take several kinds of objective examinations in their education careers. When they need to prepare to demonstrate mastery of specific skills in important examinations (e.g., for end-of-year grades or for college entrance), drill and practice courseware can help them focus on their deficiencies and correct them.

An example of an integration strategies for drill programs are shown in the Lesson Plan on page 90.

When teachers assign drill activities, they should be sure to limit the time devoted to the assignment to 10 to 15 minutes per day. This ensures that students will not become bored with the activity, so the drill and practice strategy will retain its effectiveness. Teachers should also be sure that students have been introduced previously to the concepts behind the drills; that courseware should serve mainly to debug and to help students retain their grasp of familiar concepts.

Since self-pacing and personalized feedback are among the most powerful benefits of drills, these activities usually work best for individual computer use. However, some teachers with limited technology resources have found other, ingenious ways to capitalize on the motivational and immediate feedback capabilities of drills. If all students in a class would benefit from practice in a skill using a drill program, the teacher may divide them into small groups to compete with each other for the best group scores: the class could even be divided into two groups for a "relay race" competition over which group can complete the assignment the fastest with the most correct answers.

If not all students need the kind of practice that a drill provides, the teacher may make courseware one of several learning stations to serve students with identified weaknesses in one or more key skills. The key to using drill and practice appropriately is to match its inherent capabilities with the identified learning needs of individual students.

Tutorial Activities

Tutorials: Definition and Characteristics

Tutorial courseware uses the computer to deliver an entire instructional sequence similar to a teacher's classroom instruction on the topics. This instruction is usually expected to be complete enough in itself to stand alone; the student should be able to learn the topic without any help or other materials from outside the courseware. Unlike other courseware activities, tutorials are true teaching courseware. Gagne, Wager, and Rojas (1981) stated that tutorial courseware should address all instructional events. (See the discussion of Gagne's events of instruction in Chapter 3.) Gagne et al. show how a tutorial may vary its strategies to accomplish events for different kinds of learning ranging from verbal information to complex applications of rules and problem solving.

While drill and practice activities concentrate primarily on what Hasselbring and Goin (1993) called attaining *fluency/proficiency* with skills, tutorial courseware focuses what those authors call the *acquisition stage* of learning. People may confuse drill activities with tutorial activities for two reasons. First, drill courseware may provide elaborate feedback that reviewers may mistake for the tutorial explanations required by Gagne's Events 4 and 5. Even courseware developers may claim that a package is a tutorial when it is, in fact, a drill activity with detailed feedback. Second, a good tutorial should include one or more practice sequences to address Events 5 through 7, so reviewers easily become confused about the primary purpose of the package.

Tutorials are often categorized as linear and branching tutorials (Alessi and Trollip, 1991). A simple, linear tutorial gives the same instructional sequence of explanation, practice, and feedback to all learners regardless of differences in their performance. A more sophisticated, branching tutorial directs learners along alternate paths depending on how they respond to questions and whether or not they show mastery of certain parts of the material. Even branching tutorials can range in complexity by the amount of branching they allow and how fully they diagnose the kinds of instruction that a student needs.

Some tutorials also have computer-management capabilities; teachers may both tell such a program at what level to start for a given student and get reports on each student's progress through the instruction. While a tutorial program does not need these components, data collection and management features make it more useful to teachers.

As the description of events of instruction implies, tutorials are most often geared toward learners who can read fairly well, usually older students or adults. Since tutorial instruction is expected to stand alone, it is difficult to explain or give appropriate guidance on-screen to a non-reader. However, some tutorials aimed at younger learners have found clever ways to explain and demonstrate concepts with graphics, succinct phrases or sentences, or audio directions coupled with screen devices.

Some of the best tutorial courseware activities are in packages that accompany newly purchased computers or applications software (e.g., Tour of the Macintosh or Introduction to Microsoft Works). While tutorials are found more frequently on mainframe or file server systems than on microcomputers, some good tutorials are available on standalone systems. Two examples of microcomputer tutorial sequences are shown in Figures 4.3 and 4.4.

Emulating a good teacher is a difficult assignment for any human, let alone a computer! However, courseware must accomplish this task to fulfill a tutorial function. In addition to meeting general criteria for good instructional courseware, well-designed tutorial programs should also meet several additional standards:

- **Extensive interactivity.** The most frequent criticism of tutorials complains that they are "page-turners." That is, they ask students to do very little other than reading. Good tutorials, like good teachers, should require students to give frequent and thoughtful responses to questions and problems, and they should supply appropriate practice and feedback to guide students' learning.

- **Thorough user control.** *User control* refers to several aspects of the program. First, students should always be able to control the rate at which text appears on the screen. The program should not go on to the next information or activity screen until the user presses a key or gives some other indication of completing necessary reading. Next, the program should offer students the flexibility to review explanations, examples, or sequences of instruction or move ahead to other instruction. The program should also provide frequent opportunities for students to exit as desired.

- **Appropriate and comprehensive teaching sequence.** The program's structure should provide a suggested (or required) sequence of instruction that builds on concepts and covers the content adequately. It should provide adequate explanation and examples in both original and remedial sequences. In sum, it should compare favorably to an expert teacher's presentation sequence for the topic.

- **Adequate answer-judging and feedback capabilities.** Whenever possible, programs should allow students to answer in natural language, and they should accept all correct answers and all possible variations of correct answers. They should also give appropriate corrective feedback in response to incorrect answers, supplying this feedback after only one or two

---

Example Lesson Plan
Integration of Drill and Practice Courseware

Lesson:	Traffic Officer
Developed by:	Doris Murdoch, Webster Elementary School (St. Augustine, Florida)
Courseware:	Muppet Word Book (Sunburst)
Level:	Elementary
Content Area:	Language arts: Letter cases
Instructional Resources:	In addition to computer and software, the lesson plan employs the following constructed materials: Upper/lower case letter cards mounted as necklaces for each student; stop sign and patrol belt or a large, handheld Kermit graphic; printout of traffic officer worksheet
Lesson Purposes:	• Identify given letters of the alphabet.
	• Tell whether letters are upper case or lower case.

Traffic Officer Worksheet

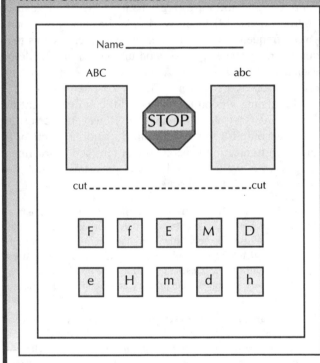

Rationale for Using Courseware. The Muppet Word Book offers excellent independent, motivational practice for letter skills that have been introduced to the class as a whole. This activity on the computer is especially helpful for students who have trouble focusing on letters as separate entities and who may be more comfortable practicing on the computer because it is nonjudgmental.

Instructional Activities.
First Day. Each student is given either an upper case or lower case letter card. Students identify letters as a full group. One child is chosen as the traffic officer. Two areas are designated as parking lots, one for upper case letters and one for lower case letters. The traffic officer directs the other students' "cars" to the correct parking lot according to letter case. Students will be guided through this activity by the teacher. After letters are sorted ("cars are parked"), the class counts the number of letters in each group. The teacher introduces the class to the parking lot activity on the Muppet Word Book disk.

Second Day. At centers or workshops, individual students or small groups of students practice classifying letter cases with the Muppet Word Book parking lot activity and the traffic officer game.

Third Day. Students complete the traffic officer worksheet as an assessment activity.

Source: From Doris Murdoch (1992). Traffic Officer. *The Florida Technology in Education Quarterly 4,* 2/(Winter). 23–24.

---

tries rather than frustrating students by making them keep trying indefinitely to answer something they may not know.

Although some authors insist that graphics form part of tutorial instruction (Baek and Layne, 1988), others emphasize judicious use of graphics to avoid interfering with the purpose of the instruction (Eiser, 1988). Eiser is among those who also recommend online evaluation and record-keeping on student performance as part of any tutorial.

Issues Related to Tutorials

Tutorials are beginning to attract the same criticism as drill and practice for "teacher-directed methods," that is, they deliver traditional instruction in skills rather than let-

ting students create learning experiences through generative learning and development projects. Also, since good tutorials are difficult to design and program, critics charge that tutorials represent trivial or even counterproductive uses of the computer. Also, a number of tutorials fail to meet criteria for good programs of this kind, thus contributing to this perception.

Tutorials are difficult to find, even for those who want to use them. Software publishers describe fewer packages as "tutorials" than any other kind of microcomputer courseware. Part of the reason for this comes from the difficulty—and, therefore, expense—of designing and developing them. A well-designed tutorial sequence emerges from extensive research into how to teach the topic well, and its requirements for programming and graphics can become

Figure 4.3   Example Software with Tutorial Activities: Welcome to Physics

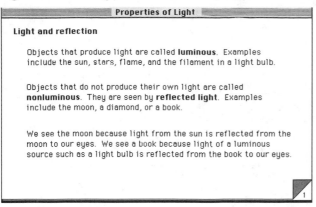

fairly involved. Designers must know what learning tasks the topic requires, what sequence students should follow, how best to explain and demonstrate essential concepts, common errors that students are likely to display, and how to provide instruction and feedback to correct those errors. Also, tutorials must be large, so they often work slowly on microcomputers. Larger tutorials must be delivered via integrated learning systems or other networked systems, making them expensive.

These problems become still more difficult because teachers frequently disagree about what they should teach for a given topic, how to teach it most effectively, and in what order to present learning tasks. A teacher may choose not to purchase a tutorial with a sound instructional sequence because it does not cover the topic the way he or she presents it. Not surprisingly, courseware companies tend to avoid programs that are problematic both to develop and to market.

How to Use Tutorials in Teaching

It is unfortunate that microcomputer tutorials are so rare; a well-designed tutorial on a nontrivial topic can be a valuable instructional tool. Since a tutorial can include drill and

practice routines, helpful features include the same ones as for drills (immediate feedback to learners, motivational aspects, and time savings). But tutorials offer the additional benefit of self-contained, self-paced substitutes for teacher presentations. This should in no way threaten teachers, since few conceivable situations make a computer preferable to an expert teacher. However, the tutorial's unique capability of presenting an entire interactive instructional sequence can assist in several classroom situations:

- **For self-paced reviews of instruction.** On many occasions, students need repeated instruction on a topic after the teacher's initial presentation. Some students may be slower to understand concepts and need additional time on them. Others seem to learn better in a self-paced mode without the pressure to move at the same pace as the rest of the class. Still others may need review before a test. Teachers can help these students by providing tutorials at learning stations to review previously presented material while the teacher works with other students.

- **As an alternative learning strategy.** Tutorials also provide alternative means of presenting material to support different learning strategies. Some students, typically advanced ones, prefer to structure their own learning activities and proceed on their own. A good tutorial can allow such students to glean

Figure 4.4   Example Software with Tutorial Activities: Welcome to Physics (continued)

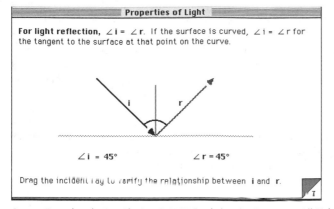
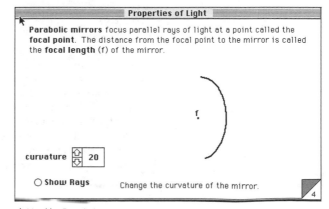

Example Lesson Plan
Integration of Tutorial Courseware

Lesson:	Following the Great Wizard
Developed by:	Gina Erickson and Kelly Foster (Adventure Learning Software, Inc.)
Courseware:	DaisyQuest and Daisy's Castle (Great Wave Software)
Level:	Elementary
Content Area:	Reading; Phonological Awareness
Instructional Resources:	Computer, software, record player that can change speeds, and a record
Lesson Purposes:	• Recognize and compare the sounds (phonemes) that make up words.
	• Segment words into component sounds.
	• Blend isolated sounds together to form words.

Rationale for Using Courseware. Educational research has shown that performance on phonological awareness is a good predictor of later reading success. Direct training in phonological awareness has been shown to support substantial improvement in a child's subsequent acquisition of reading skills. DaisyQuest and Daisy's Castle are "magical" auditory programs created especially for young children and for older children who may experience difficulty learning to read due to deficits in ability to isolate and compare sounds in words.

Instructional Activities. To determine what the students already know and to guide instruction, assess the students' blending skills by assigning Undersea Blending (whole word). Using the pretest scores, identify students who need instruction in blending. Assign these students to the computerized blending activities found in the Great Hall of Daisy's Castle.

Introduce the concept of blending by having the students listen carefully to a recorded story. Then slow the speed of the record player and let students listen to a few sentences of the "slowed speech." Ask them to tell what the slow voice said. Emphasize the importance of listening carefully to every sound in order to put together the slowed speech. The intent of this exercise is to have children recognize that words are made up of different sounds and that they can successfully join these sounds back together to form words.

Have the children form a line and place each one's hands on the shoulders of the child directly in front. The teacher is the conductor at the front of the line. Inform students that whenever you raise your arm above your head and "toot" the imaginary horn, that they are to stop marching and immediately freeze. Begin marching around the room chanting "ch, ch, ch, ch," in rhythm to the marching. Have the students repeat "ch, ch, ch, ch." Raise your arm and "toot" to practice the freezing. Proceed marching. Say a one syllable word such as *cat, bear, dog,* or *cave.* Have the students repeat the word to the rhythm of the marching. Repeat this three or four times using several different one-syllable words.

Now tell the students that the train is slowing down. Slow down your marching and say a word in slowed speech much like the speech of the slowed record player in Step 3. Freeze the train and ask the students to guess what word you just said. Proceed marching and select four or five more words to be said in slowed speech. Freeze after each word to allow the students time to guess the slowed word. The next part of this train activity is to break a single-syllable word into its component sounds (not into letters). If the target word is *cheese* you would say, "/ch/ (pause for 2 seconds), /e/ (pause for 2 seconds), /z/ (pause for 2 seconds)." Have the children repeat the sounds with the appropriate pauses. Now tell them that the train is going to gradually speed up. At this point you need to speed up your speech. Say "/ch/ (pause for 1 second), /e/ (pause for 1 second), /z/ (pause for 1 second)." Have the class repeat this. Speed up your speech and marching once again. Say the sounds again with only a slight pause in between each sound. Have the students repeat. Freeze the train. Have the children guess the word. Now repeat this process a few times until you are presenting the sounds with only a 1 second pause and the children are immediately guessing the target word.

Send the children to the computers to review with the tutorial found in the Ballroom of Daisy's Castle. When the tutorial is completed, have the students practice with the exercises found in the Throne Room and the Dining Room of Daisy's Castle. In each area they will be able to earn treasures.

To assess progress, have children do the blending (whole word) area of Undersea Challenge.

Source: Contributed by Gina Erickson, 1993.

Example Lesson Plan
Integration of Tutorial Courseware

Lesson:	Basic Concepts of Physics
Developed by:	The Brøderbund Software manual
Courseware:	Welcome to Physics (Brøderbund)
Level:	Secondary
Content Area:	Science:Physics
Instructional Resources:	Computer and Software
Lesson Purposes:	The teaching manual that accompanies this package explains that the program can be used in four different ways: initial demonstration to introduce concepts, remediation for students having trouble with teacher-introduced concepts, independent study for highly motivated students or in advanced study classes, or as a cumulative review after material is covered in class.

Rationale for Using Courseware. This program is detailed and structured enough to stand on its own as an instructional sequence or to support an inquiry approach to learning. For example, the teacher can use the animated demonstrations to illustrate points quickly and help students describe what they see and predict what will happen next.

Instructional Activities. The teacher should begin by going through the tutorial and becoming well-acquainted with its contents. He or she decides on the most helpful use of the program for the particular class and situation: initial demonstration, remediation, independent study, or review. Then do the following:

- For any strategy, be sure to introduce students to the program's operating procedures such as loading the program, using pull-down menus, moving through the frames, scrolling through listings, and stopping the program.

- Illustrate the structure of the program. (Show the table of contents and how to page through screens.) Be sure to indicate that each new chapter starts with Page 1.

- Demonstrate the Help features: Reference, Hint, and Solution.

After this introduction, proceed with the use of the program most appropriate for the class. Note that Welcome to Physics can be used in two different ways with the same class (e.g., as a device to initiate discussions and, later, as a review of the unit).

Source: Ideas taken from the Brøderbund software manuals and *Physics Teacher's Guide,* 1988.

much background material prior to meeting with a teacher or the rest of the class for assessment and further work assignments.

- **To allow instruction when teachers are unavailable.** Some students have problems when they surge ahead of their classes rather than falling behind. The teacher cannot leave the rest of the class to provide the instruction that such an advanced student needs. Many schools, especially those in rural areas, may not offer certain courses because they cannot justify the expense of hiring a teacher for comparatively few students who will need physics, German, trigonometry, or other lower-demand courses. Well-designed tutorial courses, especially in combination with other methods such as distance learning, can help meet these students' needs.

Like drill and practice functions, tutorial functions are designed primarily to serve individuals. Depending on which of the listed strategies it promotes, a tutorial may form a classroom learning station or it may be available for checkout at any time in a library/media center. Several successful uses of tutorials have been documented (Murray et

al., 1988; Kraemer, 1990; CAI in Music, 1994; Graham, 1994), but microcomputer tutorials that can fulfill the functions listed are rarely found in classroom use. Although they seem to have considerable theoretical value and have seen some popularity in military and industrial training, schools and colleges have never fully tapped their potential as teaching resources. The expense of developing them and difficulty of marketing them may be to blame for this situation. However, recent trends toward combining tutorial courseware with video media and distance education may bring tutorial functions into more common use.

Simulation Activities

Simulations: Definition and Characteristics

A simulation is a computerized model of a real or imagined system designed to teach how a certain system or a similar one works. Simulations differ from tutorial and drill and practice activities by providing less structured and more learner-directed learning activities. The person using the courseware usually chooses tasks and the order in which to do them. Alessi and Trollip (1991) identify two main types of simulations: "those that teach about something and those that teach how to do something" (p. 119). They further divide the "about" simulations into physical and process types, and they divide the "how to" simulations into procedural and situational types.

- **Physical simulations.** Users manipulate objects or phenomena represented on the screen. For example, students see selections of chemicals with instructions to combine them to see the result, or they may see how various electrical circuits operate.

- **Process simulations.** These speed up or slow down processes that usually either take so long or happen so quickly that students could not ordinarily see the events unfold. For example, courseware may show the effects of changes in demographic variables on population growth or the effects of environmental factors on ecosystems. Biological simulations like those on genetics are popular, since they help students experiment with natural laws like the laws of genetics by pairing animals with given characteristics and showing the resulting offspring.

- **Procedural simulations.** These activities teach the appropriate sequences of steps to perform certain procedures. They include diagnostic programs, in which students try to identify the sources of medical or mechanical problems, and flight simulators, in which students simulate piloting an airplane or other vehicle.

- **Situational simulations.** These programs give students hypothetical problem situations and ask them to react. Some simulations allow for various successful strategies, such as letting students play the stock market or operate businesses. Others have most desirable and least desirable options, such as choices when encountering a potentially volatile classroom situation.

These types only clarify the various forms a simulation might take. Teachers need not classify a given simulation into one of these categories. They need to know only that all simulations show students what happens in given situations when they choose certain actions. Simulations usually emphasize learning about the system itself, rather than learning general problem-solving strategies. For example, a program called The Factory has students build products by selecting machines and placing them in the correct sequence. Since the program emphasizes solving problems in correct sequence rather than manufacturing in factories, it should probably be called problem-solving activity rather than a simulation. Programs such as SimCity (Brøderbund), which let students design their own cities, provide more accurate examples of "building" simulations.

Since simulations promote such widely varied purposes, it is difficult to provide any specific criteria for selecting high-quality ones. By one frequently cited criterion, fidelity, a more realistic and accurate representation of a system makes a better simulation (Reigeluth and Schwartz, 1989). However, even this is not a criterion for judging all simulations (Alessi, 1988). Reigeluth and Schwartz (1989) also described some design concerns for simulations based on instructional theory. They listed important simulation components including a scenario, a model, and an instructional overlay that lets learners interact with the program.

Since the screen often presents no set sequence of steps, simulations need good accompanying documentation more than most courseware. A set of clear directions helps the teacher to learn how to use the program, and to show the students how to use it rapidly and easily.

Issues Related to Simulations

Most educators acknowledge the instructional usefulness of simulations. However, some people are concerned about the accuracy of the programs' models. For example, when students see simplified versions of these systems in a controlled situation, they may get inaccurate or imprecise perspectives on the systems' complexity. Students may feel they know all about how to react to situations because they have experienced simulated versions of them. Many educators feel especially strongly that situational simulations must be followed at some point by real experiences with the actual situations. Many teachers of young children feel that learners at early stages of their cognitive development should experience things first with their five senses rather than on computer screens.

Some simulations are also viewed as very complicated ways to teach very simple concepts that could just as easily be demonstrated on paper, with manipulatives, or with real objects. For example, students are usually delighted with the simulation of the food chain called Odell Lake. It lets students see what animal preys on what other animals in a hypothetical lake. However, some wonder whether or not such a computer simulation is necessary or even desirable to teach this concept. Hasselbring and Goin (1993) also point out that students can often master the activities of a simulation without actually developing any effective problem-solving skills; on the contrary, such applications can actually encourage counterproductive behaviors. For example, some simulations initially provide very little information with which to solve problems, and students are reduced to "trial-and-error guessing rather than systematic analysis of available information" (p. 156). Teachers must carefully structure integration strategies so that students will not use simulations in these inappropriate ways.

Simulations are considered among the most potentially powerful computer courseware resources. But, as with most courseware, their true usefulness depends largely on the purpose of the program and how well it fits in with the purpose of the lesson and the needs of the students. Teachers are responsible for recognizing the unique instructional value of each simulation and using it to best advantage.

How to Use Simulations in Teaching

Simulations have long been recognized for their unique teaching capabilities. Depending on the topic it addresses, a simulation can provide one or more of the following benefits (Alessi and Trollip, 1991):

- **Compress time.** This feature is important whenever students study the growth or development of living things (e.g., pairing animals to observe the characteristics of their offspring) or any other processes that take a long time (e.g., the movement of a glacier). A simulation can make something happen in seconds that normally takes days, months, or longer. Consequently, feedback is faster than in real life, and students can cover more variations of the activity in a shorter time.

- **Slow down processes.** Conversely, a simulation can also model processes normally invisible to the human eye because they happen so quickly. For example, physical education students can study the slowed-down movement of muscles and limbs as a simulated athlete throws a ball or swings a golf club in a certain way.

- **Get students involved.** Simulations can capture students' attention by placing them in charge of the program's events and asking that most motivating of questions: "What would *YOU* do?" The results of their choices can be immediate and graphic. It also allows users to interact with the program instead of just seeing its output.

- **Make experimentation safe.** Whenever learning involves physical danger, simulations are the strategy of choice. This is true any time students are learning to drive vehicles (e.g., car or airplane), or handle volatile substances (e.g., chemicals) or react to potentially dangerous situations (e.g., school violence). They can experiment with strategies that might result in personal injury to themselves or others in real life.

- **Make the impossible possible.** This is the most powerful feature of a simulation. Very often, teachers simply cannot give students access to the resources or the situations that simulations can. Simulations can show students what it would be like to walk on the moon or to react to emergencies in a nuclear power plant. They can see cells mutating or hold countrywide elections. They can even design new societies or planets and see the results of their choices.

- **Save money and other resources.** Many school systems are finding dissections of animals on a computer screen much less expensive than on real frogs or cats and just as instructional. (Animal rights activists also point out that it is easier on the animals!) Depending on the subject area, a simulated experiment may be just as effective a learning experience at a fraction of the cost.

- **Repeat with variations.** Unlike real-life situations, simulations let students run over events as many times as they wish and with whatever variations they wish. They can pair unlimited numbers of cats or make unlimited airplane landings in a variety of conditions to compare the results of each set of choices.

- **Make situations controllable.** Real-life situations are often "messy" and confusing, especially to those seeing them for the first time. When many things happen at once, students have difficulty focusing on the operation of individual components. Who could understand the operation of a stock market

Figure 4.5   Example Software with Simulation Activities: Operation Frog (1)

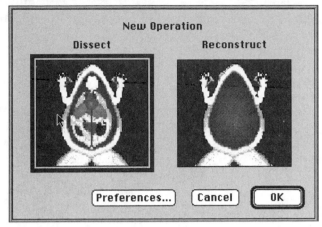

Source: Used by permission of Scholastic, Inc.

by looking at the real thing without some introduction? Simulations can isolate parts of activities and control the background "noise." This makes it easier for students to see what is happening later when all the parts together in the actual activity (see Figures 4.5 and 4.6).

Real systems are usually preferable to simulations, but a simulation can suffice when a teacher considers the real situation too time-consuming, dangerous, expensive, or unrealistic. Simulations should be considered in the following situations—keeping in mind that the real activity is preferable if it is feasible:

- **In place of or as supplements to lab experiments.** When adequate lab materials are not available or not in sufficient quantities, teachers should consider trying to locate computer

Figure 4.6   Example Software with Simulation Activities: Operation Frog (2)

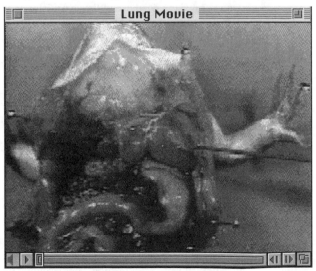

Source: Used by permission of Scholastic, Inc.

Example Lesson Plan
Integration of Simulation Courseware: Introducing the Topic of Body Systems

Lesson:	Frog cut-ups
Developed by:	Scholastic software
Courseware:	Operation Frog
Level:	Secondary
Content Area:	Biology: Anatomy
Instructional Resources:	Computer, software, and activity sheets in the software manual
Lesson Purposes:	This activity acquaints students with the structure and components of a body system. By dissecting and naming the frog body, they also become more familiar with human bodies and how they work.

Rationale for Using Courseware. This program takes the place of dissection work with a real frog. It is less expensive (since it can be used over and over again), less offensive to students (and to frogs), and more flexible (can easily be "reassembled" after dissection).

Instructional Activities. Introduce Operation Frog at the beginning of a unit on body systems. Demonstrate the parts of the program and how it is used. Place students in groups to go to various work stations. For example:

- **Group 1.** Dissect the simulated frog and then put it back together. If desired, the teacher could assign two groups to work on this: one to dissect and the other to reassemble.

- **Group 2.** Complete worksheets in the Scholastic teacher manual on organ identification.

- **Group 3.** Complete an exercise classifying various organisms according to their body structures. The teacher provides background information on how to classify organisms.

- **Group 4.** Prepare a detailed group report on a certain body system assigned by the teacher (e.g., digestive or circulatory system).

After all students have a chance to do all activities, the students present findings on their assigned body systems to the class. Teachers assess these presentations and, if desired, students ability to complete the body systems worksheets without assistance.

Source: Ideas taken from the Scholastic software manuals and *Operation Frog Teacher's Guide,* 1991.

---

simulations of the required experiments. Many teachers find simulations offer effective supplements to real labs, either to prepare students for making good use of the actual labs, or as follow-ups with variations on the original experiments without using up consumable materials. Some simulations can actually allow users to perform experiments that they could not otherwise manage or that would be too dangerous for students. (See The Earthquake lesson plan.)

- **In place of or as supplements to role playing.** When students take on the roles of characters in situations, the programs can spark students' imagination and interest in the activities. However, many students either refuse to role play in front of a class or get too enthusiastic and disrupt the classroom. Computerized simulation can take the personal embarrassment and logistical problems out of the learning experience and make classroom role playing more controllable.

- **In place of or as supplements to field trips.** Seeing an activity in the real setting can be a valuable experience, especially for young children. Sometimes, however, desired locations are not within reach of the school, and a simulated experience of all or part of the process is the next best thing.

As with labs, simulations can provide good introductions or follow-ups to field trips.

In addition to these integration strategies, which seem to represent directed methods, simulations may also promote constructivist strategies:

- **Introducing a new topic.** Courseware that allows students to explore the elements of an environment in a hands-on manner frequently provides students' first in-depth contact with a topic. This seems to accomplish several purposes. First, it is a nonthreatening way to introduce new terms and unfamiliar settings. Students know that they are not being graded, so they feel less pressure than usual to learn everything right away. A simulation can become simply a get-acquainted look at a topic. Simulations can also build students' initial interest in a topic. Highly graphic, hands-on activities draw them into the topic and whet their appetite to learn more. Finally, some software helps students see how certain prerequisite skills relate to the topic; this may motivate students more strongly to learn the skills than if the skills were introduced in isolation from the problems to which they apply (as with the *Decision!*

Example Lesson Plan
Integration of Simulation Courseware

Lesson:	Earthquake!
Developed by:	Karen Smith, McArthur Elementary School (Pensacola, Florida)
Courseware:	Science Toolkit: Earthquake Module (Brøderbund)
Level:	Elementary
Content Area:	Science: Earth concepts
Instructional Resources:	In addition to a computer and software, a laserdisc player and laserdisc: Windows on Science, Intermediate-Earth Science, Vol. 1 (Optical Data)
Lesson Purposes:	• Learn to work cooperatively with a group to perform an experiment
	• Identify fault lines and explain the movement of an earthquake

Rationale for Using Courseware. The Science Toolkit Earthquake Module simulates in an experimental setting what students could not do in real life: cause an earthquake and observe its activities. The Windows on Science videodisc helps students vividly view the movement of the earth's plates and what happens to the earth's crust in an earthquake.

Instructional Activities.
Reviewing earthquake information. To begin this lesson, the teacher should start with a basic review of an earthquake, asking questions such as: "What is an earthquake? What happens when an earthquake occurs? What causes earthquakes?" Review the layers of the earth using a visual aid showing the crust, mantle, and core. Discuss plates in the crust. Explain that the earth's crust is divided into sections called *plates*. The plates fit together like a puzzle. Show a world map with the plates defined. If possible, have a world map sectioned off showing the plates. Cut them apart and apply magnetic tape on the back to form a visual puzzle to pull apart, reinforcing the concept. Explain that the plates are constantly moving, perhaps interjecting a videodisc clip of the plates on Windows on Science: Earth Science, Volume 1. Use Frame 17984 which shows an explanation and the movement of the plates. Be sure students understand that an earthquake is the sudden moving and shaking of the earth due to pressure buildup within the plates causing plates to rise and fall.

Earthquake demonstrations. Have a child build a simulated building using paper cups and plates. Show what happens during an earthquake by gently bumping the base of the structure. Explain ways of recording movement through a machine called a *seismograph*. Explain the seismograph and its function. Talk about its importance for recording data constantly and monitoring movement for possible predictions of major earthquakes. Show the laser disc again, choosing frames related to your level. This shows an earthquake and explains what is happening. Explain the use of the Science Toolkit Earthquake Lab to create an earthquake and better understand how seismographs work.

Doing the experiment. Start up the software and set up the seismoscope according to directions in the instruction manual. Make sure that the interface box is plugged into the computer and the photocell probe is in the correct slot. The seismoscope will measure light movement as the source of earthquake activity. Set the light source (flashlight) on the table. Make sure the light is directed toward the photocell. The least movement of the light will create movement of the seismograph needle on the screen. Try three variables. First, have a student hit the table with the flashlight lightly. The computer will begin recording data as soon as it detects movement. Notice the needle jumping each time the student hits the table. Repeat the procedure with a large earthquake by pounding on the table with your fist and recording. The last part of the experiment will show a seismograph reading a severe earthquake; shake the table with the light source to create movement.

Analyzing the data. Review data gathered by using arrow keys to read the graph backward and forward. To get a hard copy of the experimental data, press Escape until you reach the Earthquake Lab menu. (You may elect to save the data and print later.) Select EXPERIMENT PLOTTER. If you want to print the whole experiment, press RETURN. Your experiment should print. You may choose to print only one section of the experiment by following the directions in the manual. End this experiment by going back to the Main menu.

Source: From Smith, K. (199), Earthquake. *The Florida Technology in Education Quarterly, 4,* 2 (Winter), 68–70.

Example Lesson Plan
Constructivist Simulation Courseware Activities: Encouraging Cooperation and Group Work

Lesson:	Community Planning Projects with SimCity
Developed by:	Marianne Teague and Gerald Teague, Northern Middle School (Calvert County, Maryland)
Courseware:	SimCity (Maxis Software)
Level:	Middle school (Seventh grade)
Content Area:	Social studies: Teaching citizenship and group cooperation in social projects
Instructional Resources:	Computer, software, and local community contacts
Lesson Purposes:	• Raising awareness of responsibility to become informed citizens and participate in local decision making
	• Learning to work cooperatively with a group to carry out a social project

Rationale for Using Courseware. The SimCity software allows students to manipulate several factors while developing a community. This helps them focus on factors they must consider when developing a local project to benefit their community.

Instructional Activities. This software and the idea for the project tied in with a current event of concern to the whole community: the development of a comprehensive plan for the township. (Many communities across the country are involved in this kind of plan.) After a representative from the county planning and zoning office talked with the students about factors of concern to community development, the social studies teacher helped introduce a group of about 50 students to the SimCity software. This program served to demonstrate the concepts that the representative had talked about with them.

The group of students formed teams of 4 to 5 students each and started their own community planning projects with SimCity. Each group met for an hour every week for about 3 months to discuss and develop their plan. As a group, they decided how to select and place features such as roads, homes, and utilities. After recording their decisions on paper, they entered them into the program and observed the results. They discussed feedback that the program gave them on areas such as taxes, crime rates, and public opinion. They videotaped their meetings, discussions, and computer work in order to document them.

After their plans were completed, each group presented its plan to the teacher and the media specialist. some of the groups that were most successful in meeting the preset criteria of low crime rates, low pollution levels, reasonable costs, and public approval were asked to prepare formal presentations. These teams presented their designs via the computer and printouts. Members of the local county planning and zoning office helped review the final products. Students had to both present their designs and explain and defend their choices.

Source: From Teague, M. and Teague, G. (1995). Planning with computers: A social studies simulation. *Learning and Leading with Technology, 23,* 1 (September), 20–22.

Decisions! software by Tom Snyder on Social Studies topics such as urbanization and elections).

• **Fostering exploration and process learning.** Teachers often use content-free simulation/problem-solving software as a motivational way to allow students to explore their own cognitive processes. Since this kind of courseware requires students to learn no specific content, it is easier to get them to concentrate on problem-solving steps and strategies. However, with content-free products, it is even more important than usual that teachers draw comparisons between skills from the courseware activities and those in the content areas to which they want to transfer the experience. For example, The Incredible Laboratory (Sunburst) brings an implicit emphasis on science process skills that the teacher may want to point out. These kinds of activities may be introduced at any time, but it seems more fruitful to use them just prior to content area activities that will require the same processes. (See the example activities above and on the next page.)

• **Encouraging cooperation and group work.** Sometimes a simulated demonstration can capture students' attention quickly and effectively and interest them in working together on a product. For example, a simulation on immigration or colonization might be the "grabber" that a teacher needs to launch a group project in a social studies unit.

Simulations offer more versatile implementation than tutorials or drills. They can usually work equally effectively with a whole class, small groups, or individuals. A teacher may choose to introduce a lesson to the class by displaying a simulation or to divide up the class into small groups and let each solve problems. Because they instigate discussion and cooperative work so well, simulations are usually considered more appropriate for pairs and small groups than for individuals. However, individual use is certainly not precluded.

The market offers many simulations, but it is often difficult to locate one on a desired topic. The field of science seems to include more simulations than any other area (Andaloro, 1991; Richards, 1992; Ronen, 1992; Smith, 1992; Mintz, 1993; Simmons and Lunetta, 1993), but use of simulations is also popular in social sciences topics (Clinton, 1991; Allen, 1993; Estes, 1994). However, more simulations are currently being developed with videodisc supplements to combine the control, safety, and interactive features of computer simulations with the visual impact of pictures of real-life devices and processes.

Instructional Games

Instructional Games: Definition and Characteristics

The category of instructional games is usually considered to include courseware designed to increase motivation by adding game rules to learning activities. Even though teachers often use them in the same way as drill and practice or simulation courseware, games are usually listed as a separate courseware activity because their instructional connotation to students is slightly different. When students know they will play a game, they expect a fun and entertaining activity because of the challenge of the competition and the potential for winning (Randel, Morris, Wetzel, and Whitehill, 1992). Naturally, classroom instruction should not consist entirely of these kinds of activities, no matter how instructional or motivational they are. Teachers intersperse games with other activities to hold attention or to give rewards for accomplishing other activities. Alessi and Trollip (1991, pp. 173–182) list ten kinds of games, but they point out that a game may actually serve more than one role:

1. **Adventure games.** Players confront hostile situations or solve mysteries. Sometimes content (e.g., mathematics or social studies) is involved; sometimes the game presents a problem-solving activity to engage students' observation and information-gathering skills.

2. **Arcade games.** Some software simulates popular games such as pinball or Pacman during which students answer questions in math or other content areas to win game points.

3. **Board games.** Players move on a computerized "board" when they answer content questions correctly.

4. **Card or gambling games.** These usually involve betting or winning money as rewards for correct responses to problems or learning tasks.

5. **Combat games.** Based on violence and confrontations with enemies, these games have sparked considerable controversy among parents and teachers because of their possible potential negative effects on students.

6. **Logic games.** These games involve logical problem-solving activities. Players usually test out hypotheses by making systematic series of moves to solve problems.

7. **Psychomotor games.** In these computerized versions of real sports such as baseball or tennis, students answer content problems to make on-screen players perform.

8. **Role-playing games.** These games let students act as on-screen characters. Many games from other categories also exhibit aspects of this kind of game.

9. **TV quiz games.** These computerized versions of typical TV game-show formats invite students to earn points and win "prizes" by answering questions correctly.

10. **Word games.** Any game that focuses on learning words or word categories falls into this classification (e.g., Word Concentration).

As with types of simulations, these categories merely illustrate the various forms an instructional game may take. Teachers should not feel that they have to classify specific games into these categories. But it is important to recognize the common characteristics that set instructional games apart from other types of courseware: game rules, elements of competition or challenge, and amusing or entertaining formats. These elements generate a set of mental and emotional expectations in students that make the instructional activities different from nongame activities.

Since instructional games often amount to drills or simulations overlaid with game rules, the same criteria for these courseware activities should apply to most games (e.g., better reinforcement for correct answers than for incorrect ones, good documentation or instructions on courseware use). When Malone (1980) examined the evidence on "what makes things fun to learn," he found that the most popular games included elements of adventure and uncertainty, and levels of complexity matched to learners' abilities. However, teachers should also examine instructional games carefully for their value as both educational and motivational tools. Teachers should also assess the amount of physical dexterity that games require of students (especially for arcade or psychomotor games) and make sure that students will not be frustrated instead of motivated by the activities. Games that call for violence or combat need careful screening, not only to avoid parent criticism, but also because girls often perceive the attraction of these activities differently than boys and because such games sometimes depict females as targets of violence.

Issues Related to Instructional Games

A classroom without elements of games and fun would be a dry, barren landscape for students to traverse. In their review of the effectiveness of games for educational purposes, Randel, Morris, Wetzel, and Whitehill (1992) found "[the fact] that games are more interesting than traditional instruction is both a basic for using them as well as a consistent finding" (p. 270). They also observed that retention over time favors the use of simulations/games. Yet many educators believe that games, especially computer-based ones, are overused and misused (McGinley, 1991). Other teachers believe games convince students that they are "escaping from learning" and that they draw attention away from the intrinsic value and motivation of learning. Critics also feel that winning the game becomes a student's primary focus, and the instructional purpose is lost in the pursuit of this goal. Observers disagree whether "getting lost in the game" is a benefit or a problem. Some teachers believe that any time they can "sneak learning in" under the guise of a game, it is altogether a good thing (McGinley, 1991). Other teachers believe that students can become confused about which part of the activity is the game and which part is the skill they are learning; they may then have difficulty transferring their skill to later nongame situations. For example, the teacher's manual for Sunburst's How the West Was One + Three × Four reminds teachers that some students can confuse the math operations rules with the game rules, and that teachers must help them recognize the need to focus on math rules and use them outside the game.

While students obviously find many computer games very exciting and stimulating, it is sometimes difficult to identify educational value. Teachers must also try to balance the motivation that instructional games bring to learning against the extra classroom time they take away from nongame strategies. For example, students may become immersed in the challenge of the Carmen Sandiego series, but more efficient ways to teach geography may be just as motivating. (On the other hand, Carmen Sandiego could make an effective reward for good performance.) Successful uses of games have been reported in all content areas (Trotter, 1991; Flowers, 1993).

How to Use Instructional Games in Teaching

Several kinds of instructional opportunities invite teachers to take advantage of the motivational qualities of games (See Figure 4.7). They might use games:

- **In place of worksheets and exercises.** This role resembles that of drill and practice.

- **To foster cooperation and group work.** Like simulations, many instructional games can serve as the basis for or introductions to group work. A game's interactive and motivational qualities can help interest students in the topic, and it can also present opportunities for competition among groups to win first or earn the most points.

- **As a reward.** Perhaps the most common current use of games is to reward good work. This is a valid role for instructional courseware, but teachers should avoid overuse of it. Otherwise, the game can lose its motivational value and become an "electronic baby-sitter." Some schools actually bar games for fear that students overemphasize the need to be entertained.

Problem-Solving Courseware

Problem-Solving Courseware: Definition and Characteristics

Teachers may find the topic of problem solving both alluring and perplexing. No goal in education seems more important today than making students good problem solvers, yet no area is as ill-defined and difficult to understand. Even scientists have difficulty defining *problem solving*. Funkhouser and Dennis (1992) quoted an earlier author as saying that "Problem solving [means] the behaviors that researchers who say they are studying problem solving, study" (p. 338). Sherman (1987–1988) was somewhat more specific, claiming that all problem solving involves three components: recognition of a goal (an opportunity for solving a problem), a process (a sequence of physical activities or operations), and mental activity (cognitive operations to pursue a solution). Sherman said that problem solving is a "relatively sophisticated mental ability which is difficult to learn" (1987–1988, p. 8) and that it is highly idiosyncratic. That is, problem-solving ability depends upon "knowledge, prior experience, and motivation, and many other attributes" (Sherman, 1987–1988, p. 8).

This definition covers a wide variety of desired component behaviors. The literature mentions such varied subskills for problem solving as metacognition, observing, recalling information, sequencing, analyzing, finding and organizing information, inferring, predicting outcomes, making analogies, and formulating ideas. Since even the definition of *problem solving* inspires an ongoing controversy in education, it is not surprising that opinions differ dramatically about the proper role of courseware and other technology products in helping to foster this important capability. The positions seem to lean toward two general ways in which teachers can view problem solving. Which of these views a teacher uses will determine the strategy for teaching problem solving and the application of technology resources to this activity.

Two views on fostering problem solving. Some teachers view problem solving as a high-level skill that can be taught directly, at least in part, by specific instruction and practice in its component strategies and subskills. Others suggest placing students in problem-solving environments and, with some coaching and guidance, letting them develop their own heuristics for attacking and solving problems. Although the purposes of the two views overlap somewhat, one seems more directed toward supplying prerequisite skills for specific kinds of problem solving, while the other seems to aim more toward motivating students to attack problems and to recognize solving problems as an integral part of everyday life. Blosser (1988) confirms this dichotomy, saying that, "Problem solving includes … an attitude or predisposition toward inquiry as well as the actual processes by which individuals … gain knowledge." Students need to combine these two elements; teachers must make ongoing adjustments to the amount of time they spend on each kind of approach in each of several content areas.

Two types of problem-solving courseware for directed instruction. Two distinct types of courseware purport to teach problem-solving skills. One is specific to contents area primarily in mathematics (e.g., The Geometric

Figure 4.7 Example Software with Instructional Game Activities: How the West Was One + Three × Four

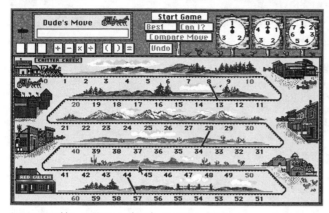

Source: Used by permission of Sunburst Communications.

Supposer by Sunburst which encourages students to learn strategies for solving geometry problems by drawing and manipulating geometric figures). The other type of problem-solving software focuses on general, content-free skills such as recalling facts, breaking a problem into a sequence of steps, or predicting outcomes (e.g., Sunburst's Memory Castle, which is designed to help students remember instructions and follow directions). Most courseware is specifically designed to focus on one of these two approaches; however, some authors point out that programs can help to teach problem solving with being specifically designed to do so (Gore, 1987–1988). Courseware implements numerous approaches to teach each of these kinds of skills. Some use challenge strategies (The King's Rule by Sunburst); others use puzzles games (Safari Search by Sunburst), adventure-games (Carmen Sandiego by Brøderbund), or simulation approaches (The Factory by Sunburst).

Issues Related to Problem-Solving Courseware

Names versus skills. As mentioned earlier, courseware packages use many terms to describe problem solving, and their exact meanings are not always clear. Terms that appear in courseware catalogs as synonyms for *problem solving* include: thinking skills, critical thinking, higher-level thinking, higher-order cognitive outcomes, reasoning, use of logic, and decision making. In light of this diversity of language, teachers can identify the skills that a courseware package addresses by looking at its activities. For example, a courseware package may claim to teach *inference skills*. One would have to see how it defines *inference* by examining the tasks it presents. They may range from determining the next number in a sequence to using visual clues to predict a pattern.

Courseware claims versus effectiveness with students. It would be difficult to find a courseware catalog that did not claim that its products foster problem solving. However, few publishers of courseware packages that purport to teach specific problem-solving skills have collected data to support their claims. When students play a game that seems to require certain skills related to problem solving, they do not necessarily learn these skills. They may enjoy the game thoroughly, and even be successful at it without learning any of the intended skills. Teachers may have to do their own field testing to confirm that courseware is achieving the results they want.

Possible harmful effects of directed instruction. Some researchers believe that direct attempts to teach problem-solving strategies can actually be counterproductive in some students. Mayes (1992) reports on studies that found problems: "teaching-sequenced planning to solve problems to high ability learners could interfere with their own effective processing" (p. 243). In a review of research on problem solving in science, Blosser (1988) also found indications that problem-solving instruction may not have the desired results if the instructional strategy does not suit certain kinds of students. For example, students with high math anxiety and low visual preference or proportional reasoning abilities will profit from instruction in problem solving only if it employs visual approaches.

The problem of transfer. Although some educators feel that general problem-solving skills (e.g., inference and pattern recognition) will transfer to content-area skills (e.g., problem-solving in geometry and math word problems), scant evidence supports this view. In the 1970s and 1980s, for example, many schools taught programming in mathematics classes under the hypothesis that the planning and sequencing skills required for programming would transfer to problem-solving skills in math. Research results never supported this hypothesis. In general, research tends to show that skill in one kind of problem solving will transfer primarily to similar kinds of problems that use the same solution strategies. Reseachers have identified nothing like "general thinking skills," except in relation to intelligence (IQ) variables.

How to Use Problem-Solving Courseware in Teaching

Directed strategies with problem-solving courseware. Integration of courseware into direct teaching of problem-solving skills places even more responsibility than usual on teachers. Usually, teachers want to teach clearly defined skills. To teach problem solving, they must decide which particular kind of problem-solving ability students need to acquire and how best to foster it. With clearly identified skills and a definite teaching strategy, the motivational aspects of problem-solving courseware has unique abilities to help focus students' attention on required activities. This kind of courseware can get students to apply and practice desired behaviors specific to a content area or more general abilities in problem solving.

Example Lesson Plan
Integration of Instructional Game Courseware: Practicing Math Rules of Operation

Lesson: Please! Please! Remember My Dear Aunt Sally!
Developed by: Gla Culpeper and Elaine Meyers (Jacksonville, Florida)
Courseware: How the West Was One + Three × Four (Sunburst Communications)
Level: Middle school
Content Area: Mathematics
Topic: Order of mathematics operations
Instructional Resources: Computer and courseware; worksheets in Sunburst, Davidson, and SRA Manuals; overhead transparency; chalkboard
Lesson Purposes:
   • Use the order of operations rule to solve mathematical problems.

   • Use correct math terminology.

Rationale for Using Courseware. This lesson was tested by a home/hospital teacher who worked one-on-one with students. (The program serves students who are sick at home or in a hospital.) HTWWO+3×4 was selected for its motivational qualities for these students. It seems to challenge them to learn the order of operations so they will do well on the game and beat the computer. This program and other game software were selected for review and practice on math concepts because students love to use them and will spend time on them that they wouldn't spend on homework.

Instructional Activities.

Reviewing prerequisite skills. Before beginning instruction, use worksheets to assure that all students know basic math operations and symbols (+,−,×,÷); reteach if remediation is necessary. Remaining students practice basic math skills with Math Blaster Plus, leaving the teacher available to work with remedial students.

Motivating students. Demonstrate the software to the students and let them play the game as a group activity. Point out how using parentheses, powers, and roots can expand the number of moves, and tell the class that they will be learning skills to help them solve mathematical problems using order of operations. Make sure the students know that the order of operations is a mathematical rule that always applies and not a game rule applicable only to this one program.

Presenting new information and learning guidance. Explain the mnemonic "Please, Please, Remember, My Dear Aunt Sally" and the meaning of the order of operations (Parentheses, Powers, Roots, Multiplication, Division, Addition, Subtraction from left to right). Point out that not all operations are used in this program. Show the rules for HTWWO on an overhead transparency. Demonstrate (+,−,×,÷) problems with and without parentheses and give examples of problems for students to as a group. Worksheets help students practice and remember these rules; the teacher checks and assists with their work.

Practice activities. When all students seem able to work the problems, demonstrate HTWWO and review the rules. Students practice one game as a group. Make arrangements for students to practice their skills on the game in the classroom or computer lab for a large group activity. Students help each other play the game against the computer. Give special recognition to students who win the most games in a period.

Assessment and followup. After students have had an opportunity to practice, give a test to evaluate their knowledge and applications of basic order-of-operations (MDAS from left to right). The teacher continues working with students who have problems and encourages them to continue using the software to get their names on the Top 10 players list and as a reward for finishing regular work.

Source: From Culpeper G., Myers, E., and Roblyer, M. D., (1991). Please! Please! Remember My Dear Aunt Sally! *The Florida Technology in Education Quarterly, 3* (2) 87–88.

Example Lesson Plan
Constructivist Integration for Problem Solving Courseware:
Fostering Exploration and Process Learning

Lesson:	Creature Capers
Developed by:	Marilyn Jussel, Kearney State College (Kearney, Nebraska)
Level:	Grades 3–12
Lesson Purposes:	Explore and practice problem-solving strategies in a number of different situations
Instructional Resources:	Computer and software

Rationale for Using Courseware: Adding computer software to the problem-solving curriculum gives students a wider variety of problem-solving practice situations. They can see how to apply the strategies they learn in one situation in a number of software packages. Teachers can also address individual differences in learning. The computer enhances motivation; students who have difficulty relating to paper-and-pencil problems especially like the computer method.

Instructional Activities: Introduce students to the process strategies they will explore and practice. For example, they may learn strategies such as pattern analysis, systematic listing, and eliminating possibilities. Exploration and use of the systematic listing strategy may be enhanced by assigning the following kinds of activities with Muppet: Mix and Match (Sunburst) or a similar software package.

1. Create two characters and record the head, body, and feet chosen to make each. Record the name given to each creature. Share results with the other students.

2. Create as many different characters as the student can in 10 minutes. Record the names. Make a complete list of the characters the class found. How many were there?

3. In a notebook, develop a plan to create all of the possible characters. Share the plan with all the students. Decide how the groups could cooperate to create all of the characters. Complete the project on the computer.

Create similar activities for older students using The Incredible Laboratory (Sunburst).

Source: From Jussel, M. (1989–1990). Teaching problem solving strategies with and without the computer. *The Computing Teacher 17*, 4 (December/January), 16–19.

These six steps can help teachers to integrate courseware for these purposes:

1. Identify problem-solving skills or general capabilities to build or foster skills in:

    a. Solving one or more kinds of content-area problems (e.g., building algebra equations)

    b. Using a scientific approach to problem solving (e.g., identify the problem, pose hypotheses, plan a systematic approach, etc.)

    c. Components of problem solving (e.g., following a sequence of steps, recalling facts)

2. Decide on an activity or a series of activities that would help teach the desired skills. (See the example lesson for eliminating confirmation bias using King's Rule and Figure 4.8.)

3. Examine courseware to locate materials that closely match the desired abilities. (Do not judge capabilities on the basis of vendor claims alone.)

4. Determine where the courseware fits into the teaching sequence (e.g., to introduce the skill and gain attention, as a practice activity after demonstrating problem solving or both.)

5. Demonstrate the courseware and the steps to follow in solving problems.

6. Build in transfer activities and make students aware of the skills they are using in the courseware.

Figure 4.8   Example Software with Problem-Solving Activities: King's Rule

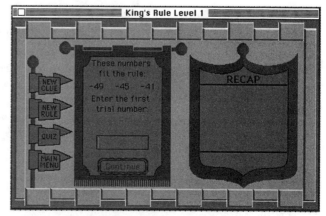

Source: Used by permission of Sunburst Communications.

Example Lesson Plan
Directed Integration of Problem-Solving Courseware: Addressing Confirmation Bias when Determining Number Rules in Mathematics

Lesson:	Could Your Answer Be Wrong?
Developed by:	James Johnson, Pennsylvania State University (University Park, Pennsylvania)
Courseware:	King's Rule (Sunburst Communications)
Level:	Middle school
Content Area:	Testing hypotheses in problem-solving strategies
Instructional Resources:	Computer and software
Lesson Purposes:	Drawing students' attention to the problem of confirmation bias in problem solving

Rationale for Using Courseware. This program is a good way to pose math problems randomly and give students nonjudgmental feedback on the hypotheses they form.

Instructional Activities. Have students play the Hidden Rule Game, in which the program generates three numbers (e.g., 1, 9, 10,) and students have to give their own sets of three numbers that follow the same rule. The program tells them whether or not their sets follow the rule.

As students consider the numbers, the teacher must ask questions to help them form alternative hypotheses about the rule and generate evidence to support their positions. For example, suppose the numbers are 16, 18, 20. Students must be encouraged to gather confirming evidence and submit their rules carefully. They may jump to the conclusion that the rule is "Jumps of two," when it may be "all even numbers" or "ascending numbers." For example:

Student enters:
4,6,8
The computer responds:
"Yes"
Student reacts:
"Numbers jump by 2."
(Teacher says, "Could it be something else? Try a set that is not numbers jumping by 2.)
Student enters:
4,8,12
The computer responds:
"Yes"
Student reacts:
"It's not numbers jumping by 2. Maybe it's ascending numbers."
(Teacher says, "What numbers could you try in order to eliminate that rule?")
Student enters:
5,7,9
The computer responds:
"Yes"
Student reacts:
"It's not ascending numbers or constant jumps. Maybe it's even numbers."
(Teacher says, "Are you sure? What can you try to find out?")

When students feel confident they have grasped good strategies well enough to take a quiz in rule-ascertaining, they select the program's quiz function. The program presents five sets of numbers for students to try.

Source: Based on Johnson, J. (1987, May). Do you think you might be wrong? Confirmation bias in problem solving. *Arithmetic Teacher.*

Constructivist strategies with problem-solving courseware. Like many technology products, some products labeled as problem-solving courseware can be employed in directed ways, but they are designed for implementation in more constructivist models. These models give students no direct training in or introduction to solving problems. Rather they place students in highly motivational problem-solving environments and encourage them to work in groups to solve problems.

Constructivists believe that this kind of experience helps students in three ways. First, they expect that students will be more likely to acquire and practice content-area, research,

Figure 4.9 Example Software with Problem-Solving Activities: Math Shop

Source: Used by permission of Scholastic, Inc.

Figure 4.10 Example Software with Problem-Solving Activities: The Factory

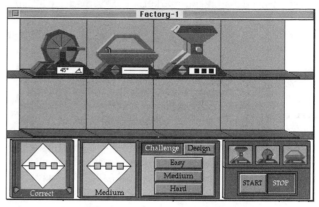

Source: Used by permission of Sunburst Communications.

and study skills for problems they find interesting and motivating. For example, to succeed in the Carmen Sandiego software series, students must acquire both some geography knowledge and some ability to use reference materials that accompany the package. Also, they must combine this learning with deductive skills to attack and solve "detective-type" problems (Robinson and Shonborn, 1991).

Second, constructivists claim that this kind of activity helps keep knowledge and skills from becoming inert because it gives students opportunities to see how information applies to actual problems. They learn the knowledge and its application at the same time. Finally, students gain opportunities to discover concepts themselves, which they frequently find more motivating than being told or, as constructivists might say, *programmed* with the information (McCoy, 1990).

Seven steps help teachers to integrate problem-solving courseware according to constructivist models:

1. Allow students sufficient time to explore and interact with the software, but do provide some structure in the form of directions, goals, a work schedule, and organized times for sharing and discussing results.

2. Vary the amount of direction and assistance depending on the needs of each student.

3. Promote a "reflective learning environment;" let students talk about their work and the methods they use.

4. Stress thinking processes rather than correct answers.

5. Point out the relationship of courseware skills and activities to other kinds of problem solving.

6. Let students work together in pairs or small groups.

7. If assessments are done, use alternatives to traditional paper-and-pencil tests.

Problem solving and simulation activities work so similarly to constructivist models that it is usually difficult to differentiate between them. Integration strategies for either are usually the same.

Integrated Learning Systems (ILSs) and Other Networked Products

Integrated Learning Systems (ILSs): Definition and Characteristics

Integrated learning systems (ILSs) are the most powerful—and the most expensive—of the courseware products, primarily because they are really more than just courseware and because they require more than one computer to run them. An ILS is a network, a combination instruction and management system that runs on microcomputers connected to a larger computer. An ILS can offer a combination of drill and practice, tutorial, simulation, problem solving, and tool courseware integrated into a total curriculum support package. In addition, it is capable of maintaining detailed records on individual student assignments and performance data and supplying printouts of all this information to teachers upon request. Bailey and Lumley (1991, p. 21) list the following general characteristics of an ILS:

*   Instructional objectives specified, with individual lessons tied to those objectives

*   Lessons integrated into the standard curriculum

*   Courseware that spans several grade levels in comprehensive fashion

*   Courseware delivered on a networked system of microcomputers or terminals with color graphics and sound

*   Management systems that collect and record results of student performance

ILS courseware and management software are housed on a computer called a *file server,* which is connected via a network to a series of microcomputers. As each student signs onto a microcomputer station, the file server sends (or "downloads") that student's assignment and courseware to the station and proceeds to keep records on what the student does during time spent on the system. The teacher makes initial assignments for work on the system, monitors student progress by reviewing ILS printouts, and provides additional instruction or support where needed.

The first ILSs on the market were primarily drill and practice delivery systems designed to improve student performance on the isolated skills measured by standardized tests. These self-contained, mainframe-based systems predated the microcomputer era, and they did not run any software besides their own. Usually housed in labs, they were designed for use in pull-out programs to supplement teachers' classroom activities. That is, students were pulled out of classrooms daily or weekly and sent to ILS labs for remedial or reinforcement work. However, these systems have evolved into multipurpose products that can run software and courseware other than their own; they can now provide a variety of instructional support from enrichment to complete curriculum. As with other media such as videodiscs, school districts are beginning to view ILSs as alternatives to traditional classroom materials such as textbooks.

ILSs compared with other networked courseware products. ILSs are characterized by their "one-stop shopping" approach to providing courseware. Each one offers a variety of instructional techniques in one place, usually as a package complete with technical maintenance and teacher training. They present strengths like prepared curricula and ease of use so that school personnel need not know a great deal about technology to use them. Consequently, they usually simplify integration decisions by defining schoolwide curriculum rather than many individual lessons. (Teachers ask "When will my class use the ILS?" Rather than "Where will I integrate this resource into my other classroom activities?")

According to Robertson, Stephens, and Company (1992), ILSs represent one of the major courseware delivery system choices for schools. Aside from standalone microcomputer products, schools can also choose so-called *multiple media systems*, which combine several kinds of media such as CD-ROMs, microcomputers, videodiscs or networked products, which deliver separate courseware packages to students via networks. According to the researchers' market surveys, "ILSs sales dominate system sales to the educational software market" (p. 6). However, this situation may change as technology becomes less expensive and more powerful and as schools become more accustomed to dealing with technology so they can integrate and tailor courseware and other media to meet their specific needs.

ILS are usually considered part of directed instruction rather than discovery learning. Companies that provide networked systems to support more constructivist approaches to learning object to the frequent classification of their products under the ILS rubric.

The courseware component of an ILS. Instructional activities available on an ILS range from simple drill and practice to extensive tutorials. Most ILSs seem to be moving toward complete tutorial systems intended to replace teachers in delivering entire instructional sequences. An ILS usually includes instruction on the entire scope and sequence of skills in a given content area. For example, it may cover all discrete mathematics skills typically presented in grades 1 through 6.

The management system component of an ILS. The capability that differentiates ILSs from other networked systems is their emphasis on individualized instruction tied to records of student progress. A typical ILS gives teachers the following kinds of information on individual performance, as well as progress reports across groups of students:

- Lessons and tests completed
- Questions missed on each lesson by numbers and percentages
- Numbers of correct and incorrect tries
- Time spent on each lesson and test
- Pretest and posttest data

An example of one report is shown in Figure 4.11. In addition, sample screens from an integrated learning system are shown in Figures 4.12, 4.13, and 4.14.

Issues Related to ILSs

The costs of ILSs. The primary criticism of ILSs centers on their expense compared to their impact on improving learning. Bentley (1991) warns that ILS benefits are largely theoretical and that most studies on their effectiveness have been done by companies with much to gain from positive reports. "The ILS could prove to be the biggest money pit in a school's budget, sapping large amounts of resources earmarked for other educational needs. ... The multitude of individual costs is staggering and each cost that is paid usually opens up further costs. ... The search for less expensive alternatives to the ILS is only logical considering how difficult it would be to find options that are *more* expensive" (p. 25).

ILS proponents, on the other hand, feel that the kinds of students who experience the most success with ILSs are those whose needs schools typically have most difficulty meeting (Bender, 1991; Bracy, 1992; Shore and Johnson, 1992). They point out the value of any system that can help potential dropouts stay in school or help remedy the deficiencies of learning disabled students. They point to studies and personal testimony from teachers over the years that attest to the motivational qualities of allowing students to work at their own pace and experience success each time they work on the system.

When Becker (1992) reported his summary of some 30 studies of ILS effectiveness, he found widely varied results with various implementation methods and systems. Generally, students tend to do somewhat better with ILSs than with other methods, and results were sometimes substantially superior to non-ILS methods. But Becker found no predictable pattern for successful and unsuccessful

Figure 4.11 Sample ILS Student History Report: Jostens Learning Co.

```
                        Green River School District
                          Jostens Elementary School

  Student History Report                                         Page 1 of 2

  Name:          Kovak, Paula
  Group:         Ms. Bartlett's Bluebirds
  Date Range:    09/01/95 – 09/29/95
  Date:          September 29, 1995      11:49 am

  Assignment:    Second Grade Math

                 Activity                 Raw Score   % Score   Time    Date

  02EM0301   Telling Time: Hours and Half    16/22       73%     0:10   09/06/95
  02EM0302   Telling Time in 5-Minute Units  15 (16)     95%     0:15   09/06/95
  02EM0303   Reviewing Hours and Minutes     100/100    100%     0:14   09/11/95
  MT-ADD21D3 Practice                        15 (16)   Mastered  0:10   09/11/95
  MT-ADD22D3 Sort                            15 (17)   Mastered  0:06   09/13/95
  MT-ADD23C3 Meanings                        100/100    100%     0:08   09/13/95
  MT-ADD24D3 Practice                        15 (16)   Mastered  0:11   09/18/95
  MT-ADD25D3 Pan Balance                     15 (16)   Mastered  0:07   09/18/95
  MT-SUB20C3 Meanings                        95/100      95%     0:05   09/20/95
  02EM0304   Days and Months                 28/32       88%     0:18   09/20/95
  02EM0305   Introducing the Calendar        - - -    Bookmrkd   0:13   09/27/95

  Average                                                92%     0:10
  Total Completed Activities:    10                     Total Time:  1:57

  Assignment:    Reading Placement

                 Activity                 Raw Score   % Score   Time    Date

  RD-PLC001  Placement Test                   3.5       Done     0:36   09/05/95

  Average                                                        0:36
  Total Completed Activities:    1                      Total Time:  0:36
```

Labels indicated on figure: 1, 2, 3, 4–6, 7, 8–9, 10

Assignment Information

1. Date range when the student worked on the assignment

2. Name of the assignment covered in this section of the report

Activity Data

3. List of activities the student completed within the date range, in order by the date of completion

4. The student's raw score for each activity, showing the number correct out of the total possible (for scored activities) or out of the number attempted (for "Mastered" activities)

5. Student's score for the activity shown as a percent or status

6. Time, in hours and minutes, that the student worked in the activity

7. Date the student last worked in the activity

Averages and Totals

8. Average percentage score and time spent to complete an activity

9. Total number of activities completed (including failures) and time spent in the assignment

10. Placement test results

Source: From *Reports Guide* by Jostens Learning Corporation. © 1995 by Jostens Learning Corp. Used by permission.

ILSs. He concluded that data were not sufficient either to support or oppose the purchase of an ILS in a given school or district.

Concerns about the role of ILSs. Another criticism of ILSs reflects a current fear rather than reality. Many educators worry that the cost of ILSs combined with the comprehensive nature of their curricula will cause schools to view them as replacements for teachers. White (1992) asks, "Will ILSs take over all teaching, as some fear?" Maddux and Willis (1993) describe a slightly different version of this problem. They warn that ILSs can have the effect of shaping or driving a school's curriculum rather than responding to it. The best way to address these kinds of concerns may be through a careful, well-planned purchasing process for an ILS. Smith and Sclafani (1989) and

Chrisman (1992) among others, have offered guidelines to potential ILS purchasers.

The list below summarizes the recommendations of Smith and Sclafani (1989):

• Clearly identify the problem the ILS is supposed to solve.

• Understand the instructional theory upon which the system is based.

• Determine whether the ILS is a closed system (one that provides 80 percent or more of the instruction for a given course) or an open system (one linked to the school's resources).

• Find out if the system's scope and sequence are matched to that of the school.

• Determine the target population for which the system was designed and whether or not it closely matches the characteristics of students with whom the ILS will be used.

Figure 4.12   Sample ILS Screen

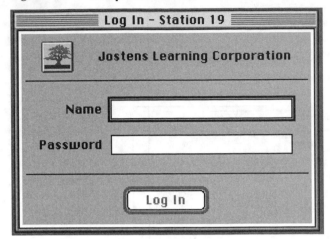

Source: Used by permission of the Jostens Learning Corporation.

- Consider the adequacy of the reporting and management system for the school's needs.
- Consider how much of its resources the school must spend on hardware and software.
- Project the educational benefits to the school from the system and compare them with the costs.

Chrisman (1992) provided another set of recommendations:

- Carefully review each vendor's educational philosophy and compare it to that of the school.
- Request that vendors inform the school on ILS updates.
- Carefully evaluate the grade-level courseware, management system, customization, and online tools and be sure they match the school's expectations.
- Set up reasonable terms of procurement.
- Calculate the personnel and fiscal impact of the ILS.

Figure 4.13   Sample ILS Screen

Source: Used by permission of the Josten Learning Corporation.

Figure 4.14   Sample ILS Screen

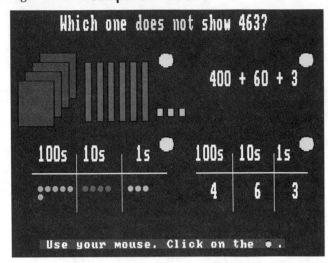

Source: Used by permission of the Jostens Learning Corporation.

- Consider all the characteristics of the system that will affect day-to-day operations (e.g., student time required to log on and off, the fit of the system with the district master plan for technology, etc.)

Factors other than quality determine the fate of an ILS. Bailey and Lumley (1991, p. 22) describe eight issues facing administrators who supervise teachers using ILSs:

1. **Use by special and regular populations.** Although many ILSs are purchased for Chapter I populations, they can be helpful for other students, as well. However, administrators must know how the technology can best serve each population.

2. **ILS integration into school/teacher culture.** Administrators and teachers should work together to decide how the ILS can benefit students in combination with other resources and curricula. Since teachers will implement an ILS, administrators should consider teacher input as vital to successful use.

3. **Research on students with varied abilities.** Administrators must keep current on ILS research with various kinds of students and be sure that teachers have this information. This will help them make the best use of ILS capabilities.

4. **Software development, selection, and adoption processes.** Adoptions of other materials must be coordinated with ILS adoption, but the ILS must not prescribe the total curriculum. Administrators and teachers must accomplish the necessary coordination without letting the ILS guide the entire process.

5. **Role and control of teachers in ILS instruction.** An ILS will require changes in the teacher's role from source of information to facilitator. Administrators must assist in this transition and help teachers become comfortable with their new roles.

6. **Financial considerations.** The expense of ILSs requires principals from individual schools to work together to develop strategic plans for purchase and implementation.

7. **Staff development.** Past research has shown neglect of teacher training in ILS use (Sherry, 1992). More and better teacher training is required to realize the potential of ILS capabilities.

8. **Technology planning.** All technology purchases must be part of the district's short-range and long-range plan for technology use. The purchase of an ILS represents only one step in a total blueprint for technology applications to improve the quality of education.

Finally, the staff of the Texas Center for Educational Technology (1991) have developed a comprehensive booklet that describes how to plan technical details of an optimal implementation of an ILS. The booklet also details how to evaluate the success of an ILS in a school setting.

How to Use ILSs in Teaching

Since an ILS creates a combination of the materials already described here (e.g., drills and tutorials), its potential benefits are quite similar. The highly interactive, self-pacing features of an ILS can help to motivate students who seem to need highly structured environments, and these activities can free up the teacher's time for students who need personal assistance. Teachers can also personalize instruction for each student by reviewing the extensive information on student and class progress provided by the ILS management system.

Successful uses of ILSs have been reported for two different kinds of teaching approaches.

• **For remediation.** While ILSs are expensive alternatives to other kinds of delivery systems, White (1992) observes that "they will probably play an increasing role in the large urban systems that have faced achievement test scores that seem intractable to the usual classroom solutions" (p. 36). However, schools still must determine how ILS functions coordinate and complement those of the classroom teacher. Most ILS uses serve target populations that have typically presented the most difficult problems for traditional classroom activities: Chapter I groups, ESOL students, special education students, and at-risk students (potential dropouts). Schools have usually tried and failed to reach these students with other methods.

• **As a mainstream delivery system.** Rather than using an ILS only as a backup system to address educational problems, a school may let an ILS do the initial job of teaching whole courses for all students in a grade level. In light of the expense of ILSs, these uses are more rare. However, some projects offered as alternatives to public schools (e.g., the Edison Project) predict that the costs of using technology in this way will amount to substantially less over time than teacher salaries. Using ILSs to increase student-to-teacher ratios has stimulated ongoing debate and study.

In either of these uses, teachers still have important roles to play. As Blickman (1992) puts it, "ILSs allow teachers a comfortable transition from the role of deliverer of instruction.... [T]eachers are still actively engaged in the teaching process but as 'guides' or facilitators as opposed to distributors of information" (p. 46). American educators generally assume that ILSs should not be seen as "teacher proof" but rather "teacher enhancing." Teachers must still assign initial levels of work, follow up on students' activities on the system, and give additional personal instruction when needed. (See Directed Example ILS)

Constructivist applications in networked environments. Just as an integrated learning system combines several kinds of courseware to create a skill-based, directed learning environment, a network can also combine several kinds of technology resources to support the goals of constructivist learning approaches. When networks provide technology resources of constructivist design and use, the resulting products are sometimes labeled with terms other than *ILS* to differentiate them from what some educators consider more traditional uses of technology. For example, they may be called *integrated technology systems (ITSs), integrated learning environments, multimedia learning systems,* or *open learning systems* (Hill, 1993, p. 29).

The technology of an ITS resembles that of an ILS, but the products themselves, as well as the ways that schools integrate them into instruction, are very different. Integrated technology systems usually provide wide varieties of unstructured tools on the same networked system. Typically, an ITS will include some kind of information bank(s) (e.g., electronic encyclopedias), symbol pads (e.g., word processing and/or desktop publishing software), construction kits (e.g., Logo or other graphic languages or tools), and phenomenaria (e.g., computer simulations and/or problem-solving resources). They also usually have data-collection systems to track student usage of the system (Mageau, 1990). Thus, this kind of networked product can provide what Perkins (1991) called a "rich environment." (Refer to "The Cat That Walked by Himself" for a constructivist example of an ILS.)

Logo Resources

Logo and Logo Products: Definition and Characteristics

In simple terms, Logo is a procedural programming language based on the artificial intelligence language LISP and designed for use with young children. But Logo has come to be much more in education. In a 1985 article, Muller characterized Logo as a philosophy, a pedagogy, a computer language for children, *and* a general-purpose language. In light of the literature, extensive classroom activity, and controversy related to Logo in education over the past 15 years, Muller's analysis seems entirely appropriate. Watt (1992) explained that Logo originated in the 1960s as a programming language designed by a Cambridge, Massachusetts research team at Bolt, Beranek, and Newman, Inc. It was based on LISP, a language designed for artificial intelligence applications. But Watts noted that the work of Papert and his colleagues at the Massachusetts Institute of Technology

Example Lesson Plan
Directed ILS

Lesson:	One-Step Linear Equations
Developed by:	Curriculum specialists at The Jostens Learning Corporation
Courseware:	Jostens Middle School Mathematics Curriculum: Algebra Expressions
Level:	Secondary
Content Area:	Mathematics: Algebra
Instructional Resources:	Computer, ILS software, chalkboard, overhead projector and erasable overheads
Lesson Purposes:	This is the third algebra lesson in Jostens' standard sequence for Grade 8. Students learn to solve and graph linear equations.

Rationale for Using Courseware. This software is part of a complete teaching sequence for mathematics and can supply the structure for all of the teachers' instruction in this area. It also is integrated with a management system that automatically tracks students' progress, identifies problems they may be having, and shows the teacher where additional explanation and practice are needed. This assists the teacher in providing individual instruction by targeting students who need special help.

Instructional Activities.

1. **Preparation.** Review on the chalkboard various algebra expressions the students have covered previously. Have students take turns solving problems involving order of operations. If necessary, have students review online assignments in MT-ALG06 - *Algebra Expressions* before going on to the next activity.

2. **On-computer.** Use the first activity in MT-ALG10C8 - *Algebra Expressions* to allow students to do intuitive equation problem solving to find the value of a variable that makes the equation true. Provide additional practice in the Learning Center for students having difficulty with this activity.

3. **Reinforcement.** The next four activities in the sequence provide various forms of practice for the skills in the previous activity.

4. **Extension.** MT-ALG15C8 is a tutorial on graphing in which students are asked both to identify the equation when given the graph of a line, and to graph the line of a given equation. Have students use an overhead projector and erasable overheads to illustrate to each other how they solved some of the problems.

Example Lesson Plan
Constructivist ILS

Lesson:	The Cat That Walked by Himself
Developed by:	Curriculum specialists at The Jostens Learning Corporation
Courseware:	Jostens Elementary Reading Expansions
Level:	Elementary
Content Area:	Reading and literature
Instructional Resources:	Computer, ILS software, and related Jostens text materials (e.g., teacher's edition, student textbooks, supplementary handouts). The *Directed Classroom Integration Activities* manual also supplies detailed notes on how to set up the classroom and how to use the software as one component in a reading/literature unit.
Lesson Purposes:	This in one of several literature-based activities, each of which is designed to be the hub of an integrated lesson.

Rationale for Using Courseware. This software helps make the stories more graphic and vivid for beginning readers and draw them into reading activities. It also helps the teacher build connections between students' own language and their reading, and between their reading and skills in other curriculum areas.

Continued

Instructional Activities

1. **Preparation.** Talk about the author Rudyard Kipling and some of the stories he wrote for children. Show the students the story in the book and explain that this on-computer activity introduces this story and helps them read a portion of it. Say that once this activity is complete, they will be doing other in-class activities related to the story.

2. **On-computer.** Let students do the on-computer reading activity.

3. **Reinforcement.** Discuss the ideas mentioned in the on-computer activity such as:

   • How would you describe the setting for this part of the story?

   • What is the meaning of the work "Nenni?"

   • What did the dog want when he visited the cave?

4. **Extension.** Use the web of activities shown here to identify components of a classroom unit centered around the story. Begin to read portions of the book to the class and use activities from the web that correspond to the portion of the book in order to extend the reading activity into other areas of the curriculum.

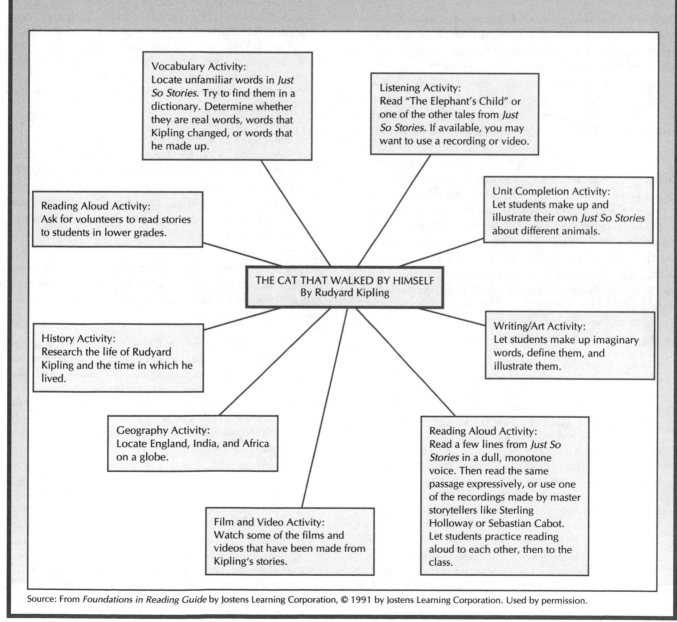

Vocabulary Activity:
Locate unfamiliar words in *Just So Stories*. Try to find them in a dictionary. Determine whether they are real words, words that Kipling changed, or words that he made up.

Listening Activity:
Read "The Elephant's Child" or one of the other tales from *Just So Stories*. If available, you may want to use a recording or video.

Reading Aloud Activity:
Ask for volunteers to read stories to students in lower grades.

Unit Completion Activity:
Let students make up and illustrate their own *Just So Stories* about different animals.

THE CAT THAT WALKED BY HIMSELF
By Rudyard Kipling

History Activity:
Research the life of Rudyard Kipling and the time in which he lived.

Writing/Art Activity:
Let students make up imaginary words, define them, and illustrate them.

Geography Activity:
Locate England, India, and Africa on a globe.

Reading Aloud Activity:
Read a few lines from *Just So Stories* in a dull, monotone voice. Then read the same passage expressively, or use one of the recordings made by master storytellers like Sterling Holloway or Sebastian Cabot. Let students practice reading aloud to each other, then to the class.

Film and Video Activity:
Watch some of the films and videos that have been made from Kipling's stories.

Source: Lesson ideas taken from the Jostens Learning Corporation's *Directed Classroom Integration Activities* manuals, 1990.

made Logo "widely used throughout the world as an introductory programming language and mathematical learning environment for students in elementary and secondary schools" (p. 615). Papert hoped that it would become "a context which is to learning mathematics what living in France is to learning French" (Papert, 1980, p. 6).

But perhaps Logo's single most important role is as a highly visible rallying point for the developing constructivist views of education. It even had its own slogan: "No threshold and no ceiling," meaning that it was simple enough for even small children to learn, but sophisticated enough to allow development of advanced concepts. For these reasons, it is perhaps insufficient to look at Logo only in terms of its descriptors (e.g., structured programming language, turtle graphics). It also seems important to look at it in terms of the hopes and aims of the educators who designed it and those who used it.

Essential characteristics of Logo in education. Papert added an ingredient to the original Logo language that proved to be a crucial link between the real (physical) world and the abstract one represented on the computer screen. This link was essential to allow children to move easily from the concrete operations of earlier stages of cognitive development to more abstract (formal) ones. In their early work with children, some members of the MIT group used a programmable mechanical robot that resembled a turtle. (See Papert, 1980, p. 218.) Children could make the "turtle" move around on the floor and draw pictures by giving it commands through a remote-control device. After a year of having children use the Logo programming language without graphics (around 1968 to 1969), Papert and his group conceived the idea of making the turtle a part of the language. In this way, children could move a turtle around the screen and draw things with Logo commands just as they had moved the turtle robot on the floor with the remote-control device.

There are now several versions of the Logo language. Distinguishing features include:

- **Logo screen devices.** The most easily recognized feature of Logo is the screen turtle, which can be either a small triangle or a figure that actually resembles a turtle. (See pictures above.) Logo programming commands make the turtle move around or appear to draw shapes and lines on the screen. When the turtle draws something, the result is called *turtle graphics.* Some versions of Logo (e.g., TI Logo) also have screen devices called *sprites* which are really several turtle-like objects that can be set in motion on the screen with specific speeds and directions after which they keep moving on their own. (This kind of animation usually requires a special hardware chip or circuit board in addition to the Logo software.)
- **Logo programming elements.** Logo programs include three kinds of programming elements. Primitives are commands to do particular actions (e.g., the primitive FORWARD moves the turtle forward one space). Procedures are series of commands (e.g., several commands together make a procedure to draw a flower). Lists are structures that hold data to be

used in the program (e.g., words in a paragraph). These data structures make Logo useful for writing computer programs to analyze language.

Primitives	A Procedure
FORWARD	TO TRIANGLE
RIGHT	REPEAT 3(FORWARD 50, RIGHT 120)
REPEAT	END

Source: Watt, D. (1992). Logo. In G. Bitter (Ed.). *Macmillan encyclopedia of Computers.* New York: Macmillan.

- **Logo program characteristics.** In technical terms, Logo is an interactive, procedural language. It is interactive because users can immediately execute commands and see what they accomplish. It is procedural because commands are structured into groups called *procedures* to accomplish tasks. For example, a person might write a procedure to draw a flower or to find how many times a word appears in a block of text. That procedure can then appear at any location within the program or in any other program to accomplish the same task. Procedures in languages such as Logo and Pascal work much like subroutines in languages such as BASIC and FORTRAN.

Although not as popular as it was in the 1980s, Logo is still used in education for a variety of purposes. It often introduces very young children to programming concepts. It has also been used successfully in instructional activities designed for students of all ages, both for programming and to explore concepts in content areas such as mathematics and language arts.

Types of Logo resources. The Logo language as it evolved through the work of Papert and his MIT colleagues became the springboard for the development of a variety of classroom technology tools:

- **Logo programming languages.** Harvey (1987) described six different versions of the original Logo programming language, each with different graphics, file handling, and other capabilities to take advantage of the hardware capabilities of Apple, Macintosh, and DOS platforms. The 1993 issue of the Logo newsletter *Logo Update* describes six sources for Logo versions and other products to support them. (See the end of this chapter for address of *Logo Update.*)
- **Logowriter.** This version of Logo combined Logo programming and word processing capabilities, thus allowing a mixture of text and graphics on the screen. This made Logo even more strongly picture-and-text-oriented for use with young children; however, it was discontinued in 1996.
- **LEGO® TC Logo.** This product is actually a kit that allows a child to build LEGO devices and write programs that direct the computer to move them. The LEGO® TC Logo kit includes a Logo software disk; computer hardware such as interface box, transformer, and circuit board; and a set of LEGO elements (e.g., plastic building pieces, motors, and sensors).

- **Microworlds packages.** A series of Logo-based products released in 1993 combine the power of Logo with the ease of use of a HyperCard environment. These products, called *Microworlds,* are used much like authoring systems, and they do not require actual Logo programming skills (Yoder and Moursund, 1993).

- **HyperStudio with HyperLogo.** This is actually an authoring package based on HyperCard software (discussed in more detail in Chapter 8). HyperLogo is a programming language used with the software.

- **Logo-controlled robots.** These robots, dubbed "Roamers" by the company that makes them, can be programmed to move around, draw designs, and even play music (see photo).

Source: Photo courtesy of Harvard Associates Inc.

Issues Related to Logo

Papert's 1980 book *Mindstorms* both fired the imaginations and raised the hopes of teachers throughout the country. In a very short time, Logo became an American success story. It was the subject of school and college courses, a popular special interest group (SIG) topic at conferences, a regular column in nearly every educational technology magazine, and hundreds of speeches and articles. A search of ERIC for articles on Logo between 1980 and 1990 yields about 600 entries. But by 1985, some educators seemed to have become disillusioned. This had more to do with unrealistic expectations and methods of implementation and research than with any deficiency in Logo itself. At least three kinds of issues surrounded Logo use.

- **Expectations versus achievements.** Observers generally acknowledge some dramatic differences between what Papert said Logo could do and what readers understood that it would do. In 1980, educators were already disenchanted with many current educational methods, especially those for teaching mathematics. When Papert's book came out, many educators and noneducators alike read a great deal into his claims. The following table summarizes some of these dramatic differences:

What Papert said:

- Logo provides an environment for children's cognitive growth.

- Logo increases children's delight in learning.

- Logo lets children create microworlds for studying mathematics concepts.

What educators heard:

- Logo will promote better mathematics problem-solving skills.

- Logo makes children become more interested in school (or learning mathematics).

- Logo will make students perform better on tests of mathematics skills.

Note the subtle but important differences between these two set of expectations. When educators' expectations proved unfounded, some began to say "Logo promised more than it has delivered" (Papert, 1987, p. 23).

- **Intended and actual implementation.** Papert (1985) warned that, while there is no "one right way" to use Logo, there are ways of using it "wrong" (Papert, 1985, 1987). Papert stated the intention of helping students to learn in a very unstructured, exploratory way, discovering both Logo rules and mathematical concepts underlying them with minimal teacher instruction in actual programming concepts. He provided several illustrations of "wrong uses" that clearly contradicts his original concept. But other educators, observing that children were not learning to program very well in Logo using Papert's approach, advocated more structured ways to teaching Logo (e.g., directed instruction and practice in Logo commands and applications). While this did, indeed, help children learn to program better in Logo, it was not the use of Logo that Papert intended.

- **Research methods and findings.** Not surprisingly, many educators set about using experimental research methods with Logo applications to substantiate the claims that had been made for them. Also not surprising was Papert's reaction to these efforts to state in behavioral terms the benefits of what are essentially nonbehavioral, Piagetian concepts. As Bracey (1988) noted, "Papert has rejected almost all research about Logo, calling it technocentric thinking" (p. 14). By technocentric, he meant that people seemed to focus on "THE effect of THE computer on cognitive development" (Papert, 1987, p. 23), rather than on determining how computers can help to create a better "culture of learning." Several respected researchers chided Papert for downplaying the importance of experimental research to validate the usefulness of Logo applications. Pea (1987) commented that "Papert gets this new field off to a bad start ... by misattributing technocentric beliefs to certain authors ..." (p. 4). Walker (1987) concurred, saying that "Papert goes too far ... when he concludes that shortcomings of conventional experiments are reason enough to rule them out ... as methods for studying the effects of educational programs such as Logo" (p. 9). Becker (1987) joined the criticism by asserting that "When Papert denigrates the 'treatment' model for evaluating Logo, he is attacking the most falsifiable methodology employed in science to validate theories" (p. 13).

Despite objections by Papert and others, research proceeded apace. Although a thorough summary of the research in this area goes beyond the scope of this book (and probably the needs of the reader), some of those who have done such summaries include Roblyer, Castine, and King, 1988; Krendl

Example Lesson Plans
Three Integration Strategies for Teaching Content-Area Skills with Logowriter

Lesson: Multiplication, Anyone?
Developed by: Judi Harris (Charlottesville, Virginia)
Level: Elementary
Content Area: Mathematics: Multiplication
Lesson Purposes: Clarify the concepts and language of multiplication

Rationale for Using Logo. Logowriter makes multiplication concepts graphic and children can use their procedures as interactive, functional multiplication translators.

Instructional Activities. Students explore multiplication representation in very simple to very complex ways. One simple example is having them choose a turtle shape at random and setting the turtle to that shape. Then position the turtle at the top left of a blank screen and draw rows of shapes according to numbers in the draw command. For example:

DRAW 7 × 4

creates

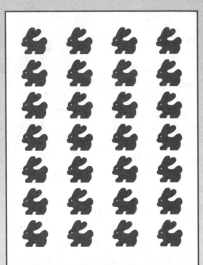

Then students can create and use routines to: (1) take the Draw command numbers and the desired shape from input lists, and (2) tally the number of shapes placed on the screen so that they can check their own computations against computer-calculated products. Writing these programs pose a challenge to older students who are more experienced with Logowriter. These students then let younger students use the programs they have written so that both can explore multiplication at different levels.

Lesson: In a Word, It's Logo!
Developed by: Judi Harris (Charlottesville, Virginia)
Level: Upper elementary, middle, and high school
Content Area: Language arts: Word structures
Lesson Purposes: Increase vocabulary through the analysis of word parts and synthesis of word component meanings

Rationale for Using Logo. Logowriter offers a uniquely creative and challenging way to focus students' attention on construction of multisyllabic words and how to determine their meanings by breaking them into their component parts.

Instructional Activities. Begin by helping students identify common prefixes, suffixes, and root words and research their meanings. Then insert these meanings into a Logo procedure that generates and outputs a random combination of three elements (one from each of the three lists: prefixes, suffixes, and roots) to form a "sniglet" (a word that doesn't exist, but should.) Students store the result-

Continued

ing words and generate and store definitions for each one. For example, if the computer generates *reportless*, the user may analyze the word for its components and store the following definition: "The state of being unprepared on the day a big report is due." Sometimes, the computer generates "sneaky sniglets" or real words. Students are encouraged to look up in the dictionary any "new" word they suspect is a sneaky sniglet.

Lesson:	Pixelated Phlowers
Developed by:	Judi Harris (Charlottesville, Virginia) and Nancy Lee Bergey (Bala Cynwyd, Pennsylvania)
Level:	Middle school
Content Area:	Science: Genetics
Lesson Purposes:	Explore basic tenets of biological inheritance or Mendelian genetics

Rationale for Using Logo. Logo procedures provide an excellent way for students to discover interactively the patterns of genetic transfer.

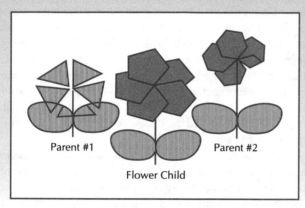

Parent #1   Parent #2

Flower Child

Instructional Activities. Use a Logo microworld called FLOWER. BREEDING (available from the authors) that specifies flower attributes for each parent and offspring according to either simple dominance or multiple levels of dominance. The following four kinds of attributes are specified: stem length, leaf size, petal color, and petal shape. Resulting flower genotypes (genetic templates) are then drawn on the screen. (The Logo procedures that draw the pictures have decision-making capabilities that determine attribute dominance.)

Use this microworld in two ways. First, have students experiment with flower phenotypes by pairing flowers and recording their offspring. With older children, teachers can use the CHALLENGE.ME! tool. The computer generates an offspring and challenges students to deduce genotype information from appearance and breeding attribute patterns.

Sources: Harris, J. (1991). Parlez-vous multiplication? *The Computing Teacher, 19* (3) 37–39, Figure used by permission of ISTE; Harris, J. (1990a). Word study, Logostyle. *The Computing Teacher, 17* (5), 29–31; Harris, J. and Bergey, N. (1990b). Pixelated phlowers: Logo-fueled phenotype and genotype investigations. *The Computing Teacher, 17* (8), 15–17, 53, Figure used by permission of ISTE.

and Lieberman, 1988; Collis, 1988–1989; Govier, 1988; Redekopp, 1989; Keller, 1990; Singh, 1992; Clements, 1993; and Walsh, 1994.

How to Use Logo in Teaching

Guidelines for teaching with Logo. In keeping with Papert's admonition to avoid technocentric thinking, it seems best to focus on applications of Logo that maximize results rather than trying to describe the unique characteristics of the language. (For detailed descriptions of Logo programming features and techniques, see Simonson and Thompson, 1994.) Keller (1990) summarized researchers' attempts to identify the characteristics of effective instruction with Logo. Some specific guidelines for employing Logo effectively arose from this summary:

• **Rules for when and how to intervene.** "Timing and degree of teacher intervention [are] critical to the effectiveness of Logo learning" (Keller, 1990, p. 57). Depending on students' needs, the role of the teacher can range from providing light intervention (e.g., acknowledging and discussing work) to strong intervention (e.g., modeling a solution or changing a student's program). Keller says that teachers typically give fewer directions and students ask more questions in Logo instruction than in other situations.

• **Techniques for (and caveats on) providing structure.** While Papert described Logo as a way to "learn without curriculum," those who have implemented Logo in classrooms agree that teachers must provide some structure. This structure may take the form of study guides and other written prompts that direct students how to do certain Logo techniques, organized group discussions and sharing sessions to provide a forum for students to reflect on what they have learned, and required work schedules. There is a clear danger that instruction will become too focused on learning to program in Logo and lose sight of the reason Logo was developed: to get students interested in Papert's powerful ideas. Teachers have a difficult job providing enough structure to keep students focused, but not so much structure that it stifles exploration and achievement.

• **Techniques for mediating (helping).** Keller finds that two kinds of "mediating" guidelines (as opposed to directed instruction-type guidelines) seem to help assure that Logo learning will transfer to problem-solving abilities. First, teachers must place more instructional emphasis on *processes* for solving problems than on the *contents* of the problems. Successful mediating techniques reported by Keller include worksheets that stress planning and reflection, teacher questioning techniques to encourage students to think about their own thinking, and teachers modeling their own reasoning processes by talking about them as they solve Logo problems. Second, teachers must take specific steps to show students how Logo skills and activities relate to other kinds of problem solving; they cannot assume that skills will transfer automatically. Keller calls this "bridging between contexts;" others refer to it as a kind of scaffolding.

- **Using peers.** Having students work in pairs or small groups on Logo problems and projects may promote their cognitive development. As students interact with each other, they tend to rely more on each other and less on the teacher to solve problems in their work.

- **Assessing progress.** Keller points out that traditional assessment methods such as tests and grades are often ineffective and counterproductive when measuring gains from Logo experiences. Alternative evaluation methods must be devised, but clear guidelines are lacking. As a tentative suggestion, teachers might make checklists of behaviors specific that they expect and then observe these behaviors before and after teaching.

Getting started with Logo. Teachers can get started with Logo in two different ways. One is to learn the Logo programming language; the other is to use a Microworlds product, which makes possible many Logo-like capabilities but requires no actual programming. Teachers who learn Logo or Logowriter gain flexibility and power from their programming skills, while microworlds offer ease of use but limited capabilities. The beginner with no programming skills may wish to try a microworlds product first, perhaps also trying the Logo-like scripting language associated with it. Those who find that they need more flexibility and power may then move on to Logo programming. In either case, educators must first purchase appropriate software: either Logo program software, or a microworlds product. (See the *Logo Update* newsletter for current information on these products.)

Examples of Logo integration. The literature on Logo use in classrooms provides many good examples of effective applications. These illustrations suggest some types of instructional needs that Logo can help meet as well as how it should be used to achieve best results. One good source of Logo integration strategies is a series of books published by the International Society for Technology in Education (ISTE) (See ISTE address in Appendix B). Example implementation plans illustrate some of these instructional needs:

- **Fostering problem solving and cooperative work.** When Logo was first introduced in Papert's 1980 book, teachers commonly used Logo programming to promote hoped-for transfers to "thinking" and problem-solving skills. These examples are not as common now; Logo applications usually seem to emphasize learning specific content area concepts and skills. However, educators also appear to be reexamining Logo's usefulness in this area. Logo may achieve better results in teaching "pure" problem-solving skills now that clearer guidelines for this use are available. (See guidelines in this chapter, page 115; and example activities for LEGO® TC Logo, pages 116–117.) Like most of the other resources used in constructivist models, this use also offers a potentially motivating environment for cooperative projects and collaborative work.

- **Illustrating content area concepts.** Logo has been used successfully in content areas from mathematics to language arts, from science to art and music. Its primary strength seems to lie in making abstract concepts more real and concrete to young learners by engaging them in highly visual, hands-on activities with these concepts.

- **Encouraging creative work.** Finally, Logo products offer attractive, rich "playgrounds" for letting students exercise their imaginations and creative skills. Its graphic and sound features allow them to experiment with self-expression in art and music in relatively nonthreatening environments.

Evaluating and Selecting Instructional Software

In the 1980s, microcomputer courseware began to flood the educational market from such diverse sources as state projects (Minnesota Educational Computing Consortium or MECC), major publishing houses, and even cottage industries. This torrent made educators increasingly aware that simply putting instructional routines on the computer did not assure that they would take advantage of its potential power as an instructional tool. Indeed, some of the products were so bad that they could be worse than no instruction at all.

Courseware quality became a major issue in education, and courseware evaluation became a popular and highly publicized practice. Many professional magazines created sections to report the results of product evaluations; indeed, whole magazines were developed to publish such evaluations *(Courseware Review)*. Organizations sprang up for the sole purpose of reviewing and recommending good instructional courseware (e.g., the Northwest Regional Lab's Microsoft Project and the Educational Products Information Exchange or EPIE).

As the field of educational technology matured and educators refined their attitudes toward computer use, the mystique of courseware faded and it assumed more of the mundane aspects of purchasing any good instructional material. During the 1980s, teachers primarily evaluated and selected their own courseware. Now, state-level and school district-level personnel increasingly control these purchases. Thus, the evaluation procedures and criteria have changed considerably from the early days of microcomputers. Regardless of who picks the products, teachers should recognize that just because courseware addresses the topics or skills they want to teach, that does not mean that it will meet their needs.

The Need for Evaluation

Courseware quality is less troublesome now than it was in the early days of microcomputers when technical soundness frequently caused problems. For example, courseware programming would not anticipate all possible answers a student might give, or it might not account for all possible paths through a sequence of instruction. Consequently, programs would frequently "break" or stop when these

Three Integration Strategies for
Problem-Solving Skills with LEGO® TC Logo

Strategy #1:	Knowledge on Wheels!
Developed by:	Marian Rosen, Ladue School District (St. Louis, Missouri)
Level:	Grades 1–8
Purposes of the Activity:	Allow children to create Lego vehicles and write Logo programs to make their vehicles complete various actions

Rationale for Using Logo. Logo provides hands-on experiences with step-by-step methods to solve problems related to building desired products. It provides concrete illustrations of many common physics principles.

Instructional Activities. Students at each grade level did some version of the basic activity. In the first grade, children were challenged to build Lego cars that would roll down a ramp and travel as far as possible. After children built their cars, they tested them to see which were most successful. Then the teacher led a discussion on how the "best cars" were similar to and different from the others (e.g., why those with larger wheels rolled farther). The discussion led students to compare cars that were different in one way only (changing only one variable at a time). Children were then allowed to hook up their cars to the computer and program them with simple Logowriter commands such as ON, OFF, and RD (reverse direction).

Older children used the same basic project, but they used increasingly complex Logo commands to make their cars do more involved actions. Students often moved on to other building projects such as elevators, toasters, and ski lifts. As they encountered unanticipated difficulties with their machines, they had to develop strategies for isolating and resolving problems. Students also learned collaborative values and techniques: "Some students are programmers, some are engineers, some are dreamers, some are artists. …The collaborative effort when they work together is wonderful to look at" (p. 57).

Problems to Anticipate. Time is never sufficient. Also, the amount of equipment available may be limited.

Strategy #2:	LEGO® TC Logo: It's Not Just for Boys!
Developed by:	Helen Faulkner and Kathleen Anderson, Montclair Kimberly Academy (Montclair, New Jersey)
Level:	Grades 4–6
Purposes of the Activity:	Give opportunities for groups of children to design and build their own Lego machines in order to learn the following skills: understanding and following directions, analyzing component parts, and integrating parts effectively. The lesson also seeks to motivate both girls and boys equally to do such projects.

Rationale for Using Logo. These cooperative projects with LEGO® TC Logo yield unique opportunities for students to get immediate, concrete feedback when they practice following directions along with the high-level skills involved in defining and carrying out a large project. LEGO® TC Logo also seems to have unique capabilities to involve both girls and boys in hands-on, technical activities.

Instructional Activities. Two important decisions were made at the outset about this after-school activity. First, sessions would be unstructured; second, girls would work separately from boys. The latter decision was made because girls and boys employed dramatically different work strategies and preferences, and girls tended to drop out of the groups when they were mixed. The materials were made available to both and they were allowed to decide what they wanted to do. Students were given their choice of creating their own product designs or following project booklets in the LEGO® TC Logo sets. Girls usually chose to begin with the booklets, while boys wanted to make their own products (usually cars). Two students from upper grades served as project "consultants" or helpers.

After the groups completed their first activities, they were challenged to move on to more complex products. After they gained confidence with the initial activities, student creativity began to emerge. Two girls built a New Jersey tollbooth, complete with coin operation. Another girl built a space station model, and a third built a home with lights, a ringing telephone, a microwave oven with timer,

Continued

and a radio. Boys' projects included a prison compound with gun towers, lights, and various vehicles; another boy built an amusement park. Having boys work separately from girls proved a key factor in allowing both groups to realize the full extent of their creativity.

Problems to Anticipate. Social organization of the classroom may have to change for this activity; teachers need specific training.

Strategy #3:	Inventor's Workshop
Developed by:	Gina Shimabukuro, All Saints Elementary School (Hayward, California)
Level:	Grade 8 (Gifted)
Purposes of the Activity:	IN a two-semester course, let students practice the problem-solving skills involved in inventing a product, building it, animating it through the computer, and testing the effectiveness of their designs.

Rationale for Using Logo. This is a unique medium for solving problems in a hands-on, realistic way while building skills in basic mechanical concepts, programming principles, and even physics.

Instructional Activities. We began by introducing students to simple, noncomputer-based activities with the Lego blocks. This allowed students to become comfortable with the medium. A large-scale project was suggested: building an amusement park. Teachers modeled part of the building and programming process. Then the entire class brainstormed all of the components in the park. They sketched a blueprint on butcher block paper. As a group, they decided which ride each student would be responsible for building. After each one was completed, it was tested by connecting it to battery boxes.

Students then programmed their rides to do various activities. They also explored business applications of their rides: cost, marketing strategies, and profits. After all projects were functional, they were placed on a large piece of plywood and aesthetic touches were added with various materials. The project was exhibited at the school's Spring open house.

Problems to Anticipate. Students never have enough time with materials. Colleagues sometimes criticize the activity as having questionable educational value. Rules must be created to ensure mutual respect for ideas. Girls sometimes need to build more self-confidence than boys.

Sources: Rosen, M. (1988). Lego meets Logo. *Classroom Computer Learning, 8* (7) 50–58; Faulkner, H., and Anderson, K. (1991). LEGO® TC Logo: Gender differences in a process-learning environment. *The Computing Teacher, 18,* (March), 34–36; Shimabukuro, G. A class act: Junior high students, LEGO, and Logo. *The Computing Teacher, 16* (5) 37–39.

unusual situations occurred. The early courseware also seemed to emphasize entertainment much more strongly (e.g., providing elaborate animations for feedback), giving less attention to educational value.

Courseware producers have obviously learned much from their early errors and problems, and overall courseware quality has improved considerably. But educators still have good reasons for spending some time reviewing and/or evaluating courseware before selecting it for use in their classrooms. Despite some general improvement in courseware quality, store shelves still stock some of the very worst instructional uses of microcomputers. Some low-quality products developed in the early days are still on the market; apparently they are still being purchased by people who don't know the difference. Even some recently developed products have surprisingly gross design or content flaws. Teachers must be aware that computerized instruction is not necessarily effective instruction, and eye-catching screen displays should not be the primary criteria for selecting materials.

Teachers should review courseware even after prescreening by committees or experts. Very often, state-level

or district-level committees are responsible only for selecting courseware that does not have gross problems and reaches the desired general level (e.g., middle school) in a general content or topic area (e.g., algebra). Each teacher must then determine which specific curriculum needs and specific grade levels the package addresses and whether or not courseware functions fit with planned teaching strategies. It cannot be emphasized enough that courseware must match clearly identified instructional needs. It should *not* be used simply because it is available at a discount or supplied free by the state or district.

Courseware Evaluation Procedures: A Recommended Sequence

Evaluation procedures and criteria vary dramatically depending on whether a teacher is selecting courseware for a single classroom or is part of a district-level committee screening materials for use by many schools. For one major difference, committees generally have to justify decisions to purchase one package over another by using weighted criteria checklists and assigning total point scores to indi-

Example Lesson Plan
Using Logo to Encourage Creative Work

Lesson: Logo Rose Windows
Developed by: Dwight Dulsky (Perkasie, Pennsylvania)
Content Area: Art
Level: Middle school
Lesson Purposes: Use Logo procedures to generate patterns for stained glass windows

Rationale for Using Logo. Logo's graphic qualities make it a natural choice to explore the design qualities of symmetry, repetition, and precision. Logo variables allow students to change their designs quickly and dramatically. With other methods, modifications would be more time-consuming and students would not be able to pinpoint the exact reasons and ways that designs were made different.

Instructional Activities. Begin by reviewing basic Logo commands and exploring the Logo programming environment. (Use simple exercises such as driving the turtle under a piece of tape placed on the computer screen.) Practice moving between the Editor and drawing screens and developing procedures. Also talk about line-by-line debugging procedures to determine how designs are drawn. Introduce and practice writing procedures to make regular, closed polygons. Divide the class into groups and assign each group a polygon to create. Analyze students' procedures as a class and allow discovery of Logo design principles and techniques.

Give students pictures of the Gothic rose window from the Notre Dame Cathedral. Ask them to analyze it, looking for patterns, shapes, and structures. Then assign them the task of drawing their own window. Initiate discussions of methods for accomplishing desired designs. Show them how to adjust basic designs by changing variable numbers. (The best moments in the project occur when a student displays a "window" for the first time and a gasp of delight fills the room.)

Source: Dulsky, D. (1993). Logo rose windows. *The Computing Teacher, 20* (7), 16–19.

vidual packages. Small groups or individual teachers can use much less formal procedures and criteria.

This section is designed primarily for individual teachers or small organizations like individual schools that (a) do not have large organizations purchasing courseware for them, (b) wish to supplement resources purchased for them by others, or (c) want to review preselected courseware to determine its usefulness for their immediate needs. These procedures are intended to help teachers anticipate and deal with problems related to courseware quality and to assist them in matching courseware to their classroom needs. The following sequence is recommended when selecting courseware for classroom use:

1. **Begin with an identified need.** Know what topics and skills you want to address and approximately how you think you will use technology. This will require some knowledge of what kinds of instructional support technology has to offer.

2. **Locate titles.** As mentioned earlier in this section, teachers should probably not base their courseware purchasing decisions on descriptive reviews. Recommendations from colleagues and professional magazines and journals should serve primarily as leads. Once teachers find out about a package they find interesting, they will probably want to use one or both of the next two general procedures to determine its usefulness for them.

3. **Complete hands-on reviews.** There is no substitute for running the courseware. Teachers should also avoid reviewing so-called *demo packages,* abbreviated versions of actual courseware. They are inadequate, frequently misleading substitutes for the real thing. A typical hands-on review consists of two or three passes through a program. A teacher usually goes through it the first time just to assess its capabilities and what it covers. During the second pass, the teacher tries to make incorrect responses and press keys that aren't supposed to be pressed in order to determine the program's ability to handle typical student use. Depending on the program's capabilities, the teacher may choose to go through the program again to review the usefulness and/or quality of particular demonstrations or presentations.

4. **Collect student reviews.** Experienced teachers can usually tell from their own hands-on reviews when instructional materials are appropriate for their students. But even they are sometimes surprised at students' reactions to courseware. Students sometimes encounter unexpected problems, or they may not seem to get out of the activity what the teacher expected they would. If at all possible, it is very beneficial to field test courseware by observing students using it, getting their reactions, and, if possible, collecting data on their achievement. Gill, Dick, Reiser, and Zahner (1992) also describe a detailed method for evaluating software that involves collecting data on student use.

Inset 4.1
Recommended Courseware Evaluation Criteria

Many sets of courseware criteria and checklists are available; they vary widely depending on the educational philosophy of the evaluator and the type of courseware being reviewed (Roblyer, 1983; Shaefermeyer, 1990). Courseware criteria may be divided into two types: those that should be considered essential and those that are sometimes applicable and sometimes not, depending on the user's needs.

Minimum criteria. Certain criteria may be used to discriminate between acceptable and unacceptable courseware materials. Criteria specific to each of the types of courseware has already been discussed in previous sections. The following additional criteria that apply to all courseware, regardless of type, are shown on the minimum criteria checklist in Table 4.1.

I. Required Instructional Design and Pedagogy: Does it teach?

• **Appropriate teaching strategy, based on best-known methods**—This covers a wide range of possible problems, from little or no interactivity to insufficient examples for concept development. One program had no graphics at all, even though it was a mathematics package intended for very young children. Learners at early stages of development are known to need concrete examples rather than text only.

• **Presentation on screen contains nothing that misleads or confuses students**—One particularly blatant error of this type was in a courseware package intended to teach young children about how the human body works. It depicted the human heart as a square box. Another, a math program, displayed a number of objects based on what the student answered, but never bothered to change the number of objects if it was a wrong answer. Thus, the student could be seeing the corrected numeral but the wrong number of objects.

• **Comments to students that are not abusive or insulting**—Programs must be sensitive to student's feelings, even if comments are intended humorously. One program based on a well-known cartoon cat with an acerbic personality belittled the student's name, saying "What kind of name is that for a worthy opponent?" It also commented on the student's "lack of mental ability" when a wrong answer was supplied. Although this was in keeping with the cat's persona, it was still inappropriate.

• **Readability at an appropriate level for students who will use it**—Although this may apply to any use of language in any program, it is particularly applicable to tutorials, which may require many explanations. For example, one tutorial for second-grade math skills had explanations at a fourth-grade reading level. This would probably not be an appropriate expectation for students who have trouble with this level of math.

• **Graphics fulfill important purpose and are not distracting to learners**—Pictures and animation are considered motivational to students, but this is not always true. For example, animated feedback may be charming the first ten times the students see it, but may achieve just the opposite effect after that. Also, some courseware attracts students' attention by flashing text or objects on the screen. This can be distracting when one is trying to focus on other screen text. Early courseware used a device called "scrolling" which had text moving up the screen as the student tried to read it, but this was quickly identified as a distracting mechanism and is rarely seen now.

II. Required for Content: Is it correct?

• **No grammar, spelling, or punctuation errors on the screen**—Even though a program may be on a nonlanguage topic, it should reflect accurate language since students learn more than just the intended skills from instructional materials. One early release on punctuation skills misspelled the word "punctuation" three different ways in the program!

• **All content accurate and up-to-date**—Many people do not associate errors such as these with courseware material; they seem to trust content presented on a computer, as if the computer would correct the text itself if it spotted a problem! Content inaccuracies have been observed in a number of packages. For example, one program referred to blood as a "red substance," which, of course, is not always true. Instructional materials in social studies should be carefully screened for inaccurate reflections of country names, which are changing rapidly.

• **No racial or gender stereotypes**—Look for diversity in names and examples used. Are they all for "Dick and Jane" and are they always in the suburbs? Also review examples for gender stereotypes. Are all doctors men? Are all homemakers women?

• **Social characteristics**—Does courseware exhibit a sensitive treatment of moral and/or social issues? For example, do games and simulations avoid unnecessary violence?

III. Required for User Flexibility: Is it "user friendly"?

• **User has some control of movement within the program**—Depending on the purpose of the program, the students should normally be able to go from screen to screen and read each screen at a desired rate. They should also have exit options available at any time.

• **Can turn off sound, if desired**—Since courseware may be used in classrooms, the teacher should have the ability to make the courseware quiet so it will not disturb others.

IV. Required Technical Soundness: Does it work correctly?

• **Program loads consistently, without error**—A common problem in early courseware, problems of this kind are not seen very often now.

• **Program does not break, no matter what the student enters**—Again, this was a more common problem in early

Continued

courseware. Programs should be designed to expect any possible answer, not just the correct or most obvious ones. When unexpected answers are entered, they should give an appropriate response to get the student back on track.

• **Program does what the screen says it should do**—If the screen indicates the student should be able to exit or go to another part of the program, this capability should be allowed as stated.

Optional criteria. Teachers reviewing courseware may consider a great many other criteria depending on their needs, the program's purpose, and the intended audience. These are detailed in Roblyer (1983), Lockard et al., 1990) and Merrill et al., (1992). Many of these criteria, which are listed below, are subjective in nature; it is up to the teacher to decide whether or not the courseware meets them.

Optional Instructional Design Criteria

• **Stated objectives**—Does the courseware state its objectives?

• **Prerequisite skills**—Are skills specified that students will need to do the courseware activities?

• **Presentation logic**—Do instructional units follow a logical sequence based on skill hierarchies?

• **Tests**—Do tests match stated skills and are they good measures of the skill?

• **Significance**—Are stated skills "educationally significant" (e.g., in the curriculum)?

• **Use of medium**—Does courseware make good use of the medium?

• **Field testing**—Is there evidence the courseware has been field-tested with students and revised based on this feedback before its release?

Optional Student Use Criteria

• **Student ease of use**—Is the program easy to use for the intended students? Does it require physical dexterity to answer items the students may not have even though they know the correct answers? Is a lot of typing required?

• **Required keys**—Are the keys required to input answers easy to remember (e.g., pressing "B" for going "back")?

• **Input devices**—Are alternate input devices allowed to make courseware more usable for special populations?

• **Directions**—Are there on-screen directions on how to use it?

• **Support materials**—Are there print support materials to support on-screen activities?

• **Optional assistance**—Is a "HELP" feature available if the student runs into difficulty?

• **Optional directions**—Can students skip directions, if they desire, and go straight to the activities?

• **Creativity**—Do materials foster creativity rather than just rote learning?

• **Summary feedback**—Are students given an on-screen summary of performance when they finish working?

Optional Teacher Use Criteria

• **Teacher ease of use**—Can teachers figure out, with minimum effort, how to work the program?

• **Management**—Does courseware contain adequate recordkeeping and management capabilities?

• **Teacher manuals**—Are clear, nontechnical teachers manuals available with the courseware?

• **Ease of integration**—Are courseware materials designed to integrate easily into other activities the teacher is doing?

• **Teacher assistance**—Does courseware improve the teacher's ability to teach the subject?

• **Adaptability**—Can teachers adapt the courseware for their needs by changing content (e.g., spelling words) or format (e.g., animated versus written feedback)?

Optional Presentation Criteria

• **Graphics features**—Are graphics, animation, and color used for instructional purposes rather than flashiness?

• **Screen layout**—Are screens so "busy" or cluttered that they interfere with reading?

• **Speech capabilities**—Is speech of adequate quality so students can understand it easily?

• **Required peripherals**—Does the program require peripherals the schools are likely to have (e. g., light pens, speech synthesizers)?

Optional Technical Criteria

• **Response judging**—Does the response judging allow for ALL possible correct answers and disallow ALL possible incorrect ones?

• **Timing**—Does the program present itself quickly so displays and responses are accomplished without noticeable delays?

• **Portability**—Can teachers transfer the courseware from one machine to another?

• **Compatibility**—Does courseware run on more than one platform?

• **Technical manuals**—Do teacher or user manuals contain technical documentation on program operation and any technical features or options?

Selecting Software for Constructivist versus Directed Uses

Nearly all references to courseware evaluation and methods in the literature seem to emphasize products that will be used with directed instruction. While many criteria are, indeed, appropriate for software designed for both uses, no one seems to give any additional detail on what to look for in software that will be used with constructivist methods. Constructivist activities seem to emphasize multimedia and telecommunications products rather than software

---

Table 4.1   Minimum Criteria Checklist for Evaluating Instructional Courseware

Title _____   Publisher _____

Content Area _____   Hardware Required _____

Courseware functions:

_____ Drill and practice   _____ Instructional game

_____ Tutorial   _____ Problem solving

_____ Simulation   _____ Other: _____

Many characteristics should be considered when selecting courseware for use in one's classroom or lab, but the following should be considered *essential qualities* for any instructional product on the computer. If courseware does not meet these criteria, it should not be considered for purchase. For each item, indicate *Y* for yes if it meets the criterion, or *N* for no if it does not.

I.   Instructional Design and Pedagogical Soundness

_____ Teaching strategy appropriate for student level and is based on best-known methods

_____ Presentation on screen contains nothing than misleads or confuses students

_____ Readability and difficulty at an appropriate level for students who will use it

_____ Comments to students not abusive or insulting

_____ Graphics fulfill important purpose (motivation, information) and not distracting to learners

Criteria specific to drill and practice functions:

_____ High degree of control over presentation rate (unless the method is timed review)

_____ Appropriate feedback for correct answers (none, if timed; not elaborate or time-consuming)

_____ Feedback is more reinforcing for correct than for incorrect responses.

Criteria specific to tutorials:

_____ High degree of interactivity (not just reading information)

_____ High degree of user control (forward and backward movement, branching upon request)

_____ Comprehensive teaching sequence so instruction is self-contained and standalone

_____ Adequate answer-judging capabilities for student-constructed answers to questions

Criteria specific to simulations:

_____ Appropriate degree of fidelity (accurate depiction of system being modeled)

_____ Good documentation available on how program works

Criteria specific to instructional games:

_____ Low quotient of violence or combat-type activities

_____ Amount of physical dexterity required appropriate to students who will use it

II. Content

_____ No grammar, spelling, or punctuation errors on the screen

_____ All content accurate and up-to-date

_____ No racial or gender stereotypes

_____ Exhibits a sensitive treatment of moral and/or social issues (e. g., perspectives on war or capital punishment)

III. User Flexibility

_____ User normally has some control of movement within the program (e. g., can go from screen to screen at desired rate; can read text at desired rate; can exit program when desired)

_____ Can turn off sound, if desired

IV. Technical Soundness

_____ Program loads consistently, without error

_____ Program does not break, no matter what the student enters

_____ Program does what the screen says it should do

Decision:

_____ Is recommended for purchase and use

_____ Is not recommended

Litchfield (1992) does address evaluation of what she calls "inquiry-based science software and interactive multimedia programs." These and other criteria and methods for multimedia products will be discussed further in Chapters 7 and 8.

Exercises

1. Identify by commonly used names each of the instructional courseware functions described below:

   _____ a. Lets students pretend they are members of the First Constitutional Congress from each of the founding states. Lets them try out various strategies for solving various problems in developing the U. S. Constitution and shows them the results of their attempts.

   _____ b. Challenges students to get through mazes in which they must remember a correct sequence of steps each time they want to progress to the next section and eventually get to the exit.

   _____ c. Lets students match up word parts (e.g., roots, prefixes, and suffixes) to form whole words. Each time they make a word, they get to see another part of the picture puzzle.

   _____ d. Provides a complete instructional sequence on fundamentals of trigonometry; it has practice items and a self-test at the end. (Software functions: step-by-step instruction and explanation, teaching as a human teacher would.)

   _____ e. Gives access to lessons on a network and software resources, includes a management system to track student progress.

   _____ f. Presents one sentence at a time, and the student puts in the punctuation that the sentence needs. If the punctuation is wrong, it explains why. (Software functions: item-by-item practice).

2. For each of the following descriptions of classroom needs for instructional materials, identify one or more types of instructional courseware functions that could help meet the need.

   _____ a. You want to let students role play as stock brokers in a stock market to make the concepts more real to them, but there is no time and few resources to support it.

   _____ b. Several students in your vocational education class missed your presentation on electrical circuits. You do not have time to deliver the same presentation again for this small group, but they need to know these concepts before they can go on to other work.

   _____ c. Students in your Brain Brawl Club have trouble remembering the states and capitals. They need this information for the upcoming meet.

3. Using the Minimum Criteria Checklist in Table 4.1, evaluate a courseware package in your area/topic of interest or expertise.

References

Introduction

Norris, W. (1977). Via technology to a new era in education. *Phi Delta Kappan, 58* (6), 451–459.

Taylor, R. (1980). *The computer in the school: Tutor, tool, and tutee.* New York, NY: Teachers College Press.

Weizenbaum, J. (1976). *Computer Power and Human Reason.* San Francisco: W.H. Freeman & Co.

Drills

Alessi, S. and Trollip, S. (1991). *Computer-based instruction: Methods and development.* Englewood Cliffs, NJ: Prentice-Hall.

Bloom, B. (1986). Automaticity. *Educational Leadership, 43* (5), 70–77.

Gagne, R. (1982). Developments in learning psychology: Implications for instructional design. *Educational Technology, 22* (6), 11–15.

Gagne, R. and Merrill, M. D. (1990). Integrative goals for instructional design. *Educational Technology Research and Development, 38* (1), 23–30.

Hasselbring, T. (1988). Developing math automaticity in learning handicapped children. *Focus on Exceptional Children, 20* (6), 1–7.

Higgins, K. and Boone, R. (1993). Technology as a tutor, tools, and agent for reading. *Journal of Special Education Technology, 12* (1), 28–37.

Merrill, D. and Salisbury, D. (1984). Research on drill and practice strategies. *Journal of Computer-Based Instruction, 11* (1), 19–21.

Okolo, C. (1992). The effect of computer-assisted instruction format and initial attitude on the arithmetic facts proficiency and continuing motivation of students with learning disabilities. *Exceptionality: A Research Journal, 3* (4), 195–211.

Robertson, Stephens, and Company (1992). *Educational technology: A catalyst for change.* San Francisco: Author.

Salisbury, D. (1988). Effective drill and practice strategies. In D. Jonassen (Ed.). *Instructional designs for microcomputer courseware.* Hillsdale, NJ: Lawrence Erlbaum Associates.

Salisbury, D. (1990). Cognitive psychology and its implications for designing drill and practice programs for computers. *Journal of Computer-Based Instruction, 17* (1), 23–30.

Tutorials

Baek, Y. and Layne, B. (1988). Color, graphics, and animation in a computer-assisted learning tutorial lesson. *Journal of Computer-Based Instruction, 15* (4), 31–35.

CAI in music. (1994). *Teaching Music, 1* (6), 34–35.

Eiser, L. (1988). What makes a good tutorial? *Classroom Computer Learning, 8* (4), 44–47.

Gagne, R., Wager, W., and Rojas, A. (1981). Planning and authoring computer-assisted instruction lessons. *Educational Technology, 21* (9), 17–26.

Graham, R. (1994). A computer tutorial for psychology of learning courses. *Teaching of Psychology, 21* (2), 116–166.

Kraemer, K. (1990). SEEN: Tutorials for critical reading. *Writing Notebook, 7* (3), 31–32.

Murray, T., et al. (1988). An analogy-based computer tutorial for remediating physics misconceptions. (ERIC Document Reproduction No. ED 299 172)

Simulations

Allen, D. (1993). Exploring the earth through software. Teaching with technology. *Teaching PreK–8, 24* (2), 22–26.

Andaloro, G. (1991). Modeling in physics teaching: The role of computer simulation. *International Journal of Science Education, 13* (3), 243–254.

Alessi, S. (1988). Fidelity in the design of computer simulations. *Journal of Computer-Based Instruction, 15* (2), 40–47.

Alessi, S. and Trollip, S. (1991). ibid

Clinton, J. (1991). Decisions, decisions. *The Florida Technology in Education Quarterly, 3* (2), 93–96.

Estes, C. (1994). The real-world connection. *Simulation and Gaming, 25* (4), 456–463.

Hasselbring, T. and Goin, L. (1993). Integrating technology and media. In Polloway and Patton (Eds.). *Strategies for teaching learners with special needs* (5th ed.). New York: Merrill.

Mintz, R. (1993). Computerized simulation as an inquiry tool. *School Science and Mathematics, 93* (2), 76–80.

Reigeluth, C. and Schwartz, E. (1989). An instructional theory for the design of computer-based simulations. *Journal of Computer-Based Instruction, 16* (1), 1–10.

Richards, J. (1992). Computer simulations in the science classroom. *Journal of Science Education and Technology, 1* (1), 67–80.

Ronen, M. (1992). Integrating computer simulations into high school physics teaching. *Journal of Computers in Mathematics and Science Teaching, 11* (3–4), 319–329.

Simmons, P. and Lunetta, V. (1993). Problem-solving behaviors during a genetics computer simulation. *Journal of Research in Science Teaching, 30* (2), 153–173.

Smith, K. (1992). Earthquake! *The Florida Technology in Education Quarterly, 4* (2), 68–70.

Instructional Games

Alessi, S. and Trollip, S. (1991). ibid

Flowers, R. (1993). New teaching tools for new teaching practices. *Instructor, 102* (5), 42–45.

Malone, T. (1980). *What makes things fun to learn? A study of intrinsically motivating computer games.* Palo Alto, CA: Xerox Palo Alto Research Center.

McGinley, R. (1991). Start them off with games! *The Computing Teacher, 19* (3), 49.

Randel, J., Morris, B., Wetzel, C., and Whitehill, B. (1992). The effectiveness of games for educational purposes: A review of recent research. *Simulation and Gaming, 23* (3), 261–276.

Trotter, A. (1991). In the school game, your options abound. *Executive Educator, 13* (6), 23.

Problem Solving

Blosser, P. (1988). Teaching problem solving—Secondary school science. (ERIC Document Reproduction No. ED 309 049)

Funkhouser, C. and Dennis, J. (1992). The effects of problem-solving software on problem-solving ability. *Journal of Research on Computing in Education, 24* (3), 338–347.

Gore, K. (1987–88). Problem solving software to implement curriculum goals. *Computers in the Schools, 4* (3–4), 7–16.

Johnson, J. (1987). Do you think you might be wrong? Confirmation bias in problems solving. *Arithmetic Teacher.*

Mayes, R. (1992). The effects of using software tools on mathematics problem solving in secondary school. *School Science and Mathematics, 92* (5), 243–248.

McCoy, L. (1990). Does the Supposer improve problem solving in geometry? (ERIC Document Reproduction No. ED 320 775)

Robinson, M. and Schonborn, A. (1991) Three instructional approaches to Carmen Sandiego software series. *Social Education, 55* (6), 353–354.

Sherman, T. (1987–88). A brief review of developments in problem solving. *Computers in the Schools, 4* (3–4), 171–178.

ILSs

Bailey, G. and Lumley, D. (1991). Supervising teachers who use integrated learning systems. *Educational Technology, 31* (7), 21–24.

Becker, H. (1992). Computer-based integrated learning systems in the elementary and middle grades: A critical review and synthesis of evaluation reports. *Journal of Educational Computing Research, 8* (1), 1–41.

Bender, P. (1991). The effectiveness of integrated computer learning systems in the elementary school. *Contemporary Education, 63* (1), 19–23.

Bentley, E. (1991). Integrated learning systems: The problems with the solution. *Contemporary Education, 63* (1), 24–27.

Blickman, D. (1992). The teacher's role in integrated learning systems. *Educations Technology, 32* (9), 46–48.

Bracy, G. (1992). The bright future of integrated learning systems. *Educational Technology, 32* (9), 60–62.

Chrisman, G. (1992). Seven steps to ILS procurement. *Media and Methods, 28* (4), 14–15.

Hill, M. (1993). Chapter I revisited: Technology's second chance. *Electronic Learning, 13* (1), 27–32.

Mageau, T. (1990). ILS: Its new role in schools. *Electronic Learning, 10* (1), 22–24.

Maddux, C. and Willis, J. (1993). Integrated learning systems: What decision-makers need to know. *ED TECH Review,* Spring/Summer, 3–11.

Perkins, D. (1991). Technology meets constructivism: Do they make a marriage? *Educational Technology, 31* (5), 18–23.

Robertson, Stephens, and Company. (1992). ibid.

Sherry, M. (1992). Integrated learning systems: What may we expect in the future? *Educational Technology, 32* (9), 58–59.

Shore, A. and Johnson, M. (1992). Integrated learning systems: A vision for the future. *Educational Technology, 32* (9), 36–39.

Smith, R. A. and Sclafani, S. (1989). Integrated teaching systems: Guidelines for evaluation. *The Computing Teacher, 17* (3), 36–38.

Texas Center for Educational Technology (1991). *ILS assessment and evaluation kit.* Denton, TX: University of North Florida, Texas Center for Educational Technology.

White, M. (1992). Are ILSs good for education? *Educational Technology, 32* (9), 49–50.

Logo

Becker, H.J. (1987). The importance of a methodology that maximizes falsifiablity. *Educational Researcher, 16* (5), 11–16.

Bracey, G. (1988). Still a storm over Logo research and Papert's ideas. *Electronic Learning, 7* (5), 14.

Clements, D. H. (1993). Young children and computers: Crossroads and directions from research. *Young Children, 48* (2), 56–64.

Collis, B. (1989). Research retrospective: 1985–1989. *The Computing Teacher, 16* (7), 5–7.

Govier, H. (1988). *Microcomputers in primary education: A survey of recent research.* (Occasional paper ITE/28a/88). Lancaster, UK: Economics and Social Research Council. (Cited in Collis, B. (1988–1989). Research windows. *The Computing Teacher, 16* (4) 7.

Harvey, B. (1987). Finding the best Logo for your students. *Classroom Computer Learning, 7* (7), 41–47.

Keller, J. (1990). Characteristics of Logo instruction promoting transfer of learning: A research review. *Journal of Research on Computing in Education, 23* (1), 55–71.

Krendl, K. and Lieberman, D. (1988). Computers and learning: A review of recent research. *Journal of Educational Computing Research, 4* (4), 367–389.

Muller, J. (1985). The great Logo adventure. In C. Maddux (Ed.). *Logo in the schools.* New York: Haworth Press.

Papert, S. (1980). *Mindstorms: Children, computers, and powerful ideas.* New York: Basic Books.

Papert, S. (1985). Different visions of Logo. *Computers in the Schools, 2* (2–3), 3–8.

Papert, S. (1987). Computer criticism versus technocentric thinking. *Educational Researcher, 16* (1), 22–30.

Pea, R. (1987). The aims of software criticism. *Educational Researcher, 16* (5), 4–7.

Redekopp, R. (1989). The significance of nonsignificance. *Journal of Research on Computing Education, 22* (2), 169–179.

Roblyer, M. D., Castine, W., and King, F. (1988). *Assessing the impact of computer-based instruction: A review of recent research.* New York: Haworth Press.

Simonson, M. and Thompson, A. (1994). *Educational computing foundations.* New York: Merrill Publishing Co.

Singh, J. (1992). Cognitive effects of programming in Logo: A review of literature and synthesis of strategies for research. *Journal of Research on Computing in Education, 25* (1), 88–104.

Walker, D. (1987). Logo needs research: A response to Papert's paper. *Educational Researcher, 16* (5), 9–11.

Walsh, T. (1994). Facilitating Logo's potential using teacher-mediated delivery of instruction: A literature review. *Journal of Research on Computing in Education, 26* (3), 322–335.

Watt, D. (1992). Logo. In G. Bitter (Ed.). *Macmillan encyclopedia of computers.* New York: Macmillan.

Yoder, S. and Moursund, D. (1993). Do teachers need to know about computer programming? *The Journal of Computing in Teacher Education, 9* (3), 21–26.

Evaluating and Using Types of Courseware

Gill, B., Dick, W., Reiser, R., and Zahner, J. (1992). A new model for evaluating instructional software. *Educational Technology, 32* (3), 39–48.

Litchfield, B. (1992). Science: Evaluation of inquiry-based science software and interactive multimedia programs. *The Computing Teacher, 19* (6), 41–43.

Lockard, J., Abrams, P., and Many, W. (1994). *Microcomputers for educators* (2nd ed.). Glenview, IL: Scott, Foresman & Co.

Merrill, P., Hammons, K., Tolman, M., Christensen, L., Vincent, B., and Reynold, P. (1992). *Computers in education* (2nd ed.). Boston: Allyn and Bacon.

Roblyer, M. (1983). How to evaluate software reviews. *Executive Educator, 5* (9), 34–39.

Roblyer, M. (1986). Careers in courseware evaluation. *Educational Technology, 26* (5), 34–35.

Instructional Software References (Company addresses in Appendix B)

Sample Software Packages with Tutorial Functions
DaisyQuest and Daisy's Castle—Great Wave Software
The Sensei Series: Calculus, Geometry, Physics, Statistics—Brøderbund Corporation

Sample Software Packages with Drill and Practice Activities
Grammar Problems for Practice—Milliken Publishing Company
Magic Spells—The Learning Company
Math Rabbit—The Learning Company
Millie's Math House—Edmark Corporation
Milliken Math Sequences—Milliken Publishing Company
Muppet Word Book—Sunburst Communications

Muppets on Stage—Sunburst Communications
Reader Rabbit—The Learning Corporation

Sample Software Packages with Simulation Activities
Operation Frog—Scholastic, Inc.
SimCity—Maxis Software
Survival Math: Hot Dog Stand—Sunburst Communications

Sample Software Packages with Instructional Game Activities
How the West Was One + Three × Four—Sunburst
    Communications
Math Blaster Plus—Davidson Software
Where in the World Is Carmen Sandiego?—
    Brøderbund

Sample Software Packages with Problem-Solving Activities
The Factory—Wings for Learning
The King's Rule—Sunburst Communications
Math Shop and Math Shop Junior—Scholastic, Inc.

Instructional Software References for Logo Products
(For more information, see the newsletter *Logo Update* (1993) *1*, (1). Available from the Logo Foundation, 250 West 57th Street, New York, NY 10107–2603.)
Logo programming languages:
•   PC Logo—Harvard Associates, 10 Holworthy Street, Cambridge, MA 02138
•   LEGO® TC Logo—LEGO Dacta, 555 Taylor Road, P.O. Box 1600, Enfield, CT 06083–1600
•   Object Logo—Paradigm Software, Inc., P.O. Box 2995, Cambridge, MA 02238
•   WIN-LOGO—Softeast Corp., Knox Trail Office Bldg., 2352 Main St., Concord, MA 01742
•   Terrapin Logo—Terrapin Software, Inc., 10 Holworthy St., Cambridge, MA 02138

HyperLogo (with HyperStudio)—Roger Wagner, Inc., 1050 Pioneer Way, Suite P, El Cajon, CA 92020
Microworlds—Logo Computer Systems, Inc. (LCSI), P.O. Box 162, Highgate Springs, VT 05460

Chapter 5

Using Word Processing, Spreadsheet, and Database Software in Teaching and Learning

This chapter will cover the following topics:

- Definition and characteristics of word processing, spreadsheets, and database programs

- Advantages of each for students and for teachers

- Example classroom uses for each kind of tool

Chapter Objectives

1. Use correct terminology to identify features and capabilities of word processing, spreadsheet, and database programs.

2. Describe specific kinds of teaching and learning tasks for both teachers and students that each kind of tool can support.

3. Develop lesson activities that integrate the functions and capabilities of each tool.

"Hence, could a machine be invented which would instantaneously arrange on paper each idea as it occurs to us, without any exertion on our part, how extremely useful would it be considered."
Henry David Thoreau, as quoted by David Humphreys, from C. Selfe, D. Rodrigues, and W. Oates (eds.) *Computers in English and the Language Arts* (1989)

Introduction to Tool Integration Strategies

This chapter and Chapter 6 describe integration strategies for common types of applications software from two perspectives: productivity and instructional applications. (See the Part II overview, pages 81–84.) Although productivity uses are not instructional in nature, they can provide powerful help for teachers who want to make improvements in their classrooms.

These chapters emphasize descriptions of the capabilities of the tools and their integration strategies, rather than steps in learning to use the tools themselves. However, for those who have not yet become skilled users of these tools, this chapter also gives a suggested sequence and some tips for learning to use word processing, spreadsheet, and database applications.

Using Word Processing Software in Teaching and Learning

Introduction to Word Processing

Word processing defined. Word processing is, simply put, typing on a computer. The term *word processor* can refer either to a computerized machine set up primarily to do word processing (e.g., an electronic typewriter) or to a general-purpose computer that can use word processing software. This chapter describes how to use microcomputers with word processing software, since this is the kind of word processing resource that educators usually use in classrooms.

Word processing can support nearly any kind of task or teaching activity that was previously done by handwriting or typewriter, but word processing offers more capability and versatility than either of these methods. Since a word processing document is prepared on screen before being printed onto paper, the writer can correct errors, insert or

delete words or sentences, and even move lines or paragraphs around before printing the document. The writer can easily change the words or appearance because the document is stored in the computer's memory and, hopefully, on a disk or a hard drive. Once stored or saved, documents can be changed or reprinted later. This chapter will help teachers understand both the general features and capabilities of word processing and its instructional applications. The following sections describe and illustrate word processing capabilities and benefits for both teacher productivity and teaching and learning tasks.

Types of word processing software. Word processing software can be classified in several ways, as Table 5.1 illustrates. One current classification depends on a document's appearance on the screen, a classification that will probably disappear in time since it concerns the capabilities of older versus newer word processing software. The first word processing programs presented documents differently on screen than the "hard copy" looked when printed out on the printer. For example, instead of showing words or phrases that would be underlined, boldfaced, or centered on the page, the screen showed symbols before and after these words or phrases to indicate what would happen on the hard copy. Today word processing software usually shows a "what you see is what you get" or "WYSIWYG" display of a document. Other word processing programs give text-based displays (WYSINWYG—"what you see is *not* what you get").

Another way to classify word processing software is according to how it is packaged. Users can interact with either a single application (e.g., WordPerfect or Microsoft Word); part of an integrated package (e.g., Microsoft Works) that also includes database, spreadsheet, and often, telecommunications and/or graphics software; or part of a total writing instruction package that also includes aids such as prewriting assistance or language analyzers (e.g., Writer's Helper). Logowriter, which combines Logo programming features with word processing capabilities, can also be considered a type of integrated package.

Teachers can also separate word processing capabilities by their intended users. Word processors have been designed to support student writing at various grade levels. For example, Kidwriter Gold is ideal for use with young children, while Research Paper Writer focuses on assisting writing instruction for older students; it provides a variety of prewriting and language analysis features. This type of software frequently includes other materials

Table 5.1	Classifications for Word Processing Software in Education	
By screen appearance	Text-based, WYSINWYG	WYSIWYG
By packaging	Single application	Integrated package
By student level	Young students (support for beginning writers)	Older students (support for later writers)

Table 5.2 Fonts and Typestyles	
Fonts	**Typestyles**
This is an example of the Palatino font.	*This is Palatino in italics.* **This is Palatino in boldface.**
This is an example of the Avant Garde font.	*This is Avant Garde in italics.* **This is Avant Garde in boldface.**

such as prepared activity files that have been developed to help teachers use word processing software for various writing exercises. Teachers must select word processing software with capabilities that match their needs. Schwartz & Vockell (1989) stated a general criterion for selecting appropriate word processing programs for classroom use: "the easier, the better."

General word processing features and capabilities. Word processing represents a significant improvement over typing on a typewriter. Although word processing capabilities and procedures vary from program to program, most programs have several features in common:

- **Storing documents for later use.** The most powerful advantage of word processing over typewriters is the ability to handle documents more than once without reentering the same text. Once created on the screen, a document can be stored on disk, re-loaded into the computer's memory later, and either modified or printed out again.

- **Erasing and inserting text.** Changes to typed documents require physical erasing or simply starting over; word processing allows a writer to insert additional letters, spaces, lines, or paragraphs into a document without trouble.

- **Search and replace.** If an error is repeated throughout a document (e.g., *work processing* instead of *word processing* or periods falling outside instead of inside quotation marks), a word processing program can easily correct the error by searching the document for all occurrences and changing them as specified.

- **Moving or copying text.** Sometimes a writer decides that a paragraph would sound better or seem more logical in a different location, or perhaps a given line will be repeated several times throughout a document. Word processing software allows the user to specify a block of text and either move it to or repeat it in the places specified (cut and paste).

- **Word wraparound.** Typists must usually place carriage returns at the ends of lines. Word processing software does this automatically with a feature called *word wraparound*. Many word processors also allow users to allow automatic hyphenation of words at line breaks.

- **Change style and appearance easily.** Word processing software allows a writer to employ a variety of fonts, typestyles, margins, line spacings, and indentations in a single document. (Table 5.2 illustrates fonts and typestyles.)

- **Justification.** The trademark feature of a word processing document is the justification of both right and left margins, sometimes referred to as *full justification*. However, a user can also easily specify that a given line or block of text be centered or right or left justified. Table 5.3 shows examples of various kinds of justifications.

- **Automatic headers, footers, and pagination.** Word processors can automatically place a title at the top (header) or

Table 5.3 Justification		
Left Justification	**Full Justification**	**Centered**
"What did I ever do before word processing? Now I am a word processing evangelist!" said the English teacher. "It has made all of my work so much easier! My students love not having to rewrite everything to make one or two corrections."	"What did I ever do before word processing? Now I am a word processing evangelist!" said the English teacher. "It has made all of my work so much easier! My students love not having to rewrite everything to make one or two corrections."	An Ode to Word Processing by Ima Convert

bottom (footer) of each page in a document with or without page numbering (pagination).

Not all word processors have the following enhanced features, but they are becoming more common:

- **Inserting text prepared on other word processors.** Each word processing program has software commands to apply formatting, and these commands are specific to the package. Therefore, one program cannot usually read a document prepared on another. However, many word processors can store documents as text or ASCII (American Standard Code for Information Interchange) files. This process removes formatting commands, so another word processor can read a file stored in this way. Some word processors also have "filters" or program functions that will accept regular files originally prepared and stored on other word processors.

- **Checking and correcting spelling.** Spell checking is a word processing capability that compares words in a document to those stored in its dictionary files. The program identifies words in the document that it cannot find in the dictionary and suggests possible corrections.

- **Suggesting words.** In addition to a spell checking dictionary, a program may include access to a thesaurus. A user can request a synonym or an alternate suffix for a given word.

- **Reviewing style and grammar.** Some word processing software can check text features such as sentence length, frequency of word use, and subject-verb agreement. This function may also suggest changes or corrections to the text.

- **Allowing insertion of graphics.** Some programs allow users to insert pictures stored as graphics files within their documents; others also have draw features that allow users to create and place their own pictures within their documents.

- **Merging text with database files.** Finally, some word processing programs can place database fields within documents such as letters. When it prints the letter or other document, the program automatically inserts information from the database as directed. Thus, a teacher could write one parent letter and merge it with a student database to print a personalized letter for each parent.

The Impact of Word Processing in Education

Advantages of word processing for education. Perhaps no other technology resource has had as great an impact on education as word processing. Not only does this tool offer a great degree of versatility and flexibility, but it is also "model-free" instructional software, that is, it reflects no particular instructional approach; a teacher can use it to support any kind of directed instruction or constructivist activity. Since its value as an aid to teaching and learning is universally acknowledged, word processing has become the most commonly used software in education. It offers many general advantages to teachers and students:

- **Time savings.** Word processing helps teachers use preparation time more efficiently by letting them modify materials instead of creating new ones. Writers can also make corrections to word processing documents more quickly than they could achieve on a typewriter or by hand.

- **Better appearances.** Materials from word processing software look more polished and professional than handwritten or typed materials. It is not surprising that students seem to like the improved appearance that word processing gives to their work (Harris, 1985).

- **Sharing methods.** Word processing allows a means of sharing materials easily among writers. Teachers can exchange lesson plans, worksheets, or other materials on disk and modify them to fit their own needs. Students can also share ideas and products among themselves.

Issues related to word processing in education. Educators seem to agree that word processing is a valuable application, but some aspects of its use in education are controversial.

- **When to introduce word processing.** The development of word processing software designed for very young children has allowed schools to introduce word processing to students as young as 4 or 5 years old. While some educators feel that word processing will free students from the physical constraints of handwriting and enable them to advance more quickly in their written expression skills, others wonder about the impact of this early use on students. It may affect their willingness to spend time developing handwriting abilities and other activities requiring fine motor skills.

- **The necessity of keyboarding skills.** Another ongoing discussion in education asks whether students need to learn "keyboarding," or typing on the computer, either prior to or in conjunction with word processing activities. Some educators feel that students will never become really productive on the computer until they learn ten-finger keyboarding. Others feel that the extensive time spent on keyboarding instruction and practice could better be spent on more important skills.

- **Effects on handwriting.** While no researchers have conducted formal studies of the impact of frequent word processing use on handwriting legibility, computer users commonly complain that their "handwriting isn't what it used to be," ostensibly because of infrequent opportunities to use their handwriting skills.

- **Impact on assessment.** Some organizations have experimented with allowing students to answer essay-type tests questions through word processing software rather than handwriting or even typing them. This practice introduces an equity issue. A word processing product may receive a different grade than a handwritten or typed one. Students who do not have access to or skills in word processing may receive different grades on the same assignment or measure than those who do.

Teachers and administrators are still deciding how best to deal with these issues. Since word processing is becoming an increasingly pervasive presence in both home and classroom writing activities, more information should soon become available to help educators make informed decisions about how best to employ its capabilities.

Benefits of word processing: Findings from research. Contradictory research findings complicate efforts to assess the benefits of word processing in education. Studies of the effects of word processing on quality and quantity of writ-

	Hawisher (26 studies) (1989)	Snyder (57 studies) (1993)	Bangert-Drowns (32 studies) (1993)
Table 5.4 Research on Word Processing Benefits			
Better quality of writing	No conclusion	No conclusion	Positive results
Greater quantity of writing	Positive results	Positive results	Positive results
More surface (mechanical) revisions	No conclusion	Positive results	No conclusion
More substantive (meaning) revisions	No conclusion	No improvement	No conclusion
Fewer mechanical errors	Positive results	Positive results	Not reviewed
Better attitude toward writing	Positive results	Positive results	No improvement
Better attitude toward word processing	No conclusion	Positive results	Not reviewed

ing have not yielded conclusive results (Bangert-Drowns, 1993). Three different reviews of research (Hawisher, 1989; Bangert-Drowns, 1993; and Snyder, 1993) found that these differences in findings may reflect differences in researchers' choices of types of word processing systems, prior experience and writing ability of students, and types of writing instruction to evaluate. Bangert-Drowns (1993) summarizes this problem with current word processing research by quoting Cochran-Smith's comment: "... research demonstrates that the answer to its bottom line question 'Do students write better with word processing?' is 'It depends'" (p. 73). Generally, word processing seems to improve writing and attitudes toward writing only if it is used in the context of good writing instruction and if students have enough time to learn word processing procedures before the study begins. Table 5.4 summarizes some of the findings of the three research reviews.

Generally, studies seem to conclude that students who use word processing software in the context of writing instruction programs tend to write more, revise more (at least on a surface level), make fewer errors, and have better attitudes toward their writing than students who do not use word processing software. Teachers who use word processing software with their students should not expect writing quality to improve automatically. Improvements of that kind depend largely on other factors such as the type of writing instruction. But the *potential* value of word processing has been established, making it one of the most validated uses of technology in education.

Word Processing in the Classroom

When to use word processing: Teacher productivity. Word processing can help teachers prepare any classroom materials that they previously typed or wrote out by hand. These include handouts or other instructional materials, lesson plans and notes, reports, forms, letters to parents or students, flyers, and newsletters. Word processing benefits these tasks by saving preparation time, especially if the teacher prepares the same documents each school year. For example, a teacher may send the parent letter

shown in Figure 5.1 every year simply by changing the dates and adding any new information. Teachers may want to keep files of templates or model documents that they can easily update and reuse with minimal effort. The figure also lists some suggestions for additions to this file of reusable documents.

When to use word processing: Teaching and learning activities. Students can also use word processing for almost any written work, regardless of subject area, that they would otherwise write by hand. Research shows that word processing alone cannot improve the quality of students' writing, but it can help them make corrections more efficiently, and this can motivate them to write more and take more interest in improving their written work. Teachers have applied word processing in various instructional ways:

* **Writing processes.** Students can use word processing software to, write, edit, and illustrate stories; to produce reports in all content areas; to keep notes and logs on classroom activities; and for other written assignments.

* **Dynamic group products.** Teachers can assign group poems or letters with various students, adding and changing lines or producing elements of the whole document in a word processing program.

* **Individual language, writing, and reading exercises.** Special word processing exercises can allow individual students to work on-screen combining sentences, adding or correcting punctuation, or writing sentences for spelling words. Word processing may also make possible a variety of reading/language-related activities ranging from decoding to writing poetry.

* **Encouraging writing across the curriculum.** A fairly recent trend in education is to encourage writing skills in courses and activities other than those designed to teach English and language arts. This practice of writing-through-the-curriculum is in keeping with the new emphasis on integrated, interdisciplinary, and thematic curricula. Word processing can encourage these integrated activities.

Example materials and integration lesson plans. Some examples of teacher-created lesson plans illustrate methods for integrating word processing into classroom activities.

Figure 5.1   Example of Word Processing Product

Mr. Belize's Sixth Grade
Happy Valley Middle School
Happy Valley, Pennsylvania 12345

August 29, 1996

Dear Parent:

I would like to welcome you and your child to my Sixth Grade classroom at Happy Valley Middle School. We have some exciting learning experiences planned for this year, and I look forward to telling you more about them later.

It would help me help your child get the most out of this year's activities if you could complete the following information and return it to me by September 7, 1996.

Thank you for your help. I look forward to meeting you in the near future! In the meantime, please feel free to call me at home at 555-1212 (until 9:00 PM) if I can be of assistance.

Sincerely,

Jorge Belize

Your child's full name   _____

Preferred first name/nickname _____

Home address   _____

                              _____

Home telephone #   _____

Emergency telephone#   _____

Special medical needs   _____

Example Word Processing Applications for Teachers

Beginning of the year welcome letter to parents	Periodic student progress letters to parents
Permission letters for field trips or other events	Frequently used work sheets, exercises
Request for fee payment letter	Student information sheets and handouts
Fundraising letter	Annual reports for the school
List of class rules	Lesson plans/notes
Flyers and announcements	Simple newsletters, letterhead stationery

Learning Word Processing: Techniques and Activities

A recommended instructional sequence. The following list of activities is recommended for first-time word processing users who prefer a structured, systematic approach to learning (as opposed to learning by doing).

Step 1   Using existing documents. Loading documents, moving in the window, page setup, printing, saving documents, and closing documents

Step 2   Changing existing documents. Selecting and deselecting text, changing font and type size, undoing changes, making style changes, aligning text, deleting text, inserting spaces and letters, replacing text, automatic search and replace, and moving sentences and paragraphs

Step 3   Beginning a new document. Setting up a new page; controlling tabs, indentations, and line spacing; and adding page numbers, headers, and footers on multipage documents.

Step 4   Using advanced features. Using ASCII (text files), spell checking, and graphics

Common mistakes and misconceptions when learning word processing. Beginning users of word processing experience some common problems. Teachers may want to review them before beginning the exercises. If students encounter problems, they may want to review this list again:

- **Forgetting to move the cursor before typing.** The computer does not automatically place text as desired. The user must indicate a spot for new text using the mouse or command keys to place the cursor at the desired location. Beginners very often have their eyes on the place they want to insert a letter or word but forget to move the cursor there first. They are surprised when their typing appears somewhere other than where they had in mind.

- **Forgetting to highlight before changing a format.** The same kind of problem occurs when beginners want to change the appearance of some text (e g., centering a title or inserting an automatic paragraph indent). The computer will not change the correct part of the text unless the user highlights the text *first* and *then* selects the option. If the selected option doesn't work, be sure to indicate which part of the text the format should change.

- **"Losing" part of the document.** Unless a computer system has a very large (e.g., 19 inch) screen, an entire document may not fit on one screen. As the user types, the top of the document scrolls up the screen and out of view. Beginners sometimes become distressed because they cannot see all of what they have typed and think it may be lost. If this happens, they can use the scroll box or scroll arrows to bring the missing part of the document into view on the screen.

- **Forgetting automatic wraparound at the ends of lines.** Word processing software helpfully sets line breaks without the user pressing Return or Enter; the text automatically wraps around to the next line. Beginners sometimes forget this fact and use the typewriter convention of hitting the Return key at the end of each line. This will make large spaces appear in the text when the document prints out on paper. Remember to press Return or Enter *ONLY* at the ends of paragraphs.

- **Problems with naming and saving files.** Word processing is frequently the first application that beginning computer users learn. Thus, this may provide their first experience with the concepts of storing and replacing documents in computer memory. The most common error is forgetting to save a document before closing it. The novice user may think he needs to save only once. He is surprised when he opens a document later and finds that some of what he typed is missing. Remember to save a document before closing it.

- **Incorrect spacing at the top or bottom of the document.** Beginners are sometimes surprised when their printed documents have different top and bottom margins than what they saw on a supposedly WYSIWYG screen. This can be caused by two kinds of problems:

- **Setting paper in the printer.** The printer is a separate machine from the computer. When it receives a command to print from the computer, it will begin printing wherever the paper is positioned, even if the paper is already positioned halfway up the page. If unexpectedly large or small spaces

Three Integration Strategies for
Teaching Writing Skills with Word Processing

Strategy #1:   Using Word Processing for First Grade Writing
Developed by: Nancy Kuechle, Horace Mann School (Beverly Hills, California)
Level: Grade 1
Purposes of the Activity: First grade students in an experimental writing program used word processing at two of five learning stations to support writing activities.

Instructional Activities. Students attended the computer lab 4 days a week for 45-minute periods. The 12-computer lab was divided into five learning stations, three with computers and two with other activities. Students were divided into 5 groups and each attended a different station each visit. Before beginning the activities, students were allowed to experiment with the software and familiarize themselves with it. The stations consisted of phonics-based software activities, word processing and graphics, listening activities with headphones, and paper-based writing activities with pencil and crayon. At the word processing station, a student selected graphics to create a scene and then used the simple word processor to write a story about it. Students learned how to write, delete, and insert characters. They were given sentences to copy with blanks to fill in: "My name is _____. I am _____ years old." After completing the sentences, each student added another sentence telling something else about himself or herself. As their ability to use the word processor increased, students were given less direction until they needed only a topic sentence. Topics included holidays, classroom events, or subject areas such as social studies or science. Teachers corrected misspelled words and collected stories from these stations into book form for all to enjoy. At the end of each session, students read their stories to the class and shared their illustrations.

Problems to Anticipate. Students tend to get caught up in the graphics, and writing time can suffer unless limits are set on the number of graphics.

Strategy #2:   Process Writing and Word Processing
Developed by: Lee VerMulm, Cedar Falls High School (Cedar Falls, Iowa)
Level: High school juniors and seniors
Purposes of the Activity: Students used word processing for all aspects of the writing process in a one-semester writing elective course. A networked lab was used for the writing activities.

Instructional Activities. Students wrote several small assignments and one 500 to 700-word essay in each of four units. In the final unit, they wrote longer (25-page) research papers that combined text and graphics. In addition to writing and revising their work, students maintained writing logs in which they recorded what they accomplished each day. They also completed lessons on the basic skills of analytical and persuasive writing. These lessons were downloaded to students from the file server as needed. Students had access to several word processing programs, graphics programs, a bibliographic citations database, a note-taking program, several writing analysis programs, and online help in grammar, word usage, and punctuation in the form of a thesaurus and various desk accessories. They could also use an electronic bulletin board and fax machine to support their information searches. Students' writing was reviewed daily by the teacher and their writing groups. They reviewed completed drafts with a teacher and decided on appropriate revisions. When a draft was revised and completed, the teacher evaluated it based on the criteria and made notes on any areas that needed improvement.

Problems to Anticipate. Students often have to be coaxed away from their computer station to make room for the next class. Students quickly become so captivated by the word processor that they want access to the computers during non-school hours.

Strategy Lesson #3:   Reading, Reporting, and Reviewing
Developed by: Ellen Barker, Webster Elementary School (St. Augustine, Florida)
Level: Grade 4
Purposes of the Activity: Students will learn to prepare in word processing programs summaries of books they have read. They will put book reports in a format that can be shared with others (HyperCard stacks).

Continued

Instructional Activities. After students have learned the basics of word processing, a HyperCard activity provides for extensive use of language arts skills. Students are required to use handwriting, typing, reading, and editing. The standard format connected by "buttons" on each card allows them to learn more easily about books read by other students, thus motivating them to enjoy other books and learn from peer models. The teacher will need to procure or create a HyperCard book report form into which students can type elements of their reports. Students will then be assigned to read various books. Following the reading but before use of the computer, students will analyze, summarize, and evaluate the book(s), depending upon the particular questions the teacher may have included in the HyperCard stack. Once students are ready to use the computer, the teacher will need to boot or turn on the computer and locate the book report program.

Students write their book reports with word processing software, using the same formatting as they will encounter when using the HyperCard book report. Once students have done the prewriting, they go individually to paste their reports onto the HyperCard book report form. A peer editor reads the book report and corrects mistakes of grammar, syntax, semantics, and punctuation, depending upon the requirements of the teacher. When both students are satisfied with the report, it is saved and graded. When all students have had the opportunity to enter one book report, the students will learn how to enter a second report behind the original. This will enable them to see how many books they have read throughout the year. This activity can continue to the end of the year.

For extension activities, students can read one another's book report stacks to see if they would like to read books summarized and recommended by classmates. Students can work in small groups to prepare composite reports on books they have all read and discussed. A panel discussion involving two or more students with quite different opinions about a book can serve as a forum for mini-debates. The class can serve as the panel of judges to determine which side does the best job of presenting arguments and supporting them with information taken from the text. Points can also be awarded for well-supported personal opinions.

Problems to Anticipate. It is generally a good idea for the teacher to have a colleague work through the HyperCard book report stack to locate glitches that may occur when students attempt to do book reports. If student reports are to be saved, the program needs to have some provision for them to be saved onto the hard disk or diskette.

Source: Kuechle, N. (1990). Computers and first grade writing: A learning center approach. *The Computing Teacher, 18* (1), 39–41; VerMulim, L. (1993). The Christa McAuliffe writing center: Process writing with a networked Mac lab. *The Computing Teacher, 20* (7), 48–53; Barker, E. (1992). Reading, writing, and reporting. *Florida Technology in Education Quarterly, 4* (2) 79–81.

fall at the top of a document, check to see if the paper is positioned in the printer at the top of the sheet.

- **Extra blank lines in the document.** Even though blank lines do not always appear on the word processing screen, the computer knows they are there and allows space for them on the printed page. Beginners do not always realize that when they press Return, they insert a blank line that is as real to the program as text lines. If the paper is set at the top of the page and unexpected blank spaces still appear at the end of a printed document, see if blank lines appear in the document itself. If so, delete them by highlighting them and pressing the Delete key.

• **Problems with search and replace** This very handy feature is easy to misuse, and it can result in unexpected changes to the document. Seasoned users joke about the accountant who wanted to change wording in all his letters from "TO CUSTOMERS" to read "BY CUSTOMERS." He instructed the word processor to automatically search and replace all text from "TO" to "BY." The computer changed everything to "BY CUSBYMERS." Before telling the word processor to change all instances of some text, be sure you can predict the result accurately.

Using Spreadsheet Software in Teaching and Learning

Introduction to Spreadsheet Software

Spreadsheet defined. Electronic spreadsheets programs organize and manipulate numerical data. The term *spreadsheet* comes from the pre-computer word for an accountant's ledger: a book for keeping records of numerical information such as budgets and cash flow. Unlike the term *word processing,* which refers only to the computer software or program, the term *spreadsheet* can refer either to the program itself or to the product it produces. One teacher may say, "I use a spreadsheet program to keep my grades" and "Here is my grading spreadsheet for Period 3." Spreadsheet products are sometimes also called *worksheets.* Information in a spreadsheet is stored in rows and columns. Each row–column position is called a *cell.* Cells can contain numerical values, words or character data, and formulas or calculation commands.

A spreadsheet helps users manage numbers in the same way that word processing helps them manage words.

**Three Integration Strategies for
Dynamic Writing with Word Processing Software**

Strategy #1:	A Class Literary Paper
Developed by:	Mary Schenkenberg, Nerinx Hall High School (St. Louis, Missouri)
Level:	High school
Purposes of the Activity:	The teacher used word processing to let students develop a literary paper as a group in order to model the process.

Instructional Activities. The computer was attached to a large monitor so all students could see. They created a class essay on William Faulkner's "A Rose for Emily," which they had been reading and discussing. They first brainstormed a list of ideas and the class selected "Emily" as their theme. They looked through the story searching for images or words illustrating Emily's connection with the Old South, and filled the monitor with their ideas. By the end of the class, students had developed a thesis, an outline, and a substantial amount of content for the essay. More importantly, they had a clearer grasp of the process for writing such a paper.

Strategy #2:	A Class Poem
Developed by:	Joan Hamilton, Emerson School (Bolton, Massachusetts)
Level:	Grade 8
Purposes of the Activity:	A connection between reading and writing is built by having students write a class poem modeled after Walt Whitman's "Song of Myself."

Instructional Activities. After students read the original poem, the teacher set up six stations in the computer lab. Each was labeled with a beginning line from the poem: I hear ..., I understand ..., I saw ..., I want ..., Injustices ..., and Who are you? Students were asked to move from computer to computer adding one idea to each category. For example, at the "Who are you?" station, students were encouraged to add words, phrases, or ideas that describe a typical 8th-grade student. The resulting lists were then saved, printed, and made available to all students for use in writing their own poems. When the individual poems were completed, students had the option of combining their efforts into a class poem. Groups of students decided which parts of individual poems should be included and in what order. The final poem was printed, used as a choral reading, and displayed on bulletin boards.

Strategy #3:	A Class Novel
Developed by:	Lee Sebastiani (University Park, Pennsylvania)
Level:	High school
Purposes of the Activities:	The class explores the genre of science fiction by creating its own science fiction novel on the word processor.

Instructional Activities. The class began by imagining possible sites for the novel. After creating a fictional planet and city, they decided to focus on one period in the planet's history. They then developed character outlines, with each student contributing one character for the novel. The teacher encouraged students to let the characters reflect their own personalities and interests. Each student was asked to imagine an incident in the planet's history and describe it in a 3-to-5 page story. Word processing was used for all phases of this process. Students also used graphics programs to illustrate their stories and provide detail for the imaginary planet and its events.

Source: Wresch, W. (1990). Collaborative writing projects: Lesson plans for the computer age. *The Computing Teacher, 18* (2), 19–21.

Bozeman (1992) described spreadsheets as a way to "word process numbers" (p. 908). Spreadsheets were the earliest application software available for microcomputers. Bozeman identified them as largely responsible for creating the microcomputer revolution since people began to buy microcomputers so they could run the first spreadsheet software, Visicalc.

Today, teachers typically use electronic spreadsheets for any work that involves keeping track of and calculating numerical data such as budgets and grades. Spreadsheets process calculations faster, more accurately, and with more visual feedback than other tools such as calculators. For example, if a worksheet is set up to add a column of expense items, the cell showing the sum will change automatically in

Integration Strategies for
Language and Writing Exercises with Word Processing Software

Strategy #1:   Investigating Sentence Variety
Level: Grade 5 and up
Purposes of the Activity: This exercise helps students focus on making sentences vary in length and structure. The word processor is used to set off sentences in paragraphs so they can be more easily read and reviewed.

Instructional Activities. Have students select an essay they have already written and revised. Ask them to load the file into the computer, select their longest single paragraph, and set it off from the others by inserting blank lines before and after it. If the word processor allows a Return or Enter in its search and replace function, search for periods in the paragraph and replace them with a period and a Return or Enter. (Otherwise, do a Find function for periods, and press Return or Enter after each one.) Sentences will be isolated from each other in a list. Encourage students to determine whether or not all sentences have the same structure and/or length. If so, then edit in the following ways:

- Combine sentences, especially those that have overlapping meanings. Omit ones that do not communicate meaning clearly.
- Add new sentence beginnings.
- Rearrange word order within sentences to vary rhythm and structure.

Strategy #2:   Creating Coherence between Sentences
Level: Grade 5 and up
Purposes of the Activity: This exercise gets students to focus on the relationships between sentences. The word processor is used to scramble sentences and then put them back together in order to examine how they fit together to communicate meaning.

Instructional Activities. Students begin the exercise in the same way as in the lesson above on Investigating Sentence Variety. They isolate the sentences in a list. Then they use cut and paste functions to scramble the sentences in random order in the list. Now have them look carefully at the sentences like pieces of a puzzle and look for clues that signal where they go together. Which words "link back" or "forecast ahead" (p. 132). In the best test of coherence, each student should have a partner use the cut and paste function to rearrange the sentences to their intended order. If the person cannot put them in the desired order, the links between the sentences may not be explicit enough. If this is the case, students will have to make the connections clearer.

Strategy #3:   Appreciating Punctuation
Level: Grade 5 and up
Purposes of the Activity: This exercise encourages students to look at the important role of punctuation in writing. The word processor is used first to remove punctuation and capitalization and then to restore them.

Instructional Activities. Students load or type a paragraph onto the screen. They are asked to remove all punctuation marks and all capital letters, as well as all extra spaces at the ends of sentences. The teacher points out how difficult it is to read without these "clues" to meaning. Students trade computers and try to restore all the punctuation marks, capital letters, and spaces in their partners' paragraphs. They save their files to different names, load the original versions at the same time, and compare their versions with the originals. If the versions differ, students examine whether or not other punctuation or capital is necessary for meaning and change it as necessary.

Source: Elder, J., Schwartz, J., Bowen, B., and Goswami, D. (1989). *Word processing in a community of writers*. New York: Garland, 117–132.

response to any change to one of the expense items. If a worksheet is set up to calculate a student's grade average, the cell showing the cumulative average will be updated if the points change for any one of the grades. These capabilities allow both teachers and students to play with numbers and see the results. This section of the chapter helps teachers to define both the capabilities and the classroom applications of spreadsheets. It will describe capabilities and benefits of spreadsheets for both teacher productivity and teaching and learning.

Integration Strategies for
Reading and Language Exercises with Word Processing Software

Strategy #1: Writing Poetry to Develop Literacy
Developed by: Marguerite Nelson
Level: Any age (The article discussed 7-to-13-year-old LD students.)
Purposes of the Activity: Students use the word processor to write several styles of poems that focus on sounds. These poetry exercises become a way of practicing and improving fluency in decoding skills.

Instructional Activities. The students are introduced to a variety of forms of "experimental poetry," which is "an attempt to express writers' frustration with the limitations of language to describe the rapid changes in culture.... The writers tried to obliterate meaning by replacing traditional rules for writing poetry with randomness" (Nelson, 1994, p. 39). The result are forms of poetry which emphasize sound as opposed to meaning. Writing these poems gives students motivational opportunities to practice decoding without relying on context, while they exercise their own creativity. In each of the following forms, students use the unique capabilities of the word processor to facilitate the development of their poems:

- **Dada.** Students use banks of newspaper articles or headlines saved on disk. A student loads a story and uses the move function to go to random places in the document. Each word on which the cursor stops is cut and pasted until the whole article is rearranged randomly. The resulting Dada is saved and printed out:

Example: Day Spending Work Flexible Striking (Stephanie, Grade 2)

- **Sound poems.** These are created with nonsense syllables that make certain desired sound effects. Students use sources of vowel combinations, consonant blends, and digraphs saved on disk. They use the word processor's replace function to replace a specified vowel or blend. The result is a sound poem:

Example: ick   phick   snick
          twick   blick   click   (Anthony, Grade 3)

- **Optophonetic poems.** Sounds take on a visual form in these poems. The word processor is used to generate letters with various typefaces which the students then read out loud.

- **Oulipo.** Students use various alphabetical algorithms to specify which vowels will be included or left out or to arrange words alphabetically. The word processor's move function can be used to alphabetize words as desired.

Example—Only I or E Vowels: I like mice/In white pies   (Daniel, Grade 4)

- **Snowballing iceograms.** Poems are written by starting with a single letter and then adding one letter to each subsequent line to make series of real words.

Example: O
          No
          Now
          Know
          Known   (Reid, Grade 3)

- **Iterative poetry.** Students begin with a famous line or quote and replace the nouns or other parts of speech in it according to a certain algorithm (e.g., replace a noun with the word that falls seven words after it in the dictionary). The word processor's replace function is used for the replacing activity after the word is located in the dictionary.

Example: I heard it through the grapevine / I heard it through the grassland (Tony, Grade 5)

Continued

- **Transformations.** In these poems, words are replaced with their definitions. The replace or edit functions are used to delete and insert text.

Example: I saw a girl with a bow in her hair. I saw a female child with a knot with loops in it in her fine threadlike structure growing from the skin of most mammals. (Laura, Grade 3)

After writing the poems, students are allowed to read them aloud and share with each other.

Source: Nelson, M. (1994). Processing poetry to develop literacy. *The Computing Teacher, 22* (3), 39–41.

Integration Strategies for
Using Word Processing to Encourage Writing across the Curriculum

Strategy #1:	Math, Social Studies, and Writing
Developed by:	Marilyn Burns
Level:	High school
Purposes of the Activity:	This lesson integrates writing into a unit applying mathematics, which looks at the problems faced by families during the westward migration in the 1840s to 1850s.
Strategy #2:	Writing and Mathematics
Developed by:	Nancy Brown
Purposes of the Activity:	Students create word problems related to arithmetic problems they are doing in math, they then edit the problems in their English classes.
Strategy #3:	The Role of Word Processing in EST
Developed by:	Ken Hyland
Purposes of the Activity:	Students use word processing to facilitate writing as they learn a second language.
Strategy #4:	Word processing adds to experiencing science
Developed by:	Heidi Imhof
Purposes of the Activity:	Hands-on science experiences in the outdoors lead to writing activities with a word processor.
Strategy #5	Art, Reading, and Writing
Developed by:	Nancy Scali
Purposes of the Activity:	Visual arts and crafts offer opportunities for students to communicate and interpret what they are learning in science, social studies, math, language arts, and literature using the word processor.

Source: Burns, M. (1992). Math and the westward migration. *Writing Notebook: Creative Word Processing in the Classroom, 9* (3), 30–31. Brown, N. (1992). Writing mathematics. *Arithmetic Teacher, 9* (3), 40. Hyland, K. (1990). Literacy for a new medium. *System 18, 3,* 335–343. Imhof, H. (1991). Writing as a reflection of students' thinking: One field experience. *Writing Notebook: Creative Word Processing in the Classroom, 9* (1) 12–15. Scali, N. (1992). Using art to enrich reading and writing. *Writing Notebook: Creative Word Processing in the Classroom, 9* (4), 42–43.

Types of spreadsheet programs and products. Like word processing software, a spreadsheet program can form an application of its own or be part of an integrated package such as Microsoft Works that also contains word processing and database software. Sometimes spreadsheet capabilities are also combined with a database program to create a powerful, multipurpose product such as Lotus 1-2-3 or QuattroPro. Teachers usually select a program like Lotus 1-2-3 to present business education concepts to high school students or to handle more complex recordkeeping tasks than simple gradekeeping. These combination spreadsheet/database programs have more capabilities than self-contained spreadsheets or integrated packages and are more complicated to learn and use.

Teachers also use spreadsheet derivatives. Gradebooks or gradekeeping packages amount to special-purpose spreadsheets designed exclusively to store and calculate grades. Some software publishers also sell spreadsheet templates, predesigned worksheets for special instructional purposes such as demonstrating concepts of budgeting.

Figure 5.2 Charting Spreadsheet Data

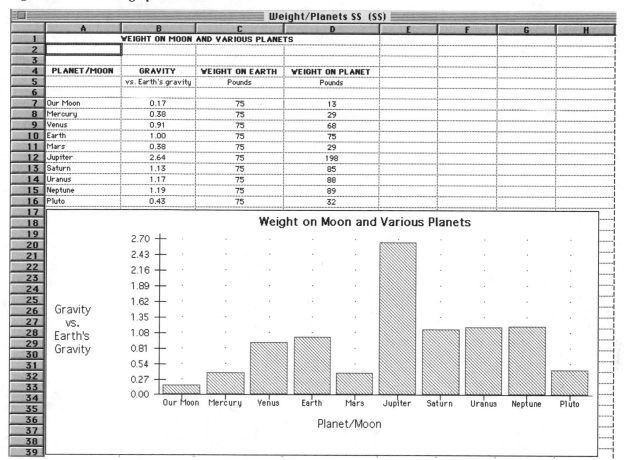

General spreadsheet features and capabilities. Spreadsheet packages offer significant improvements over calculating values by hand or with a calculator. Like word processing documents, spreadsheets can be easily edited and stored for later use. Although spreadsheet capabilities and procedures vary from program to program, most programs have the following features in common:

*   **Calculations and comparisons.** Spreadsheets can calculate and manipulate stored numbers in a variety of ways through formulas. In addition to adding, subtracting, multiplying, and dividing, specified in formulas, spreadsheets can also manipulate data in many more complex ways through function commands. These include mathematical functions such as logarithms and roots, statistical functions such as sums and averages, trigonometric functions such as sines and tangents, logical functions such as Boolean comparisons (if … then), and financial functions such as periodic payments and rates. Most spreadsheets also offer special-purpose functions such as lookup tables. These are sets of numbers that are automatically compared with those in the spreadsheet and assigned a value if they match. For example, a teacher might have a lookup table to assign letter grades based on students' final numerical grades. Formulas can also allow users to "weight" given grades.

*   **Automatic recalculation.** This is the most powerful advantage that spreadsheets offer. When any number changes, the program updates all calculations related to that number.

*   **Copying cells.** Once a user enters a formula or other information into a cell, it can be copied automatically to other cells. This can save time, for example, when placing a long formula at the end of each of 20 rows; the user can simply copy the information from the first row to other rows.

*   **Line up information in columns.** Spreadsheets store data by row-column positions, a format that makes information easy to read and digest at a glance.

*   **Create graphs that correspond to data.** A spreadsheet program can display entered and calculated data in a chart or graph (e.g., a pie chart or bar graph). Figure 5.2 shows an example spreadsheet and a bar chart derived from its data.

*   **Use worksheets prepared on other programs.** Spreadsheet programs have software commands, invisible to the user, that perform formatting features such as centering text in columns. Since these program commands vary from package to package, one spreadsheet program cannot usually read and manipulate a worksheet prepared on another program. However, many spreadsheet programs allow a document to be saved as a text, ASCII, or SYLK file. When stored in this way, a whole worksheet or specific parts can be brought in (imported) and used in another spreadsheet program.

The Impact of Spreadsheets in Education

Advantages of spreadsheets. Spreadsheet programs are in widespread use in classrooms at all levels of education. Teachers use them primarily to present mathematical topics

Figure 5.3   Sample Spreadsheet for Keeping Grades

	A	B	C	D	E	F	G	H	I	J	K	L	M	N	O	P
1	Fall 1994 Grades - PERIOD 4 Social Studies															
2																
3				Individual Assignments					Prod.	Prod.		Test	Test		Final	FINAL
4				1	2	3	4	5	1	2		*1	*2		Avg.	GRADE
5				8%	8%	8%	8%	8%	20%	20%		10%	10%			
6																
7	Adams, Alma			89	92	84	96	80	88	95		54	70		84.28	B
8	Betts, Lee			95	84	81	77	90	91	95		90	100		90.36	A
9	Bradley, Brindell															
10	Brush, Jason			86	95	96	90	90	91	95		87	45		86.96	A
11	Dirk, Dwan			80	97	90	83	90	97	100		75	100		92.10	A
12	Gretsky, Gerald			72	75	90	97	90	77	76		81	50		77.62	C
13	Howard, Kay			84	89	79	97	100	91	90		78	80		87.92	B
14	Johnson, Betty			89	98	96	96	90	94	90		94	100		93.72	A
15	Jones, Natalie															
16	Lane, Michael			83	85	72	50	76	91	95		76	80		82.08	B
17	McBur, Yolanda			96	100	96	96	77	91	90		81	100		91.50	A
18	McClellan, Will			45	100	92	95	100	97	100		99	100		93.86	A
19	Morrison, Addie			97	93	92	96	90	88	95		93	100		93.34	A
20	Moultrie, Fred			98	91	88	89	100	91	90		56	80		87.08	B
21	Sanders, Lillie			97	90	88	85	100	94	90		83	100		91.90	A
22	Shepherd, April															
23	Williams, Peter			93	65	92	82	56	80	70		71	100		78.14	C
24																
25																

but sometimes for other purposes. They can help teachers and students in several ways:

• **Time savings.** Spreadsheets save valuable time by allowing teachers and students to complete essential calculations quickly. They save time not only by making initial calculations faster and more accurate, but their automatic recalculation features make it easy to update products such as grades and budgets. Entries such as grades also can be changed, added, or eliminated easily, with formulas automatically recalculating final grades.

• **Creating charts.** While spreadsheet programs are intended for numerical data, their capability to store information in columns makes them ideal tools for designing informational charts that may contain few numbers and no calculations at all (e.g., schedules and attendance lists).

• **Answering "what if" questions.** Spreadsheets help people visualize the impact of changes in numbers. Since values are automatically recalculated when changes are made in a worksheet, a user can play with numbers and immediately see the result. This capability makes it feasible to pose "what if" questions (e.g., what weight should I give this grade?) and answer them quickly and easily.

• **Motivation.** Many teachers feel that spreadsheets make working with numbers more fun. Collis (1988) described spreadsheets as "sufficiently enjoyable and interesting in themselves that students can sometimes be experiencing the *pleasure of exploring* math at the same time as they are *doing math*" (p. 264). Students sometimes perceive mathematical concepts as dry and boring; spreadsheets can make these concepts so graphic that students express real delight with seeing how they work.

Issues related to spreadsheets. Few observers disagree about the applications of spreadsheets in classrooms. One of the few questions related to spreadsheets in education

emerges when teachers decide whether to use them to keep grades or to rely instead on gradekeeping packages (gradebooks) designed especially for this purpose. Spreadsheets usually offer more flexibility in designing formats and allowing special-purpose calculation functions, while gradebooks are simpler to use and require little setup other than entering students' names and assignment grades. Teachers appear to be about evenly divided on which is better; the choice comes down to personal preference.

Since spreadsheet use creates no "researchable issues," there are no research results to report in this area. Studies do show, however, that spreadsheets can be useful tools for teaching topics ranging from problem solving (Sutherland, 1993) to statistical analysis methods (Klass, 1988). The literature contains numerous testimonials by teachers who have used spreadsheets successfully in teaching topics ranging from mathematics (Baugh, 1995) to social studies (Voteline, 1992).

Spreadsheets in the Classroom

Applications of spreadsheets: Teacher productivity. Teachers can use spreadsheets to help them prepare classroom materials and complete calculations that they would otherwise have to do by hand or with the aid of calculators. An example spreadsheet a teacher might use for keeping grades is shown in Figure 5.3. They have helped with many activities and products in education, including:

• Gradekeeping (for a teacher's own records and to prepare grade charts for posting)

• Club and/or classroom budgets

• Computerized checkbooks for clubs or other organizations

- Attendance charts
- Performance assessment checklists

Applications of spreadsheets: Teaching and learning activities. The literature reflects an increasing variety of applications for spreadsheets. While their teaching role focuses primarily on mathematics lessons, they have also effectively supported instruction in science, social studies, and even language arts. Teachers can use spreadsheets in many ways to enhance learning:

- **Demonstrations.** Whenever concepts involve numbers and some kind of concrete representation can clarify the ideas, spreadsheets can contribute to effective teaching demonstrations. Spreadsheets offer an efficient way of demonstrating numerical concepts such as multiplication and percentages, for example. A worksheet can make a picture out of abstract concepts and provide a graphic illustration of what the teacher is trying to communicate.

- **Student products.** Students can use spreadsheets to create neat timelines, charts, and graphs as well as products that require them to store and calculate numbers.

- **Support for problem solving.** A spreadsheet takes over the task of performing arithmetic functions so students can focus on higher-level concepts (Ploger, Rooney, and Klingler, 1996). By answering "what if?" questions, spreadsheets help teachers to encourage logical thinking, develop organizational skills, and promote problem solving.

- **Storing and analyzing data.** Whenever students must keep track of data from classroom experiments, spreadsheets can help them organize these data and perform required descriptive statistical analyses.

- **Projecting grades.** Students can be taught to use spreadsheets to keep track of their own grades. They can do their own "what if?" questions to see what scores they need to make on their assignments to project desired class grades. This simple activity can play an important role in encouraging them to take responsibility for setting goals and achieving them.

Example materials and integration lesson plans. The Integration Strategies on pp. 141–144 describe some teacher-created plans that integrate spreadsheets into classroom activities.

Learning to Use Spreadsheets: Techniques and Activities

A recommended instructional sequence. The following list of activities is recommended for first-time spreadsheet users who prefer structured, systematic learning (as opposed to learning by doing).

Step 1   Manipulating worksheets. Loading spreadsheet files, moving around in the spreadsheet, setting up and printing spreadsheet documents, and closing documents

Step 2   Changing existing worksheets. Highlighting cells, changing font and type size, undoing changes, changing column width, formatting cells for numerical and character data, entering and replacing data, inserting and deleting rows and columns, moving rows and columns, and sorting data

Step 3   Creating and changing formulas. Entering functions such as "sum" and "average," entering constructed formulas, editing formulas, and copying formulas (or other data) down columns or across rows to other cells

Step 4   Creating and using worksheets. Creating gradebook and budget spreadsheets and using the "what if?" capability

Step 5   Advanced features. Importing ASCII or SYLK files (text files) and creating graphs and charts with spreadsheet data

Common mistakes and misconceptions when learning spreadsheets. Many beginning spreadsheet users encounter some common problems. If you are having problems, look over this list to see if you can find a solution:

- **Forgetting to highlight cells to be formatted.** The computer cannot format a cell in the spreadsheet (e.g., centered or boldface text with dollar signs) until the user highlights the affected cells *first* and *then* selects the option. If a chosen formatting option doesn't work for the cells you thought you were formatting, be sure you first indicate the part of the worksheet where the format should apply.

- **Difficulties in developing formulas.** Perhaps the most common problems in spreadsheet use have to do with creating formulas. The most frequent problem results from failing to complete the first step in the procedure for creating formulas: placing the cursor in the correct cell. The next most common problem results from pressing the right arrow key (instead of the Return or Enter key) to leave the cell while creating a formula. Rather than moving the pointer, this action adds something to the formula. Many students become very confused when they see a formula grow as they struggle to leave the cell! A final common problem with formulas is accidentally including the formula cell itself in the formula's calculation. This is sometimes called a *circular reference error*. Even if an error message does not appear, this error can usually be spotted quickly because the formula results in much larger numbers than expected.

Using Database Software in Teaching and Learning

Introduction to Database Software

Database defined. Databases are computer programs that allow users to store, organize, and manipulate information, including both text and numerical data. Database software can perform some calculations, but their real power lies in allowing the user to locate information through keyword searches. Unlike word processing software (which can be compared to a typewriter) or a spreadsheet (which can be compared to a calculator), a database program has no electronic counterpart. It is most often compared to a file cabinet or a Rolodex card file. Like these precomputer devices, the purpose of a database is to store important information in a way that makes it easy to locate later. This capability becomes increasingly important as the society's store of essential information continues to grow in volume and complexity.

Integration Strategies for Spreadsheets

Strategy #1:	Using Spreadsheets for Student Products
Developed by:	John Beaver, SUNY College at Buffalo (Buffalo, New York)
Level:	Various middle school and high school grades
Purposes of the Activity:	Students use a spreadsheet to generate charts and graphs of data they have gathered.

Instructional Activities: The teacher should introduce graphing concepts by having students interpret some commercially produced charts and graphs. Then they should assign charts to do without benefit of spreadsheet software. (This gives concrete experience that teaches both the benefit of using a spreadsheet to produce these items and the procedures for generating them.) Then the teacher demonstrates how to use spreadsheet software to create charts and graphs.

Students generate and conduct a brief class survey on a topic of interest. They use the spreadsheet to collect the data and display it in chart or graph form. The products are displayed on a bulletin board as "A Profile of the Class." If additional surveys of other classes are done, the results can be posted on the bulletin board side-by-side and compared across classes.

Problems to Anticipate. Simplify students' introduction to spreadsheet charting procedures by creating a file for them with predesigned charting specifications. All the options can overwhelm them.

Strategy #2:	Using Spreadsheets to Anticipate Grades
Developed by:	John Beaver, SUNY College at Buffalo (Buffalo, New York)
Level:	Various middle school and high school grades
Purposes of the Activity:	Students use a spreadsheet to store grades and to project results of potential scores on assignments.

Instructional Activities. High school students are usually very "grade-conscious" and concerned about what effect a given test or other score could have on their class grades. The teacher can let students create their own spreadsheets for keeping and projecting grades. They begin by entering headings and formulas. When they complete their spreadsheets, the teacher demonstrates the spreadsheet's modeling power by letting students enter their worst-case and best-case grades to see the effects on their total GPAs in each situation. They can use this technique to determine the effect of individual scores on their final grades, thus preventing any overly drastic reactions to one poor score (or any unwarranted optimism as a result of one good score).

Problems to Anticipate. Letter grades must be translated to numerical equivalents for the spreadsheet.

Strategy #3:	Using Spreadsheets to Teach Problem Solving
Developed by:	Richard Sgroi, SUNY (New Paltz, New York)
Level:	Middle school and junior high school
Purposes of the Activity:	Students use spreadsheets to do calculations as they learn a four-step sequence for problem solving.

Instructional Activities. The teacher prepares students to learn the four-step problem-solving approach by reviewing the contradiction strategy. This procedure attacks a problem by identifying clues to aid in the solution and then answering questions about which solutions are possible and not possible (e.g., "Could 65 be a possible solution? Why or why not?"). When students are comfortable with this approach, the teacher introduces spreadsheet use in the context of two types of problems: coin problems and ratio problems. In a coin problem, the student knows a certain number of coins and a total sum, and must determine how many of each coin could give the sum. In the ratio problem, the student must find three 3-digit numbers that use all of the digits 1–9 only once; the ratio of the first number to the second number must be 1:2 and the ratio of the first number to the third number must be 1:3 (a 1:2:3 ratio). For each kind of problem, the teacher presents a prepared spreadsheet to support the necessary calculations. Then the four-step problem-solving procedure is introduced: understand the problem, devise a plan, carry out the plan, and look back. The class works through these problems in small groups and then discusses procedures and solutions together.

Problems to Anticipate. Some teachers may want students create their own spreadsheets, but for this lesson, teachers should probably supply them so students can focus on the problem-solving process.

Continued

Strategy #4: Using Spreadsheets to Help with Calculations during Problem Solving
Developed by: James R. M. Paul
Level: Elementary school
Purposes of the Activity: Students use a spreadsheet to help do numerical calculations involved in problem solving.

Instructional Activities. "Primary school children think big" (Paul, 1995, p. 65). They talk about and attack many math-oriented problems that require arithmetic skills that they lack. Spreadsheets can help with the calculations involved in solving such problems and can support the conceptual development related to these activities.

With the first problem, students visited a pizza parlor with three different size pizzas (small, medium, and large) divided them into 4, 6, and 8 portions. Students discussed whether individual portions were equal across all three sizes. They talked about whether it was more cost effective to buy two small pizzas of four portions each, or one large pizza with eight portions. They made initial estimates and then measured the pizzas. After discussion of the best methods to use and after making some initial calculations on a calculator, they formed initial hypotheses, looked at the kind of calculations necessary to solve the problems, and entered the data and formulas into the spreadsheet. After determining the solution to their problem, they changed the pricing parameters to answer "what if?" questions. Then they changed the radii of the pizzas until the cost for all sizes was equal.

A variation on this problem arose from observing the spaghetti measuring stick, which had holes representing various size servings of spaghetti. The cook has to fit as much dry spaghetti into the hole as possible, and the resulting serving size will be appropriate for a designated number of people. The problem and methodology was similar to the pizza problem. The class discussed how the stick worked and concluded that the size (area) of each hole would have to be proportional to the serving. The diameters of the holes were measured, and the spreadsheet was set up and used to determine whether the measuring stick was accurate and whether the spreadsheet could calculate exact diameters.

Strategy #5: Using Spreadsheets to Demonstrate Concepts
Developed by: Kenneth Goldberg, New York University, (New York)
Level: Grade 5
Purposes of the Activity: A spreadsheet is used to display data from a U.S. presidential election to show how popular votes and electoral votes differ, and how it is possible for a person to win the popular vote and still lose the election.

Instructional Activities. This lesson was used just prior to the 1988 U.S. presidential election, but it can also illustrate the concepts involved in any U.S. presidential election. The class held a mock election and assigned electoral votes to each class in the school based on enrollment numbers (e.g., a class with 10 or fewer students got one vote, one with 11 to 20 students got two votes, and one with 21 or more got three votes). The spreadsheet was set up to match the list of classes and their popular and electoral votes. Data on election results were entered after the election was held, and the spreadsheet was displayed on a large monitor so the whole class could see the results as they were entered. Some facts became clear when students began to see the data. It was evident that: Bush would win; the popular vote was fairly close; the electoral vote was not close; and if very few of the popular votes in key areas were changed, the results of the election would be reversed. The class discussed these results, as well as the possibility that a candidate could win the popular vote and lose the electoral vote.

Strategy #6: Using Spreadsheets to Store and Analyze Data
Developed by: Scarlet Harriss, Webster Elementary School (St. Augustine, Florida)
Level: Grade 3
Purposes of the Activity: Students use a spreadsheet to record the results of their M&M count and do various activities with estimation and prediction.

Instructional Activities. The teacher must prepare for this activity by developing a paper M&M Record Form for each student and a spreadsheet for use by the whole class. Each student pair receives a bag of M&Ms and a form. They count and separate by color the M&Ms in their bag. Given what they know about the contents of their own bags, students use estimation and prediction to project what they will find by color and number for the whole class. Each student pair fills out their M&M Record Form. The pairs take turns entering their data into the spreadsheet. When all data are in the spreadsheet, the students reevaluate their predictions. The teacher shares the spreadsheet sums and averages

Continued

using an overhead projector and LCD panel, and the class compares predictions with the actual totals and averages. The teacher ends by summarizing the prediction and estimation processes the class has used. As an extension activity, the class may discuss how these processes can be used in various ways in everyday life.

Problems to Anticipate. Teachers will need to make sure students understand concepts such as sums, averages, estimating, and predicting before beginning this activity.

Source: Beaver, J. (1992). Using computer power to improve your teaching, Part II: Spreadsheets and charting. *The Computing Teacher, 19* (6), 22–24; Sgroi, R. Systematizing trial and error using spreadsheets. *Arithmetic Teacher 39* (7), 8–12; Paul, J. R. M. (1995). Pizza and spaghetti: Solving math problems in the primary classroom. *The Computing Teacher, 22* (7), 65–67; Goldberg, K. P. (1990). Bringing mathematics to the social studies class: Spreadsheets in the electoral process. *The Computing Teacher, 18* (1), 35–38. Harriss, S. (1992). "M&M count." *Florida Technology in Education Quarterly, 4* (2), 60–61. Another popular lesson based on this data-storage approach is described in Edwards, J. (1992). What's in our trash? *Florida Technology in Education Quarterly 4, 2* (Winter).

People often use the term *database* to refer both to the computer program and the product it creates. However, database products are also sometimes called *files*. While a spreadsheet stores an item of data in a cell, a database stores one item of data in a location usually called a *field*. Although each field represents one item of information in a database, perhaps the more important unit of information is a record, since it relates directly to the designated purpose of the database file. For example, in a database of student records, each record corresponds to a student, and it consists of several fields of information about the student (e.g., name, address, age, parents' names). In a database of information on a school's inventory of instructional resources, then each record represents one resource and consists of

several fields describing the resource (e.g., title, publisher, date published, location).

Database software packages vary considerably in the format and appearance of information. Figure 5.4 shows an example database file on the regions of Florida. Designed by Barker (1992) to teach geography research skills, the database holds a record for each region, and each item about the region (e.g., topography, rivers) appears in a field.

The importance of databases. Technology advocates face a challenge in trying to do justice to the usefulness of database software in a classroom or curriculum. Unlike word processing or spreadsheet programs, a database may not enhance or take the place of noncomputer resources. In fact,

Figure 5.4   Database for Florida's Five Regions

Region	Topography	Rivers	Lakes	Bays	Forests
North Florida	rolling hills and coastal lowlands	Apalachicola River St. Johns River St. Mary's River Suwanne River	Lake Seminole		Apalachicola National Forest Ocala National Forest Osceloa National Forest
West Coast Florida	lowlands of the coast barrier islands plains, marshes, and sandy beaches	Caloosahatchee River Manatee River Myakka River	Lake Tsala Apopka	Charlotte Harbor Pine Island Tampa Bay	Big Cypress National Preserve
Central Florida	rolling hills of the Central Highlands lowlands near coasts sinkholes thousands of lakes plains	Caloosahatchee River Kissimmee River St. Johns River Withlacoochee River	Lake Apopka Lake Kissimmee Lake Okeechobee		Ocala National Forest
East Coastal Florida	forested lowlands plains marches barrier islands lagoons	Indian River Matanzas River St. Johns River	Blue Cypress Lake Lake Harney Lake Okeechobee		
South Florida	big cities in east swamps in west in between are suburbs		Lake Okeechobee	Biscayne Bay Florida Bay	Everglades National Park Big Cypress National Preserve

Source: Barker, E. (1992). Florida's five regions. *Florida Technology in Education Quarterly 4, 2,* 76–77.

Table 5.5 Classifications for Databases in Education		
	First Type	**Second Type**
By packaging	Single application (dBase, Microsoft Word)	Integrated package (ClarisWorks, Microsoft Works)
By purpose	Database software (dBase, Microsoft Works)	Prepared database (ERIC on Disc)
By filing type	Flat filing system (ClarisWorks, Microsoft Works)	Relational filing system (dBase, Fox Pro, Oracle)
By capability	Nonprogrammable systems (ClarisWorks, Microsoft Works)	Programmable DBMS systems (dBase, Fox Pro, Oracle)

instruction using databases may require a fairly dramatic shift in the way a teacher thinks about and teaches working with information. Some cite the database program's potential for facilitating new, constructivist teaching strategies as the source of its reputation as an indispensable classroom tool. Heine (1994) said that "A database is one of the computer tools that students should be able to use by the end of elementary school" (p. 39). Teachers have long recognized the unique capabilities of database software to support instruction in problem solving, research skills, and information management. Teachers have also found databases useful for teaching higher-level concepts such as classification and keyword searching to very young students (Hollis, 1990; Jankowski, 1993–1994). Students as young as 6 or 7 years old can begin to learn how to classify and group people or animals according to characteristics, and how to locate entries that match a certain description.

The productivity uses of databases have not earned them the same widespread popularity with teachers as tools such as word processing. Teachers may not have any existing filing system that a database could facilitate. For example, they may not have any special system (other than looking along shelves) to search for instructional resources that could help them teach a topic. They might not keep any personal classroom records on student performance, relying instead on those in the main office for vital information. Those who do not use databases may view the work required to create and/or use personal databases for these purposes as more time-consuming and difficult than existing mechanisms for dealing with information. However, as examples will show later in this chapter, a classroom database also gives teachers new capabilities and options for information management that they would not otherwise have.

Types of database programs and products. Table 5.5 presents several ways to categorize database programs.

The first classification separates packages according to purpose. As with word processing and spreadsheet programs, a database program can be an application on its own like dBase or part of an integrated package like Microsoft Works or ClarisWorks that also contains software for word processing, spreadsheets, graphics production, and/or telecommunications. The table also notes a difference

between database software intended to allow people to prepare their own database files to store their own data and prepared database files such as Dialog's ERIC on Disc or Sunburst's Animals Data Bases, which give access to existing collections of information. Prepared databases are usually designed to support learning of certain curriculum topics or make certain research tasks more efficient. Sometimes these collections are available on media such as CD-ROM or microdisk, and sometimes they are available online via telecommunications.

Another way to categorize database products is according to their schemes for storing and organizing information. A flat file database program produces a single file consisting of records with several fields each. Another kind of database program, called a *relational database,* links or relates separate files through a common field called a *key field.* For example, a student database containing personal background data may be linked to another student database with course and grade histories. These files are linked through a common field such as student name or Social Security number. The database program can draw reports from either file through the key field.

A final way to categorize database products is according to their levels of capability. More capable systems are also more complex to use, but some complex systems provide Database Management Systems (DBMSs) that give users helpful interfaces to the databases. This interface is a language of some kind, either a structured query language (SQL) or an actual programming language, which makes it easier to select items from the database for reports of various kinds. Most, but not all, relational database systems also have DBMSs.

General database features and capabilities. Database programs offer people several kinds of capabilities for handling information that they could not use for information stored in a noncomputer format such as on paper in file folders in a file cabinet. Although capabilities and procedures vary from program to program, most programs have the following features in common:

* **Allowing changes to information.** To make changes to paper documents, one must locate documents from each of the file folders that contain copies of the information. One

must then retype or otherwise alter the sheets of paper. Computer users can access information stored in a database from a number of locations (either via disk or at a terminal), no matter where it is physically stored. Changing data usually means simply calling up the file and editing information on-screen in one or more fields or giving a command to search and update all the information that meets certain criteria.

- **Sorting information alphabetically or numerically.** The computer's ability to put data in order comes in handy when information is stored in a database. The program can sort or order records according to data in any one of the fields. For example, student records could be printed out alphabetically according to the Last Name field, or the same information could sort student information from youngest to oldest by the Age field.

- **Searching for information.** All database programs allow users to search for and compare information according to keywords. For example, a teacher might want to locate the records of all teenaged students in a certain grade. If the information were stored in file folders, the teacher would have to go through each one, look up the person's birthday, and check to see if it fell before a certain date. With a database, the teacher could simply give a command to display all records whose Age field contained a number higher than 12 or whose Birth Date field contained a year lower than a certain year.

- **Automatically retrieving reports or summaries of information.** Storing information in databases makes it easier to prepare summaries across all data elements. For example, a teacher may want to group students for work outside class on group projects; the database could indicate when each student could meet outside class time. The teacher could search for all students who had free periods at the same time and assign these students to work together.

- **Merging with word processing documents.** A user can insert information stored in a database automatically in several letters or other word processing documents simply by preparing one document and putting field names in it instead of actual names or other information. The information stored in those fields *for each record* will automatically be inserted as each copy of the document is printed out. This process is called *merging* the database with the word processing document.

The Impact of Databases in Education

Advantages of databases. Database programs and products are in widespread use in classrooms at all levels of education. They can help teachers and students in the following ways:

- **Reducing data redundancy.** In education, as in business and industry, many different organizations need access to the same kinds of information on the same people or resources. In precomputer days, each organization had to maintain its own stores of information that were often identical to those of other organizations. For example, each school's office and the school district office might have duplicate files on teachers and students. Nowadays, since databases can be accessed from multiple locations, an organization needs to keep only one actual copy of these kinds of information. This cuts down on both the expense and the physical space needed to store the information.

- **Saving time locating and/or updating information.** People need to locate information and keep it accurate; time is money. It takes time to find the information and keep it up to date for everyone who needs it. Since a database stores information in a central computer instead of in several different file folders in various offices, users can find information more quickly, and they can more easily make changes whenever updates are needed. For example, if a student's address or legal guardian changes, changing the information in a database is both quicker and easier than locating and changing it in many file folders.

- **Allowing comparisons of information through searches across files.** Electronic databases also offer an important capability of locating information that meets several criteria at once. For example, a teacher may want to locate all of the resources in video format at a certain grade level that focus on a certain topic. A database search would make locating these materials an easy task as compared to a search of library shelves. For a large collection of information, this kind of search is possible *only* if information is stored in a database.

Issues related to databases. Databases are permanent and pervasive parts of life in the information age. They allow users to locate bits of important data in a landscape crowded with information; they support decisions with confirmed facts rather than assumptions; they put the power of knowledge at our fingertips. Yet this power is not without its dangers, and knowing how to find information is not the same as knowing what to do with it.

- **Simplifying access versus safeguarding privacy.** Each of our names is listed, along with much of our personal information, on literally dozens of databases. This cataloging begins when we are born, even before we are named, when we appear on the hospital's patient database. Our doctors and schools—even our places of worship—have our names and other notes about us in their databases. Whenever we apply for credit cards, driver's licenses, or jobs, we enter still more databases. These information entities reside on computers that can communicate and exchange notes, so information in one database can be shared with many information systems. Education, like other systems in our society, has come to depend on ready access to these information sources.

  However, easy access to information about people has long been recognized as a threat to personal privacy. If information is easy to get to, it may also be easy for unauthorized people to obtain and possibly misuse it or for organizations to use it in ways that violate basic human rights. In the Privacy Act of 1974, the U.S. Congress formally recognized the problems presented by government access to information on private citizens. This law requires that federal agencies identify publicly the records they maintain on U.S. citizens. It also limits the kinds of information that can be kept and requires that people be told what information the government keeps on them. As teachers begin to keep student information in classroom databases and use information from school and district databases, they must recognize their responsibility to safeguard this private information and protect it from unauthorized access. Sometimes this means keeping disks in secure places; sometimes it means making sure passwords remain secret. It may also mean deleting information if parents or students request it.

Figure 5.5   An Example Database: An Instructional Materials Inventory

Title	Content Area	Topic	Grade Level	Type	Docs	Publisher
Compton's Multimedia Encyclopedia	Misc.	Encyclopedia	All grades	Tool		Compton's NewMedia
Computer Inspector	Misc.	Computer Programs	All levels	Utility Tool	X	MECC
Conduit: Algebra Drill & Practrice II	Math	Algebra	Middle School	Drill & Practice	X	Conduit
Conduit: Coexist	Social Studies	Population Dynamics	Secondary	Simulation	X	Conduit
Conduit: Discovery in Trigonometry	Math	Trigonometry	9th & up	Tutorial	X	Conduit
Conduit: Evolut	Science	Evolution/Natural Sci.	Secondary	Simulation	X	Conduit
Conduit: Surfaces for Multi-Variables Calculus	Math	Calculus	College level	Tutorial	X	Conduit
Conquering Decimals (+,–)	Math	Decimals (+, –)	4th – 8th	Drill/Game	X	MECC
Conquering Decimals (x, /)	Math	Decimals (x, /)	3rd – 8th	Drill/Game	X	MECC
Conquering Fractions (+, –)	Math	Fractions (+, –)	4th – 8th	Drill/Game	X	MECC
Conquering Fractions (x, /)	Math	Fractions	5th – 8th	Game	X	MECC
Conquering Math Worksheet Generator	Math	Worksheets	Teachers (3–8)	Utility	X	MECC
Conquering Percents	Math	Percents	5th – 8th	Drill/Game	X	MECC
Conquering Ratios & Proportions	Math	Ratios & Proportions	5th – 8th	Drill/Game	X	MECC
Conquering Whole Numbers	Math	Whole Numbers (+, –, x)	3rd – 6th	Drill & Practice	X	MECC
Coordinate Math	Math	Charts, diagrams	4th – 9th	Drill & Practice	X	MECC
Counting Critters	Math	Counting	Pre K – K	Drill & Practice	X	MECC
Creative Writer	Lang. Arts	Writing	Ages 8 – 14	Word Processing		Microsoft
Database in the Classroom: Dataquest Sampler	S.S./Science	Database	5th – 12th	Database	X	MECC
Decimal Concepts	Math	Decimals	3rd – 6th	Game	X	MECC
Decision, Decisions: Immigration	Social Studies	Problem Solving	8th – 12th	Game/Simul.	X	Tom Snyder Productions

It always means being sensitive to who is looking at screens or printouts with students' personal information.

• **Instructional uses of databases.** The literature reflects extensive applications of databases for instructional purposes. There are several reasons for this popularity. First, databases are a completely "philosophy-free" technology resource. They can be used in teacher-directed ways or to support student-directed projects. Second, databases are a relatively inexpensive type of software, with a wide range of capabilities; thus, they can be used effectively in ways ranging from very simple to very complex. Hunter (1983) and Collis (1988) were among the first to document the uses of microcomputer databases for instructional purposes. Their books contained many varied lesson plans and demonstrated that databases could underlie teaching and learning activities for many content areas and grade levels. Early database applications were frequently offered to teachers as good ways to teach problem-solving skills (Watson, 1993). However, some 10 years of use and research on the instructional applications of databases has shown that simply having students use them does not ensure learning of desired research and problem-solving skills.

Collis (1990) summarized six different studies on instructional uses of databases. She found that students can use databases to acquire useful skills in searching for and using information, but they need guidance to ask relevant questions and analyze results. If allowed to proceed on their own, students may regard a simple printed list of results a sufficient measure of success. Studies by Maor (1991) in science and Ehman, Glenn, Johnson, and White (1992) in social studies yielded essentially the same results. Databases seem to offer the most effective and meaningful help when they are embedded in a structured problem-solving process and when the activity includes class and small-group discussion of search results.

Using Databases in the Classroom

When to use databases: Teacher productivity. Teachers can use databases to help them prepare classroom materials and to do tasks that they would otherwise have to do by hand or could not do at all. The appearance of the screen display may change greatly depending on the software package, so it is not possible to show a standard example. However, Figure 5.5 shows an example database display from one software package that a teacher might use. Teachers rely on databases to make their work more efficient and productive in several ways:

• **Inventorying and locating instructional resources.** Teachers can store titles and descriptions of instructional resources in a database to help them identify materials that meet certain instructional needs. If a school has a large collection of resources that are used by all of the teachers, the school's library/media center probably catalogs this collection on a database designed for this purpose. Some teachers like to keep databases of their own materials so they can match available resources with specific instructional needs quickly and easily.

• **Using information on students to plan instruction and enhance motivation.** Whether a teacher designs and keeps a personal information database or uses one from the school or school district office, this information can suggest many ways to meet students' individual needs. For example, the teacher might keep information on students' reading levels or particular learning problems. Information on each child's favorite sport or hobby could influence designs of motivating activities and selection of materials that would attract each child's attention. Personal touches for each student are easier to accomplish when all the information is on a database. A teacher might begin each week with a "birthday search" of the student database to give a special congratulations or banner for each student with a birthday.

• **Using information on students to respond to questions or perform required tasks.** Teachers are often asked to supply personal information on students or deal with situations relating to their personal needs, yet it is difficult to remember everything on dozens of students. For example, some students require special medication, and the teacher is responsible for reminding them to take it. A teacher might need to decide quickly whether a particular adult is authorized to take a child from the school. A database with these kinds of information can be very helpful.

- **Sending personalized letters to parents and others.** The capability to merge database information with a word processing document comes in handy whenever the teacher wants to send personalized notes to parents or to the students themselves. The teacher can create only one letter or note, and the database program takes care of the personalizing.

When to use databases: Teaching and learning activities. When database software became available for microcomputers in the 1980s, instructional activities with these tools quickly became popular. Thus, the literature reflects an increasing variety of applications for databases. The heaviest uses seem to be in social studies, but effective applications have been designed for topics in content areas from language arts to science. The following list summarizes the ways teachers are using databases to enhance learning:

- **Teaching research and study skills.** Skills in locating and organizing information to answer questions and learn new concepts have always been as fundamental as reading and writing skills. Students need good research and study skills not only for school assignments, but also to help them learn on their own outside school. As the volume of information in our society increases, the need to learn how to locate important information quickly also grows. Before computers, each family strived to buy reference tools such as a good dictionary and a set of encyclopedias so that children could do research for their school reports and other assignments. Today, these and other sources are stored on electronic media, and students need to know how to do computer searches of these references.

  In one sense, electronic formats make it easier to get to information, and, thus, they support students' learning and make it easier for them to acquire study skills. However, because of the unique capabilities offered by databases, looking up information in electronic formats is a far different activity than doing research in books. To take advantage of the new information resources available to them, students must learn new skills such as using keywords and Boolean logic (e.g., look for references with the keywords "Macintosh" AND "Apple" but NOT "fruit"). They must also learn skills in operating computer equipment and using procedures and commands required by database software. Many classroom activities with databases are designed to introduce these kinds of searches along with information for a report or a group project.

- **Teaching organization skills.** Students also need to understand concepts related to handling information. To solve problems, they must locate the right kinds of information and organize it in such a way that they can draw relationships between isolated elements. One way to teach these skills is to have students develop and use their own databases, and many examples of this kind of activity have been reported. Even very young students can learn about organizing information by creating databases of information about themselves: birth dates, heights, weights, eye colors, pets, parents' names. They can then do simple searches of their databases to summarize information about the members of their class. These kinds of activities help them understand what information is and to use it. In later grades, students can design and create databases related to content areas they are studying. For example, a class might create a database of descriptive information on candidates running for office across their state or in a national election.

- **Understanding the power of information "pictures."** Students need to understand the great persuasive power of information organized into databases. Sometimes a database can generate an information "picture" that may not be visible in any other way. While these pictures may or may not be completely accurate, many people make decisions based on them. For example, the U.S. government uses databases of information on people who have been convicted of past income tax offenses to generate descriptive profiles of people who may be likely to try to defraud the government in the future. Students can learn to use database information to generate these pictures, either with existing databases or their own.

- **Posing and testing hypotheses.** Many problem-solving activities involve asking questions and locating information to answer them. Therefore, using databases is an ideal way to teach and provide practice with this kind of problem solving. Students can either research prepared databases full of information related to a content area (e.g., presidents, animals), or they may create their own databases. Either way, these activities encourage them to look for information that will support or refute a position. In lower grades, the teacher may pose the question and assign students to search databases to answer it. Later, the activity may call for students to both pose and answer appropriate questions. For example, the teacher may ask students to address popular beliefs concerning artistic or gifted people. The students formulate questions, form debate teams, and design searches of databases on famous people to support their positions.

Example materials and integration strategies. The examples of Integration Strategies on pp. 149–151 present some teacher-created plans that integrate databases into classroom activities.

Learning to Use Databases: Techniques and Activities

A recommended instructional sequence. The following list of activities is recommended for first-time database users who prefer a structured, systematic approach to learning (as opposed to learning by doing).

Step 1   Manipulating database files. Loading database files, moving around in the database, and closing files

Step 2   Changing existing database information. Changing information in fields, inserting and deleting fields and records, changing field names, and sorting data by fields

Step 3   Creating a new database. Creating a database structure, naming fields, entering new information in fields

Step 4   Searching for and printing information from a database. Searching by keyword, structuring a report, printing a report

Step 5   Merging a database file with a word processing document. Creating a word processing letter with database fields, printing merged letters

Common mistakes and misconceptions when learning databases. Beginning users of databases often encounter some common problems. If you are having problems, look over this list again to see if you can find a solution:

Integration Strategies for
Teaching Research and Study Skills with Databases

Strategy #1: Locating Geographical Information
Developed by: Doug Magee, Lord Alexander School (Kemano, Canada)
Level: Grades 7–8
Purposes of the Activity: Locate geographical information needed to play Carmen Sandiego quickly and efficiently by using the World Geograph database

Instructional Activities. Let the class play the Carmen Sandiego game in teams of two or three students. One team member keeps a record of clues about the country and the time period. Once each team catches the criminal, they use World Geograph to locate the countries they "traveled" in the game and develop fact sheets, maps, graphs, and questions on them. The teams can also make graphs to show comparisons among countries on selected categories (e.g., area, population, language, religion).

Limit the number of categories available from World Geograph in order to limit the comparisons later. The 55 categories of information available in the database can overwhelm middle school students.

Strategy #2: Using Databases of Sources to Support Research
Developed by: Michael Coe and Thomas Butts, Plano Senior High School (Plano, Texas)
Level: Grades 11–12
Purposes of the Activity: Students use a database that lists resources by topic to locate information for their research projects.

Instructional Activities. Assigning research projects helps keep both teachers and their students up-to-date on new developments. Keeping and updating a database of sources for this research not only helps a student learn skills in locating information, it also provides information on the most important sources for a field. A student selects a topic of interest in the field, reviews the sources in the database, and researches the topic using materials outside the classroom. He or she writes a report based on this research and presents it orally to the class.

Problems to Anticipate. The teacher needs to provide considerable guidance to the student researchers so they do not become too frustrated during the process of learning research skills. At first, the teacher may want to assign specific articles or give a list of several from which students can choose.

Source: Magee, D. (1991). Carmen Sandiego and World Geograph. *The Computing Teacher, 19* (3), 31–32; Coe, M. and Butts, T. (1991). Keeping up with technology, *The Computing Teacher, 18* (5), 14–15.

Integration Strategies for
Teaching Organization Skills with Databases

Strategy #1: A Database Yearbook
Developed by: Ruth Hollis, Peru Central School (Plattsburg, New York)
Level: Grade 2
Purposes of the Activity: Information on all students in the second grade is placed in a database in order to organize it for a class yearbook and support various classroom activities.

Instructional Activities. The second grade teachers introduced the yearbook project to students by talking about the district's curricular theme ("Beginnings") and about the things students would like to know about each other. Each class brainstormed the 10 most important kinds of information about themselves. Then all the second grade teachers consolidated the areas into the final 10 to be included in the database. The computer teacher created a database template and modeled for the teachers how to collect the information with their classes by using butcher paper on the wall. Once the classes collected the information, students worked in pairs to enter it into the database in their own classrooms. When the database was completed, teachers helped the students use it to answer questions about themselves, e.g., "How many kids like pizza?" and "What is the class favorite color?" Teachers then created the yearbooks by printing out the database records for their

Continued

classes and adding a picture for each student, a graphic cover, and an introduction. Many kinds of learning activities were designed based on this information.

Problems to Anticipate. When students enter information in the database, encourage them to have a doer and a helper and exchange roles halfway through the task. The helper should help by holding the paper rather than touching the keys.

Strategy #2:	Let the Computer Do It
Developed by:	Leo Jankowski (Dunedin, New Zealand)
Level:	Children ages 8–9
Purposes of the Activity:	Students learn how to use databases to store, organize, and locate information.

Instructional Activities. Each pair of students creates an on-paper data file. Each sheet of paper is a record containing three fields of data: last name, first name, and address. Students are asked to do searches on one and two variables using this on-paper data; they quickly find out how unwieldy this system becomes. They are then shown how to place the information in a database. The teacher discusses what a sort is and demonstrates sorting on a field name. Then search techniques are demonstrated. The class is divided into groups, and each group is assigned to create its own databases. Each group practices sorting data on various fields, printing the sorted lists of records, and searching for records. The teacher shows the students how to add three new fields: age, height, and weight. Students gather this new information, add it to the database, and use the completed data summary to answer questions about their class, (e.g., "Who is the shortest person?") Next, the students load a prepared database of information on a topic they need to study (e.g., a country). They learn to sort and search this database in much the same way as the one they designed, but they use specific questions that focus on important characteristics of the country. Finally, they add to this database other information about the country (e.g., 20 more towns and cities).

Problems to Anticipate. It is helpful to place required database program commands on big sheets on the wall so everyone can see them readily. Also, be sure to save often to prevent loss of data.

Source: Hollis, R. (1990). Database yearbooks in the second grade. *The Computing Teacher, 17* (6), 14–15; Jankowski, L. Getting started with databases. *The Computing Teacher, 21* (4), 8–9.

Integration Strategy for
Using Databases to Create Information Pictures

Strategy #1:	Getting a Clearer Focus on History
Developed by:	Rick Thomas (Eugene, Oregon)
Level:	High school
Purposes of the Activity:	Students use a database of 1835 census information on Cherokee citizens of the United States to gain a more accurate historical perspective on the Cherokee lifestyle and culture of the time than is available from traditional written sources.

Instructional Activities. This activity is designed to impress social studies students with the power of database information for finding a truer picture of people or events than they can get in sources such as public papers, letters, and diaries. For example, much information about the lives of long-dead people can be gleaned from information such as tax rules, census reports, and church and business records—information that is difficult to manage without a database.

The example in this lesson is a study of the Cherokee nation in the 1800s. The documents of the time paint a picture of a savage and uncultivated people. Even speeches by famous politicians of the time such as Andrew Jackson support this view. Students begin by reviewing this traditional information and discussing the rationale for the removal of the Cherokees from Georgia, Tennessee, North Carolina, and Alabama by 1839 in what came to be known as the "Trail of Tears." Then students begin to pose questions such as "What kind of people were the Cherokee?" and "How will our understanding of their life affect our perspective on their removal?"

Continued

Students begin their review of the database of 1835 census information on the Cherokee nation by exploring the contents and organization of the file. Then they begin to pose questions that evaluate the generalizations about the Cherokee that are evident in documents of the time. Ask them to formulate conclusions about the Cherokee lifestyle. Make an assignment for students to write a summary of their findings and whether or not their beliefs about the events of the time have changed as a result of what they found.

Problems to Anticipate. Teachers may need to motivate students to get started on their exploration questions by suggesting they examine individual names and how they reflect personalities and interests. Ask what they say about the Cherokee nation. For example, note the complete absence of names denoting violence or extreme action.

Source: Thomas, R. (1991). A focus on history. *The Computing Teacher, 18* (6), 30–33. The database activity is described in *Classworks: Microsoft Works for the Classroom* (Eugene, Oregon: ISTE).

Integration Strategies for
Posing and Testing Hypotheses with Databases

Strategy #1: Investigating Science with Databases
Developed by: A.W. Strickland, University of Maryland, and Terry Hoffer, Eastern Montana College
Level: High school
Purposes of the Activity: Students use a database to store data from a science experiment and then use it to form and test hypotheses about chemicals.

Instructional Activities. Science classes focus on the question, "What physical properties would be useful for identifying pure substances?" Students list properties such as physical state, melting point, color, solubility, and density. They use these properties to create a database template. The teacher uses a large-screen monitor to display the template so the students can see the development of the database. Then students look up the physical properties of the sample compounds and enter them into the database. Each student is assigned three pure substances to research and enter into the database. The teacher then assigns an exercise to give students an opportunity to develop database query skills. After they complete the exercise, they set up and print a report on their findings.

Problems to Anticipate. Students will need to develop strategies to focus their searches and make them a more efficient use of time. For example, they can enter criteria that search for a range of values for a variable such as melting point, rather than for an exact value.

Strategy #2: A Rock Database Project
Developed by: Thalia Hartson (Alberta, Canada)
Level: Grades 3–8
Purposes of the Activity: A unit of study on rocks forms the basis for learning science inquiry skills. A database of information on rocks helps organize the information for use in supporting these problem-solving activities.

Instructional Activities. The National Geographic filmstrip *Rocks and Minerals* supplies background information on the unit. This stimulates students' natural interest in rocks and minerals and they spontaneously begin to collect rock samples to sort, observe, and classify. The teacher introduces the three main types of rocks and the rock formation cycle. Laserdisc pictures supplement reference book pictures. Students each select one mineral to research. They enter the information they find into a database. The teacher and student then develop questions and use the database to answer them. Typical questions are: "Which crystal structures are the most common?" and "What makes certain minerals valuable?" A printed set of questions can be used later as a consolidating exercise. They use the MECC simulation Murphy's Minerals to follow-up this technique of testing and identifying minerals according to characteristics.

Problems to Anticipate. Students should enter their data in separate files which the teacher later consolidates. This seems to prevent errors and unintentional deletions.

Source: Strickland, A. W., and Hoffer, T. (1990–1991). Integrating computer databases with laboratory problems. *The Computing Teacher 18* (4), 30–32; and Hartson, T. (1993). Rocks, minerals, databases, and simulations. *The Computing Teacher, 21* (1), 48–50.

• **Confusing spreadsheet and database features.** Depending on the software package, a spreadsheet or worksheet may look very similar to a database listing. Sometimes learners get confused about which they are seeing on the screen and which of their activities call for which kind of software. Some spreadsheet software (e.g., QuattroPro) has many database-like capabilities in the same package. But other packages (e.g., Microsoft Works) have separate software components for spreadsheet and database functions, and they produce separate outputs for each one. When a spreadsheet is a separate package or component, it can normally be identified by its characteristic row numbers and column letters. Use a spreadsheet for anything that calls for extensive calculations; to organize information so it can be searched, you need database capabilities. A spreadsheet can produce a simple chart with nicely formatted straight rows and columns more easily than a database.

• **Difficulties with keyword searches.** The most useful feature of databases also presents the most difficulties for new users. A keyword search automatically reviews the whole database and selects only those records that contain certain words, phrases, or numbers. A user can combine several different keywords using the principles of Boolean logic. New users make the most mistakes with database searches when they do not understand how this logic works. When constructing a search, it is sometimes helpful to create a diagram or picture that shows graphically the results expected. See the example below for Apple AND Macintosh BUT NOT fruit:

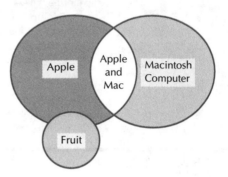

• **Decisions about which fields to include.** Most people who have never before used databases have difficulty conceptualizing just what fields to include to make a helpful database file. For example, a teacher who will want to do mail merges in order to send personal letters to students may want to include a separate field for first name only. The teacher may find it helpful to discuss with others the purpose of the database and how it will be used when it is complete. The teacher may also want to review some examples of similar databases done by others.

Exercises

Word Processing Exercises

1. **Word processing terms.** Fill in the word processing term for each of the following features:

    _____ 1. What is a file called when you save it in a form so that it can be used by a different word processing program than the one that created it?

    _____ 2. What word processing feature changes all instances of an incorrect word to the correct word throughout an entire document?

    _____ 3. What optional feature in some word processors tells the user if words are misspelled?

    _____ 4. What is another word for a choice of typeface (e.g., Courier, Chicago)?

    _____ 5. What is the name for the blinking bar within a word processing document that indicates where text may be inserted?

    _____ 6. What word processor command saves a document under a different name?

    _____ 7. What name describes text flush against both right and left margins?

    _____ 8. What is it called when the text goes to the next line automatically without pressing Return or Enter?

    _____ 9. What three-word, hyphenated term describes taking something out of a document and putting it in another place in the document?

    _____ 10. Name the text item that appears at the top of every page of a word processing document.

2. **Word processing applications.** Describe the ways you plan you to use word processing to enhance your students' learning activities and your productivity as a teacher.

3. **Word processing development.** Develop a classroom lesson activity that integrates word processing functions into instruction in one or more of the following ways:

    • Writing processes
    • Dynamic group products
    • Language and writing exercises

    The lesson activity will be evaluated according to the following criteria. It should:

- Be planned for a realistic time frame, allowing for students to learn word processing before applying it
- Use the unique capabilities of word processing to support instructional activities
- Save instructional time and/or assist with the logistics of activities involved in the lesson

Spreadsheet Exercises

1. **Spreadsheet terms.** Fill in the spreadsheet term for each of the following features:

_____ 1. What is location in a spreadsheet, identified by a row–column position, that can store one item of information?

_____ 2. Give a name for the item in a spreadsheet that does calculations on other data.

_____ 3. What is another name for the kind of data in a spreadsheet on which calculations can be done?

_____ 4. What is another name (besides *spreadsheet*) for the product of a spreadsheet program?

_____ 5. Name special-purpose functions that automatically compare numbers with those in the spreadsheet and assign labels or values (e.g., letter grades) for those that match.

2. **Spreadsheet applications.** Describe the ways you plan you to use spreadsheet software to enhance your students' learning activities and your productivity as a teacher.

3. **Spreadsheet development.** Develop a classroom lesson activity that integrates spreadsheet functions into instruction in one or more of the following ways:

- Presenting demonstrations
- Developing student products
- Supporting problem solving
- Storing and analyzing data
- Projecting grades

The lesson activity will be evaluated according to the following criteria. It should:

- Be planned for a realistic time frame, allowing for students to learn spreadsheet use before applying it
- Use the unique capabilities of a spreadsheet to support instructional activities
- Save instructional time and/or clarify the concepts involved in the lesson

Database Exercises

1. **Database terms.** Fill in the database term for each of the following features:

_____ 1. Name a type of database program that relates two or more files through a shared key field.

_____ 2. What type of database program stores information as separate, rather than related, files?

_____ 3. What is the smallest unit of information in a database?

_____ 4. Give the name for all items of information related to a database entry (e.g., in a database of teaching resources, all items of information about one resource).

_____ 5. State another name besides *database* for database program.

_____ 6. What is a term for the process of looking for information that matches certain keywords?

_____ 7. Name a term for the word or words used to obtain a subset of information from the database.

_____ 8. What process automatically combines the entries in a database with a single word processing document?

2. **Database applications.** Describe the ways you plan you to use database software to enhance your students' learning activities and your productivity as a teacher.

3. **Database development.** Develop a classroom lesson activity that integrates database functions into instruction in one or more of the following ways:

- Teaching research and study skills
- Teaching information organization skills
- Creating information pictures
- Posing and testing hypotheses

The lesson activity will be evaluated according to the following criteria. It should:

- Have a realistic time frame, so students can learn database use before applying it
- Use the unique capabilities of databases to support instructional activities
- Save instructional time, support the logistics involved in the lesson, and/or make possible instructional activities that would be impossible otherwise

References

References for Word Processing

Bangert-Drowns, R. (1993). The word processor as an instructional tool: A meta-analysis of word processing in writing instruction. *Review of Educational Research, 63* (1), 69–93.

Barker, E. (1992). Reading, writing, and reporting. *Florida Technology in Education Quarterly, 4* (2), 78–81.

Elder, J., Schwartz, J., Bowen, B., and Goswami, D. (1989). *Word processing in a community of writers.* New York: Garland.

Harris, J. (1985). Student writers and word processing. *College Composition and Communication, 36* (3), 323–330.

Hawisher, G. (1986, April). The effects of word processing on the revision strategies of college students. Paper presented at the Annual Meeting of the American Educational Research Association, San Francisco, CA (ERIC Document Reproduction Service No. ED 268 546).

Hawisher, G. (1989). Research and recommendations for computers and compositions. In G. Hawisher and C. Selfe (Eds.). *Critical perspectives on computers and composition instruction.* New York: Teachers College Press.

Humphries, D. (1989). A computer training program for teachers. In C. Selfe, D. Rodrigues, and W. Oates (Eds.). *Computers in English and the Language Arts.* Urbana, Illinois: National Council of Teachers of English.

King, R. and Vockell, E. (1991). *The computer in the language arts curriculum.* Watsonville, CA: Mitchell McGraw-Hill.

Kuechle, N. (1990). Computers and first grade writing: A learning center approach. *The Computing Teacher, 18* (1), 39–41.

Levin, J. A., Boruta, M. J., and Vasconcellos, M. T. (1983). Microcomputer-based environments for writing: A writer's assistant. In A. C. Wilkinson (Ed.). *Classroom computer and cognitive science.* New York: Academic Press.

Levin, J. A., Riel, M., Rowe, R., and Boruta, M. (1984). Muktuk meets Jacuzzi: Computer networks and elementary school writers. In S. W. Freeman (Ed.). *The acquisition of written language: Revision and response.* Hillsdale, NJ: Ablex.

Lockard, J., Abrams, P., and Many, W. (1994). *Microcomputers for 21st century educators* (3rd ed.). New York: HarperCollins.

Mehan, H., Miller-Souviney, B., and Riel, M. (1984, April). Knowledge of text editing and the control of literacy skill. Paper presented at the Annual Meeting of the American Educational Research Association, New Orleans, LA.

Morehouse, D., Hoaglund, M., and Schmidt, R. (1987, February). *Technology demonstration program final evaluation report.* Menomonie, WI: Quality Evaluation and Development.

Nelson, M. H. (1994). Processing poetry to develop literacy. *The Computing Teacher, 22* (3), 39–41.

Roblyer, M., Castine, W., and King, F.J. (1988). *Assessing the impact of computer-based instruction: A review of recent research.* New York: Haworth Press.

Schwartz, E. and Vockell, E. (1989). *The computer in the English curriculum.* Watsonville, CA: Mitchell.

Schramm, R. (1989). The effects of using word processing equipment in writing instruction: A meta-analysis.

Unpublished doctoral dissertation, Northern Illinois University, DeKalb.

VerMulm, L. (1993). The Christa McAuliffe writing center: Process writing with a networked Mac lab. *The Computing Teacher, 20* (7), 48–53.

Wresch, W. (1990). Collaborative writing projects: Lesson plans for the computer age. *The Computing Teacher, 18* (2), 19–21.

References for Spreadsheets

Baugh, I. (1995). Tool or terror? *The Computing Teacher, 22* (5), 14–16.

Beaver, J. (1992). Using computer power to improve your teaching, Part II: Spreadsheets and charting. *The Computing Teacher, 19* (6), 22–24.

Bozeman, W. (1992). Spreadsheets. In G. Bitter (Ed.). *Macmillan encyclopedia of computers.* New York: Macmillan.

Collis, B. (1988). *Computer, curriculum, and whole-class instruction.* Belmont, CA: Wadsworth.

Edwards, J. (1992). What's in our trash? *Florida Technology in Education Quarterly, 4* (2), 86–88.

Goldberg, K. (1990). Bringing mathematics to the social studies class: Spreadsheets and the electoral process. *The Computing Teacher, 18* (1), 35–38.

Harriss, S. (1992). M&M count. *Florida Technology in Education Quarterly, 4* (2), 60–61.

Klass, P. (1988, April). Using microcomputer spreadsheet programs to teach statistical concepts. (ERIC Document Reproduction Service No. ED 293 726).

Paul, J. R. M. (1995). Pizza and spaghetti: Solving math problems in the primary classroom. *The Computing Teacher, 22* (7), 65–67.

Ploger, D., Rooney, M., and Klingler, L. (1996). Applying spreadsheets and draw programs in the classroom. *Tech Trends, 41* (3), 26–29.

Sgroi, R. (1992). Systematizing trial and error using spreadsheets. *Arithmetic Teacher, 39* (7), 8–12.

Sutherland, R. (1993). A spreadsheet approach to solving algebra problems. *Journal of Mathematical Behavior, 12* (4), 353–383.

Voteline: A project for integrating computer databases, spreadsheets, and telecomputing into high school social studies instruction. (1992). Report of a statewide North Carolina project. (ERIC Document Reproduction Service No. ED 350 243).

References for Databases

Coe, M. and Butts, T. (1991). Keeping up with technology. *The Computing Teacher, 18* (5), 14–15.

Collis, B. (1988). *Computer, curriculum, and whole-class instruction.* Belmont, CA: Wadsworth.

Collis, B. (1990). *The best of research windows: Trends and issues in educational computing.* Eugene, OR: International Society for Technology in Education (ERIC Document Reproduction No. ED 323 993).

Ehman, L., Glenn, A., Johnson, V., and White, C. (1992). Using computer databases in student problem solving: A study of eight social studies teachers' classrooms. *Theory and Research in Social Education, 20* (2), 179–206.

Heine, E. (1994). The world at their fingertips. *The Florida Technology in Education Quarterly, 7* (1), 38–42.

Hollis, R. (1990). Database yearbooks in the second grade. *The Computing Teacher, 17* (6), 14–15.

Hartson, T. (1993). Rocks, minerals, databases, and simulations. *The Computing Teacher, 21* (1), 48–50.

Hunter, B. (1983). *My students use computers.* Alexandria, VA: Human Resources Research Organization.

Jankowski, L. (1993–1994). Getting started with databases. *The Computing Teacher, 21* (4), 8–9.

Magee, D. (1991). Carmen Sandiego and world geography. *The Computing Teacher, 19* (3), 31–32.

Maor, D. (1991 April). Development of student inquiry skills: A constructivist approach in a computerized classroom environment. Paper presented at the Annual Meeting of the National Association for Research in Science Teaching, Lake Geneva, WI, April 7–10, 1991 (ERIC Document Reproduction No. ED 326 261).

Strickland, A. and Hoffer, T. (1990–1991). Integrating computer databases with laboratory problems. *The Computing Teacher, 18* (4), 30–32.

Thomas, R. (1991). A focus on history. *The Computing Teacher 18,* 6, 30–33.

Watson, J. (1993). *Teaching thinking skills with databases* (Macintosh version). Eugene, OR: ISTE.

Wetzel, K. and Painter, S. (1994). *Microsoft Works 3.0 for the Macintosh: A workbook for educators.* Eugene, OR: ISTE.

Software References (For company addresses, see Appendix B)

Sample Integrated Software Packages

This list includes both single packages with several integrated parts (e.g., Microsoft Works), as well as packages that are actually several programs designed to work well together (e.g., Smart Suite).

ClarisWorks—Claris Corporation
Microsoft Works—Microsoft Corporation
Smart Suite—Lotus
Corel Office—Corel, Inc.
WordPerfect Suite—Corel, Inc.

Sample Word Processing Software

Bank Street Writer—Scholastic Inc.
Kidwriter Gold—Spinnaker
Logowriter—Logo Computer Systems Inc.
MacWrite Pro—Claris Corporation
Microsoft Word—Microsoft Corporation
Magic Slate—Sunburst Communications
Research Paper Writer—Tom Snyder
Talking TextWriter—Scholastic Inc.
WordPerfect—Novell, Inc.
Word Pro—Lotus, Inc.
The Writing Center—The Learning Company
Writer's Helper—CONDUIT

Sample Spreadsheet Software

Excel—Microsoft Corporation
Lotus 1-2-3—Lotus Development
QuattroPro—Novell Corporation

Sample Database Software and Ready-made Files

Database programs
    Access—Microsoft
    Approach—Lotus
    Bank Street School Filer—Sunburst Communications
    dBase—Borland
    Filemaker Pro—Claris Corporation
    4th Dimension—STI?
    Fox Pro—Fox Software
    Oracle—Oracle, Inc.
    Paradox—Borland
Ready-made databases:
    Teacher's Idea and Information Exchange, P. O. Box 6229, Lincoln, NE 68506

Chapter 6

Using Various Technology Tools to Support Teaching and Learning

This chapter will cover the following topics:

- Definitions and characteristics of a variety of technology tools

- Unique advantages of these tools for various classroom activities

- Example classroom uses for each tool

Chapter Objectives

1. Use correct terminology to identify the names and features of several technology support tools.

2. For each of several teaching and learning tasks, identify one or more technology tools that can enhance and support the activity.

"... [T]]he notion that computers are neutral and just another technological tool may seem quite a reasonable one.... Computers are just powerful tools that people will use as they see fit. But ... think of the influence the automobile and television have had on our culture. The evidence is accumulating that computers are having a decided impact on the way our schools and society organize, communicate, and make decisions.
Joe Nathan (From Micro-Myths, 1989)

Introduction to Technology Support Tools

Why Use Technology Support Tools?

Chapter 5 described productivity applications and example classroom integration strategies for the three most common technology support tools: word processing, spreadsheet, and database programs. Chapter 6 focuses on the remaining tools, a wide variety of computer-based products that can support teachers and students in a multitude of teaching and learning tasks. These tools vary greatly in their purposes, the kinds of benefits they offer, and their utility for teachers. Some, such as electronic gradebooks and CMI tools, are designed to organize and analyze information; they are fast earning an image as indispensable aids for teachers struggling to cope with increasing amounts of data related to student performance and achievement. Other tools, such as certificate makers and clip art packages, serve merely to improve the appearance of instructional products and make it easier for teachers and students to produce attractive, professional-looking materials that inspire pride.

The tools described here range in importance from nearly essential to nice to have, and in function from presenting instruction itself to supporting background tasks that make a classroom function smoothly. However, each one has unique and powerful features. As Nathan (1985) emphasized, these tools, if used wisely and creatively, have the potential not only to support classroom activities, but also to transform the very nature of the way people learn and work.

Each tool described in this chapter requires both additional classroom resources and time to learn and to implement. Teachers should choose them for the qualities and benefits they bring to the classroom, rather than simply because they are available on the market. Depending on the capabilities of a particular tool and the needs of the situation, a technology support tool can offer several kinds of benefits:

- **Improved productivity.** Many of the tools described in this chapter can make it faster to get organized, produce instructional materials, or accomplish paperwork tasks that teachers must do anyway (e.g., create tests or calculate grades). Using a technology tool to do these tasks can free up valuable time that can be rechanneled toward working with students or designing learning activities.

- **Improved appearance.** Many tools help teachers to produce polished-looking materials that resemble the work of professional designers. In fact, these tools are frequently the same ones used by professional designers. The quality of classroom products is limited only by the talents and skills of the teachers

and students using the tools. Students appreciate receiving more attractive-looking materials, and they also find it rewarding and challenging to produce handsome products of their own.

- **Improved accuracy.** Several tools make it easier to keep more precise, accurate records of events and student accomplishments. More accurate information can support better instructional decisions about curriculum and student activities.

- **More support for interaction.** Some products have capabilities that promote interaction among students or allow input from several people at once. These qualities can encourage many creative, cooperative group-learning activities.

Types of Technology Support Tools

This chapter divides support tools into six general categories that describe their functions. Example software products available for each type are listed at the end of the chapter:

- **Materials generators.** Tools that assist in teachers' production of instructional materials

- **Data collection and analysis tools.** Resources that help teachers collect and organize numerical information that indicates student progress

- **Graphics tools.** Tools that allow production of pictures and illustrated written products

- **Planning and organizing tools.** A variety of tools that help teachers and students conceptualize their work before they actually begin it

- **Research and reference tools.** CD-ROM versions of encyclopedias, atlases, and dictionaries

- **Tools to support specific content areas.** Tools that assist with activities associated with certain content areas

Using Materials Generators

Desktop Publishing Software

Definitions: Desktop publishing versus desktop publishing software. It is perhaps ironic that one of the most useful and widely used of the technology tools is one that communicates information in a traditional medium: the printed page. By allowing teachers and students to design elaborate printed products, however, desktop publishing tools give them the very important advantage of complete control over a potentially powerful form of communication. This control over the form and appearance of the printed page defines the activity of desktop publishing. Norvelle (1992) reported that the term *desktop publishing* was coined in 1984 by Paul Brainerd, founder of the Aldus Corporation, to focus on the role of personal computers and laser printers in making the individual-as-publisher a viable concept.

Like word processing software, desktop publishing software allows manipulation of text, and the capabilities and classroom applications of these two products overlap. According to many accounts in the educational technology literature, one can perform the tasks of desktop publishing with many available word processing software packages; the key element in such an activity is control over the design and production of a document. Both kinds of tools allow users to mix text and graphics on each page.

The primary difference between software *designed* for word processing and that *designed* for desktop publishing is that the latter is designed to display documents page by page. It also allows more easy flexibility over the placement and formats of both text and graphics on individual pages. Word processing is designed to "flow" text from page to page as it is typed in. Text boxes can appear anywhere on a desktop-designed page. As a result, classroom products created with desktop publishing software can be as eye-catching and professional-looking as those produced by the most prestigious ad agency in New York or San Francisco. The impact of a desktop published product is limited only by the capabilities of the person creating it. Just as the quality of a word processing article depends on the skill of the writer, the quality of a newsletter or brochure from a desktop publishing system depends on the creativity and expertise of the designer. Thus, desktop publishing software focuses on designing communications through a combination of written words and page appearance, while word processing communicates the message primarily through words (McCain, 1993; Williams, 1994).

Elements of desktop publishing. Desktop publishing software gives power to users by allowing them to control three elements: page setup, text format, and graphics. They can create text boxes of any size and place them anywhere on the page. Users can manipulate text format by changing type font, size, style, and color. As noted in Chapter 5, software offers fonts in an ever-increasing array from very plain, typewriter-looking Courier to very fancy *Boulevard*. Font size, measured in points (1 point equals about 1/72 inch), ranges from very small (6 or 7 points) to as large as will fit on a page. (The type size for text in most books is about 10 points.) *Type style* refers to appearance changes such as **boldface**, *italics*, outlining, and underlining. Desktop publishing software also allows users to choose colors of type. For example, many documents use the special effect of white type on black backgrounds. Graphics of any kind can be designed using drawing tools provided by the software, or the user can bring existing pictures and diagrams into the document (import them).

Types of desktop publishing software. Teachers can choose from a wide range of desktop publishing software for classroom use. Several packages are designed especially for education or for simple newsletter production (e.g., Microsoft's *Publish It!*). For more sophisticated work, they may select one of the higher-end packages used in professional design offices (e.g., *Quark Xpress* or *PageMaker*). Naturally, the more capable packages take more time to learn. Schools usually choose them for more complicated design tasks that require advanced page design features

such as designing complex graphics and rotating graphic and text elements on the page.

Making the most of desktop publishing software: Skills and resources. Like other technology tools, desktop publishing contributes the most if the user knows something about the activity before applying the tool. Designing effective print communications is an entire field of expertise in itself with its own degree programs. Graduates are frequently in high demand in business and industry. As Knupfer and McIsaac (1989) observe, many aspects of page design can influence reading speed and comprehension. They describe four different categories of variables that have been researched: graphic design, instructional text design, instructional graphics, and computer screen design. While the last category focuses on reading from a computer screen, Knupfer and McIsaac note that "certain of its features apply to both electronic and print-based research" (p. 129), including factors like text density and uppercase versus lowercase type.

However, even with this in mind, teachers and students need not be professional designers to create useful desktop publishing products, and their skills will improve with practice. According to Parker (1989) and Rose (1988), desktop publishing products will have greater impact and communicate more clearly if they reflect some fairly simple design criteria. Beginners may want to keep the following suggestions in mind:

- **Select and use typefaces (fonts) carefully.** Unusual typefaces can help direct the eye toward text, but too many different fonts on a page are distracting and *some fancy fonts can be difficult to read*. A serif font (a font with small curves or "hands and feet" that finish off the ends of the letters) is easier to read in paragraphs; use it for text in the main body of the document. Use sans serif type (a font without "hands and feet") for titles and headlines. Make type large enough to assist the reader. For example, younger readers usually need larger point sizes than older ones.

- **Use visual cueing.** When certain information on the page is very important, attract readers toward it by cueing them in one of several ways. Desktop publishing allows users to employ cueing devices like frames or boxes around text, bullets or arrows to designate important points, shading the part of the page behind the text, and changing the text itself in some way such as **boldface** or *italic* type. Captions for pictures, diagrams, and headings also help to guide the reader's attention.

- **Use white space well.** There is a saying in advertising that "white space sells." Don't be afraid to leave areas in a document with nothing in them at all; this will help focus attention on areas that do have information.

- **Create and use graphics carefully.** Use pictures and designs to focus attention and convey information, but remember that too many elaborate pictures or graphic designs can be distracting.

- **Avoid common text format errors.** Parker (1989) describes ten "desktop design pitfalls" to avoid. These include irregularly shaped blocks of text, angled type, excessive underlining, widows and orphans (leftover single words and phrases

---

**Integration Strategy for
Desktop Publishing**

Strategy: In the News
Developed by: Connie Skinner, Gla Culpepper, and Marsha Wiggam (Duval County, Florida); Elaine Myers, Florida
                       Diagnostic Learning Resources System; NEFEC; and Jack Wright, Florida Diagnostic Learning
                       Resources System/Crown Region
Level: Middle school
Lesson Purposes: These activities show students the components of a news publication and how to produce a text and
                       graphics layout for a publication.

Instructional Activities. Begin by discussing the project with students and showing them examples of the kind of publication they will produce. Focus on the parts of a news story by having them preview a local TV broadcast looking for specific facts in each news story (e.g., who, what, when, where, why). Go on field trips to the local newspaper and TV and radio stations to talk to the professionals at these locations. Have guest speakers from the local area such as newspaper reporters and TV and radio news reporters. Discuss the terminology of the newspaper (which students should have been hearing about in their field trips and from speakers). Do a hands-on exploration of example newspapers, identifying the parts of each one. Teach students how to use the desktop publishing program to produce each part of one of the example layouts. Then have students move on to designing their own newspaper. They form small groups, with each group working on a section of the publication, writing their own stories, laying them out, and editing the final product. Students who have artistic talent produce the graphics, emulating the examples they have seen in other newspapers.

Source: Skinner, C., Culpepper, G., Wiggam, M., Myers, E., and Wright, J. (1991). In the news. *The Florida Technology in Education Quarterly, 3* (2), 73–74.

---

at the tops of pages), unequal spacing, excessive hyphenation, exaggerated tabs and indentations, grammatical errors, cramped logos and addresses, and too many typefaces.

To take full advantage of desktop publishing software, a system needs the right hardware and additional software. Remember some helpful resource hints:

* **Recommended hardware.** With any design elements, desktop publishing products are always more professional looking if they are printed out on a laser printer rather than a dot matrix one. An optical scanner can also be handy to import existing pictures and other graphics into the computer for use in a document.

* **Recommended software tools.** Some predesigned graphic tools can be very useful. One such tool is clip art packages, or computer files of drawings and logos. These are available on disk or CD-ROM and can be combined with each other or with original drawings to produce a desired picture. Some people also buy clip art books and scan the pieces of art into files as needed. Whenever a document will be reused with modifications, it is helpful to save it in a file of templates or predesigned products.

Example classroom applications. Desktop publishing software can be used for many of the same classroom activities and products as word processing software. Desktop publishing is the tool of choice, however, to produce more elaborate, graphic-oriented documents. Teachers can use desktop publishing software to help them produce notices and documents for parents, students, or other faculty members.

But desktop publishing can also support some highly motivating classroom projects. Hermann (1988), McCarthy (1988), Newman (1988), and Beeken (1992) have reported instructional benefits of these uses for both students and teachers. These include increases in children's self-esteem when they publish their own work, heightened interest in and motivation to write for audiences outside the classroom, and improved quality of instruction through teacher collaboration. A list of common classroom applications and ideas for implementing them include:

* **Letterhead.** Teachers can design their own individual stationery and encourage students to do the same for themselves or for the class. Students can design letterheads for their clubs or even for the school itself.

* **Flyers and posters.** Whenever teachers or students must design announcements, desktop publishing can turn the chore into an instructional adventure. A simple notice of some upcoming event can become an opportunity to learn about designing attractive and interesting communications. Teachers can smuggle in instruction in grammar and spelling.

* **Brochures.** To communicate more extensive information than a simple flyer can convey or to provide information on more than one occasion, student projects can revolve around creating brochures. Popular examples of such classroom projects include travel brochures that report on students' explorations during field trips, descriptions of their local region, and creative descriptions of organizations or activities.

* **Newsletters and magazines.** The literature has reported many examples of classroom projects built around student-designed newsletters or magazines. Sometimes this activity

Figure 6.1　Desktop Publishing Product/PageMaker

Source: Courtesy of the Florida Design Initiative, Florida A&M University, School of Architecture.

represents the culmination of a large project such as a series of science experiments or a social studies research unit; sometimes it is simply a way to get every student to contribute writing for a class product. All these projects are reported to be highly motivational to students, and they attract "good press" for the teacher and the school. Figure 6.1 shows a cover page from a teacher-produced newsletter.

- **Books and booklets.** Even the youngest students are thrilled to produce and display their own personal books. Sometimes these products present work produced by students over the course of a school year; sometimes they show creative works that result from a competition; frequently they collect examples of the students' best work for a particular topic or time period. Students have often sold their publications as fund-raising activities, but this kind of project is reported to reap other real benefits for students at all ability levels. One teacher of low-achieving students found that, "Making their books has really turned these kids around. … Getting published … had an enormous positive impact on … self confidence and self-esteem" (McCarthy, 1988, p. 25).

Test Generators and Test Question Banks

Computer software is available to help teachers with what many consider one of the most onerous and time-consuming of instructional tasks: producing on-paper tests. The teacher creates and enters the questions, and the program prepares the test. The teacher may either print out the required number of copies on the printer or print only one copy and make the required copies on a copy machine. These tools have several advantages, even over word processing programs (Gullickson and Farland, 1991). For one, they produce tests in a standard layout; the teacher need not worry about arranging the spacing and format of the page. For another, such as program can automatically produce various forms or versions of the same test upon request. Also, changes, deletions, and updates to questions are also easy to accomplish, again without concern for page format. The features of test generators vary, but the following capabilities are common:

- **Test creation procedures.** The software prompts teachers to create tests item-by-item in formats such as multiple-choice, fill-in-the blank, true/false, matching, and (less often) short answer and essay.

- **Random generation of questions.** Test items are randomly selected from an item pool so that a different version of the test can be created with each request. This is especially helpful when the teacher wants to prevent the "Wandering Eye Syndrome" as students are taking the test.

- **Selection of questions based on criteria.** Programs usually allow teachers to specify criteria for generating a given test. For example, items can be requested in a specific content area, matched to certain objectives, or in a certain format (e. g., short answer items only).

- **Answer keys.** Most programs automatically provide an answer key at the time the test is generated. This can be helpful when it comes to grading, especially if there are different versions of the test to grade.

- **On-screen testing and grading.** Most test generators offer only on-paper versions of tests, but some allow students to take the test on-screen after it is prepared as Figure 6.2 illustrates. These programs usually also provide for automatic grading of the test and summary statistics on performance on each question.

Many test generators allow use of existing question pools, or test question banks, and some offer these banks for purchase in various content areas. Some programs can also import question banks prepared on word processors.

Worksheet Generators

Teachers can also use software to produce worksheets that is very similar in many ways to test generators. Worksheet generators help teachers produce exercises for practice rather than tests. Like the test generator, the worksheet generator software prompts the teacher to enter questions of various kinds, but it usually offers no options for completing exercises on screen or grading them. The most common worksheet generators deal with lower-level skills such as math facts, but other programs are available to generate activities such as Cloze exercises. In many cases, test generator software and worksheet generator software are similar enough to be used interchangeably, and some packages are intended for both purposes.

Puzzle Generators

Tools that automatically format and create crossword and word search puzzles fall under a general category of tools called *puzzle generators*. The teacher enters the words and/or definitions, and the software formats the puzzle. Children are often fascinated by word search puzzles, but these materials may have little instructional value other than reviewing spelling of new words. Crossword puzzles, on the other hand, can be used as exercises to review words and definitions or even low-level concepts.

Figure 6.2 Example Test Generator Software On-screen Questions: Test Designer Plus

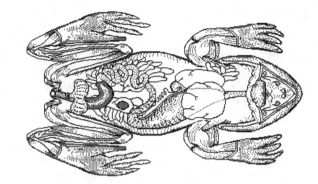

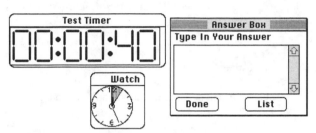

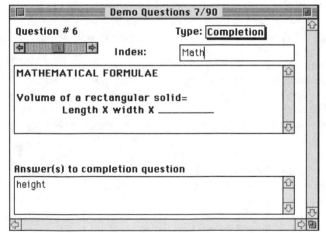

Source: Used by permission of SuperSchool Software.

Bar-code Generators

One technological advancement that has provided great benefits to education is the ability to store large numbers of pictures and documents on optical media such as videodiscs and CD-ROMs. As Chapter 7 describes collections of still and motion video images and pictures of important documents are available in a wide variety of topics ranging from science and history to art and music. However, teachers need more than mere availability to incorporate these resources into instruction, teachers must have a way to access them quickly as needed. This is the function of bar-code generating software. A bar-coded frame number can identify the

Figure 6.3 Example Bar Codes for Images of Mother and Baby Animals

loon	lion	chicken
frame 48745 to 49013	frame 16764	frame 16312
loon	hartebeest	chicken
frame 49017 to 49174	frame 16803	frame 16313
loon	impala	chicken
frame 52630 to 52987	frame 16804	frame 16314
gorilla	raccoon	chicken
frame 49361 to 49764	frame 16795	frame 16314
gorilla	elephant	tortoise
frame 49769 to 49956	frame 16838	frame 16112
macaque	chicken	frog
frame 49176 to 49358	frame 16310	frame 16042
lemur	chicken	frog
frame 39195 to 39321	frame 16311	frame 16043

Source: Murdoch, N. (1992). BioSci babies. *The Florida Technology in Education Quarterly,* 4 (2), 30–32.

location of each of the items stored on optical media. These bar codes look like the UPC codes on grocery items, as Figure 6.3 shows. Bar-code generating programs allow teachers to prepare pages of bar codes for pictures and information that relates to a given topic.

A science teacher might use this technology to show students examples of various transformations of energy by displaying images stored in various places on a physical sciences videodisc. Before class, the teacher could use a bar-code generator program to prepare a page of bar codes designating the locations where the examples were stored. During class, the teacher would run a bar-code reader across each bar code and the desired picture would display on the screen. Students can also use this technique to generate their own project presentations or reports.

IEP Generators

The current restructuring movements in education have brought increasing emphasis on school and teacher accountability. With this comes an increase in paperwork on student progress. Teachers of exceptional students, however, still seem to hold the record for the most paperwork requirements. Federal legislation such as PL 94–142 and the Americans with Disabilities Act require that schools prepare an individual educational plan, or IEP, for each exceptional student. These IEPs serve as blueprints for each special student's instructional activities, and the teacher must provide documentation that such a plan is on file and it governs classroom activities. Software is available to assist teachers in preparing IEPs (Lewis, 1993). Like test and worksheet generators, IEP generators provide on-screen prompts that remind users of the required components in the plan. When the teacher

Integration Strategies for
Bar-code Generator Software

Strategy #1: BioSci Babies
Developed by: Doris Murdoch, Webster Elementary School (St. Augustine, Florida)
Level: Kindergarten
Purposes of the Activity: Students identify various animals by name, compare them, and sort them according to whether or not they are hatched from eggs or born from their mothers.

Instructional Activities. The class reads the book *Baby Animals* (by Podendorf) or another, similar book on baby animals. Using prepared bar-code sheets, show clips from the BioSci II Elementary videodisc (Videodiscovery) of various animals that hatch from eggs. Then show clips of animals that were born from their mothers. Ask, "How do these animals differ?" "How are they alike?" Pass out example stuffed animals to each of the children. Have them identify the animals and tell whether they think the animals were hatched or born. When all animals have been discussed, ask the children to draw a general conclusion about which animals are hatched and which give birth to their young. Follow up the lesson with other books such as *Are You My Mother?* (Eastman), *Horton Hatches an Egg* (Dr. Seuss), and *Baby Animals* (Kuchalla).

Strategy #2: Let the Earth Creatures Live!
Developed by: Jack Edwards, Webster Elementary School (St. Augustine, Florida)
Level: Elementary to middle school
Purposes of the Activity: Students work in small groups to design convincing presentations designed around clips found on videodiscs.

Instructional Activities. Present the following scenario to the whole class:

The military on Earth is in a state of readiness. It seems that a group of alien spaceships are hovering above the planet. They are Naclites from the planet Nacl, creatures who require tremendous amounts of salt to live. Since the resources on their home planet are close to depletion, they have been out searching the neighboring galaxies for new sources of salt. Realizing that the Earth's seas are rich with the element they need, they have made known their intentions to extract the salt from the oceans. You are a committee sent from Earth's leaders to persuade the Naclites that taking the salt from our waters would kill off many beautiful species of sea life. You must select and present film clips, slides, and information to persuade the Naclites that the Earth's seas are too valuable to be used for salt supplies.

Assign parts of the overall task to small groups of students. Have each group scan through the discs available to them looking for convincing evidence of the number of beautiful and unique kinds of sea life in the oceans. Have each group use a mapping form to record the bar-code numbers, then use a bar-code program to produce a bar-coded sheet of the example frames and clips. After students select their examples, they design a 5-to-10-minute presentation. Each group presents its findings to the rest of the class.

Source: Murdoch, D. (1992). BioSci babies. *The Florida Technology in Education Quarterly, 4* (2), 30–32; strategy #2 based on an idea provided by Jack Edwards.

finishes entering all the necessary information, the program prints out the IEP in a standard format. Some IEP generation programs also accept data updates on each student's progress, thus helping the teacher with required recordkeeping as well as IEP preparation.

Certificate Makers

Recognizing achievement is a powerful means of motivating people. Teachers have found certificates to be a useful form of this recognition. Certificates congratulate students for accomplishments, and the students can also take them home and share them with parents and friends. Certificate makers simply provide computerized help with creating such products. Most certificate makers include templates for various typical achievements. The teacher selects one appropriate for the kind of recognition desired (e.g., completing an activity, first place winner) and enters the personalizing information for each student. This software helps teachers produce certificates quickly and easily, so they can award them frequently.

Form Makers

Like desktop publishing, form design is a special area of expertise that software can facilitate. Teachers must frequently create forms to collect information from students, parents, or other faculty. Sometimes these forms are as simple as permission for students to participate in a class event, but they can become much more complicated. For example, a teacher may need forms to collect personal information from students for student records or to enter information on software packages as they are evaluated. Formatting even the simplest form can be time-consuming on a word processor. Form maker software structures the process and makes the design simpler to accomplish. Most such packages have some graphic abilities, allowing users to add lines and boxes for desired information. As teachers create these forms, they can store them as templates for later use, perhaps with revisions.

Groupware Products

Groupware is one of the newest technology tools. The term itself has been in the educational technology lexicon only since around 1991. It refers to software products that are designed to promote cooperative learning among groups of students by helping them document their work as they progress. Groupware usually resembles a special-purpose word processor that allows students to enter the results of work sessions and each student's contribution to the development of a product. Some packages allow links via modem, so students in more than one location can work on a cooperative project. Pearlman (1994) says that these products "stimulate group activity in a one-computer or sometimes no computer classroom" (p. 1). Cowan (1992) describes three groupware products and gives some suggestions for using them effectively with students:

- Set and communicate ahead of time well-structured rules and procedures for using the groupware. These guidelines keep the students on task and prevent collaborative activities from becoming chaotic. Teachers should participate in and monitor sessions to enforce these rules.

- Make sure that all students have opportunities to express themselves so that one or two students do not control the process of entering information.

- Groupware takes teacher time and computer resources to use. Identify ahead of time places in the curriculum that seem likely to profit from groupware as opposed to word processing software. Use groupware to improve communication and collaboration skills.

Using Data Collection and Analysis Tools

Gradebooks

Although many teachers prefer to keep their grades on flexible spreadsheet software, some also prefer special software designed exclusively for this purpose. A gradebook

(electronic gradekeeping) program allows a teacher to enter student names, test/assignment names, data from tests, and weighting information for specific test scores. The program then analyzes the data and prints out reports based on this information. Some gradebooks even offer limited-purpose word processing capabilities to enter notes about tests. The software automatically generates averages and weighted averages for each student and averages across all students on a given test. Gradebooks require less teacher set up time than spreadsheets, but they also allow less flexibility on format options. Wager (1992) describes the process and criteria one group of teachers used to select a gradebook for use throughout the school. Important criteria included:

- Capacity to track many tests/assignments
- Flexibility in report formats (e.g., sorting by name, individual as well as group reports)
- Wide range of peripheral support (e.g., use with various printers, networks)
- Ease of setup and use
- Use on multiple platforms

Stanton (1994) also reviewed currently available gradebook packages according to a list of criteria. He noted some new capabilities to look for in an electronic gradebook. These included generating graphs, making seating charts, and tracking attendance.

Statistical Packages

As Gay (1993) joked, many teachers believe that the field of statistics should be renamed *sadistics*. Yet several kinds of instructional situations may interest teachers in statistical analyses. Brumbaugh and Poirot (1993), among others, maintain that teachers should take advantage of opportunities to do research in their classrooms. If teachers do choose to do classroom research, they must follow data collection with data analysis. Depending on the type of research, several typical analyses can yield helpful information, including descriptive statistics such as means and standard deviations to inferential statistics such as t-tests and analyses of variance. Software can also help with qualitative data collection and analysis. Teachers may perform statistical analysis of students' performance on tests. Question analysis procedures help them analyze test questions that they intend to use more than once. By changing and improving questions, teachers can make their tests more accurate and reliable. Finally, teachers may have to teach beginning statistics to their own students, for example in a business education course.

Statistical software packages perform the calculations involved in any of these kinds of procedures. Naturally, a teacher must have considerable knowledge of the proper applications of various statistical procedures; the software merely handles the arithmetic. But this alone can save considerable time. Webster (1992) reviewed several statistical software packages for use in business education courses, and she found that the packages varied considerably in their usefulness for this purpose.

Data Management (CMI) and Testing Tools

Definitions: CMI versus data management. The term *computer managed instruction,* or CMI, is an old one—some feel it is archaic—left over from the time when nearly all education technology software could be classified as CMI or CAI (see Chapter 1). As early as 1978, Baker admitted that a precise description of CMI did not exist since "the definitions are as diverse as the number of existing systems" (p. 11). Baker gave examples of several such systems that were popular at the time, noting widely varying characteristics. Among these systems were the Teaching Information Processing System (TIPS), the Sherman School System, and the Program for Learning in Accordance with Needs (PLAN). At that time, there was a burgeoning interest in mastery learning, in which teachers specify a sequence of objectives for a student to learn and prescribe instruction to help the student master each objective. Clearly, the teacher must keep track of each student's performance on each objective—a mammoth recordkeeping task. CMI systems running on large, mainframe computers were designed to support teachers in these efforts.

Today's teachers emphasize mastery learning less than keeping students on task and monitoring their progress to make sure their work challenges them without frustrating them. Some educators feel the term CMI has always been a misnomer for a recordkeeping routine within an instructional system rather than a type of instructional delivery. However, software tools are still available—now on microcomputers and networks—to store and analyze data on student progress during instruction and to provide summary reports on progress. The purpose for collecting the data may be different, but the recordkeeping task is still considerable. While some people still call this software CMI, others refer to it simply as *data management software.*

Types and functions of data management tools. Roblyer (1992) identified three different kinds of computerized data management tools:

- **Components of computer-based learning systems.** These tools allow teachers to enter names and other information on students. When each student types in his or her name, the system presents a sequence of activities tailored to that student's needs. The system also collects data as students go through the instruction. Reports show the teacher what students have accomplished and point out areas where they may still need assistance and off-line work. While some stand-alone microcomputer-based packages have these tools built into the software, these systems are more commonly seen as components of networked integrated learning systems (ILSs) and integrated technology systems. These systems have the power and capacity to handle large amounts of data on student performance.
- **Computerized testing systems.** With these tools, students receive actual instruction elsewhere, usually by noncomputer media. Computers facilitate on-screen testing and recordkeeping after instruction. Sometimes known as *computer-assisted testing* (CAT), these tools both generate test forms and process performance data. They differ from test generators, which allow on-screen testing but do not give detailed reports on results for individuals and groups. Some of the major standardized tests such as the SAT and GRE are now given on computerized testing systems. These systems offer many benefits, including immediate knowledge of results. Tests can also be shorter, since the systems can assess each person's ability level with fewer questions. This is because the software continuously analyzes performance and presents more or less difficult questions based on the student's performance, a capability known as *computer adaptive testing,* also shortened to CAT (Strommen, 1994). CAT is used more and more frequently for testing in professional courses like those in nursing education. The capabilities of computerized testing systems let educators go beyond the limits of multiple choice tests and they make possible alternative assessments. These systems also simplify scheduling of tests, since everyone need not take tests at the same time.
- **Test scoring and data analysis systems.** These types of data management tools accept test data input either through the keyboard or by optically scanning bubble sheets. Tests are automatically scored and reports on the results are generated. Both devices and sheets are available from companies such as National Computer Systems (NCS) or Scantron.

All of these data management systems serve two primary purposes. First, they provide clerical support for all of the calculation and paperwork tasks required to track student progress. Second, they help teachers to match instruction to the needs of each student. Each type of system provides various kinds of reports about student progress. ILS management systems usually provide the most extensive reporting. They can give a wealth of feedback ranging from the number of test questions answered correctly and incorrectly on a given student's test to summary data on the performance of whole classes in a topic area. One example of these reports is shown in Figure 6.4.

Using Graphics Tools

Print Graphics Packages

These software tools have a very limited purpose, but one that many teachers seem to find indispensable. They are essentially simple word processors designed especially for quick and easy production of one-page signs, banners, and greeting cards. One of the graphics programs, Print Shop Deluxe (Brøderbund) has become one of the best-selling software packages in education. Teachers can find hundreds of uses for the products of print graphics software ranging from door signs to decorations for special events. Some schools have even had Print Shop Deluxe contests to design the most creative signs or banners. The graphics and other options available in the program are selected from menus, making the programs so easy to use that anyone can sit

Figure 6.4 Sample ILS Group Pre/Posttest Report: Jostens Integrated Learning Corporation

Green River School District
Jostens Elementary School

Group Pre/Posttest Report   Page 1 of 4

Group: Ms. Zapata's First Grade Math
Objectives: JCAT Math Computation 1A
Score Used: Last
Date Range: 09/22/95 - 10/5/95
Date: October 6, 1995   2:21 pm

AM1 JCAT Math Grade 1
AM1-A STRAND - Study Skills
AM1 - A1 GOAL - Applications
AM1 - A1101 AM1 - A1102

Student		AM1-A1101 %	Time	AM1-A1102 %	Time	AM1-A1103 %	Time	AM1-A1 Summary %	Time
Allred, Leigh J.	PRE	60 −	1:27	100 +	0:59	100 +	0:49	85 +	3:15
	POST	100 +	1:03	100 +	0:33	75 +	1:22	92 +	1:58
	DIFF	+40		+0		−25		+7	
Kim, Young Fik	PRE	40 −	0:22	25 −	0:45	75 +	0:50	46 −	1:57
	POST	80 +	0:42	75 +	0:55	75 +	0:55	77 +	2:32
	DIFF	+40		+50		+0		+31	
Tanner, John C.	PRE	80 +	1:32	100 +	1:21	100 +	0:48	92 +	3:41
	POST	100 +	1:00	100 +	0:59	100 +	1:01	100 +	3:00
	DIFF	+20		+0		+0		+8	
Group Average	PRE	52	1:07	65	1:10	68	1:03	61 +	3:20
	POST	86	0:49	74	0:51	80	0:55	77 +	3:25
	DIFF	+20		+9		+12		+16	

+ = Objective, Goal, or Strand was mastered
− = Objective, Goal, or Strand was not mastered

Objectives

1. Objectives on this report shown in their hierarchical structure

Individual Student Data

2. Names of selected students

3. For each student, percentage scores and mastery status on two administrations of a test, including total time in minutes and seconds (optional)

4. For each objective, pretest scores (PRE), posttest scores (POST), and the difference between them (DIFF)

Group Data

5. Group averages showing percentage correct, time, and difference between pretest and posttest for each objective

Legend

6. Codes used on the report:
   + for mastered objectives
   − for non-mastered objectives

Source: *Reports Guide* by Jostens Learning Corporation. © 1991 by Jostens Learning Corp. Used by permission.

down and create a product in a matter of minutes. One example of a print product is shown in Figure 6.5.

Draw/Paint Programs

To produce more complex hard-copy graphics, draw/paint programs are the technology tool of choice. These tools are usually used to create designs and pictures that are then imported into desktop publishing systems or desktop presentation tools, as described in the following section. Just as print graphics programs are known for their simplicity and limited options, draw/paint programs are known for their sophistication and wide ranging capabilities. Many of these packages (e.g., Aldus Freehand or CorelDraw) require considerable time to learn and implement. But some, like Kid Pix, in Figure 6.6, are designed specifically for children to use with-

out formal training. Draw/paint programs have many instructional uses, as the lesson strategies on p. 167 illustrate.

Presentation Software and Computer Projection Systems

Definitions and types. Presentation software packages help users create on-screen descriptions, demonstrations, and summaries of information. Presentation tools represent a notable example of a technology that migrated from business and industry to education. These tools were first adopted by business executives and salespeople who used them to give reports at meetings and presentations to clients. Their capabilities to demonstrate, illustrate, and clarify information became evident, and presentation tools began to make their way into K-12 and university classrooms.

Figure 6.5 Sample Product Using Print Shop Deluxe

The programs allow more flexibility than print graphics programs in the selection of features, allowing a user to choose from an array of text, graphics, and animation features. Hoffman (1994) reported that presentation tools began exclusively as "electronic slide shows," but they have evolved into an additional kind of presentation product: multimedia authoring tools, which allow users to incorporate motion sequences from CD-ROM and other video media into their presentations. Hoffman classifies some software that combine the capabilities of both slide show and multimedia presentations as "hybrid packages." (See Figure 6.7.) Hypermedia software (discussed in detail in Chapter 8) may also be used as presentation software.

Making the most of presentation software: Skills and resources. As with desktop publishing, the effectiveness of a presentation tool depends largely on the communications skills of the presenter. Ferrington and Loge (1992) gave several helpful tips for designing effective presentations to groups. For large classes and other groups, presentation software products are usually used in conjunction with computer projection systems. These may be devices such as LCD panels that fit on top of overhead projectors or systems that operate as standalone devices. All of these devices enlarge the image produced by the software by projecting it from the computer screen onto a wall screen.

Example classroom applications of presentation software. These tools are especially useful for teacher presen-

Figure 6.6 Sample Product Using Kid Pix

tations to whole classes of students, but they can also be used to enhance conference presentations or talks to large groups. They can enhance instruction in traditional large-group lecture courses (Bolduc, Hale, and Webb, 1994) or increase the effectiveness of a distance learning curriculum (Joiner and Alvarez, 1994). Presentation software is not just for teachers; Granning (1994) describes how student skills with presentation tools have become the focal point for a high school program to develop communications skills. Every student in the program is required to put together presentations that combine computer-generated graphics, audio, and video.

Charting/Graphing Software

Charting and graphing software tools automatically draw and print out desired charts or graphs from data entered by users. The skills involved in reading, interpreting, and

Figure 6.7 Classifications of Presentation Software

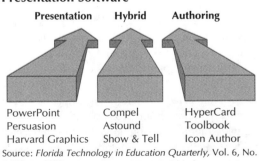

Presentation	Hybrid	Authoring
PowerPoint	Compel	HyperCard
Persuasion	Astound	Toolbook
Harvard Graphics	Show & Tell	Icon Author

Source: *Florida Technology in Education Quarterly,* Vol. 6, No. 2, Winter 1994.

Integration Strategies for Draw/Paint Programs

Strategy #1:	Map Skills with Kid Pix
Developed by:	Dennis Day, Westhill Elementary School (Bothell, Washington)
Level:	Grade 3
Purposes of the Activity:	Students use the Kid Pix drawing program to design a map as an introduction to a unit on basic geography skills.

Instructional Activities. This activity is the beginning of a 4-week unit on geography skills. It begins with a whole-class demonstration of Kid Pix on the computer projection system. Students are given handouts describing the requirements of the project they will do and giving an example product. The map they produce should include a compass rose with an arrow and an *N* for North, a map title, a date, and seven map symbols of their choice. After a session for hands-on exploration of Kid Pix, they are given time to design and produce their own maps. If certain students have trouble getting organized, the teacher might try assigning the elements and symbols for them to use in their maps. Students may also find it better to use the mini-word processor in Kid Pix to label their maps rather than pointing and dragging individual letters from the alphabet menu.

Strategy #2:	Drawing Kids In
Developed by:	Bonnie Meltzer (Portland, Oregon)
Level:	Pre-K to Grade 1
Purposes of the Activity:	Students with hearing and vision impairments use a basic drawing program to develop drawings and illustrated books as a means of self-expression. Computer drawing activities also help develop skills in shape and object identification, eye/hand coordination, controlled movements, visual and language stimulation, and creative expression.

Instructional Activities. The teacher may want to select a simple drawing program with limited options that does not scroll automatically. This limits children to drawing so they can see the whole screen and total image at any time. An enlargement program is necessary for visually impaired children. Also, printing on a laser printer makes the drawings darker and easier to see. The teacher begins by introducing basic mouse skills, working with each child individually by holding the child's hand on the mouse and directing the movements until the child is ready to work independently.

Children first concentrate on using a brush shape, then move on to other patterns, followed by circles and rectangles. As children become more comfortable, the teacher gradually introduces new tools, techniques, and ideas. If they like and are able to do so, children get to label their own drawings before printing them. These drawings can be combined and bound into a booklet for each child to display.

Strategy #3:	Drawing in the Art Curriculum
Developed by:	Rick Wigre, Everett High School (Everett, Washington)
Level:	High school
Purposes of the Activity:	Students are taught how to use drawing programs in an introduction to art course to pave the way for integrating art into other curriculum areas.

Instructional Activities. Students are first introduced to draw/paint programs in an introduction to art course taught in the freshman year. This course is also a good base for the computer art courses, which are designed to integrate with other disciplines. Students use the visual communication skills they learn through the draw/paint programs to develop products and assignments for many other courses. For example, they may use computer images to present reports in history or science. The computer art course begins with 2 weeks of introduction to an array of computer art and graphics in which students write and talk about what they are seeing. They refer back to visual concepts taught in the introduction to art class. During this time, students also become familiar with looking up visual images in the library and using them as resources. They spend the rest of the course learning basic drawing features such as tiling (creating a design and covering a page with it) and use of clip art, as well as more advanced skills which combine a variety of drawing features. They also learn use of other equipment to design presentations, e.g., the camcorder and laserdisc. The art teacher works closely with teachers in the other disciplines to develop assignments that make use of these drawing/presentation skills.

Source: Days, D. (1994). Active mapping. *The Computing Teacher, 21* (5), 27–28; Meltzer, B. (1990). Who can draw with a Macintosh. *The Computing Teacher, 17* (7), 21–23; Wigre, R. (1993). Developing a computer art course in the art curriculum. *The Computing Teacher, 21* (3), 14–16.

producing graphs and charts are useful both to students in school and adults in the world of work. However, those with limited artistic ability face special challenges in learning and using these skills. Fortunately, charting and graphing software can take the mechanical drudgery out of producing graphs and charts. If students do not have to labor over rulers and pencils as they try to plot coordinates and set points, they can concentrate on the more important aspects of the graphics: the meaning of the data and what they represent. As Duren (1990–1991) observed, this kind of activity supports students in their efforts at visualizing mathematical concepts and engaging in inquiry tasks. Graphing activities in science, social studies, and geography can also profit from applications of these kinds of software tools.

Clip Art Packages, Video Collections, and Sound Collections

Clip art packages are collections of still pictures drawn by artists and graphics designers and placed in a book or on a disk for use by others. When teachers prepare presentations with desktop publishing or presentation tools, they need not use draw/paint programs to draw original pictures. A wealth of pictures is available in the form of clip art. Most word processing and desktop publishing systems can import clip art stored on disk into a document. Draw/paint programs can also import clip art for use in designing other pictures and graphic images. Clip art from a book may be optically scanned into a computer file and stored on a disk for use with these programs. When more realistic depictions are desired, actual pictures stored on video media such as CD-ROM and videodisc may be imported in the same way.

Clip art packages and video collections are invaluable tools to help both teachers and students illustrate and decorate their written products. Teachers find that such pictures help make flyers, books, and even letters and notices look more polished and professional. Some teachers feel that students are more motivated to write their own stories and reports when they can also illustrate them.

For teachers and students who want to develop their own multimedia presentations, collections of sound effects and movie clips are also becoming more common. A system may need special hardware (e.g., sound cards or synthesizers) or software (e.g., Quicktime, a movie-making software) in order to incorporate these elements.

Digitizing Systems and Video Development Systems

Several other kinds of tools help users prepare graphics for both print and on-screen presentations. Digitizing programs are software tools that handle pictures scanned into the computer and stored as picture files. Users can edit these pictures as needed or include them as they are. With Apple's QuickTake camera and similar devices, teachers can take pictures as they would with any camera, storing the images on disk instead of on film. They can then import the photos into video presentations or desktop publishing products. Kodak's Photo CD is a unique product that digitizes images from 35mm film and stores them on CDs (Jones, 1994). The number of products that assist with graphics production is on the increase, and new capabilities to develop and import images are constantly being developed. Teachers and students can even use movie-making software such as Quicktime and FusionRecorder in their graphics productions (See Chapters 7 and 8.)

Using Planning and Organizing Tools

Outlining Tools and Other Writing Aids

Several kinds of technology tools are available to help students learn writing skills or to assist accomplished writers in setting their thoughts in order prior to writing. Outlining tools are designed to prompt writers as they develop outlines to structure documents they plan to write. For example, the software may automatically indent and/or supply the appropriate number or letter for each line in the outline. Outliners are offered either within word processing packages or as separate software packages for use before word processing.

Other writing aids include software designed to get students started on writing reports or stories. One type of writing aid is sometimes called a *story starter.* This kind of program provides a first line and invites students to supply subsequent lines. Other tools give students topic ideas and supply information about each topic that they can use in a writing assignment. Sometimes a software package combines outlining tools and other writing aids.

Brainstorming Tools

These might also be called *conceptualizing tools,* since they help people think through and explore ideas or topics. One such tool, Idea Fisher, provides semantic connections with given words. For example, if a user were to click on the word *whale,* the program might display several words associated with whales such as *ocean* and *sea life,* as well as some less obvious connections such as *endangered species.* The program can also ask probing questions to provoke thoughts about each topic selected. Another program, Inspiration by Inspiration Software, assists brainstorming by helping people develop concept maps. Concept maps are visual outlines of ideas that can offer useful alternatives to the strictly verbal representations provided by content outlines. (See Figure 6.8.) Anderson-Inman and Zeitz (1993) give some good examples of classroom applications of concept mapping tools.

Lesson Planning Tools

Most teachers do not rely heavily on written lesson plans to guide their teaching activities. However, many occasions demand some form of documentation to show what teachers

Figure 6.8   Concept Map

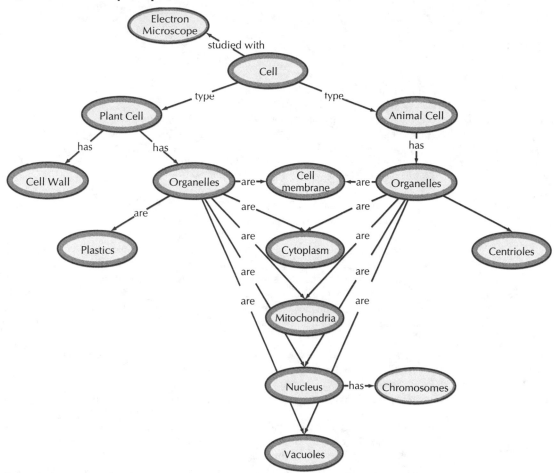

Source: Anderson-Inman, L. and Zeitz, L. (1993). Computer-based concept mapping: Active studying for active learners. *The Computing Teacher, 21* (1), 6–8, 10–11. Used by permission of ISTE.

are teaching and how they are teaching it. Tools that help teachers develop and document their descriptions of lessons are sometimes called *lesson makers* or *lesson planners.* Most of these programs simply provide on-screen prompts for specific lesson components (e.g., objectives, materials, activity descriptions). They also print out lessons in standard formats, similar to the way test generators format printouts of tests.

Schedule/Calendar Makers and Time Management Tools

Several kinds of tools have been designed to help teachers organize their time and plan their activities. Schedule makers help formulate plans for daily, weekly, or monthly sequences of appointments and events. Calendar makers are similar planning tools that actually print out graphic calendars of chosen months or years with the planned events printed under each day. Other time management tools are available to help remind users of events and responsibilities. The teacher enters activities and the dates on which they will occur. Then, when he or she turns on the computer each day, the software displays a list of things to do on the screen. Some integrated packages combine all these tools.

Using Research and Reference Tools

Encyclopedias

For many years, American families kept sets of encyclopedias to support their children's education. Young people used these books for research on school projects, and parents used them to take advantage of "teachable moments" when their children require more than quick answers. Now most major encyclopedias come on CD-ROM with some kind of database structure. CD-ROM encyclopedias have several advantages over books. Users can search to locate one specific item or all references on a given topic; they usually offer multimedia formats that include sound and/or film clips; they also offer hypertext links to related information on any topic. Chapters 7 and 8 present more information on and examples of disc-based encyclopedia applications.

Atlases

Like encyclopedias, atlases are popular educational reference tools for families as well as schools. They summarize geographic and demographic information ranging from

Integration Strategies for Presentation Software

Strategy #1: Presentation Technologies in Social Studies
Developed by: Fred Peterkin, Eastside High School (Gainesville, Florida)
Level: High school
Purposes of the Activity: A teacher uses presentation technologies to keep information current and to integrate visual images into teaching current events.

Instructional Activities. With a configuration set up for classroom presentations, the teacher of current events can combine text with still and motion graphics and integrate both into daily classroom instruction. An optimal configuration combines presentation software with photo CDs, videodisc images, stereo equipment, and OCR software. With this setup, captured images, frames, or video clips from current events can be presented as needed. These presentations can show office holders and other important players in their actual environments. Structural concepts of government can be taught, and the system can also illustrate current events. For example, when teaching about the executive office of the president and the cabinet, the teacher might show captured clips from the NAFTA discussion, a C-Span tally of the House vote on NAFTA, and a picture of Secretary of Commerce Ron Brown interspersed with a description of the president's preparation for a trade and economic conference. All of this information is projected in living color and action on a scale that makes it easy for students sitting in the back of the room to see and discern details in text and graphics.

Strategy #2: Adding Pizazz to Writing
Developed by: Karen Parker, Iowa City Community School District (Iowa City, Iowa)
Level: Grades 5–6
Purposes of the Activity: Students use Slide Shop presentation software to present their illustrated rewrites of popular folk tales.

Instructional Activities. The teacher begins by presenting the project idea to the students and demonstrating the capabilities of the software. Students are divided into groups and each group decides on a folk tale to rewrite. Students work in groups to write first drafts of their folk tales. Stories are kept short (100 words or fewer) to make them easy to adapt to the visual format. The teacher discusses storyboarding techniques, giving each group 4 × 6 cards with a specific format to follow. Each card (storyboard) represents a frame on the software. Students "chunk" their text into frames and enter storyboard pictures and text into the software. Printouts are generated and reviewed, and text editing is done on-screen. When all stories are typed and saved on disk, a special viewing day is arranged to let each group show off its work. Some innovative titles from the past include: The "Three XL Pigs" and "Snow White and the Seven Radical, Cool, Tubular Dudes."

Source: Peterkin, F. (1992). Presentation technology: A necessity for the modern classroom. *The Florida Technology in Education Quarterly, 6* (2) (Winter), 25–35; Parker, K. (1992). Add pizazz to writing projects. *The Computing Teacher, 20* (3), 22–23.

population statistics to national products. CD-ROM versions of these atlases are especially helpful since they are so interactive. Students can see information on a specific country or city or gather information on all countries or cities that meet certain criteria. (See Figure 6.9.) Some atlases even play national songs on request!

Dictionaries

Sometimes called *word atlases,* CD-ROM dictionaries specify pronunciation, definition, and example uses for each word entry. They also offer many search and multimedia features similar to those of encyclopedias and atlases. Many CD-ROM dictionaries can play an audio clip of the pronunciation of any desired word, a capability of special help to young users and others who cannot read diacritical marks.

Figure 6.9 Example Atlas Software: MacGlobe

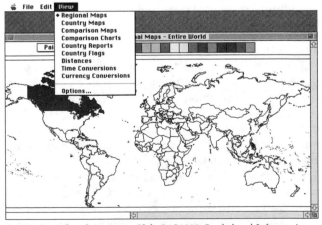

Source: Reproduced using, MacGlobe®; ©1992, Brøderbund Software, Inc. All Rights Reserved. Used by Permission.

Integration Strategies for Charting/Graphing Software

Strategy #1: Graphing in the Second Grade

Developed by: Joan Martin (Watertown, Massachusetts), Mei-Hung Chiu (Pittsburgh, Pennsylvania), and Anne Dailey (Quincy, Massachusetts)

Level: Grade 2

Purposes of the Activity: Students used graphing software to learn abstract concepts about time and distance relationships and learn how to produce and interpret line graphs with this and other kinds of information.

Instructional Activities. The teacher begins by marking off a line on the floor with masking tape to make it easy for students to determine the exact distance they walk from a sensor (connected to a microcomputer-based lab or MBL station). The students are introduced to the computer set-up, the sensor, and the markings on the floor. The sensor has been placed adjacent to the computer monitor so the students can see what happens. The instructor shows how the graph is created by walking along the floor marks, and students are asked to "walk through" simple directions: Start at the 3-meter mark, stand there for 4 seconds, walk to the 4-meter mark in 2 seconds, then run to the 2-meter mark and stand. The students then graph their motion over a 10-second interval as they walk the line. Since graphs are constructed on the monitor in real time, students need to realize how long 10 seconds is. They practice counting elapsed time by saying, "1-Mississippi, 2-Mississippi," and so on. Students help one another by counting out the time and verifying the correct meter marks. Next, students make up verbal instructions to produce different effects on the computer screen. Instructors ask specific questions about the physical shapes of the graphs concerning the concept of slope, direction of motion, determining movement for a certain period of time. Students then examine a variety of graphs and discuss which ones could be duplicated by the computer as they walked the line in various ways. This kind of activity seems successful in teaching young children graphing concepts that are normally difficult to understand for students this age.

Strategy #2: Inquiry Skills and Graphing Tools

Developed by: Philip Duren, California State University

Level: High School

Purposes of the Activity: Graphing software facilitates an inquiry-oriented, problem-solving learning environment for algebra instruction.

Instructional Activities. Graphing software replaces traditional paper-and-pencil methods for practice in solving equations. With the teacher's help, students begin by exploring the relationship of solutions to $f(x) = 0$ and the x=intercepts of $f(x)$. They verify these relationships by making up their own examples. They also investigate the conditions under which polynomial or rational functions that cannot be factored have roots or x-intercepts. The teacher gives the students additional functions and has them predict the x-intercepts or the maximum possible roots. Students use the graphing software to graph the functions and check their predictions. Students work with partners or in cooperative groups to generate more data and ideas and to encourage more negotiation and discussion about the meanings of the observations and findings. Each student assumes a role within the group: facilitator, questioner or devil's advocate, computer operator, and recorder. They switch roles for each set of observations.

In another activity, students might explore expressions such as:

$$\frac{x^3 - 11x^2 + x = 50}{x - 2}$$

They discuss questions. (Is there a relationship between the number of real roots of the expression and the number of zeros of the numerator's polynomial? What are the zeros of the expression? What does the graph do with the vertical lines [asymptotes] that appear on the screen?) Before they graph a rational expression, students make conjectures about where it will have asymptotes and how many real zeros there will be. Given a graph, students try to predict the degree of the rational expression by investigating graphs from a greater distance. After several example graphs of equations of varying degree, students develop generalizations to prove or disprove.

Source: Martin, J., Chiu, M., and Dailey, A. (1990). Graphing in the second grade. *The Computing Teacher, 18* (3), 28–32; Duren, P. (1990–1991). Enhancing inquiry skills with graphing software. *The Computing Teacher, 10* (1), 23 26.

Integration Strategies for Brainstorming Tools

Strategy #1:	Concept Mapping
Developed by:	Lynn Anderson-Inman and Leigh Zeitz
Level:	Grade 11
Purposes of the Activity:	This activity introduces electronic concept mapping as a way to represent a specific knowledge domain in graphic form. Concept-mapping software eases the logistics of constructing and modifying the maps (see example map on p. 169).

Instructional Activities. Student-created concept maps (also called *semantic maps* or *webs*) can help learners construct knowledge by providing a vehicle for integrating new information with that learned previously. Because the learner plays an active role in creating and modifying the concept map, this study strategy promotes active learning and student involvement. Prior to reading an assigned chapter, students are asked to create a concept map for a set of terms. Then while reading the chapter, the students revise their maps to include newly learned information and clarify newly understood relationships. After classroom instruction and discussion, they might modify their maps again. Depending on the complexity of the material, this process might be modified several times. An example is an assignment in biology. Students are asked to learn about the structure of cells; they are given 11 important vocabulary words. They create a concept map showing how they feel these terms are related. Using Inspiration concept mapping software, each term is placed in a node and the lines linking the nodes are labeled. After reading the chapter on "Cell Structure," students revise their maps. After a classroom discussion of the topic, students are given another opportunity to revise their maps. On the last day of the unit, the teacher suggests augmenting the concept maps once again in preparation for the test. More detailed information for some concepts is placed into note windows linked to each node.

Source: Anderson-Inman, L., and Zeitz, L. (1993). Computer-based concept mapping. *The Computing Teacher, 21* (1), 6–11.

Using Tools to Support Specific Content Areas

CAD and 3D Modeling/Animation Systems

A computer-assisted design (CAD) system is a special kind of graphics production tool that allows a user to prepare very sophisticated, precise drawings of objects such as houses and cars. Like presentation tools, CAD systems began to appear in classrooms after their introduction in business and industry. This kind of software is usually employed in vocational-technical classrooms to teach architecture and engineering skills. However, some teachers use CAD software to teach drawing concepts in art and related topics. More advanced graphics students will probably want to use 3D modeling and animation software systems to do fancy visual effects such as morphing. (See Figure 6.10 for a sample of CAD software.)

Music Editors and Synthesizers

Music editor software provides blank musical bars on which the user enters the musical key, time, and individual notes that constitute a piece of sheet music. This software is designed to help people develop musical compositions on-screen, usually in conjunction with hardware such as a musical instrument digital interface (MIDI) keyboard and music synthesizer. This hardware allows the user to either (a) hear the music after it is written or (b) create music on the keyboard and automatically produce a written score.

Steinhaus (1986–1987) explained that music editors offer powerful assistance in the processes of precomposing, composing, revising, and even performing. Forest (1993) offers examples of these activities in a school setting, as well as a list of good music-related software and media. (Also see Chapter 14 for more classroom uses of these tools.)

Reading Tools

Both reading teachers and teachers of other topics occasionally need to determine the approximate reading level of specific documents. A teacher may want to select a

Integration Strategies for Probeware MBLs

Strategy #1:	Exploring the Relationship between Science and Mathematics
Developed by:	Louis Nadelson
Level:	High school
Purposes of the Activity:	Students use the line equation in the software that is intended for calibrating the probes in the MBL probeware to explore mathematical concepts underlying the general linear equation $y = mx + b$ (where m = slope and b = the y intercept).

Instructional Activities. Many different probeware setups require users to calibrate prior to using it for data collection. This process establishes a relationship between the input to the computer from the probe and the value that the measurement represents. The software also gives instruction on how to do the calibration. For example, if the MBL is measuring voltage, the probe is calibrated by connecting it to AA battery previously confirmed by a multimeter to output 1.5 volts. The computer registers 215 counts for 1.5 volts. The beginning point of 0 counts and the ending point of 215 define a line. This line is used to calibrate the probe.

After the probe is calibrated, an equation is developed for the line (0, 0 and 215, 1.5). Students sketch the graph of the line defined by these points using either paper or graphing software. They predict values for counts given specific voltages using their graph by locating the value on the x axis and finding the corresponding y value for counts. They use the general linear equation $y = mx + b$ to discuss the slope of the line as the rise (vertical distance between the two points) divided by the run (horizontal distance between the points). The teacher has the students calculate the slope for their line. They enter the value for the slope into the general linear equation and do the final equation for the example calibration. Once the equation has been determined, students calculate counts for specific voltages. When the probe is ready to be used, students bring in samples to be tested.

Source: Nadelson, L. (1994). Calibrating probeware: Making a line. *The Computing Teacher, 21* (6), 46–47. See also Flick, L. (1989). Probing temperature and heat. *The Computing Teacher, 17* (2), 15–24.

story or book for use in a lesson or to confirm that works are correctly labeled as appropriate for certain grade levels. Several methods are available for calculating the reading level of a written work; all of them are time-consuming and tedious to do by hand. Readability analysis software automates calculations of word count, average word length, number of sentences, or other measures of reading difficulty.

Another software tool related to reading instruction, Cloze software, provides passages with words missing in a given pattern (e.g., every fifth word, every tenth word). Students read the sentences and try to fill in the words. Cloze passages have been found to be good measures of reading comprehension. Some teachers also like to use them as exercises to improve reading comprehension.

Many books for children as well as adults are available in interactive CD-ROM versions (Truett, 1993). Some of these allow children to hear narrations in English or Spanish. Others, like the Living Books series (Just Grandma and Me), let children explore the screen, activating animations and sounds when they click in various locations. These books are designed to provide an interesting, interactive way to read and increase reading fluency.

Figure 6.10   Sample CAD Software

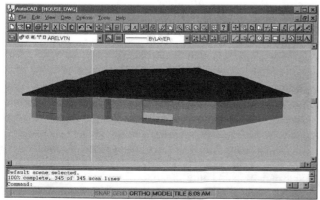

Source: Courtesy of Bill Wichelie.

MBLs (Probeware)

A technology tool that has proven particularly useful in science classrooms is the microcomputer-based labs (MBL), sometimes referred to as *probeware*. These packages consist of software accompanied by special hardware probes designed to measure light, temperature, voltage, and/or speed. The probes are connected to the microcomputer, and the software processes the data collected by the probes. Bitter, Camuse, and Durbin (1993) asserted that microcomputer probeware can actually replace several items of lab equipment such as oscilloscopes and volt-meters because MBLs outperform this traditional equipment. Ladelson (1994) pointed out that probeware achieves a dual purpose of gathering empirical data and revealing the relationship between science and math. Stanton (1992) described a variety of MBLs, covering their capabilities and prices along with grade levels and science subjects they can help teach.

Exercises

1. **Terms used with technology support tools.** Fill in the correct term for each of the following features:

   _____ 1. Files of predrawn pictures that can be imported into word processing or desktop publishing documents

   _____ 2. Software to allow students to develop and play back their own musical compositions

   _____ 3. The part of an ILS that automatically tracks and reports on each student's progress through computer-based instruction

   _____ 4. Nonspreadsheet software that allows teachers to enter, maintain, and calculate grades on the computer

   _____ 5. A combination of hardware and software that helps collect and analyze data during science experiments

2. **Support tool applications.** What kind of software support tool(s) would help meet each of the following kinds of instructional needs:

   _____ 1. A teacher is doing a workshop for a professional conference. He wants to use a medium that will highlight and dramatize his main points in a way that will focus the attention of his audience on them. He also wants to play video clips from the classroom to illustrate what he is talking about. What should he use?

   _____ 2. A vocational education teacher is presenting a unit on planning and designing structures (e.g., buildings and houses). During this unit, she wants to illustrate how architects use computers to assist them in this work. What software would she need?

   _____ 3. A fourth grade teacher wants to begin his unit on sea life by showing his students pictures of sea creatures that are either mammals or fish. He has two different videodiscs with many slides and video clips, but he wants a way of getting to these pictures quickly during his presentation and the subsequent class discussion. What software will he need?

   _____ 4. A teacher wants to prepare her students to take upcoming standardized tests. She develops several hundred questions that will provide practice in taking these tests, and she wants to produce a sample test for each of her students by randomly selecting items and printing out a different version of the test for each student. What software would be helpful?

   _____ 5. Students in an English class want to develop their own newspaper complete with various kinds of typefaces, fancy page layouts, and graphics. What kind of software would help them carry out this activity?

References

Anderson-Inman, L., and Zeitz, L. (1993). Computer-based concept mapping: Active studying for active learners. *Computing Teacher, 21* (1), 6–8, 10–11.

Baker, F. (1978). *Computer-managed instruction: Theory and practice.* Englewood Cliffs, NJ: Educational Technology Publications.

Beeken, L. (1992). Ideas for teacher collaboration: What happens when teachers collaborate? *NCRVE Professional Development Bulletin, 1* (1).

Bitter, G., Camuse, R., and Durbin, V. (1993). *Using a microcomputer in the classroom* (3rd ed.). Boston: Allyn and Bacon.

Bolduc, R., Hale, M., and Webb, J. (1994). Multimedia for presentations in large-lecture classrooms. *The Florida Technology in Education Quarterly, 6* (2), 65–68.

Brumbaugh, K., and Poirot, J., (1993). The teacher as researcher: Presenting your case. *The Computing Teacher, 20* (6), 19–21.

Cowan, H. (1992). The art of group communication. *Electronic Learning, 11* (8), 38–39.

Duren, P. (1990–1991). Enhancing inquiry skills with graphing software. *The Computing Teacher, 20* (3), 23–25.

Ferrington, G., and Loge, K. (1992). Making yourself presentable. *The Computing Teacher, 19* (5), 23–25.

Forest, J. (1993). Music and the arts: Keys to a next-century school. *The Computing Teacher, 21* (3), 24–26.

Gay, L. R. (1993). *Educational research: Competencies for analysis and application* (4th ed.). Columbus, OH: Merrill.

Granning, M. Presentation technologies enhance students' communication skills at Lakewood High School. *The Florida Technology in Education Quarterly, 6* (2), 37–43.

Gullickson, A., and Farland, D. (1991). Using micros for test development. *Tech Trends, 35* (2), 22–26.

Hermann, A. (1988). Desktop publishing in high school: Empowering students as readers and writers. ERIC Document Reproduction No. ED 300 837.

Hoffman, E. (1994). Overview of presentation software: What every teacher needs to know. *The Florida Technology in Education Quarterly, 6* (2), 11–15.

Joiner, D., and Alvarez, D. (1994). Presentation software in distance learning: Marion County's MacAir project. *The Florida Technology in Education Quarterly, 6* (2), 16–24.

Jones, P. (1994). Photo CD: Implications for Florida education. *The Florida Technology in Education Quarterly, 6* (2), 44–49.

Knupfer, N., and McIsaac, M. (1989). Desktop publishing software: The effects of computerized formats on reading speed and comprehension. *Journal of Research on Computing in Education, 22* (2), 127–136.

Ladelson, L. (1994). Calibrating probeware: Making a line. *The Computing Teacher, 21* (6), 46–47.

Lewis, R. (1993). *Special education technology.* Pacific-Grove, CA: Brooks-Cole.

McCain, T. (1993). *Designing for communication: The key to successful desktop publishing.* Eugene, OR: ISTE.

McCarthy, R. (1988). Stop the presses: An update on desktop publishing. *Electronic Learning, 7* (6), 24–30.

Nathan, J. (1985). *Micro-myths: Exploring the limits of learning with computers.* Minneapolis: Winston Press.

Newman, J. (1988). Online: Classroom publishing. *Language Arts, 65* (7), 727–732.

Norvelle, R. (1992). Desktop publishing. In G. Bitter (Ed.). *The Macmillan encyclopedia of computers.* New York: Macmillan.

Parker, R. C. (1989). Ten common desktop design pitfalls. *Currents, 15* (1), 24–26.

Pearlman, B. (1994). Designing groupware. *ISTE Update, 6* (5), 1–2.

Roblyer, M. (1992). Computers in education. In G. Bitter (Ed.). *The Macmillan encyclopedia of computers.* New York: Macmillan.

Rose, S. (1988). A desktop publishing primer. *The Computing Teacher, 15* (9), 13–15.

Stanton, D. (1992). Microcomputer-based labs. *Electronic Learning Special Edition (Buyers Guide), 12* (1), 16–17.

Stanton, D. (1994). Gradebooks, The next generation. *Electronic Learning, 14* (1), 54–58.

Steinhaus, (1986–1987). Putting the music composition tool to work. *The Computing Teacher, 14* (4), 16–18.

Strommen, E. (1994). Can technology change the test? *Electronic Learning, 14* (1), 44–55.

Truett, C. (1993). CD-ROM storybooks bring children's literature to life. *The Computing Teacher, 21* (1), 20–21.

Wagner, W. (1992). Evaluating grade management software. *The Florida Technology in Education Quarterly, 4* (3), 59–66.

Webster, E. (1992). Evaluation of computer software for teaching statistics. *Journal of Computers in Mathematics and Science Teaching, 11* (3/4), 377–391.

Williams, R. (1994). *The non-designer's design book: Design and typographic principles for the visual novice.* Berkeley, CA: Peachpit Press, 1994.

Instructional Software References

Example Materials Generators
Desktop publishing
    PageMaker—Adobe
    Publish It!—Microsoft
    Quark XPress—Quark
    Ready, Set, Go!—Manhattan Graphics
    The Writing Center and The Children's Writing and Publishing Center—The Learning Company
    (See also references for Integrated Software Packages in Chapter 5.)
Test generators and test questions banks
    Create-a-Test—Educational Resources
    Test Designer Plus—SuperSchool Software
    Test Bank 4.0—Microsystems
    Test It! Deluxe—EduSoft
    Test Writer—Micro Media
Worksheet generators
    Reading Strategy Series—Prentice-Hall
Puzzle generators
    Crossword Magic—Mindscape
    Designer Puzzles—MECC
Certificate makers
    Certificates and More—Mindscape
    Certificate Maker—Spinnaker
Form makers
    DB FORMAKER—Data Blocks
    Filemaker Pro—Claris
    Formtool—BLOC Development Corp.
Bar-code generators
    Bar n' Coder 3.01—Pioneer New Media Technologies
IEP generators
    Talley Goals and Objectives Writer—Curriculum Associates
Groupware products
    Aspects—Group Technologies
    GROUPwriter—Wings for Learning/Sunburst
    Realtime Writer—Realtime Learning Systems

Example Data Collection and Analysis Tools
Gradebooks
    Gradebook Deluxe—Edusoft
    Grade Machine—Misty City Software
    Gradebook Plus—Mindscape/SVE
Statistical packages
    The Data Collector—Intellimation
    SAS—SAS, Inc.
    SPSS—Statistical Packages for the Social Sciences, Inc. (SPSS)
    SYSTAT—SPSS

Data management (CMI) and testing tools
    NCS Abacus Instructional Management Software—NCS
       Education
    Grady Profile—Aurbach & Associates
    (See also Test generators (above) and references for
       Integrated Learning Systems in Chapter 4.)

Example Graphics Tools
Print graphics packages
    Bannermania—Brøderbund
    Click Book—Bookmaker, Inc.
    PrintShop Deluxe—Brøderbund
    SuperPrint—Scholastic
Drawing and painting programs
    Adobe Illustrator—Adobe
    Freehand—Macromedia
    Canvas—Deneba Software
    CorelDraw—Corel
    DazzleDraw—Brøderbund
    Kid Pix—Brøderbund
Presentation software
    Astound—Gold Disk
    Kid Pix SlideShow—Brøderbund
    Persuasion—Aldus
    PowerPoint—Microsoft
Charting/graphing software
    Cricket Graph—Cricket Software
    Harvard Graphics—Software Publishing
    MECC Graph—MECC
    See also spreadsheet software in integrated packages such as
       ClarisWorks, Microsoft Works, and WordPerfect Works.
Clip art packages
    Click ART—T/Maker
    Desk Gallery Clip Art Library—Zedcor
Digitizing systems
    Adobe PhotoShop—Adobe
    Ofoto—Light Source
    Photo-CD—Kodak
    QuickTake camera—Apple
Movie-making systems
    FusionRecorder—Video Fusion, Inc.
    QuickTime—Apple

Example Planning and Organizing Tools
Outlining tools and other writing aids
    See word processors such as
    Writer's Helper—CONDUIT and Research Paper Writer—
       Tom Snyder
Brainstorming tools
    Idea Fisher—Idea Fisher Systems Inc.
    Inspiration—Inspiration Software
Lesson planning tools
    Plan to Teach—Teaching Inc.
    Daily Plan-It—SmartStuff
Schedule/calendar makers and time managers
    Calendar Creator—Power Up
    Calendar Maker—CE Software

Example Research and Reference Tools
Encyclopedias
    Compton's New Media Interactive Encyclopedia—Compton
    Encyclopedia Brittanica (EB)
    Mammals: A Multimedia Encyclopedia—National
       Geographic
    Encarta Multimedia Encyclopedia—Microsoft
    New Grolier Multimedia Encyclopedia—Grolier
    WorldBook Multimedia Encyclopedia—World Book
Dictionaries
    American Heritage Talking Dictionary—Softkey
    Oxford English Dictionary—Oxford University Press
    Random House Unabridged Dictionary—Random House
Atlases
    PC Globe—PC Globe, Inc.
    MacGlobe—Brøderbund
    MECC Geograph—MECC
    Street Atlas USA—DeLorme Mapping
    World Vista Atlas—Applied Optical Media

Example Tools to Support Specific Content Areas
CAD systems
    ArchiCAD—Graphisoft
    AutoCAD and AutoSketch—AutoDesk
    DataCAD—CADKey
    Micro Station and GenericCAD—Bentley Systems
3D modeling and animation systems
    3D Studio—AutoDesk
    StrataVision 3D—Strata
    Form Z—AutoDessys
Music editors and synthesizers
    ConcertWare +, Kid's Notes Version 1.3, and ConcertWare
       + MIDI—Great Wave Software
    Music Construction Set—Electronic Arts
    Music Studio—Activision
    Piano Partners Music Learning System—M. Waldman
    StickyBear Music—Weekly Reader Family Software
Reading tools (readability and Cloze)
    Readability Machine—Prentice-Hall
    Cloze-Plus—Milliken Publishing
Electronic books on CD-ROM
    Bravo Books—Computer Curriculum Corporation
    Discis book series for children—Discis
    Disney Interactive animated storybooks
    Living Books: *Just Grandma and Me* and *Arthur's Teacher
       Trouble* (Brøderbund)
    Talking Classic Tales—Orange Cherry New Media
Probeware and MBLs
    LabVIEW 1—National Instruments
    LEAP System—Quantum Technology
    Science Toolkit series—Brøderbund
    SensorNet—AccuLab Products Group
    Universal Lab Interface (ULI)—Vernier Software
    Whales and Their Environment—Wings for Learning

Part III

Using Technology Media and New Technology Tools: Principles and Strategies

The chapters in this part will help teachers learn:

1. To identify the unique capabilities of each of the following resources:
   - Optical technologies—Interactive videodisc, CD-ROM, and CD-I
   - Hypermedia authoring systems
   - Telecommunications and distance learning
   - Emerging technologies—Personal Digital Assistants, Artificial Intelligence, and Virtual Reality

2. To recognize the various teaching and learning functions that these technologies can fulfill

3. To develop integration strategies for each of these technologies that match their capabilities to classroom needs

Introduction

As we approach the magical year 2000 A. D. in the history of human activity on the earth, people also look forward to marking about a half-century of computer use in society. Computers have been used in education for almost all of their existence, but thanks largely to microcomputers, the last 15 years have seen an explosion of computer technology-related activity in schools and classrooms. In only about the last 10 years, video technologies and telecommunications have shifted the spotlight from standalone microcomputers to more recent developments such as interactive video and the Internet.

Some of the oldest technology tools (e.g., word processing programs) and software resources (e.g., simulations) are still among the most useful technologies for teachers. Still, more recent developments generate the real excitement and innovation in educational technology. Much more is being written about multimedia applications than about microcomputers, and telecommunications stirs much more excitement than tutorials. The chapters in this section describe the current and emerging roles of four categories of these fresh technologies that have taken the lead in the post-microcomputer era.

Each of the chapters in this section follows a similar structure. Each begins with a description of the technology resources considered in the chapter including the background, characteristics, and capabilities of each one. Instructional advantages and disadvantages are considered, and types of integration strategies are outlined. Finally, each chapter presents sample directed and constructivist integration plans that are linked to curriculum.

Chapter 7: Using Optical Technologies in Teaching and Learning

This chapter deals with three recent technologies that resemble television. Interactive videodiscs have been around since about 1972, but only in about the last decade have educators begun to take advantage of the capabilities and practical applications of Level I and Level III videodisc technology. Publishers and other curriculum developers have recognized the potential power of interactive video and CD-ROM. The use of Level I videodisc technology, a noncomputer-driven medium, has become increasingly popular, and integration strategies for this technology are among the most powerful and the easiest to implement. Compact disc-interactive (CD-I) technology is currently still in its infancy, but it is predicted to play a more high-profile role in educational technology in the near future.

Chapter 8: Using Hypermedia in Teaching and Learning

A combination of audio, video, and text-based media under the direct control of teachers and students has long been a dream and a goal in education. Tools such as HyperStudio and Linkway Live have helped fulfill this dream of practical, easy-to-use classroom multimedia applications. This chapter describes the new multimedia/hypermedia authoring tools and integration strategies that employ them.

Chapter 9: Linking to Learn

The information superhighway has become a pervasive metaphor for technology in modern culture, but educators are just beginning to develop integration strategies for the Internet. Distance learning, long a means of providing courses and workshops for industry training and higher education, continues to increase in popularity, and the range of applications still grows. This chapter reviews a variety of telecommunications and distance learning resources and methods to link learners with each other and with needed resources in distant sites.

Chapter 10: Emerging Technologies

Technologies such as artificial intelligence and virtual reality make the future seem right around the corner. This chapter describes three technologies whose practical applications are just emerging, but whose potential power seems almost limitless. These include personal digital assistants (PDAs), artificial intelligence (AI), and virtual reality. Integration strategies for these technologies are still experimental, but some of the elements of these technologies (e.g., voice and handwriting recognition, multimedia-based simulations) promise to change the future of educational technology, as well as the face of society in general, more than any other set of tools.

Chapter 7

Using Optical Technologies in Teaching and Learning

This chapter will cover the following topics:

- Definitions and characteristics of three kinds of optical technologies

- Unique advantages of interactive videodisc, CD-ROM, and CD-I technology for various classroom activities

- Example classroom uses for each technology

Chapter Objectives

1. Identify the characteristics of interactive videodisc, CD-ROM, and CD-I technologies.

2. Design directed and constructivist classroom activities for each kind of optical technology.

"By mixing things up and making the classroom a more multisensory environment, you take advantage of multiple paths to learning."
David Thornberg

Introduction to Optical Technologies

Some Background

The belief that "a picture is worth a thousand words" has been an important part of American culture for some time. Since it was first introduced in the 1950s, television has had an incalculable role in shaping both people's knowledge of the world and their preferred mode of learning about it. The emergence in recent years of optical technologies has only added to this dependence on visual images and reinforced the cultural expectation of pictures, preferably with movements, to accompany descriptions of important concepts. This chapter describes the emergence of three of these optical technologies and their current—and potential—influence on educational practices.

Optical disc technology was first introduced by the Dutch firm NV Philips in 1972, leading to the release of a commercial product in 1976. Videodiscs were initially marketed as media to show movies at home. Soon after they got started in the consumer market, however, relatively low-cost VCR technology became available and emerged as the medium of choice for consumers.

In recent years, the combination of optical technologies with personal computers has begun to affect the education market significantly. First, videodisc players paired with computers gave students and teachers high-quality, interactive video and audio media. More recently, audio CD technology was extended to create CD-ROMs. The CD-ROM medium provides a low-cost method for storing an enormous amount of data, giving educational institutions easy access to large quantities of information. Most recently, compact disk-interactive (CD-I) technology has offered yet another option. CD-I promises to simplify the interface between people and the machines that serve them.

Definition of Visual Technologies

Optical storage devices include several forms of classroom tools: videodiscs, CD-ROMs (compact disc-read only memory), CD-I (compact disc-interactive), and several other technologies including DVI (digital video interactive), and photo CDs (photographic compact discs). This chapter will focus on videodisc, CD-ROM, and CD-I technologies, since they have the most powerful current impact on K-12 education.

Optical technologies all share a single core technology: laser beams heat light-sensitive material on a disc so that a chemical reaction causes the area to either remain opaque or reflect light, thus revealing encoded information. In contrast, magnetic storage units, such as hard disks and tape devices, store data as magnetic pulses, which are read or altered by the disk drive (Mathisen, 1991; Walkenbach, 1992).

Generally accepted style in the industry has adopted the spelling *disk* magnetic storage systems and *disc* for optical devices.

Issues Related to Optical Technologies in Education

Since optical media have enabled developers to produce multimedia and hypermedia programs, perhaps the biggest issue concerning optical technologies is how to use them effectively. No doubt, many programs produced with optical technologies are technological wonders. But, educators need to ask what these breakthroughs do for the intellectual development of students. It seems evident that teachers must do more than just sitting students down in front of engrossing programs.

Blissett and Adkins (1993) suggest the following strategies for teachers to employ when integrating interactive multimedia into the classroom:

- Provide guidance and further explanation on the nature of the task when a group gets stuck or, worse, misunderstands what to do.
- Check that the software package provides advance organizers for its conceptual content.
- Individualize the learning experience by assessing learning as it occurs and then intervening to link and relate or extend and consolidate concepts to meet the needs of particular pupils or groups.
- Ask open-ended questions that require pupils to verbalize their thought processes and review their understanding of the conceptual subject matter.
- Challenge and provoke thinking, leading to more abstract and conceptual discussions.
- Help the group to review its problem-solving strategies and direct them toward more powerful ones.
- Conduct training in group-work skills.

Interactive Videodisc Technology

Videodiscs are optical storage media for random-access storage of high-quality audio and analog information. Laser videodiscs store text, audio, video, and graphics data in analog format. Interactive videodisc (IVD) technology was first released in the 1970s and it has built applications in both the education and business worlds. A videodisc resembles an audio CD disc, except it is generally larger in diameter. Most videodiscs are 12 inches in diameter (Figure 7.1) and hold 54,000 still frames or the equivalent of 675 carousel slide trays. They represent an extremely durable medium for storing and displaying visual information. Videodiscs are read by a laser beam, and the mechanism allows random access to any part of the disc. The random access feature is important because it avoids the need to

Figure 7.1 Laser Videodisc

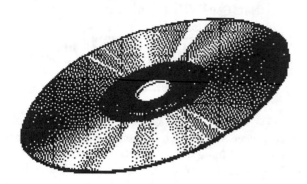

Table 7.1 Attributes of CLV and CAV Videodisc Formats		
Attribute	**CLV**	**CAV**
Random access	yes	yes
Still frame	no	yes
Frame search	no	yes
Time search	yes	no
Chapter search	yes	yes
Scan	yes	yes
Multispeed	no	yes
Minutes per side	60	30
Straight play	yes	yes
Two audio tracks	yes	yes

fast forward or rewind to find a particular image as would be necessary with a videotape.

Although not generally considered cutting-edge technology, videodiscs are becoming very popular teaching tools for many educators. The quantity, and particularly the quality, of applications are expanding rapidly. The wide variety of software linked to videodiscs encompasses programs that should appeal to just about any teaching style. The ease of use of Level I programs (described below) encourages adoption by all teachers.

IVD Formats

Currently teachers can use videodiscs in two formats: CAV and CLV. CAV (constant angular velocity) discs permit interactive applications by enabling users to access randomly any of 54,000 frames on each side in only a few seconds. In addition, this format permits the teacher to display a single frame continuously, treating images as individual slides. To make this process easy, each frame on the disc has its own number ranging from 1 to 54,000. Laser videodisc players offer the following features with CAV discs: frame search, chapter search, single frame stepping, scanning (essentially fast forward or fast reverse), slow motion, and triple speed. All of these features can function in forward or reverse mode. The total video capacity of a CAV disc is 60 minutes (30 minutes each side).

CLV (constant linear velocity) videodiscs are used for linear applications, sometimes called *extended play applications,* such as movies or television specials. The CLV discs have limited interactive possibilities; for example, the players do not provide random access to individual frames and freeze frames images. CLV discs are indexed by time rather than frame numbers, so players provide random access to exact seconds on CLV discs. The total running time of video on a CLV disc is approximately 60 minutes per side. Table 7.1 summarizes the features of CLV and CAV formats.

All videodisc players can accommodate both CLV and CAV formats. Of the two formats, CAV is considered more appropriate for education since it offers more options for access and interactivity.

Levels of Interactivity

The level of interactivity of a videodisc program refers to the amount of control the program gives the user and what kinds of software and hardware the user needs to achieve control. Videodisc programs are referred to as Level I, Level II, Level III, and Level IV.

- **Level I.** In Level I interactive video, the user controls the program directly without the involvement of a computer. The user can control the videodisc player in three different ways: (1) with the controls on the front of the player itself, (2) with the handheld remote control keypad, or (3) with a handheld bar code reader and printed bar codes.

- **Level II.** Level II videodiscs are not used very much in education. The videodisc player needs an internal microprocessor that reads the computer program stored along with other data on the videodisc. This level of interactivity will be replaced by a digital technology such as CD-ROM or CD-I.

- **Level III.** Level III interactivity requires the use of a computer connected to a videodisc player. A computer program operates the player via a cable by running appropriate "driver" software. Level III software programs include drivers for most videodisc players. This combination offers the user the advantage of access to computer data and videodisc data simultaneously.

- **Level IV.** This term refers to Level III interactivity with a single monitor to display all material. Some industry experts think of Level IV interactivity as programs that apply artificial learning techniques to create expert learning systems. Level IV systems often include sophisticated devices for user interfaces, such as touch screens, speech recognition apparatus, and virtual reality peripherals. In a broad sense, *Level IV* also refers to future developments in interactive multimedia (Skolnik and Kanning, 1994). Because they remain relatively undeveloped at this time, Level IV applications will not be covered in this text.

Hardware Requirements for IVD Instruction

A videodisc player connects to a television set or computer monitor in much the same way as a VCR does. Both are generally configured to use either RF or RCA connectors. Remote controls and bar-code readers are not difficult to use, but they do require some training and practice. To take full advantage of the interactivity in IVD technology, the teacher must handle the mechanics of its operation smoothly.

- **Level I hardware.** The necessary hardware for Level I interactive video includes a videodisc player, a monitor, and a method of controlling the player—either the control panel on the machine, a remote control device, or a bar-code reader. (Not all videodisc players are equipped for bar-code control.) The user also needs a computer and appropriate software to make customized bar codes.

- **Level III hardware.** Level III interactivity requires an appropriately configured videodisc player, a monitor, a computer, and an RS-232 cable. (Some videodisc players will function only at Level I.)

Software Requirements for IVD Instruction

- **Level I software.** The videodisc provides the only software needed for this level. The user also needs a special software package to generate customized bar codes. (Bar'n'Coder, Lesson Maker, and My Report Generator do this, as Figure 7.4 illustrates.)

- **Level III software.** This level gives the user several software options: a curriculum program like Tom Snyder's The Great Ocean Rescue, a slide show generating program such as Videodiscovery's Media Max, or a hypermedia authoring program like Roger Wagner's HyperStudio or Alchemidia's Multimedia Scrapbook.

Classroom Applications of Videodiscs

When schools introduced IVD technology a number of years ago, its advocates focused their excitement on the potential of Level III interactivity. It's easy to see why. Smooth interaction between the user and the technology developed an almost "magical" quality. Some even predicted that Level III interactive technology would revolutionize education. They tended to dismiss Level I technology as inferior to Level III, mainly because it involved no computer.

As often happens, the experts did not read accurately the pulse of the schools or the teachers. Level III interactivity required more hardware than schools could afford to implement on any scale. In addition, one station in a classroom did not fit in well with the directed instruction styles of most teachers. Instead, Level I usage has proven to be very popular with teachers, and software developers have begun to take notice. As more compatible technology becomes available in schools, Level III usage is increasing along with Level I. In order to try to meet the needs of both sets of customers, developers often ship videodisc programs with bar codes for Level I and software for Level III.

Figure 7.2 **Sample Bar Code**

Nat. Geo. Wild Chimps
Chimps Use Intelligence

1644 to 1729

Bar-code access. Users can scan bar codes to access chapters, individual frames, or segments of video on both CLV and CAV discs. The bar codes look like the UPC codes on commercial-use products (Figure 7.2). The bar code reader that interprets the patterns of stripes resembles the technology in many stores (Figure 7.3).

Bar codes are considered the easiest way to access information from videodisc. They simplify control of the technology so the teacher can concentrate on interacting with the students without worrying about entering frame or chapter numbers with a remote control device. Vendors have taken advantage of these capabilities and now include bar codes with most Level I programs. A good example of the power of bar codes emerged when Optical Data Corporation upgraded its Windows on Science program. By indexing discs with bar codes, the company turned a potentially cumbersome program into one that teachers find both exciting and easy to use.

Bar-coding software. A number of software packages for Macintosh, Windows, and DOS systems (Bar'n'Coder 3.0, Barcode Maker, Lesson Maker) enable users to create bar codes for videodiscs. Most of these programs work similarly. The user enters the frame or time numbers to display (see Figure 7.4) and then types in a descriptor for the code. At he user's command to make the code, the program either prints it out or exports it to a word processing document.

Once the bar code is generated, a teacher can use it many ways. A teacher might create bar codes that access data on a videodisc dealing with geography and then glue the bar codes on appropriate parts of maps. This becomes an exploration center for students. Another instructor might use the computer to paste the bar code into a word processing file and then scan it during a teacher-directed activity (See the Example Lesson Plan.)

Figure 7.3 **Handheld Bar-code Reader**

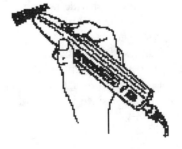

Figure 7.4 Screen Display for Pioneer's Bar'n'Coder 3.0

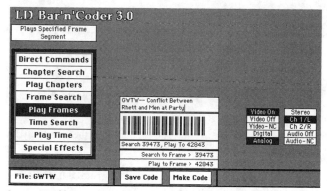

Source: Used by permission of Pioneer New Media Technologies.

Bar code software can also make bar codes for an audio CD. Videodisc players sold since 1992 support the LB2 standard. They can play audio CDs under control of bar codes that match the standard. The "address" for the audio CD shows up on the TV monitor hooked up to the videodisc player. The user can enter this information into the bar-code program as for a videodisc frame to gain access to a specific part of the audio CD.

Level I Applications for Education

- **Curriculum packages.** Perhaps the most ambitious use of videodisc technology comes from Optical Data Corporation. This firm has developed entire curriculum packages around Level I technology. It has followed its Windows on Science, now in wide use, with Windows on Math and Windows on Social Studies. Some states now allow school districts to adopt these programs in lieu of textbooks. D. C. Heath's Interactions program uses video technology to bring interactive examples of math applications in real-world settings.

- **Problem solving.** Videodiscovery's Science Sleuths and Math Sleuths (also available in CD-ROM versions) provide students with mysteries to solve. The videodisc program offers clues in the form of interviews, textual and numeric data, pho-

tographs, and diagrams. The teacher plays a very active role in guiding the students via questioning techniques and just-in-time teaching.

- **Simulations.** The Adventures of Jasper Woodbury teaches middle-school math through a series of simulated stories. The students solve real-world applied math problems by retrieving mathematical data embedded in the stories. Bar-code technology comes in very handy, since it enables students to review segments from the story as needed. It is designed so that the teacher can stop the action and teach certain math techniques just when students need those techniques to solve techniques problems.

- **Visual databases.** Visual databases are collections of individual pictures and short video segments. BioSci II was one of the earliest examples of this type of program. These databases are perhaps most useful in the areas of science and art. They provide a wealth of resources for both teacher presentations and student projects.

- **Movies and documentaries.** Schools can choose from thousands of movies, documentaries, and other general-use videodiscs at very reasonable prices. These resources can yield tremendous educational benefits. Through the search feature controlled by the remote control, the teacher can access any frame or segment of a disc almost instantly. The random access capability of the technology holds great promise for encouraging and facilitating a sound pedagogical use of video as opposed to mundane uses such as playing movies straight through. This category of videodiscs includes a whole series of National Geographic Specials, PBS specials such as Nova and The Civil War, and Disney's Sing Along Songs.

Repurposing videodiscs for Level I. Videodisc producers and developers have specific uses in mind for it. Depending on the program, these can range from home entertainment to very structured, directed-instruction lessons.

Repurposing refers to an originally unintended application of a videodisc via a control program (Barron and Orwig, 1993, p. 43). In other words, the user, perhaps a teacher or student, can display a portion of a videodisc to serve a completely different purpose than what the developers had in mind. This represents one of the most powerful uses of IVD in the schools. It opens up the door to tremendous opportunities for students and teachers to develop their creative potentials. Also, it gives teachers and students a perspective on video through the eyes of a director, and it helps to prepare teachers and students for the time in the near future when they will have tremendous editing capabilities with digital video.

Repurposing also can enable schools to get a lot more value for the money they spend on educational resources. It is very expensive for a team of instructional designers to develop an educational videodisc program and, naturally, customers must cover that cost. Schools can expect to pay $200 to $500 for fully developed programs. On the other hand, movies, documentaries, and general-interest videodiscs with no teacher support materials sell for well under $100. These relatively inexpensive discs offer tremendous resources for creative educators and students.

Example Lesson Plan
Repurposing a Videodisc

Lesson: Observation Skills
Level: Grade 5
Lesson Purposes: In this lesson, the teacher has repurposed a high school/college level videodisc to meet the needs of a fifth grade class.

Materials. Videodisc version of Smithsonian's Insects: The Little Things That Run the World by Lumivison, videodisc player with bar-code reader

Teacher Note. The bar code segment displays a visually rich segment that accompanies a narration dealing with how ancient civilizations viewed insects. Don't tell the students the subject matter of the segment. This activity provides a good lead-in to a study of insects.

OBSERVATION SKILLS ACTIVITY

insects/silent

6135 to 6639   Play   Pause

insect/audio on   insect/video off

6135 to 8639   6135 to 8839

Instructional Activities.

1. Have the students pair up and assume roles as No. 1 and No. 2.

2. Tell them that the first time through the clip, No. 1 will watch it and No. 2 will face away from the screen. (Use the bar code marked "insects/silent.")

3. Instruct No. 1 to watch the clip very carefully, because when it ends, he or she will describe it in detail to No. 2.

4. When the clip is completed, give the pairs time to converse. After 4 to 5 minutes, stop the conversations and let No. 2 watch the clip. (No. 1 can watch it a second time.)

5. No. 2 now gives No. 1 feedback on the match between expectations after the description and the actual clip.

6. Play the segment with sound for all participants. Discuss the activity. Encourage students to reflect on the experience.

Level III Capabilities

Hypermedia resources. In probably its most powerful use, Level III technology can incorporate videodisc resources into interactive hypermedia programs. These resources can include video, audio, (including song and speech), animation, photographic, and text data. The most common resources accessed from a videodisc are video and still images.

Figure 7.5 gives an example screen from a hypermedia program that utilizes Level III interactive video. At the bottom of the card, the question "Can War Be Justified?" represents a hypermedia link. If the user clicks on this area with the mouse, the videodisc player will play a video segment. If the user clicks on "Support materials," the system will access a section of the videodisc that provides text documentation, maps, and still images.

Presentation tool. Level III interactive videodiscs can also act as components of presentation tools that are integrated into educational programs. Some curriculum pro-

grams, like those in the ABC Interactive series and BioSci II, have built-in tools that allow teachers and students to develop their own presentations by drawing on the videodisc resources.

Certain control programs allow users to access any CAV disc to create slide shows and/or presentations. Media Max from Videodiscovery (see Figure 7.6) and the Voyager VideoStack from Voyager can also add text overlay onto a slide displayed on the TV monitor.

Level III Applications for Education

• **Databases.** Level III software provides an easy way to access the contents of visual databases. The Visual Almanac provides a collection of stills, short video segments, and sounds for users to include in their own multimedia productions. The software is used primarily to access the information on disc. The Living Textbook series from Optical Data offers the teacher a vast collection of visual resources to enhance science instruction.

• **Hypermedia exploration.** A number of Level III programs are designed to let users choose their own pathways through

Figure 7.5 Sample Hypermedia Screen (ABC News InterActive's *Lessons of War*)

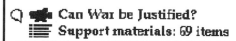

Can War be Justified?

Virtually all of the world's great philosophies declare that people should treat each other with dignity, justice and respect. To Christians, Muslims and Jews, who follow faiths that consider the Old Testament to be a sacred text, "Thou Shalt Not Kill" is believed to be one of the **ten commandments** given by God to Moses on Mount Sinai.

Despite the moral and religious prohibitions against violence, war and violence have continued throughout human history. One of the challenges of philosophers has been to explain why wars continue, and to determine under what circumstances war can be justified. The concepts of fairness and justice that all cultures share have also resulted in a body of customs and agreements that have become **international law**. These laws of nations also establish the circumstances under which nations may claim that waging war is justified.

Can War be Justified?
Support materials: 69 items

Source: Used by permission of Capital Cities/ABC, Inc.

large bodies of information. IBM's Illuminated Manuscript, STV: Human Body and GTV: A Geographic Perspective on American History are examples of this type of program. Some educators believe that hypermedia represents the best and highest use of videodisc technology.

- **Simulations.** IBM's Exploring Chemistry series simulates an interactive laboratory that a provides a safe, efficient, and effective way for students to do experiments. Tom Snyder's The Great Ocean Rescue and The Great Solar System Rescue discs invite teams of students to analyze visual data and solve problems. Both Snyder programs are for whole-group instruction in a one-computer classroom.

- **Problem solving.** Apple Computer's TLT: Teaching, Learning, and Technology (1991 version) is a Level III tool intended to help schools work through the school improvement process. The videodisc component provides interviews with experts in the field and numerous examples of teachers and students in action. A presentation program interfaces with CD-ROM drives as well as videodisc players.

Advantages of IVD Technology

- **Flexibility.** This resource supports a variety of applications for teachers and students. The control options, as well as the unlimited creative potential of repurposing videodiscs, allow this tool to integrate effectively into most curriculum areas.

- **Dual audio tracks.** Developers often use the second track to play program audio in a second language (usually Spanish) or to include a simpler or more detailed version of the presentation. The teacher can then let a student or students listen to one track through headphones while the rest of the class listens to the other track through the monitor.

- **Ease of use.** Most teachers and students feel comfortable enough to begin using this technology with only a few hours of training. However, they may become fully comfortable only after many hours of use. Some schools have encouraged teachers to check out videodisc players for practice at home, making them much more likely to use the equipment in their classrooms.

- **Quality.** High-resolution video images and audio clips stored on videodiscs provide much higher quality than

Source: Joseph H. Bailey, © 1991 National Geographic Society. Used by permission.

Figure 7.6 Level III Remote Control Screen

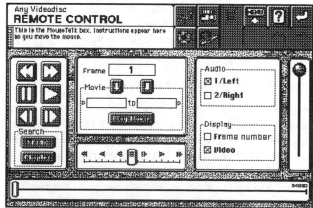

Source: MediaMax by Videodiscovery. Used by permission.

videocassette tapes can achieve. Videodiscs also beat the quality of digitized video displayed directly on computer monitors.

- **Durability.** Since videodiscs are read with a laser beam, they suffer no real wear and tear in normal use. Discs are coated with plastic that generally protects them from small scratches or fingerprints. They are certain to last longer than videotapes or films.

- **Standards.** Videodisc manufacturers have adopted a worldwide standard for the technology. As a result, any videodisc will play on any player. They also adhere to the LB and LB2 bar-code standards, which means that all videodisc bar codes work with all bar-code readers.

Disadvantages of IVD Technology

- **Cost.** The cost of some videodisc programs can be quite high, $300 to $500 or more. This presents a big problem when a school's media purchasers see these programs simply as "videos." Someone who can pay $49.95 for a videotape will often balk at the cost of a videodisc. Schools should evaluate these higher-end videodisc programs as curriculum packages, however, rather than just discs.

- **Read only.** The videodisc is a read-only technology, which means that the user cannot record information. This is an obvious disadvantage compared to the VCR.

- **Lack of interface standards.** No industry standards regulate the interface between the videodisc player and the computer. This means that different brands of players require different cables, and some software programs are incompatible with some players. To guard against incompatibility, it would be wise for a school to settle on one brand of player.

- **Hardware intensive systems for Level III.** A Level III setup includes a videodisc player, monitor, computer, and cables. Teachers often find it a burden to gather all of this equipment in one place at one time. Some schools have opted to create portable multimedia stations that can easily be transported from one classroom to another. Level III interactivity also presents a cumbersome interface for the user who must pay attention to two monitors (the computer screen and the television screen).

- **Maintenance costs.** Videodisc players can be quite expensive to repair, and users should handle them with care. However, the industrial type of players sold to schools are designed to be more durable than home players.

- **Limited video capacity.** Although videodiscs provide tremendous storage capacity for still frames, when played as continuous video, the CAV format offers only 30 minutes per side. A complete movie can fill three, four, or even five separate discs.

Issues Related to IVD Use in Schools

Some people have been predicting the demise of the videodisc as a result of the explosion of CD-ROM technology, but their dire predictions seem to have been premature. The installed base of videodisc players and software in many schools ensures their survival, and teachers still find Level I IVD technology an easy-to-use and effective tool (Shields, 1994). Schools will not likely decide simply to drop videodiscs and switch to CD-ROMs, especially since

Level I use requires no computer and videodisc players can easily be moved and set up.

The future of Level III is somewhat uncertain. Because of the awkward hardware configuration for Level III versus the simplicity of CD-ROM and CD-I systems, schools will likely prefer the simpler interactive multimedia platform. Level III may be relegated to the role of a presentation tool. This use does take advantage of its real strength—delivering high-quality, full-motion video on a TV screen, without the necessity of a scan converter. The choice between videodisc and CD technology need not settle on one over the other; instead, schools must choose how to prioritize expenditures.

CD-ROM Technology

Introduction to CD-ROMs

Commonly called *compact discs* or *CDs,* CD-ROMs are made of the same material as videodiscs, but they are smaller in size, just 4.72 inches (12 cm) in diameter. In appearance, CD-ROMs look identical to audio CDs. The main practical difference between an audio CD and a CD-ROM is what they can store and how they are used. Audio CDs store music in digital form, while CD-ROMs store text, audio, video, animation, and graphics information, also in digital form. CD-ROMs are known for their huge storage capacity, up to 650 MB of data, which equates to the equivalent of 250,000 pages of text, or five hundred 500-page novels. Computer systems access CD-ROMs through internal or external CD-ROM players. Users of these discs need to understand some important technical issues:

- **Hardware.** Internal, or built-in CD-ROM drives are housed inside computers' cases. External drives connect to computers by cables. In recent years, more and more machines have been sold with built-in CD-ROM drives. Some CDs play audio information only through external speakers or headphones. Another hardware consideration is the size of the computer's monitor; some programs that include video will require 13-inch or larger monitors. Be sure to read the fine print in the catalog when ordering CD-ROM programs. That's where hardware requirements are usually listed.

- **Software.** A computer must run special software to communicate with a CD-ROM player. This software is commonly referred to as CD-ROM *drivers.* This is usually not an issue with a built-in CD-ROM drive, since the manufacturer would have installed the software along with the hardware. Some CD-ROM programs require users to install additional software on the hard drive in order to run those specific CDs. This expedites the use of the program.

- **Drive speed.** Many of the newer CD-ROMs include video segments. In order to play video segments successfully, a CD-ROM player should have access time of 280 milliseconds (ms) or less. A quicker access time (smaller number of milliseconds) is faster and more desirable.

- **Platform.** CD-ROMs typically fit under one of four platform types: Macintosh, MS-DOS, Windows, and MPC. MPC (multimedia PC) is a multimedia standard agreed upon by some hardware and software producers in the PC industry. It provides enhanced audio and video, but for only MS-DOS computers. When ordering, it's important to request the correct type of disc for your machines.

In addition, some CD-ROM producers create what is known in the industry as *hybrid* CDs. This type of CD-ROM contains software for two or more platforms on a single disc. For example, a disc may contain software for the Mac, Windows, and MPC platforms.

Classroom Applications of CD-ROMs

The number of titles in the CD-ROM market has increased at a phenomenal rate over the past few years. With the expected increase in sales of home and school-based multimedia computers, the number of titles will continue to grow. Users can choose from titles in several categories:

- **Interactive multimedia programs.** The multimedia storage capabilities of CD-ROM technology have made interactive multimedia products one of the most frequently touted uses. While researchers still need to study the effectiveness of these programs as learning tools, there is no doubt about their appeal. These programs often organize information in the hypermedia model. Users usually do not follow directions, instead, they choose their own pathways through the information. Some examples of this type of program include, The San Diego Zoo Presents: The Animals; A.D.A.M. Essential, An Exploration of the Human Body; and Smithsonian's America.

- **Interactive storybooks.** These on-screen stories have become extremely popular with primary teachers and students. On the audio tracks, narrators read pages as the words are highlighted on screen. If a student needs to hear a word again, just clicking on it with the mouse pointer will activate the audio. Some teachers prefer the straightforward approach of the Discus Books series, while others are more drawn to the Brøderbund titles because students tend to find them more engaging. As educators learn more about the cognitive value of these books, development methodology is sure to show great advancements.

- **Computer software.** A single CD-ROM can store the equivalent of 800 3½-inch micro disks. This makes CD-ROM a wonderful technology for distributing software. Some companies have taken advantage of this added capacity by enhancing successful programs with new multimedia features (Oregon Trail by MECC and The Adventures of Carmen Sandiego).

- **Reference materials.** CD-ROM technology meets the needs of the students and teachers in this area more than any other. A plethora of reference materials are available at very reasonable cost. To add still more value, these resources are accompanied by software that makes searching for information both easy and efficient. Below are just a few of the categories and titles:

  - **Encyclopedias:** *Compton's Encyclopedia, Encarta* (Microsoft's multimedia encyclopedia), *The Aircraft Encyclopedia, Encyclopedia of Science and Technology*

  - **Almanacs:** *Illustrated Facts, The CIA World Factbook, The KGB World Factbook, The Time Almanac*

  - **Atlases:** *U. S. Geography, Picture Atlas of the World, U.S. Atlas, Small Blue Planet: The Real Cities Atlas*

- **Collections of resources.** A wide variety of resources are now shipped on CD-ROM. These include collections of clip art, sound effects, photographs, video clips, fonts, and document templates. Some major conferences, like the Florida Educational Technology Conference (FETC), distribute proceedings, presenter handouts, vendor samples, and shareware on CD-ROM to each registrant.

Advantages of CD-ROM Technology

- **Large storage capacity.** CD-ROMs provide an excellent way of storing large quantities of text, audio, and video information. Floppy disks just do not have the capacity to take advantage of the potential of multimedia.

- **Durability.** The discs are strong and hold up well under constant handling by students.

- **Cost.** Once the content is developed, the cost of replicating CD-ROMs is very low.

- **Search speed.** Although CD drives read information more slowly than hard drives, they provide a fast way of searching through large collections of information.

- **Availability.** Thousands of titles are available with many more to come.

- **Audio CD capability.** CD-ROM players will play tracks from audio CDs. This can come in handy for hypermedia authoring. Students can add segments of music or sound effects to a hypermedia product by accessing a disc in the CD-ROM drive. (See Chapter 8.) This has the advantage of alleviating copyright concerns. Heed a word of caution, though: Never attempt to play a CD-ROM disc on an audio CD player; it will ruin the CD-ROM disc.

Disadvantages of CD-ROM Technology

- **Access speed.** CD-ROM drives access data slowly compared to hard drives.

- **Installation.** Due to a lack of a standard in the MS-DOS world, connecting a CD-ROM drive to a PC can be very frustrating. The industry is aware of this problem and steps are being taken to correct it.

- **Shortage of hardware.** The aging computers of many schools prevent them from taking advantage of CD-ROM technology. As time passes, this should become a less and less serious problem. If at all possible, schools should buy machines with built-in CD-ROM drives.

Compact Disc-Interactive (CD-I) Technology

Introduction to CD-I Technology

CD-I was developed primarily as a consumer electronics product that combines optical discs with a computer to provide a home entertainment system that delivers music, graphics, text, animation, and video output in the living room. However, the emergence of "edutainment" software and CD-I's easy-to-use interface have interested schools in this product.

Unlike a CD-ROM drive, a CD-I player is a standalone system that requires no external computer. It plugs directly into a TV or monitor and comes with a remote control to allow the user to interact with software programs sold on discs. The hardware looks and feels much like an audio CD player, but runs interactive multimedia programs. Like the CD-ROM player, it can also play audio CDs.

The user interacts with the player via an input device such as a remote control or a joystick. Recently, a new device called the Roller Controller has become available. This large, colorful trackball device is designed to let children as young as 2 1/2 years old use the player. The combination of the Roller Controller and CD-I technology has become very popular with prekindergarten and special education teachers. In fact, it provides the only "higher" technology that some children can use.

Hardware Requirements for CD-I Technology

Source: Courtesy of Philips Consumer Electronics Company.

Classroom Applications of CD-I Technology

The number of CD-I titles is increasing, but it still remains small compared to the CD-ROM market. Many CD-I uses overlap with those of CD-ROMs. One big difference is that this technology is not usually used as an output device. Although all CD-I units have some RAM and many models have floppy drives to import and export data, there are no CD-I resource discs. The drives are usually used strictly as play machines.

Advantages of CD-I Technology

• **User-friendly interface.** CD-I was designed for young children to use it; this is a very nonintimidating technology. With the addition of a touch screen, it may model the interface of tomorrow.

• **Ease of use.** Because the computer and drive are all in one box, this is an extremely easy technology to use and transport.

• **Economical.** The cost of the player is dropping and the programs usually sell for around $50. This is certainly an advantage for education. In addition, CD-I players run on TVs or monitors that many schools already have. The cost of a CD-I player is considerably less than that of a computer with a CD-ROM drive.

• **Versatility.** CD-I drives can play both audio CDs and photo CDs.

Disadvantages of CD-I Technology

• **Shortage of software titles.** At this time, CD-I is still finding its niche in the market. Until it is more widely used, the number of titles will be limited.

• **Incompatibility with CD-ROMs.** A CD-I disc cannot play on a CD-ROM player. This obviously presents a problem in schools, which often accumulate varied hardware with no real sense of standard.

Using Optical Technologies in the Classroom: Some Examples

Directed Instruction Activity: Stretching the Mind
Level: Grades 4–12
Purpose of the Activity: To develop students' creative and critical thinking skills

Instructional Activities: Students need a piece of paper and something to write with. The television screen displays an image and a word. The students must write down one way in which the word and picture are alike and one way in which they are different. This can be repeated any number of times. When done, the students can share answers. As a culminating activity, students receive a list of the class's words and are asked to write a story using at least half of them. If the school cannot provide CD-ROM or CD-I technology, allow students to use resources like atlases, almanacs, and encyclopedias to search for ideas. When they have completed the activity, they can read their stories read to the class.

Suggestions for Teacher: Some of the videodisc control programs, like MediaMax and Voyager Videostack enable the user to display a slide and overlay words on the screen. If this overlay feature is not available, the words can be written on the chalkboard. The teacher may use Level I IVD for this activity if Level III technology is not available.

Continued

A Constructivist Activity:	Geography and Travel
Level:	Grades 7–12
Purposes of Activity:	To enhance students' knowledge of geography through a study of the airline industry, to broaden students' knowledge of how the airline industry operates, and to provide students with an opportunity to utilize visual technologies in the development of a project
Instructional Activities:	Each group studies one assigned airline. The groups are subdivided into study groups dealing with hub airports, main routes, commuter routes, and international routes. Some individuals may research airline history and the FAA. Students should utilize CD-ROM, videodisc, CD-I, and telecommunications resources when possible. As a final product, each group puts together a promotional presentation designed to get people to fly on its airline. Level I or Level III videodisc, CD-ROM, or CD-I technology may be used in the presentation.
Suggestions for the Teacher:	This activity is aimed at geography teachers, but with adaptations it could suit economics classes, as well. Optical resources are designed primarily to reinforce the travel aspect of the presentation. Numerous videodiscs are available that highlight geographical aspects of many locations. Either video segments or slides can be extremely effective components of a presentation. The presentations do not necessarily have to target large groups with CD-ROM, CD-I, or Level III videodisc technology; a kiosk approach may be better. Written directions could be given to the users to guide them through construction of a persuasive presentation.

Other Sample Integration Activities

Activity 1:	Persuasive Presentation
Level:	Grades 5–12
Media:	Level I videodisc
Purpose of Activity:	To use Level I interactive videodisc technology to create a persuasive presentation
Instructional Activities:	Pass out a copy of the following scenario to the students. Ask them to read it.

Scenario: The Salt Quest. Military commands all over the world are in a state of high alert. A fleet of alien spacecraft has been lurking just outside earth's atmosphere for the past 2 days. An organization called What's Up, a think tank composed of highly educated men and women who for years have been convinced that extraterrestrial life exists, was asked to contact the aliens to see what they wanted.

The spacecraft are from a planet called Nacl, which is in another solar system. The inhabitants of Nacl require tremendous amounts of salt to exist. They are getting close to depleting their own supplies and have been on a long search for new supplies. They have noticed that earth's oceans are full of salt, so they thought that they would help themselves to this huge supply.

You are a committee of What's Up members that must persuade the Naclites that it would not be a good idea to mine the salt from the oceans. The Naclites are reasonable beings who, it is thought, will move on to other places if they realize that mining the oceans would kill off millions of beautiful creatures.

What's Up is charged with putting together a presentation that will convince the Naclites that the sea life on earth is too magnificent to be sacrificed just to satisfy their need for salt. The members decide that the presentation must have visuals, and Level I interactive video would be the best media to use.

To tap the creative resources of as many members as possible, the group has divided into subgroups that will each develop a 5-minute presentation. The group will then decide which presentation to use with the Naclites. The individual presentations will be evaluated according to the following criteria: persuasiveness, creativity, and entertainment value.

Divide the students into groups of four. Provide the groups at least one videodisc that contains images of sea creatures. The groups then need to scan through the videodisc and select still frames or video segments that they can use in their presentation. They will then develop the actual presentations, keeping in mind the criteria for evaluation.

Continued

Suggestions for Teacher: Encourage the students to be as creative as they would like. The videodisc player should be used as a resource to embellish the presentations. Some groups will lean toward simply playing series of video segments; this should be discouraged. Teacher should intervene during the development of the project to make sure that the students are actively engaged in developing a quality product.

Activity 2: Tracking Down Trivia
Level: Grades 3–12
Media: CD-ROM
Purpose of Activity: To improve students' research skills

Instructional Activities: Pass out copies of the following list of questions to the class. Have the students work in pairs. Using the CD-ROM as a resource, look up the answers to the questions:

1. What is the capital city of Sweden?
2. What was E. B. White's first name?
3. Who won the Nobel Prize for Peace in 1985?
4. When did the Japanese bomb Pearl Harbor?
5. How did John Paul Jones die?
6. What is the state flower of Utah?
7. What does *illiterate* mean?
8. What gases make up the atmosphere of Jupiter?
9. What movie won the Academy Award for best picture in 1985?
10. What is a synonym for the word *supercilious?*
11. What is the chemical formula for sulfuric acid?
12. If Babe Ruth had lived 20 years longer, how old would he have been when he died?

Suggestions for Teacher: This activity works well in a center or as an activity that can be completed at the media center. This type of activity involves considerable reading and can be rewritten to include problem-solving skills and math.

Activity 3: Who's Got Smart Ears?
Level: Grades Pre-K–Grade 1
Media: Level I CD audiodisc
Purpose of the Activity: To develop audio discrimination skills

Instructional Activities: This directed-instruction activity takes advantage of the random access capability of the videodisc player using the LB2 standard. The teacher plays a frequently heard sound (door closing, dog barking) and the students try to identify it. The teacher uses bar codes and audio compact discs with collections of sound effects.

Suggestions for Teacher: Record stores usually carry collections of dozens of sound effects on audio CD at reasonable prices. A bar code generating program that utilizes the LB2 standard can be used to make bar codes for the lesson. The bar codes give the teacher a very efficient way of playing sounds again and again.

Activity 4: Research Detectives
Level: Grades 7–12
Media: CD-ROM
Purpose of the Activity: To develop students' research skills

Instructional Activities: Pass out copies of the following scenario to small groups of students. Provide access to a CD-ROM player and a variety of CD-ROM reference tools. Microsoft's Bookshelf is excellent for this activity since it has an encyclopedia, dictionary, almanac, atlas, and other resources all on one CD-ROM. Some high-level vocabulary and ambiguous facts have been deliberately included in the scenario. This strategy encourages students to use the reference resources.

Scenario: The Case of the Missing CD-ROM. The staff, parents, and students of Luke Bean High School have worked for the past year to develop a comprehensive technology plan for their school. They have collected a vast quantity of data to present to the school board in the hope that the board will fund their proposal. In an attempt to make the data as user friendly as possible, the LBHS technology

Continued

plan has been transferred to a CD-ROM master. This CD-ROM presents the school's plan in a multi-media/hypermedia format that should enable the board members to sort easily through the plethora of resources. The resources include interviews, sample lessons, and research reports all intended to inform and persuade the school board.

A major glitch developed last night, however, when somebody stole the only CD-ROM master. The police have narrowed down the list of suspects to three individuals: a teacher, a student, and a parent who, for different reasons, are opposed to the use of technology in the schools. Summaries of the police interviews with the three suspects follow. Your job is to find out who stole the CD-ROM. If you find out which one is telling lies in the interview, then you will find the thief. Therefore, you must fully research their statements made during the interviews.

Suspect 1. Beeso Bidmee has been a faculty member at Luke Bean H.S. for the past 19 years. He was born on July 14, 1923, as Mr. Bidmee pointed out, the same date as Senator Bob Dole. He teaches advanced math and driver education. Mr. Bidmee does not believe that schools need to spend money on advanced technology. He sees nothing wrong with the way schools have been teaching; the problem is the kids. He would like to use the money earmarked for technology to hire a full-time staff person to staff an ongoing in-school suspension program. "There is no way that I could have taken that CD-ROM last night," said Beeso. "My consort Brenda and I were watching a theatrical representation about the 21st president of the United States, Teddy Roosevelt." Plus the police said that the disc was stolen between the hours of 10:00 p.m. and 9:00 a.m. and I didn't get up this morning until after the time that the Japanese bombed Pearl Harbor in 1941."

Suspect 2. Alice B. Cokeless is a child of the 1960s. She was born on May 9, 1945, the day after V-E Day. She has matured over the years and now is a responsible parent who has put her wild years behind her. She still holds onto some of her antiestablishment beliefs, however, and she steadfastly opposes the use of instructional technology in schools. She complains that big business is pushing the use of technology in schools as a way to make money; in the process, it is depleting schools of precious fiscal resources. Alice denies any involvement in the theft of the CD-ROM. She was at the library working on her book about the civil rights movement. Last night she was researching an incident in 1972 when the governor of Alabama was shot.

Suspect 3. Raul Fernandez, a student at LBHS, is one of the police suspects in the case of the missing CD-ROM. Raul is a football player who thinks that technology is taking money away from more important needs such as athletic equipment. Raul was born during the time that the 1980 Olympics were being held in Lake Placid, New York. He was busy last night doing schoolwork; he was working on a report that covered the United States invasion of Grenada in 1983. This subject interests Raul since the U.S. troops invaded in order to evict the troops of a noted antireligious dictator, Fidel Castro.

Suggestions for Teacher:	As a twist on this type of project, students could write the actual activity. This might be suitable for a high school class. The products could then be published and shared with other classes to develop research skills.
Activity 5:	Understanding U.S. Culture
Media:	CD-I
Level:	Grades 9–12
Purposes of the Activity:	To give students an opportunity to reflect on, synthesize, and explain the culture of the United States through an examination of Smithsonian collections
Instructional Activities:	The CD-I disc Treasures of the Smithsonian provides the major resource for this activity. This disc offers the viewer a glimpse of a wide variety of Smithsonian exhibits. The resource presents audio, video, and text information to explain the backgrounds of the collections. The students' objective is to put together a booklet that guides the user through the CD-I disc. The audiences for the booklet are immigrant or foreign exchange students who would like to become more familiar with the culture of the United States. This assignment is intended to provide some structure and additional background that will make the CD-I a useful tool for those unfamiliar with the cultural background of this country.
Suggestions for Teacher:	Students could work in small groups with each group taking on a different aspect of the Smithsonian. At the end, they could combine their sections into one booklet.

Continued

Activity 6:	Slide Show Presentation
Media:	Level III interactive videodisc
Level:	Grades 5–12
Purpose of the Activity:	To provide students an opportunity to develop and deliver a presentation using slides from a CAV videodisc

Instructional Activities: The teacher assigns a pair of students the task of constructing a presentation utilizing slides or short video clips from a relevant videodisc. If available, the students can use a program like MediaMax which enables them to develop a computer-controlled slide show (Level III). If a computer interface is not available, the students may use Level I technology (with or without bar codes). When completed, the students present the project to their classmates.

Suggestions for Teacher: This type of presentation can also use a number of Level III videodiscs. The ABC Interactive series, BioSci II, and many other programs come with software that enables users to develop slide shows. Some of the programs let users incorporate text overlays on the TV monitor.

Activity 7:	Critical/Creative Listening
Media:	Level I Interactive
Level:	Grades 3–12
Purpose of the Activity:	To develop the creative and critical listening skills of the students. Specifically, students will determine a speaker's purpose or point of view.

Instructional Activities: The teacher selects a videodisc that may be suitable for this activity. The academic level of the students and the availability of discs should determine this choice. The teacher chooses some exemplary segments from the disc that model speakers expressing points of view or demonstrating clear purposes. The teacher then plays a segment and asks students to surmise the speaker's purpose or point of view. The videodisc player enables the teacher to show the segment over and over again. This activity often leads to lively discussion in the class. The videodisc player can also support student assessment. For example, a segment could be shown, and students could write their responses.

Suggestions for Teacher: At the highest listening and thinking level is critical-creative listening. Here, listeners really engage the brain as they process spoken language. At the critical-creative level, they listen with emotional and intellectual involvement. The higher mental processes are in full gear to interpret what people hear. The teacher can select a student objective from the following list:

- Determine the main idea of what the speaker is saying.
- Identify details that support the main idea.
- Determine the speaker's purpose or point of view.
- Adapt information to one's own needs or experiences, or both.
- Indicate cause and effect relationships among the speaker's ideas.
- Evaluate continuity or logic in what the speaker says.
- Visualize a scene that the speaker describes.
- Judge the validity or veracity of what the speaker says.
- Relate what the speaker says to one's own experiences.
- Determine whether additional meanings are hidden behind the speaker's words.
- Anticipate what the speaker will say next.
- Sort out relevant information from the irrelevant and fact from opinion.
- Identify propaganda techniques that the speaker may be using.

Exercises

Exercise 7.1. Fill in the table at the end of these exercises with check marks in appropriate places to identify the characteristics on the left that apply to interactive videodisc, CD-ROM, and/or CDI technology.

Exercise 7.2. View a CLV videodisc and develop an index or map of the content of the disc. The following steps are suggested:

- Select a CLV videodisc from the resources that are available to you and view the program.
- Play the program a second time. This time, begin to record the addresses of relevant segments. This process will involve a lot of scanning back and forth with the remote control in order to get precise addresses. When you have completed the index, use a word processor or database program to type up the final product.

Exercise 7.3. Use videodisc from Exercise 7.2 to create a lesson for your class. The activity should be designed to develop a creative/critical thinking or listening skill. The index that you created will help guide you to relevant por-

tions of the disc. Select the specific time segments that you need and record their addresses. If the resources are available, create bar codes for your segments. If no bar-code software is available, the lessons may be delivered by using the remote control. Teach the lesson to your class.

Exercise 7.4. Develop a lesson plan for a project that requires your class to use an optical technology to deliver a persuasive presentation. See the example lesson plans included in this chapter.

Exercise 7.5. Develop a scavenger hunt type of activity that utilizes an IVD, CD-ROM, or CD-I resource geared to a specific program. The number and complexity of the questions should match the developmental level of the intended audience.

Exercise 7.6. Develop a lesson plan that integrates a Level III IVD resource into your class. This may involve a whole-group, directed-instruction setting; individual projects; or small-group classroom centers.

	Interactive Video (Level 1)	Interactive Video (Level III)	CD-ROM	CD-I
Come in CAV and CLV format				
Repurposing for various lessons				
Used with bar-code software				
Costly				
Relatively cheap				
Stores text, audio, and video information				
Used mainly for audio/video information				
Needs a computer				
Used without a computer				

References

Barron, A. (1993). Optical media in education. *The Computing Teacher, 20* (5), 6–10.

Barron, A. and Orwig, G. (1993). *New technologies for education: A beginner's guide.* Englewood, CO: Libraries Unlimited.

Brooks, R. and Perl, B. (1993). Interactive technology for education. *Principal, 71* (2), 20–22.

Bruder, I. (1991). Multimedia: How it changes the way we teach and learn. *Electronic Learning, 11* (1), 22–26.

Blissett, G. (1993). Are they thinking? Are they learning? A study of the use of interactive video. *Computers and Education, 21* (1–2), 31–39.

Bunnel, D. (1992). Let's start a revolution: Bring multimedia to education. *New Media,* (1) 5.

Cockayne, S. (1991). Effects of small group sizes on learning with interactive videodisc. *Educational Technology, 31* (2), 43–45.

Cohen, K. (1993). Can multimedia help social studies teachers? Or are videodiscs worth the expense? *Social Studies Review, 32* (2), 35–43.

D'Ignazio, F. (1992). Are you getting your money's worth? *The Computing Teacher, 19* (5), 54–55.

Dyrli, O. E. and Kinnaman, D. E. (1994). Preparing for the integration of emerging technologies. *Technology and Learning, 14* (9), 92–98.

Galbreath, J. (1994). Multimedia in education: Because it's there? *TechTrends, 39* (6), 17–20.

Gustafson, K. and Smith, M. (1995). Using a bar-code reader with interactive videodiscs. *TechTrends, 40* (1), 29–32.

Karlin, M. (1994). Videodiscs and CD-ROM: Impact on a history lesson. *Media and Methods, 30* (5), 12–15.

King, J. (1990). Optical disc technology: Education trend of the future? *Technology Teacher, 49* (8), 25–29.

Lewis, P. (1991). The technology of tomorrow. *Principal, 71* (2), 6–7.

Luskin, B. (1993). CD-I from boob tube to teacher's assistant— The smart TV. *Journal of Instructional Delivery Systems, 7* (1), 3–5.

MacKenzie, S. (1992). Beating the book: Megachallenges for CD-ROM and hypertext. *Journal of Research on Computing in Education, 24* (4), 486–498.

Madian, J. (1995). Multimedia—why and why not? *The Computing Teacher, 22* (7), 16–18.

Mathisen, R. (1991). Interactive multimedia and education: Specifications, standards, and applications. *Collegiate Microcomputer, 5,* 93–102.

Moore, M., Myers, R., and Burton, J. (1994). What multimedia might do … and what we know about what it does. In *Multimedia and learning: A school leader's guide.* Alexandria, VA: NSBA.

Padgett, H. (1993). All you need to know about videodiscs: One easy lesson. *Media and Methods, 29* (4), 22–23.

Parham, C. (1995). CD-ROM storybooks revisited. *Technology and Learning, 15* (6), 14–18.

Porter, S. (1995). Waving the magic wand: Making books using laserdisc bar codes. *The Computing Teacher, 22* (7), 21–23.

Shade, D. (1994). Here we go again: Compact disc technology for young children. *Day Care and Early Education, 22* (1), 44–46.

Shields, J. (1994). Getting the big picture on videodiscs. *Technology and Learning, 15* (2), 48–52.

Sullivan, J. (1995). Exciting ways to use videodiscs. *Media and Methods, 31* (3), S8–S10.

Thorpe, B. (1993). Kids can create videodisc reports. *The Computing Teacher, 20* (2), 22–23.

Truett, C. (1994). CD-ROM, videodiscs, and new ways of teaching information and research skills. *The Computing Teacher, 21* (6), 42–43.

Walkenbach, J. (1992). Optical storage comes up to speed. *New Media, 10,* 26–30.

Resource References

Bar-code Software

Lesson Maker—Optical Data
My Report Generator—Ztek, Inc.
Bar'n'Coder 3.0—Pioneer
Barcode Maker—Voyager
MediaMax—Videodiscovery
Voyager VideoStack—Voyager

Level III IVD programs

The Great Ocean Rescue, The Great Solar System Rescue— Tom Snyder
ABC Interactive: The Lessons of War, Communism and the Cold War, AIDS—ABC
Illuminated Manuscript, Exploring Chemistry—EduQuest
STV: Human Body, GTV: A Geographic Perspective on American History—National Geographic Society
TLT: Teaching, Learning, and Technology, Visual Almanac— Apple

Level I IVD Programs

Windows on Science, Windows on Math, Windows on Social Studies, The Adventures of Jasper Woodbury, Living Textbook—Optical Data Corp.
Interactions—D. C. Heath
Science Sleuths, Math Sleuths, BioSci II—Videodiscovery
National Geographic Specials—National Geographic Society
Nova, The Civil War—Public Broadcasting Corp.
Disney's Sing Along Songs—Image
Smithsonian's Insects: The Little Things that Run the World— Lumivison

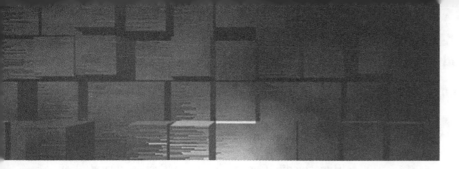

Chapter 8

Using Hypermedia in Teaching and Learning

This chapter will cover the following topics:

- Importance of hypermedia in education

- Features and capabilities of hypermedia programs

- Examples of hypermedia programs

- Integration strategies for hypermedia programs

- Sample activities for hypermedia integration

Chapter Objectives

1. Identify by name components of a hypermedia program.

2. Develop activities that integrate hypermedia authoring into the curriculum.

   - Develop a program for a directed instruction activity.
   - Develop a program for a constructivist activity.

3. Identify the advantages and disadvantages of hypermedia applications in education.

"Technology is the campfire around which we will tell our stories."
Laurie Anderson

Introduction to Hypermedia

Hypermedia Defined

Considerable confusion surrounds the exact meanings of the terms *multimedia, hypermedia,* and *hypertext* since there is no generally accepted definition for any of them (Tolhurst, 1995). Frequently, working definitions depend on the point of view of whoever is doing the defining. Computer scientists may define them based on user interfaces, while cognitive psychologists will consider effects on human learning. To a great extent, the terms and their applications are still evolving. This chapter offers descriptions and applications that put them in perspective for teachers:

• **Hypertext.** The basic concept of nonsequential, computer-based text was first envisioned more than 40 years ago by Vannevar Bush. Today, Tripp and Roby (1990) define *hypertext* as a "nonlinear, multidimensional, semantic structure in which words are linked by associations." These links resemble connections in a concept map in contrast to a traditional page-by-page progression through a textbook. For example, an electronic encyclopedia might link certain words with sections of text in another part of the encyclopedia. When the user clicks the mouse pointer on a word, let's say *Boston,* the screen then displays information from another part of the program about that city. The hypertext model has no real ending point; the section on Boston may allude to the Revolutionary War, and these words could also be a "hot spot" with links to another section about that war, and so on. Some have broadened the definition of *hypertext* to include sound, video, animation, and pictures (Landow, 1990; Foss, 1989; Megarry, 1988). Tolhurst

(1995) suggests that references to video, audio, and animation displays as *text* are inaccurate; she assets that textual information can include diagrams, pictures, and tables. Generally speaking, static media are acceptable in the hypertext format: photographs, sketches, tables, and diagrams.

• **Hypermedia.** Current literature reflects a great deal of overlap between the definitions of *hypertext* and *hypermedia.* A useful interpretation of hypermedia includes all the characteristics of hypertext plus the capabilities of video, audio, and animation displays.

• **Multimedia.** Multimedia has become such a common term in commercial usage that its definition has become particularly vague. Many tailor its definition to suit their own particular needs (Blattner and Dannenburg, 1990). Kozma (1991) suggests that the term has actually been in use for several decades, only recently being linked to the use of computers. Since some definitions of *multimedia* do not include nonlinear links between information, a suggested definition would specify the use of multiple media formats to present information.

Tolhurst (1995, p. 25) offers a useful diagrammatic view of the distinctions between hypertext, hypermedia, and multimedia (Figure 8.1).

Teachers integrate hypermedia into the curriculum in two primary ways:

• Present existing software like the videodisc-based program The Lessons of War or the CD-ROM Microsoft Bookshelf.

• Both teachers and students use hypermedia authoring programs such as HyperStudio and SuperLink to develop instructional materials or projects.

Chapter 7 on visual technologies reported on the first use. Chapter 8 will focus on the second. Hypermedia authoring offers one of the most exciting and promising areas of instructional technology currently available to edu-

Figure 8.1　Overlapping Domains of Hypertext, Hypermedia, and Multimedia

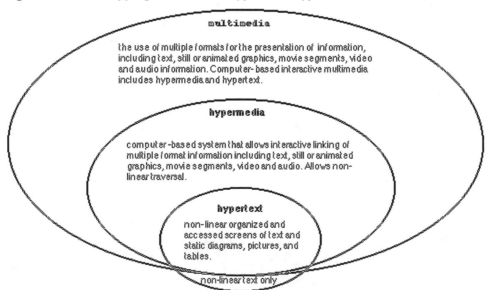

Source: Tolhurst, D. (1995). Hypertext, hypermedia, multimedia defined? *Educational Technology, 35* (2), 21–26. Used by permission.

cators and students. The chapter will cover the basics of hypermedia authoring along with an overview of the most popular programs for this purpose. Both directed-instruction and constructivist uses of hypermedia authoring will be discussed, and a variety of classroom integration strategies and activities will be provided.

Why Is Hypermedia Authoring Important to Schools?

The use of hypermedia tools in the learning process has a number of important implications for schools. Dede (1994) sees uses of hypermedia tools beyond simple presentations of information as having the following implications:

- They offer new methods of structured discovery.
- They address varied learning styles.
- They motivate and empower students.
- They accommodate nonlinear exploration, allowing teachers to present information as a web of interconnections rather than a stream of facts.

These tools may also permit sophisticated evaluations of learning. In the process of using hypermedia, people are said to "leave a track" (Simonson and Thompson, 1994). Future hypermedia systems might apply pattern-recognition techniques from the field of artificial intelligence to help schools assess students' mastery of higher-order cognitive skills (Dede, 1994).

Perhaps most importantly for schools, hypermedia authoring may play a major role in preparing students for the information-intensive world of the future. In tomorrow's digital world, readily available, powerful personal computers and ubiquitous electronic networking will allow people to incorporate a variety of media into the writing process (Gates, 1995). Indeed, hypermedia publishing may eventually supersede paper publishing in importance. Those who work to transfer information and knowledge, and in the information age that will include most people, will choose among many media, including video, animation, music, graphics, and sound effects. In perhaps their biggest challenge, these people will have to craft effective blends of technology, information, and creativity to develop valuable products.

Hypermedia authoring will very likely have a major impact on learning, business, and entertainment (Negroponte, 1995). Dyrli and Kinnaman (1995) contend that continued increases in processing power, storage capacity, CD-ROM technology, graphics software, and communications bandwidth will soon make digital video commonplace, and computer-based hypermedia with rapidly become as fundamental to the way we live, work, play, and learn as telephones, televisions, and automobiles are today. To prepare for this information-intensive environment, students and teachers alike will need skills in hypermedia authoring that go far beyond text-based writing.

Figure 8.2 General Hypermedia Icons

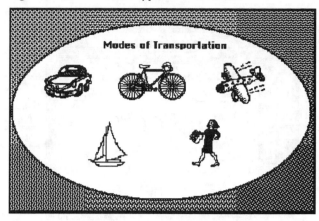

Features and Capabilities of Hypermedia

Hypermedia enables users to search or explore a collection of data in a nonlinear fashion. For example, the first screen of a hypermedia project on transportation might illustrate a variety of transportation modes. (See Figure 8.2.)

Each picture, or icon, representing a mode of transportation, might represent an invisible button or hot spot on the screen. When the user clicks the mouse pointer on an icon, the screen changes and a menu specific to the chosen transportation mode then appears. The new menu might consist of a number of buttons that link to still more specific categories. (See Figure 8.3.)

By clicking these subcategorical icons, users could display topical information using a variety of different media, such as video, audio, text, or animation. This system offers users a broad choice of pathways, so they may choose to work their way through a body of information in many different ways. For example, one student may choose to scan the data on all modes of transportation, while another may decide to focus on one type and study it in detail.

Although different hypermedia programs may employ different terminology, most follow a similar metaphor. As this chapter presents its general overview of hypermedia

Figure 8.3 More Specific Hypermedia Icons

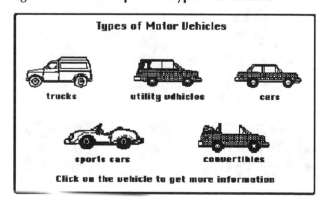

Figure 8.4   Hypermedia Cards and Stack

A group of "cards" is a stack

Figure 8.6   Text Field

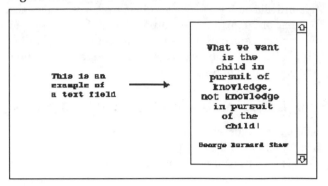

authoring programs, it cannot show all of the programs currently offered. Therefore, this text will use the terminology of HyperCard and HyperStudio. Readers should focus on the general concepts behind this terminology.

Basic Structure of a Hypermedia Program

Programs vary in their procedures and capabilities, but most share some defining characteristics:

Characteristic 1: Stacks made up of cards. Each screen of information is called a *card* (see Figure 8.4). A group of cards is collectively called a *stack*. The hypermedia author need not arrange these cards in any conventional manner, such as alphabetically or chronologically. Due to the nature of the hypermedia/hypertext metaphor, stacks don't have clearly defined beginning or ending points.

Cards contain graphic items, text fields, and buttons. Graphics items, or computer drawings, may be placed on a card using the paint tools included in the hypertext authoring program, by importing pre-made computer-generated graphics, or by using a scanner to digitize a paper copy. Figure 8.5 shows an example. Large collections of graphics covering many subjects are commercially available on both floppy disk and CD-ROM at reasonable prices. Text may also be added to a card using

the text tool within the program's graphics function. This feature is useful for labeling items, but the difficulty of editing these elements makes this an inconvenient way to add large blocks of text.

Instead, a hypermedia author can place text on a card by creating a text field. (See Figure 8.6). A text field is essentially a small word processing screen that can be formatted, sized, and placed on a card as the author sees fit. One card can contain multiple text fields. The author can easily edit text placed in a text field.

A button is another type of object that can be added to a hypermedia card. Buttons define hot spots on the screen that initiate actions when clicked with the mouse pointer. (See Figure 8.7.) Since invisible buttons can be placed on top of graphics or text, buttons can adopt almost any appearance. Buttons represent the most powerful component of the hypermedia metaphor; they are the engines that drive hypermedia and hypertext interactivity.

Depending on the complexity of a program, the author can control button actions in two ways: through pre-made buttons or scripting. In the most common control method, the author selects from a number of preprogrammed options for the most common button actions. This feature enables the author to develop hypermedia stacks very efficiently. It has also proved a necessary help to allow elementary-age students to become hypermedia authors. The HyperStudio screen in Figure 8.8 indicates a number of actions that a pre-made button can initiate.

Figure 8.5   Graphic Items and Graphic Text

Source: Used by permission of Roger Wagner, Inc.

Figure 8.7   Hypermedia Buttons

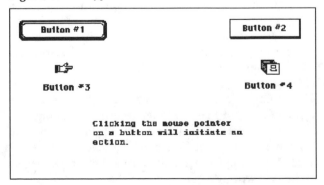

Figure 8.8   Some Possible Button Actions on a HyperStudio Card

```
================ Button Actions ================
┌─ Places To Go: ──────┐  ┌─ Things To Do: ──────┐
│ ○ Another card...    │  │ ☐ Play a sound...        │
│ ○ Next card          │  │ ☐ Use HyperLogo...       │
│ ○ Previous card      │  │ ☐ New Button Actions...  │
│ ○ Back               │  │ ☐ Play a Movie or Video..│
│ ○ Home stack         │  │ ☐ Play animation...      │
│ ○ Last marked card   │  │ ☐ Magic Buttons™...      │
│ ○ Another stack...   │  │ ☐ Testing functions...   │
│ ○ Another program... │  └──────────────────────┘
│ ● None of the above  │  ┌─────────┐ ┌──────┐
└──────────────────────┘  │ Cancel  │ │ Done │
                          └─────────┘ └──────┘
```

Source: Used by permission of Roger Wagner, Inc.

Characteristic 2: Scripting. In another common feature, most hypermedia authoring programs allow the author to program button actions in simple programming languages such as Hypertalk or Logo. Once the only way to select button actions, this method is now used primarily by advanced authors to develop more sophisticated projects. However, some teachers choose to have their students script regularly as a way to encourage development of basic programming skills. Figure 8.9 illustrates scripting for a very basic command in HyperCard.

Overview of Hypermedia Options

As hypermedia programs have evolved, they have become more powerful and more user friendly. Consequently, authors can now draw on wide variety of resources. This section describes some common features that are available in hypermedia authoring programs.

Audio resources. Hypermedia authoring programs offer the user a number of ways to incorporate audio clips:

*   **CD Audio.** Cards can include segments of audio CDs in CD-ROM drives or videodisc players that use the LB2 standard. CDs can provide digitized music, speech, or sound effects.

*   **Videodiscs.** Rather than playing both the video and audio tracks of a videodisc, authors may choose to leave the video and access the audio track alone.

*   **Recorded sounds.** Hypermedia programs usually allow authors to record sound into their programs. This can include voice, as when an author records his/her reading of a poem.

*   **Prerecorded sounds.** Many hypermedia authoring programs come with built-in selections of sound effects. Authors can also add sounds from packaged collections stored on floppy disks or CD-ROMs.

Video resources. Video clips can add a whole new dimension to a program and provide authors with many new communication possibilities. As with audio, authors can incorporate video displays into a program in many ways:

*   **Digitized videos.** By using a video digitizer, a hypermedia author can import video images from external sources such as

Figure 8.9   Simple Scripting Commands in HyperCard

```
on mouseUp
   go to next card
end mouseUp
```

a VCR, a videodisc, or a camcorder. Programs such as Quicktime allow authors to create and edit their own short video clips (movies) and place them on cards. Teachers and students need to observe copyright laws when importing video in this manner. Digitized video also consumes a great deal of hard drive storage space, thus limiting the amount of video that an author can realistically incorporate into a program. An external hard drive offers one solution to this problem; removable cartridge systems are very effective in this role, although the costs for the individual cartridges can be quite high. In the future, more efficient video compression routines promise greater latitude for hypermedia authors.

*   **Videodiscs.** Most hypermedia programs enable authors to access either individual slides or segments of video from videodiscs. This can prove a big advantage due to high quality of audio and video recordings from videodiscs.

*   **Prerecorded videos on CD-ROMs.** Authors can buy collections of short video clips on CD-ROM for the purpose of incorporating them into hypermedia programs. No copyrights inhibit use of these images, giving authors much more leeway in their presentations.

Photographs. "A picture is worth a thousand words" in hypermedia, as much as elsewhere. Photographs provide a powerful resource for authors in all subject areas.

*   **Scanned photos.** Authors can digitize traditional photographs using scanners and then incorporate these images into hypermedia stacks.

*   **Captured from video sources.** By using a video digitizer, an author can freeze images from a VCR, camcorder, or videodisc player and then import them.

*   **Digital cameras.** These cameras take digitized, color pictures that authors can add to hypermedia cards. Pictures can be downloaded directly from the camera to the computer's hard drive.

*   **Imported from CD-ROMs.** Collections of photographs on CD-ROMs are marketed expressly for inclusion in hypermedia programs.

Graphics. Graphics or drawings offer another tool for authors to communicate their ideas. Often an illustration will make a point that is very difficult to get across with words. This aspect of hypermedia authoring is particularly appealing to artistically inclined users.

*   **Created by authors.** Virtually all hypermedia programs offer basic collections of tools that let users draw or paint graphics. These tools enable users with very limited artistic talents to create credible designs and drawings.

• **Imported from clip art collections.** A vast array of clip art collections are available for purchase. These pre-made graphics cover a wide assortment of subject areas.

• **Scanned images.** As another alternative for accessing graphics, an author can scan an image from either a book of clip art or a drawing done in conventional art media such as pencils or paintbrushes. Since computer access is often limited, some teachers prefer to have students draw their pictures off the computer and then digitize them using scanners.

Animation. Animation can be a highly effective tool for illustrating a concept; a student might create an animation of a seed germinating as part of a project on plants. The sources of these displays are familiar:

• **Imported from CD-ROMs.** Collections of animation are also available on CD-ROM. Like some of the other media, these premade collections allow authors to rapidly add effective and professional animations to a project.

• **Created using animation tools.** Hypermedia programs have improved dramatically in their animation capabilities. A novice animator can now relatively easily generate quite sophisticated and effective animations.

Text. In spite of the attention paid to all the other components of hypermedia, text still remains one of the most powerful ways of communicating ideas.

• **Writing as project develops.** All hypermedia programs offer standard word processing features that enable users to write text. In addition, text may also be added as a graphic item. This feature lets the user easily drag text around the screen, and it is very handy for adding labels to pictures.

• **Importing from word processing files.** Most programs also let authors import text created separately in word processors. This can be a boon for an author who has saved a great deal of writing in a large collection of word processing files.

Hypermedia Tools

This section discusses the tools available in HyperStudio, shown in Figure 8.10. The top six tools on the palette show the types of tools that are typically available in hypermedia programs. The bottom 15 are common to most paint programs and are usually components of hypermedia programs. Figure 8.11 describes the functions of these tools.

Title page. Figure 8.12 shows the title page from a sample hypermedia stack. Each topic is a button that links to a different media use.

• **Music of the Movement.** This button links to a CD-ROM player that holds an audio CD entitled "Protest Songs of the 60s." The user can choose to hear a variety of short segments of music.

• **Vivid Images.** This links to a series of photographs with powerful visual images of the civil rights era that have been scanned and imported into the stack. A text field with each photo provides background information. Another collection

Figure 8.10 HyperStudio Tools

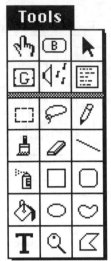

Source: Used by permission of Roger Wagner, Inc.

of images includes drawings done either on the computer or by hand and then scanned into the stack.

• **Historical Background.** This button takes the user to a section of the stack that provides background information on the civil rights movement. This text-oriented section branches off to a database of important people and places.

• **Timeline.** Graphics tools and animation features have been combined to create an animated timeline. Maps were scanned and animation was added to show the routes of civil rights marches.

• **The Major Players.** This button displays a section of the stack with biographies of major historical figures of the era. These biographies are in turn linked to a videodisc that plays short video segments on some of these people.

• **Some Personal Reflections.** For this section, the student used a tape recorder to interview people in her neighborhood who lived through the civil rights movement of the 60s. She then edited the tape and imported sections into her hypermedia project.

• **How Was It Reported?** This section links to a videodisc player with the videodisc "Eyes on the Prize." Buttons are linked to specific sections of the disc that show how news media reported relevant events. The user is then given some thought-provoking questions about the video segment.

• **What Does It Mean Today?** The student used a camcorder to record responses from fellow students and community leaders to the question, "What does the civil rights movement of the 60s mean to today's generation?" Certain poignant sections of video were then digitized and imported into the stack.

A model linking card. Suppose that the user decides to begin his/her browse through the stack by clicking on the button entitled "The Major Players." The linking card for that button may look like Figure 8.13.

This card allows the user to choose to view video images of a number of historical figures. It also gives the

Figure 8.11   Hypermedia Paint and Tool Functions

The **Browse tool** is used to click on buttons.

The **Button tool** edits or changes the buttons created.

The **Arrow tool** edits buttons, graphic items, and text items.

The **Graphics tool** edits graphics.

The **Sound tool** enables user to find and edit the buttons with sound.

The **Text tool** shows all the text items and enables user to edit them.

The **Selection Tool** captures rectangular areas.

The **Lasso tool** is used to select an irregularly shaped object.

The **Pencil tool** is used for freehand drawing.

The **Paint brush** tool paints an area using the selected brush shape.

The **Eraser tool** erases the area under the eraser.

The **Line tool** creates a straight line at any angle

The **Spray can tool** sprays an area with a dotted pattern.

The **Rectangle tool** produces a rectangle. Holding down the shift key while drawing produces a perfect square.

The **Rounded rectangle** tool produces a rectangle with rounded corners.

The **Paint bucket tool** is used to fill an area with a color or pattern

The **Oval tool** produces an oval. If the shift key is held down while drawing, a circle is created.

The **Freehand tool** closes the shape that you make automatically.

The **Paint text** tool creates text as a graphic item.

The **Magnifier tool** lets the user zoom in and work at a closer level.

The **Polygon tool** draws a closed polygon.

Source: Used by permission of Roger Wagner, Inc.

user the option of proceeding to answer some questions about the person. The video buttons are programmed to show selected video clips from a disc in the videodisc player. The buttons with question marks link to other cards that pose the questions.

Hypermedia Authoring Programs

Since the late 1980s, teachers have had access to hypermedia authoring programs like HyperCard for the Macintosh, LinkWay for MS-DOS machines, and TutorTech for the Apple II. These early programs certainly represented a

Figure 8.12   Title Page from a Student-produced Hypermedia Stack

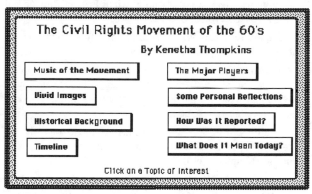

major jump forward in technology, but their use was limited. The author had to make a major time commitment to learn the software, and developing a project of any length also took a long time. Their major limitation resulted from the need to include extensive scripting, or programming commands. Although the scripting language was easy to learn compared to more traditional computer languages, its complexity still limited its popularity.

Things began to change when Roger Wagner's Hyper-Studio was released for the Apple IIGS. This program used the same basic metaphor as HyperCard, but it eliminated much of the need for scripting. In recent years, a number of programs have become available that emulate HyperStudio's easy-to-use format. A brief summary of some currently available hypermedia programs includes:

- **HyperCard.** Apple has only slowly upgraded this product. However, at press time a major upgrade was planned. HyperCard still has a large user base and remains one of the most powerful programs.

- **LinkWay Live.** A major upgrade of the original IBM program simplified the integration of multimedia inputs.

- **HyperStudio.** One of the most innovative products on the market for Mac and Windows computers, this product has developed a large following among teachers. Roger Wagner Inc. has focused its efforts on HyperStudio and upgrades it regularly.

Figure 8.13   Linking Card Model: "The Major Players"

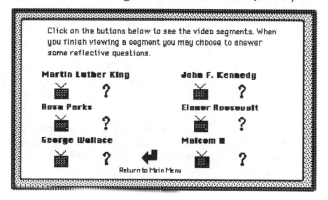

- **MicroWorlds Project Builder.** The unique emphasis on Logo gives the user experience at simple programming. Separate packages are also sold as Math Project Builder and Language Arts Project Builder.

- **SuperLink.** This Windows program offers a wide variety of hypermedia options, plus ease of use.

- **Multimedia Toolbook.** This Windows program features video editing software, support for many file formats, and a Media Packager that gathers and compresses multimedia elements for a given product.

- **Digital Chisel.** A powerful product for the Macintosh, this program supports full text-to-speech capability and includes question templates for developing tests.

- **Multimedia Workshop.** This program is very rich in resources and offers a large collection of built-in clip art, images, sounds, and video clips. It is designed primarily as a presentation tool.

- **HyperWriter.** This Windows and DOS program has a "Wizards" feature that guides the user through processes such as linking cards and designing an interface. Its files can be saved in the HTML (hypertext markup language) format for easy uploading to the Internet's World Wide Web.

- **Media Text/My Media Text Workshop.** This easy-to-use product for the Macintosh amounts to a multimedia word processor. It gives students excellent practice at linking sound, animation, and video with text.

As computer power increases and becomes more affordable, even more sophisticated and easy-to-use programs will become available.

Hardware Requirements for Hypermedia Authoring

This text presents a limited look at both hardware and software for hypermedia applications. Rather than attempting to cover all aspects of the field, it offers a look at what might be useful and available to the typical classroom teacher, either now or in the near future. Hardware resources for such a system may vary:

- **Computer with keyboard and monitor.** Full-blown development platforms require a minimum computer configuration that includes a DOS machine with at least a 486 microprocessor and a Macintosh system with a 68030 or better processor. Either system needs a large hard drive and 12 to 16MB of RAM. The minimum requirements to utilize programs like HyperStudio or LinkWay Live are much more realistic for the average classroom.

  - HyperStudio: 4MB of RAM for System 7.0 or above
  - LinkWay Live: 2MB of RAM with a 286 or above processor

  Although newer computers and software offer users much more elaborate options for hypermedia authoring, many older systems still have a great deal of educational value and should not easily be dismissed as obsolete.

- **Digital camera.** These cameras, like the Canon XapShot or Apple QuickTake, let the user take digital photographs and store them as digital files. The images can then be incorporated into hypermedia projects. Students of all ages enjoy using their own photographs in projects.

- **Scanner.** Scanners can capture still images such as those from magazines or books. Used in conjunction with a video digitizer card, this is an excellent way to incorporate still images into a hypermedia project.

- **Video digitizer.** Video digitizers, also known as *digitizing boards,* capture full-motion video from video cameras, VCRs, videodisc players, or live TV. The video segments are then stored as computer files, and they can be edited using software like Adobe Premier. Both teachers and students should recognize copyright restrictions when digitizing and editing video.

- **Video input.** To implement motion or still video in a production, the author needs access to the source of these images. Video cameras, VCRs, videodiscs, or CD-ROMs are among the possible sources.

- **Audio card.** To incorporate sound, an audio capture and playback card is needed. Many computer systems sold in recent years have had built-in audio cards. An audio source such as a microphone will also be needed.

- **CD-ROM drive.** CD-ROMs have become essential elements in multimedia technology. Because of its huge storage capacity, CD-ROM is the only technology for storing large quantities of digitized video or audio. Hypermedia authors can also buy large collections of digitized video, audio, and still image resources on CD-ROM.

- **Audio speakers.** In order to monitor quality and simply to hear the audio parts of a program, speakers are a must for hypermedia development. Many newer computers are shipped with either external or internal speakers.

- **Videodisc player.** Videodisc players with Level III capabilities provide excellent resources for hypermedia authors. High-quality video or audio input can be easily accessed from any videodisc. With thousands of videodisc titles on the market and existing videodisc players in many schools, this technology will prove to be a valuable resource for years to come.

Impact of Hypermedia in Education

Advantages of hypermedia in education. Capabilities and strengths of hypermedia include:

- **Motivation.** Hypermedia programs offer such varied options that most people seem to enjoy using them. Students who usually struggle to complete a project or term paper will often tackle a hypermedia project enthusiastically. McCarthy (1989) expressed the belief that the most important characteristic of hypermedia is its ability to encourage students to be proactive learners.

- **Flexibility.** The user of a hypermedia program can draw on such diverse tools that the technology truly offers something for everybody. For example, a student who may not excel at expressing ideas in writing can record what he/she wants to say orally.

- **Development of creative and critical-thinking skills.** The tremendous access to hypertext and hypermedia tools opens up a multitude of creative avenues for both students and teachers. Marchionini (1988) referred to hypermedia as a "fluid" environment that constantly requires the learner to make decisions and evaluate progress. He asserted that this process forces students to apply higher-order thinking skills. Turner and Dipinto (1992) reported that the hypermedia environment seems to encourage students to think in terms of metaphors, to be intro-spective, and to give free rein to their imaginations.

- **Improved writing and process skills.** Turner and Dipinto (1992) also found that exposure to hypermedia authoring tools helps students in the following areas:

  - It gives students a new and different perspective on how to organize and present information.

  - It gives students a new insight into writing; instead of viewing their writing as one long stream of text, they now see it as "chunks" of information to be linked together.

- **Forward thinking.** The hypertext/hypermedia seems likely to persist. A review of the number of World Wide Web pages on the Internet gives ample evidence that linking information together via hypertext and hypermedia is indeed an effective way to present and add value to large bodies of information. In time, millions of people may publish multimedia documents on the information highway in the hope of attracting viewers, readers, and listeners (Gates, 1995).

Disadvantages of hypermedia in education. Limitations of hypermedia include:

- **Hardware intensity.** To take full advantage of the benefits of hypermedia technology, students need ample on-line development time. This presents a problem in most classroom settings due to an insufficient number of computers. The problem is further exacerbated when available computers are not configured for hypermedia authoring. For example, they may lack the capacity to digitize sound or input video.

- **Lack of training.** Although hypermedia programs are becoming quite easy to use, they still require extensive training. Unfortunately, training does not seem to be a priority in most school districts. Recent surveys show that staff development makes up only about 8 percent of technology budgets (Siegel, 1995). The toughest challenge that instructional personnel face is not learning to use a particular program, but rather learning to integrate it within the curriculum. To help alleviate this problem, hypermedia training needs to go beyond just learning how to make an authoring program work. Training must also give serious consideration to effective curriculum integration. In addition, to ensure quality products, hypermedia training should extend to the areas of media, design, and the arts.

- **Difficult of projecting.** Teachers often want to project students' hypermedia projects onto large screens so that others can see the results. This is still a somewhat cumbersome task that requires the teacher to hook up an LCD panel or a video projector. Both of these pieces of hardware also are quite expensive, so not every classroom can have a projection setup (Barron and Orwig, 1993). A compromise solution may be to use a converter that can project the computer signal onto a television/monitor.

- **Integration problems.** As mentioned earlier, integration of hypermedia technology into the curriculum presents some major problems. To assure quality projects, students need sufficient time to focus, build, and reflect. The conventional school schedule, often chopped up into 50-minute blocks, does not lend itself to serious project development. If hypermedia authoring is to have a major impact on learning, educators will need to look at ways of infusing more flexibility into students'

daily schedules. One step in the right direction might be more integration of subject matter.

- **Data compression.** A hypermedia project can fill a tremendous amount of storage space on computer's hard drive. Digitized video and sound files are the major culprits. Until compression techniques improve and become more cost effective, this problem will persist (Malhotra and Erickson, 1994). Another component of this problem is the difficulty of transferring a file from one computer to another, since even very small hypermedia files will exceed the capacity of a single data disk.

There are some ways of getting around these problems, however. For one, students can store files on external hard drives; the removable-cartridge type is most effective. The cartridges, which resemble large data disks, can store 200MB or more. These add another cost element, but the prices are dropping quite rapidly. Another way of dealing with large files is to employ a program like Disk Doubler, which can split a large file into a number of smaller ones each of which will fit on an 800K disk. Using this method, a student can copy a project from the hard drive onto a number of $3^1/2$-inch microdisks.

Using Hypermedia in the Classroom

Integration Strategies and Tips for Developing Hypermedia Authoring Skills

Whether teachers are developing their own skills or those of their students, they must remember that the hypermedia authoring process involves two distinct phases. Initially, authors need to learn the mechanics of the programs and develop their understanding of the concept of hypermedia. No one could develop a quality product without first being reasonably comfortable with the tools. However, to move to the next level, hypermedia authors must develop their awareness of the complexities of the different media as well as their knowledge of learning design. While this is a long-term process that will emerge through a great deal of experience, a number of strategies can aid the classroom teacher in helping students to focus on producing quality over quantity:

- **Media literacy.** Given the complexities and proliferation of different media, an understanding of media basics will become a fundamental skill for the information age. Since most people will have tremendous capabilities to adapt and alter existing media in the near future, a critically important part of instruction in hypermedia authoring will focus on how to be critical and ethical consumers and producers of media (Lloyd-Kolkin and Tyner, 1991).

- **Music and art.** Visual art and music often play major roles in the effectiveness of hypermedia products. As students gain more knowledge in the theory and aesthetics of music and art, they will more likely utilize them productively in the authoring process.

- **Principles of design.** Many principles of desktop publishing layout also apply to hypermedia designs. Students will often initially overindulge in the face of the tremendous choices with hypermedia tools. They should observe design principles such as those offered by Litchfield (1995). (See Table 8.1.) Brunner (1996) and Clark (1996) also offer criteria for assessing students' hypermedia work.

Table 8.1   Computer Project Standards for Students

Language

- Use numbers or noncontroversial names to identify groups. Do not use ethnic, slang, or rude names.
- Use correct English punctuation and grammar. Make sure that all of your screens are error free.
- Do not use questionable vocabulary, slang terms, or curse words.

Type and Font

- Use no more than two different fonts and three different type sizes.
- Limit the project to just two different typestyles.
- Use bold or plain text fonts for main text. Shadow and outline typestyles are too difficult to read for more than a few words.

Graphics and Visuals

- Use graphics and visuals that are appropriate, that add to the project, and that are directly related to the topic.
- Do not use obscene or rude graphics or visuals. When in doubt, leave it out!
- Decide on a couple of styles of screen changes (e.g., wipes, zooms, fades) and use those only. Too many different styles will be distracting.

Content

- Make certain that content is accurate.
- Content should include text, graphics, visuals, and sound.
- Check and recheck information to make sure that everything is correct.
- Make sure the project matches storyboard instructions.
- Every group member must review and sign off on the project before submitting it.

Source: Litchfield, B. (1995). Helping your students plan computer projects. *The Computing Teacher, 22* (7), 37–43.

- **Creativity and novel thinking.** When assessing students projects, look for and encourage creative uses of the potential of hypermedia. Too many student projects resemble glorified paper-based projects; they do not take advantage of the true power of this medium. Classroom activities that encourage creative and critical thinking in all subject areas will help develop skills and a mindset that will naturally enhance the authoring process.
- **Storyboarding.** Storyboarding will help students make better use of valuable computer time. On index cards, students can lay out what they want on individual cards. Many students will resist this, preferring to develop only when they are on line. It will help to explain that professional media creators practice storyboarding. Even famous movie directors plan their scenes with storyboards.

- **Benchmark.** An effective way of developing authoring skills is to see what others have done. This is particularly true in the area of scripting. Evaluating the scripts of existing programs can give insight into how to write a script for a new project. Through the Internet or commercial on-line services, teachers can now download stacks. This opens up a wide variety of low-cost or free hypermedia resources for teachers and students alike. It is also helpful to examine some effective uses of media; Ken Burns's series on the Civil War, for example, demonstrates the power of images and sound when melded together in the context of a story.
- **Audience.** Whenever possible, the teacher should try to give students an opportunity to display their projects. Students will be much more motivated if they believe that others value their work. Research on writing has shown that students will invest more effort in the writing process when they know that others will read their writing. Turner and Dipinto (1992) have observed that this sense of audience seems to carry over to hypermedia authoring. However, teachers often find that components of a student's project make sense only to the author. Younger students in particular should be constantly reminded that they need to think of their project from the point of view of the user. Encourage them to test out their projects on other students, family, or friends.
- **Assessment.** Teachers or fellow students may look for several signs of effective authoring in the assessment process:
  - Was the information accurate?
  - Were the objectives of the program clear?
  - Was creativity demonstrated?
  - Was the design aesthetically pleasing?
  - Were storyboards or flow charts used?
  - Were the hypertext and hypermedia links effective?
  - Were media used efficiently and effectively?
  - Was the program easy to use?
  - Does the project demonstrate novel thinking?

Dipinto and Turner (1995) suggest that student self-assessment of hypermedia projects may be the most important component of the assessment process. They offer the following thoughts on self-assessment: "Perhaps it is this notion that knowledge gained through reflection on what was learned and how it was learned enables students to construct a microworld where assessment becomes a feedback mechanism that leads to further exploration and collaboration" (p. 11).

Hypermedia projects make valuable additions to student portfolios. Many programs include player files that can run program files without the application itself. This is particularly useful when a student wants to take a project home and run it on the family's computer.

Directed-Instruction Example

Scenario. Mr. Chung's sixth grade class has just finished reading the novel *The Black Stallion.* Although many of the students said that they enjoyed the book, Mr. Chung found it very difficult to engage them in thoughtful discussion. In fact, they seemed unwilling to answer any questions other than those that required direct factual recall. Chung has encountered this problem before, and this time he decided to try something different. He recently took in service training in hypermedia authoring and thought that a hypermedia program based on the videodisc of the film *The Black Stallion* might get his students more engaged. He decided on several objectives for the hypermedia program:

1. Explain how films are constructed.

2. Compare and contrast the film with the novel on which it is based and evaluate the major differences and similarities.

3. Identify and describe the setting for the story.

4. Identify the major plot of the story: problem (conflict), continuing action, climax (crisis), and conclusion (solution).

5. Identify at least one subplot.

6. Identify some of the various types of camera shots and angles used in film composition.

Mr. Chung found that he could easily link segments of the film to buttons in his hypermedia project. He was also able to provide notes to his students when he projected his stack using an LCD panel. When he presented the program to the students, he found out that they grasped things about the story that previous classes had not been able to comprehend. The class was motivated and rated the unit as one of the best that they had ever studied. Students also scored better on the unit test than any other class had ever done. The students did especially well on the essay part of the exam.

Constructivist Example

Scenario. The faculty at Jane Goodall High School have recently begun to team teach across curriculum areas. Two faculty members pioneered this approach at Goodall: Ms. Sanchez, a world history teacher and Mr. Bono, an English teacher. They have taught as a team for the past 2 years and find that the areas of history and English are well-suited for integration. Sanchez and Bono have been busy for the past month guiding their 11th grade classes through a study of World War II. They now feel that the students have sufficient background to proceed to the next level, which is the development of hypermedia projects. The teachers envision the students working in groups of three or four, and they expect the project to take approximately a month to complete. The projects will, by necessity, involve extensive work outside class. They believe that the following description will get their students excited about the projects:

As the country approaches a new millennium, those who lived during WWII are rapidly passing on. With them will pass their unique and personal knowledge of a world at war. In an effort to preserve some personal accounts of WWII, an historian from a local university has decided to try to capture the reflections of local residents who lived through this momentous

period. The historian wants this project to move beyond the traditional method of conducting interviews and then reporting on them in a paper or book. He worries that all the work of the students and the interviewees would just end up on a bookshelf somewhere. He would like to have the high school students collaborate with some of the WWII-era citizens to develop a series of hypermedia projects that would help capture the essence of life in that era. He wants to give the developers sufficient latitude to cover a broad range of topics. Some possibilities might include the home front, women at work, race relations during the war, rationing, when the soldiers came home, etc. Upon completion, the projects could be displayed on Veterans Day and Memorial Day in local libraries or malls, used by local schools for teaching, uploaded to the Internet, put onto a CD-ROM and sold for fund-raising purposes.

The students were very excited about developing these projects knowing that they were helping to preserve local history as well as providing a service to the community.

Other Sample Activities

Teachers must prepare students carefully for hypermedia authoring activities. First of all, students need to understand well-defined parameters of the project. Younger students especially need guidance in developing their projects. Students can also benefit from working cooperatively to develop hypermedia projects.

In addition to project ideas, this section includes some activities designed to foster the development of the students' aesthetic judgment in the use of different media.

Additional Activities to Develop Hypermedia Skills

Animation and science. To provide students an opportunity to develop animation skills, this activity offers an opportunity to incorporate science into the animation process. Students view a video segment that illustrates an aspect of science, for example a time-lapse view of a plant flowering. The students then develop an animation of the sequence. An animation program or a hypermedia program with animation capabilities may be used. If possible, use a videodisc player to show the video segment; this allows the students to easily view the segment as many times as they need. If technology is not available for the students to use, they can still create the animation using pencils and paper to make flip books. These skills are directly transferable to the computer animation process.

Choral reading. This activity is designed to develop students' skills at effectively utilizing sound effects or music in a presentation. Sound can play an important role in determining the quality of a hypermedia product, and teachers should not assume that students will automatically know how to properly utilize sounds. For this activity, students use a videodisc player configured for the LB2 standard, which enables the user to access segments from an audio CD. In conjunction with a bar-code program, these tools give students a great deal of editing capability with sounds or music.

Tell the students that they will be doing an activity to read a poem and add sound effects or music to embellish the reading. Students may either write a poem or use a piece of existing poetry. They then need to select segments from a audio CD that somehow add to the reading. Music stores often carry compact discs with short segments of sound effects. For example, a CD of horror sounds could be used with a Halloween reading. Once the segments have been selected, a bar-code program can be used to create bar codes for the audio clips. It is advisable to have students work in pairs on a project like this, so one can read while the other operates the bar codes. A tape recording can be made of the presentation for the purpose of authentic assessment.

Hypertext database. This activity gives students practice at effectively linking information together. This is a good activity to use in situations that lack access to many peripherals such as video and audio sources. It is also suitable when the teacher wants to focus on effective and creative linking without distracting students with flashy media. Students can work most effectively in pairs. A partner provides support and a simple checks-and-balances system. Students either choose or are assigned topics on which to collect data. This might be something like the rosters of their favorite sports teams. The students then work to create links within their beginning body of information. For example, the developer might link a player's year of birth with a list of the major news stories of that year. The linking possibilities are almost limitless. In the course of this project, students will demonstrate imagination and gain a great deal of research experience.

Autobiography. This activity begins with students researching their backgrounds. This activity should employ a questionnaire that the students help generate. They then take it home and get help from relatives with filling out details of their lives. The teacher may suggest or require that the projects contain some or all of the following information:

- Events that happened the year that they were born (These may be drawn from newspapers, almanacs, magazines, or parents' recollections.)
- Their interests and hobbies
- Information about the town where they were born
- Family tree diagrams
- Top-ten lists (This may include books, foods, movies, songs, people, sports teams, etc.)
- Scanned-in photos

This activity might culminate in an open-house where relatives can come and view the class projects.

Activity 1:	Presentation Program
Level:	Grades 4–12
Purpose of Activity:	To provide students an opportunity to develop a hypermedia stack to assist in a presentation
Instructional Activities:	In this activity, the student groups will develop and deliver multimedia presentations. The students must be familiar with a multimedia authoring program. They begin by agreeing with the teacher on a subject for the presentation. The possibilities are almost limitless, although the teacher may decide to relate the topic to a specific curriculum concept. When they have finished, the students need opportunities to deliver their presentations. Depending on the grade level, the audience could range from another class (in the case of elementary students) to a group of local business people (in the case of a high school class).
Suggestions for Teacher:	Emphasize to the students that content is most important; they should not simply play with the technology. Help guide them through the storyboarding phase, where much of the thinking takes place. Stress proper design theory; by keeping screens clean, simple, and consistent, the stacks will probably communicate their points most effectively. Students may tend to use too many fonts or graphics; encourage moderation. For older students, it might work well to bring in a guest speaker who could discuss proper presentation design. A local business or college may be able to provide a good resource person.
Activity 2:	Hypermedia Story
Level:	Grades 3–12
Purpose of Activity:	To provide students with an opportunity to write a story using hypertext and hypermedia
Instructional Activities:	Tell the students or student groups that they will be writing a story using a hypermedia program. They should use some of the many media tools. This activity offers a wide variety of possibilities:

Continued

- Older students may want to offer multiple endings to a story.
- If a student writes a story about "fairies," she may create a hyperlink that provides more information about the origin of this idea.
- Dramatic sound effects or music could accompany a story.
- An author may create an audio link that plays a recording of himself explaining how he came up with the idea for the story.
- Paint tools can be used to create illustrations for the story. Students might collaborate, with one writing and the other illustrating.

Suggestions for Teacher: The quality of the students' work will benefit if they use storyboards to lay out the story before they actually produce it on the computer. Index cards work very well for this purpose. Encourage some hyperlinks; some students may tend to create simple linear stories.

Activity 3:   States Report
Level: Grades 4–8
Lesson Purposes: To provide students an opportunity to learn about individual states by constructing hypermedia projects

Instructional Activities: A class could approach this activity in a number of ways. For example, a map of the United States could serve as the main menu with each state's outline a button that would link to additional information about the state. The teacher might want to set up a computer center where students periodically develop presentations for one or two states. Depending on the grade level, the information for each state could range from simple demographic data to a more detailed discussion of the state's history or economy. If a student is not motivated by traditional reports, he/she might want to report on something like the sports teams or major universities of each state.

Suggestions for Teacher: The teacher should encourage students to compile their data and lay out their basic plan before they start authoring. They could find much of their data in CD-ROM atlases or encyclopedias. If telecommunications tools are available, online services or the Internet offer a wealth of current information. The chance for students to present their finished product to an audience will boost motivation. Ask the media specialist in the school if the project can be put on display in the media center and subsequently added to the media collection. The finished product could be included in the students' portfolios.

Activity 4:   Media Literacy
Level: Grades 9–12
Purpose of the Activity: To improve students' media literacy and hypermedia skills

Instructional Activities: This is a fairly complex activity; students could work most effectively in pairs or small groups. The students choose a videodisc to examine critically its contribution to media literacy. This project would be a good followup to a series of lessons that exposed the class to the concept of media literacy. The chosen videodisc should follow a documentary format. National Geographic and Nova specials work well for this type of activity. Once the group has had sufficient time to examine the videodisc program and its varying media literacy perspectives, members should then plan a hypermedia project. The project should show examples from the videodisc and illustrate a number of points about media. It might examine production techniques, for example the use of fades, pan shots, zoom shots, and sound effects. Buttons could link to specific segments of video. The ultimate goal is to make both the project developers and its users much more literate consumers of media.

Suggestions for Teacher: A number of excellent books address the issue of media literacy. It would be helpful to have some available as resources. To expedite the process, it might be helpful to have a VCR copy of the program available; public libraries might be a resource. This would allow some flexibility in the preliminary viewing of the show. Students could even watch it at home and bring notes to school.

This is a major project that might well produce a very useful product. Students would feel empowered if they knew that their project would be used by other classes. Try to facilitate this feeling of pride by offering the media literacy projects to middle school or upper elementary school teachers.

Activity 5:   Critical Viewing of a Film
Level: Grades 9–12
Purpose of the Activity: To develop a hypermedia project that critically examines a film

Continued

Instructional Activities:	This activity requires a videodisc version of a film and a connection to a videodisc player. The individual or group working on this project should be thoroughly familiar with the film. They will also need access to basic information about film appreciation. The project can focus on any aspect of critical film viewing. One potential project might guide the user through a primer on film appreciation. Another might compare and contrast the book and the movie versions of a story. Segments of the film could be accessed via buttons, thus providing almost instant examples to accompany the program text.
Suggestions for Teacher:	This is a time-consuming project that is not for all students. It could be incorporated into a number of subject areas including literature, social studies, and even foreign language. Encourage the students to use online resources for this project. Film reviews and even comments from people in the field might be accessed by posing questions on electronic bulletin boards. The following list of criteria may be helpful in evaluating a film:

- Use of technical elements: sound, music, lighting, composition of shots, movement, special effects, color, camera angles, editing
- Type and extent of emotion generated
  - Character development
  - Interest and unity of plot
  - Credibility of plot and characters
  - Theme or message
  - Abilities of actresses or actors
  - Extent of realism
  - Relation to one's own life
  - Ease of understanding

Activity 6:	Talking Colors
Level:	Grades 1–3
Purpose of the Activity:	To introduce the students to the concept of hypermedia authoring
Instructional Activities:	This teacher-guided activity is intended to introduce students to hypermedia. As a prerequisite, students should be familiar with the computer and basic drawing tools.

Begin by drawing a series of rectangles on the first card of the stack. These rectangles then can be filled in with the primary colors. Create an invisible button and place it over one of the rectangles, then link the button to another card. On the new card use the graphic tool to write out the name of the color. Also on that card, create a sound button and have the students record the name of the color. (Older students could use the name of the color in a sentence.) Create another button that will take the user back to the first card. Repeat this process for each of the colored rectangles. |
Suggestions for Teacher:	Develop a model project to show the students what they will be doing. The students' finished products might then be used by pre-K, kindergarten, or first grade students. This same type of project might also be done with letters of the alphabet. The linking card might show an object or series of objects beginning with that letter.
Activity 7:	Many Multimedia Marvels: Student-Designed Projects
Purpose of the Activity:	Students research and develop their own multimedia presentation to inform and persuade their classmates about ecology issues.
Instructional Activities:	Students are paired with classmates of their choice, and each pair is taught how to use the technology involved (e.g., remote control and methods of accessing frames on the videodisc player). After reviewing the laserdiscs, they decide on a project topic. Some topics that have been done successfully include: Toxic Wast, Landfills, and Whales. Each pair of students selects frames that correlate with their topic, and collect other information by researching resources such as the *Grolier's Encyclopedia*. They then compose HyperCard or HyperStudio stacks to present their findings and to test their classmates' understanding. To combine their stack with videodisc materials, they must place buttons on the frames that will access the videodisc sequences. Some students may opt to add special effects or sound features to their presentation to give it additional power. After all groups are finished, they present their products to the class. The class discusses and comments on the findings and the presenters' techniques.

Source: Activity 7 is from McKinney, K. (1991). Students develop their own multimedia programs with laserdiscs and HyperCard. *The Florida Technology in Education Quarterly, 4* (1), 45–51.

Exercises

Exercise 8.1. Identify the parts of the following hypermedia card:

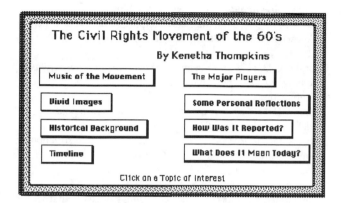

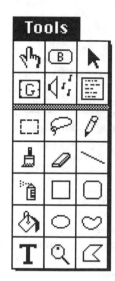

_____ **1.** Topics are really these places on the screen you can click.

_____ **2.** The whole frame above is one of these.

_____ **3.** Several frames together are usually called one of these.

_____ **4.** The words "Click on a topic of interest" are actually one of these.

_____ **5.** The "coloring" tools on the tools palette are called these.

Exercise 8.2. Plan a hypermedia project that can be used in a whole-group, directed-instruction lesson. Use index cards to develop a storyboard for your project. Consult with your instructor regarding the content and developmental level of the lesson.

Exercise 8.3. Using a hypermedia authoring program of your choice, develop the lesson that you planned in Exercise 8.2. If feasible, teach the lesson to students. Plan and deliver a presentation to your classmates on what you have learned from this exercise.

Exercise 8.4. Develop an activity that involves your students in the creation of a hypermedia project. The activity should be designed to have your students work in pairs or small groups. The activity may include the use of a scenario if you like.

Exercise 8.5. Put yourself in the place of one of your students in Exercise 8.4. Find at least one classmate to work on the project with you. Plan the project, use index cards, and lay out the project in storyboards. Meet with your instructor when the planning stage is complete. Then develop the hypermedia project and demonstrate it to your class.

Exercise 8.6. Create a hypermedia stack to teach the user about the components of hypermedia authoring. Try to plan this stack so that it could be used either by an individual at a workstation or by the teacher for large-group instruction.

References

Ambron, S. and Hooper, K. (1990). *Learning with interactive media: Developing and using multimedia tools in education.* Redmond WA: Microsoft Press.

Ambrose, D. W. (1991). The effects of hypermedia on learning: A literature review. *Educational Technology, 31* (12), 51–55.

Barron, A. E. and Orwig, G. W. (1993). *New technologies for education: A beginner's guide.* Englewood, CO: Libraries Unlimited.

Blattner, M. M. and Dannenburg, R. B. (1992). *Multimedia interface design.* Reading, MA: Addison-Wesley.

Bruder, I. (1991). Multimedia: How it changes the way we teach and learn. *Electronic Learning, 11* (1), 22–26.

Brunner, C. (1996). Judging student multimedia. *Electronic Learning, 15* (6), 14–15.

Bunnel, D. (1992.) Let's start a revolution: Bring multimedia to education. *New Media, 6* (1), 5.

Bush, V. (1986). As we may think. In Lambert and S. Ropieque (Eds.). *CD-ROM: The new papyrus.* Redmond WA: Microsoft Press. [Reprinted from *The Atlantic Monthly,* (1945 July), 101–108.]

Carlson, E. (1991). Teaching with technology: It's just a tool. Paper presented at the annual meeting of the American Educational Research Association, Chicago.

Clark, J. (1996). Bells and whistles ... but where are the references: Setting standards for hypermedia projects. *Learning and Leading with Technology, 23* (5), 22–24.

Dede, C. (1994). *Making the most of multimedia. Multimedia and learning: A school leader's guide,* Alexandria, VA: NSBA.

Dipinto, V. and Turner, S. (1995.) Zapping the hypermedia zoo: Assessing the students' hypermedia projects. *The Computing Teacher, 22* (7), 8–11.

Dyrli, O. and Kinnaman, D. (1995.) Moving ahead educationally with multimedia. *Technology and Learning, 15* (7), 46–51.

Ellis, D., Ford, N, and Wood, F. (1993). Hypertext and learning styles. *The Electronic Library, 11* (1), 13–18.

Foss, C. L. (1989). Tools for reading and browsing hypertext. *Information Processing and Management, 25* (4), 407–418.

Galbreath, J. (1994). Multimedia in education: Because it's there? *TechTrends, 39* (6), 17–20.

Gates, W. (1995). Multimedia revolution is here. Life on Line. *Gainesville (FL) Sun,* 5/15/95, 7.

Kalmbach, J. A. (1994). Just in time for the 21st century: Multimedia in the classroom. *TechTrends, 39* (6) 29–32.

Kozma, R. B. (1991). Learning with media. *Review of Educational Research, 61* (2), 179–211.

Landow, G. P. (1990). Popular fallacies about hypertext. In D. H. Jonassen and H. Mandl (Eds.). *Designing hypermedia for learning* (pp. 27–37). Germany: Springer-Verlag.

Landow, G. P. (1992). *Hypertext: The convergence of contemporary critical theory and technology.* Baltimore: John Hopkins Press.

Litchfield, B. (1995). Helping your students plan computer projects. *The Computing Teacher, 22* (7), 37–43.

Lloyd-Kolkin, D. and Tyner, K. (1991). *Media and you.* Englewood Cliffs, NJ: Educational Technology Publications.

Malhotra, Y. and Erikson, R. (1994). Interactive educational multimedia: Coping with the need for increasing data storage. *Educational Technology, 34* (4), 38–46.

Marchionini, M. (1988). Hypermedia and learning: Freedom and chaos. *Educational Technology, 28* (11), 8–12.

McCarthy, R. (1989). Multimedia: What's the excitement all about? *Electronic Learning, 8,* 26–31.

Megarry, J. (1988). Hypertext and compact discs: The challenge of multimedia learning. *British Journal of Educational Technology, 19* (3), 172–183.

Milone, M. N. (1994). Multimedia authors, one and all. *Technology and Learning, 14* (9) 25–31.

Negroponte, N. (1995). *Being digital.* New York: Alfred A. Knopf.

Postman, N. (1993). *Technopoly: The surrender of culture to technology.* New York: Vintage Books.

Saettler, P. (1990). *The evolution of American educational technology.* Englewood, CO: Libraries Unlimited.

Sharp, V. F. (1994). *HyperStudio in one hour.* Eugene, OR: ISTE.

Siegel, J. (1995). The state of teacher training. *Electronic Learning, 14* (8), 43–53.

Simonson, M. R. and Thompson, A. (1994). *Educational computing foundations.* New York: Merrill.

Skillen, P. (1995). ThinkingLand: Helping students construct knowledge with multimedia. *The Computing Teacher, 22* (7), 12–15.

Thornburg, D. (1991). *Education, technology, and paradigms of change for the 21st century.* San Diego: Starsong Publications.

Tolhurst, D. (1995). Hypertext, hypermedia, multimedia defined? *Educational Technology, 35* (2), 21–26.

Tripp, S. D. and Roby, W. (1990). Orientation and disorientation in a hypertext lexicon. *Journal of Computer Based Instruction, 17* (4), 120–124.

Turner, S. V. and Dipinto, V. M. (1992). Students as hypermedia authors: Themes emerging from a qualitative study. *Journal of Research on Computing in Education, 25* (2), 187–199.

Vandergrift, K. E. (1988). Hypermedia: Breaking the tyranny of the text. *School Library Journal, 35,* 30–35.

Wilson, K. (1991). New tools for new learning opportunities. *Technology & Learning, 11* (7), 12–13.

Resources

Hypermedia Programs/Products
The Lessons of War—ABC Interactive
Microsoft Bookshelf—Microsoft
LinkWay Live—EduQuest
HyperCard—Apple
HyperStudio—Roger Wagner
MicroWorlds Project Builder—LCSI
Multimedia Toolbook—Asymetrix
Digital Chisel—Pierian Springs Software
Multimedia Workshop—Davidson
HyperWriter—Ntergaid
Media Text/My Media Text Workshop—Sunburst
Multimedia Scrapbook—Washington Computer Services
Quicktime

Digital Cameras
QuickTake—Apple
XapShot—Canon

Other Resources
Risk Doubler—Connectix
Adobe Premier—Adobe Systems
The Black Stallion—MGM/UA

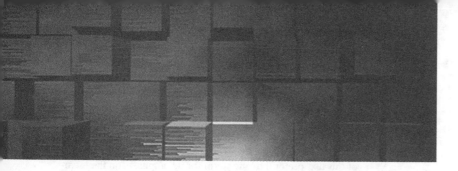

Chapter 9

Linking to Learn: Using Technology to Connect People and Resources

This chapter will cover the following topics:

- Definitions and descriptions of various linking options available to educators

- Benefits of linking for learning

- How to select and implement linking resources

- Teaching and learning activities that link learners

- Implementation issues that affect linking activities

Chapter Objectives

1. Identify illustrative examples for three types of linking options that can support educational activities:

   - Technology for course delivery
   - Technology to facilitate communications
   - Technology to locate and use educational resources

2. Interpret terms associated with these types of links.

3. Describe a distance learning program or telecommunications resource that could help meet a given instructional need.

Reaching Out to a World of Resources

Technology has changed no aspect of society more quickly and dramatically than its communications capabilities. Today's children regard revolutionary technologies such as the fax machine and the Internet as normal, everyday parts of the electronic landscape in which they live. Even in the rapid environment of technological evolution, these remarkable changes in communications have come about with incredible speed; some resources have developed from possible to pervasive in only a few years, and these changes are by no means completed or even slowing down. The primary reason for this breathtaking revolution in communications is our society's recognition of the importance of ready access to people and resources. If knowledge is power, as Francis Bacon said, then communication is freedom—freedom for people to reach information they need in order to acquire knowledge that can empower them.

These rapid developments in communication technologies have made possible three kinds of activities that allow educators and their students to reach out from their classrooms to bring in people and material resources. These links for learning employ many of the same communications technologies to assist learning through:

- Course delivery systems
- Communication among teachers, students, and others
- Locating and consulting people and information resources

This chapter begins with a comparison of the definitions and purposes of these types of applications, the benefits each offers to education, and some implementation concerns. Subsequent sections explain how teachers can integrate these linking activities into school and classroom activities.

Definitions of Key Terms

All linking activities share one feature: the use of some kind of network to communicate. The user may reach the network through technical mechanisms like broadcasts (satellite, microwave, or instructional television), cable systems, fiber optics, or a combination of these technologies. The purpose of the link may range from sending messages to taking a college course. Each can be considered an aspect of linking to learn.

As with nearly every activity related to educational technology, users must understand a variety of terms that describe linking activities. They must overcome a decided lack of agreement on which terms refer to which activities. Teachers should be aware of the commonly accepted usages for the two most common umbrella terms, as well as the areas of overlap between them:

- **Distance learning.** The terms *distance education, remote learning,* and *distance learning* all refer to learning situations in which the instructor and learner are separated over distance and/or time. Consequently, instruction relies on electronically transmitted educational or instructional programming and print materials (Holmberg, 1981; Sewert, 1982; Keegan, 1983; Wagner, 1988). While many authors use the terms *distance education* and *distance learning* interchangeably, the authors of this text prefer the latter term for its emphasis on the learning process. In this textbook, *distance learning* refers primarily to course delivery systems, but some people apply it to any linking activity to promote learning.

- **Telecommunications.** While the most precise definition of this term is communication at a distance, many people (and most educators) think of it as using a computer to communicate. Kearsley, Hunter and Furlong (1992) offered the term *telecomputing* as a substitute for *telecommunications* to describe activities in which classroom computers and modems link groups with each other and with educational resources. In this textbook, *telecommunications* means primarily (a) facilitating communications among teachers, students, and others and (b) locating and using people and information resources. Telecommunications can help teachers keep track of attendance, keep in touch with colleagues, ask questions of content experts, locate resources not readily available locally, or keep track of recent events. Students communicate with others in remote locations, share data, conduct joint experiments, collaborate on projects, or access remote resources from school or home. As with the term *distance learning,* some people encompass all linking activities under *telecommunications,* but many educators use the term to denote all activities except formal courses or workshops.

Who Makes the Decisions about Linking Options?

The number and types of linking methods used in education seem to be growing daily. This is partly because communication technologies are expanding rapidly, but also because teachers and students are constantly finding new ways of using these technologies to solve problems and meet needs that arise from distance or lack of accessible resources. However, not all these options are always available to a given teacher. Distance learning (DL) activities (that is, course delivery systems) usually replace classroom instruction rather than supplementing it; these activities involve expensive resources and time-consuming implementation strategies that require changes to the traditional classroom structure. Therefore, the decision to integrate distance learning activities is usually made by a school, district, or state department rather than an individual teacher. Many telecommunications activities, on the other hand, are less expensive and require less restructuring of the classroom delivery system. These activities are usually initiated by teachers to supplement other classroom instruction.

Although teachers do not normally initiate the selection and use of DL technologies, they do significantly influence district and school decisions. In order to do this, they must recognize the varied ways in which this technology can improve educational opportunities for all students. This chapter provides information on DL to build a basic understanding of its applications, their contributions to education, and implementation issues.

Present and Potential Advantages of Linking to Learn

Teachers can assess the benefits of various linking activities in two ways: by studying research evidence of effectiveness in and impact on learning and by evaluating expected changes in the nature of the learning environment.

Research Evidence: The Effectiveness of Linking Activities

More actual studies and evaluations have focused on the effectiveness of course delivery via distance learning activities than on other kinds of linking activities. Research conducted by scores of professional educators and evaluators from 1954 through 1994 has consistently found no significant difference between instruction delivered through traditional classroom methods and instruction delivered over one or more remote technologies in which a teacher and students are separated by distance (Schramm, 1977; Johnston, 1987; Russell, 1992; Schlosser and Anderson, 1994). In fact, so many researchers have studied instructional television over the years—Russell reviewed 21 research summaries covering data obtained from over 800 separate studies—that no one should doubt the efficacy of instruction delivered over television. The findings of comparative studies also attest that students learn equally well with any video-based delivery technology, with or without interaction. Although no current studies have compared the effectiveness for different types of students of instruction delivered via distance learning (e.g., general, adult, K-12, exceptional students), evaluation from the federally funded Star Schools Program indicates that distance learning students enrolled in high school courses function slightly better than comparable students in traditional classes (Withrow, 1992). Existing research, project evaluation, and anecdotal evidence support distance learning as an effective means for delivering formal instruction (Lane, 1993, p. 179).

State departments of education have noticed the usefulness of distance learning to increase access to effective instruction. In 1987, fewer than ten states were involved in distance learning. By 1989, virtually every state was offering some form of distance learning program (St. Onge, 1992). Current surveys show that almost half of all U.S. school districts are implementing distance learning programs in their schools (Quality Education Data, Inc., 1993). Many of these states and school districts are also evaluating the performance of new technologies, the experiences of teachers in these environments, and effects on student achievement and dropout rates. In the next few years, results from this research will help to guide and improve the applications of this technology. It seems logical that the added speed and capabilities of newer communications technologies will only increase the instructional impact that has already been demonstrated for distance learning.

Less documentation addresses the impact on learning of modem-based telecommunications activities such as research via the Internet or using e-mail for classroom development projects. Most reviews of such activities are descriptive rather than experimental, and they provide descriptive data on the characteristics of linking activities rather than evaluations of usual concerns such as achievement or dropout rates (Honcy and Henriquez, 1993). Some studies have reported on the effects of telecommunications-based activities on writing (Cohen and Riel, 1989) and on cultural awareness (Roblyer, 1992), but no extensive reviews have detailed the effectiveness of resources like word processing programs.

How Linking Can Change the Nature of Education

Linking activities appear to hold great promise for improving the quality of education in many ways. Educators believe they can provide several different kinds of assistance. They can provide ready access to a variety of people and information resources, create opportunities for collaboration between classrooms, support learner-initiated study, offer advanced or otherwise unavailable courses, and deliver staff development programs with minimal restrictions of time or place. The Telecommunications Act of 1996 has three provisions dealing with education's access to telecommunications resources (Bybee, 1996). Clearly, the expectation is that these will be key to raising and maintaining the quality of American education. In fact, at a time when American schools are receiving less and less money to cope with larger and more pressing problems, linking technologies offer new hope for the entire educational system. The following sections describe some of these anticipated improvements.

Hypothesized improvements in instructional quality. Distance learning technologies can remarkably expand the quantity and quality of information resources that students can access from the classroom. In the most fundamental sense, these technologies connect the teacher with the student and/or provide access to remote resources for both when time, distance, or expense prevent face-to-face interaction. But these technologies can also do much more:

- Both teachers and learners can work with others at remote sites.

- The community of learners can expand to include virtually anyone who wishes to obtain information and who is not excluded by policy or cost.

- They can provide real access to experts in universities, research laboratories, the business community, government agencies, and political offices.

Hypothesized structural changes in educational systems. The technologies of distance learning offer many opportunities to restructure the teaching–learning process. In schools, computers, modems, televisions, and telephones can easily transform learning by offering alternatives to teacher-provided information, access to virtually unlimited resources, and opportunities for real-world communication, collaboration, and competition. For example, in his book about the demise of network television and the coming transformation of media and American life, futurist George Gilder states that distance learning technologies may have the power to transform public education by making the nation's best teachers available to students anywhere (Gilder, 1992). The promise of technology extends beyond a substantial increase in available teacher or information resources, however. Instead, the application of technology to education presages significant changes in curriculum or content, the physical learning environment, and the roles of both teacher and learner.

- **Changes related to curriculum/content.** Installing sophisticated technologies makes sense only if updated curriculum can utilize them effectively. Effective educational programs designed to accompany distance learning technologies combine the best characteristics of the traditional classroom with the latest advances in learning theory and interactive technologies. New technologies allow program designs to add content and process changes not possible with standard, textbook-based curricula. These changes will undoubtedly alter methods for developing and disseminating curriculum to teachers and learners; they include increased flexibility, interactivity, and access to additional resources. Also, with the global human knowledge base literally doubling every 3 to 5 years, no education system can possibly hope to enclose under one roof all the information that's available to students on every subject. Education's most important goal is to teach students how to locate, access, and apply information (Purcell, 1993). Technology for on-site storage, retrieval, and manipulation of data allows teachers to customize learning and to move toward individual and small-group collaboration. Personal computers, display technologies, optical memory systems, facsimile (fax) machines, and graphics scanners also enable learners to practice locating, accessing, and applying information in the contexts of the classroom and the real world.

- **Changes related to the physical learning environment.** Technology has the potential to alter significantly the physical environment in which learning occurs. A technology-rich classroom may group students in learning centers around computer monitors or television screens rather than in more traditional rows of desks directed toward the instructor. Groups of students may be working on completely different subjects, at different levels, and in different media—all simultaneously. A single student may be part of a multistate virtual community in which members know one another only through e-mail addresses, computer-transmitted biographies, or faxed photographs; a student may also join a homogeneous work group concentrating on solving an assigned or chosen problem. Conversely, the student's "classroom" may be at home, the office, a hospital, a correctional institute, or a hotel far from the originating point of the course or information. The potential consequences for both public and private schools are substantial, affecting everything from the number of schools and teachers to the mechanisms for funding education.

- **Changes in the roles of teacher and learner.** Changes in technology create changes in culture, and one effect of the increasing use of educational technology has been a change in the culture of the classroom. In fact, distance learning tech-

nologies may have radical effects on the relationship of teacher and student. For example, teacher-participants in the Apple Classrooms of Tomorrow (ACOT) project expressed some reluctance to learn how to use available classroom technological tools (Ringstaff, 1993). Teachers felt discomfort knowing only a little more than their students about these tools. At the same time, students were eager to assume responsibilities to provide assistance to one another and to teachers. The result was the gradual evolution of a "student expert" structure in which teachers began to capitalize on the technical expertise of their students. In turn, the students themselves realized several unforeseen benefits. Slower students blossomed, less popular students were sought out for advice and assistance, and formerly unmotivated students became excited and involved. Teachers were themselves surprised at the many benefits to students, including improved academic performance, increased self-esteem, and students' greater acceptance of responsibility.

Creating learning communities. Long a goal of K-12 education programs, community involvement plays a new role in education through remote delivery system. For example, no geographic boundaries may restrict attendance at a specific educational institution, allowing unlimited learner access to programs of choice and eliminating age as a criterion for enrollment. At the same time, the length of time that learners spend on a given topic or in a particular course comes to depend on their ability to demonstrate achievement or competency. Individual institutions could determine the specific timing, lengths, and schedules of their unique learning environments to best serve the needs of the learners in their (extended) communities.

In a learning community, education plays a critical part in the lifelong transformation of the human personality, helping to develop children into adults who can participate freely and successfully in the social, economic, spiritual, and civic activities of society. In today's ever-changing and increasingly global world, that transformation is continual, necessary for economic survival, and an important part of our cultural growth. A community of learners fosters a fundamental commitment to learning and a recognition that the most important asset of any community is the people within it (Education Imperative, 1992). Organizations become successful within a community by expanding and building upon the knowledge of their workers. It is this capacity for continuous learning that will, in turn, determine the quality of the future.

Selecting and Using Linking Resources

The Nature of Links: Some Technical and Conceptual Background

In the last few years, the technological options for delivering or supplementing education over distance have increased dramatically. Advances in telecommunications and distance learning technologies, coupled with steadily decreasing costs for transmission and reception equipment, have created an affordable and effective option for remote delivery of

Figure 9.1 Microcomputer with Modem

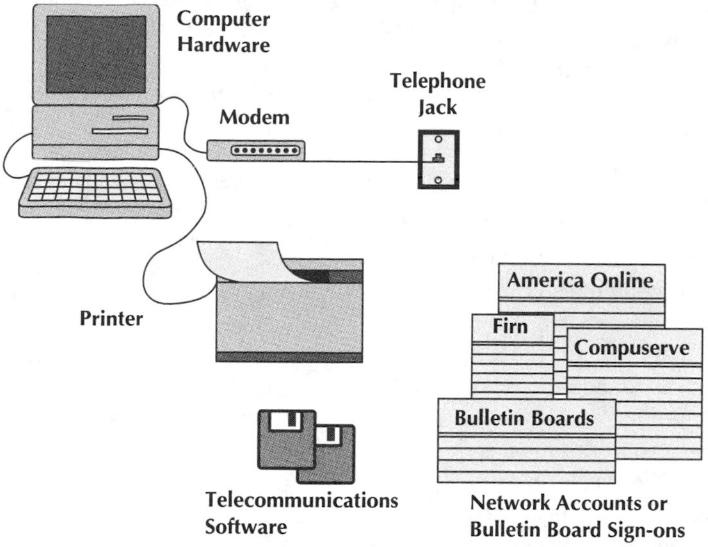

instruction and electronic links between students and teachers. These technologies allow users to transmit, receive, create, store, and combine information in unprecedented ways, improving the breadth and quality of educational offerings and services and allowing students and teachers to access even the most volatile of information resources.

Networks: The common elements. Any linking activity needs three basic components: people who seek information, a source of information, and a transmission technology that links the two. Teachers may find it helpful to conceive of these elements as forming a kind of loosely defined network.

Transmission technologies. Several delivery or transmission systems allow teachers and students to establish links with people and material resources, including broadcasts (by satellite, microwave, and Instructional Television Fixed Service (ITFS); the Public Switched Telephone Network (PSTN), cable systems, and fiber optic systems. An institution can employ these technologies independently or combine them to form a "hybrid" system that meets its special needs. No single technology is best for all situations or applications. Each technology has different capabilities and limitations. Using these technologies effectively requires a careful match between technological capabilities and educational needs. Satellite systems and other broadcast technologies, for example, do not allow the teacher to see students, so they limit options for interaction.

One fairly recent development in transmission technologies is the Integrated Services Digital Network (ISDN). Fox, Loutsch, and O'Brien (1993) described ISDN as "a new and exciting technology that will revolutionize education and allow you to integrate [technology into] classroom curriculum in ways you never imagined before" (p. 18). Before ISDN, a system needed a separate telephone line for each application (e.g., voice calls, faxes, cable TV). Since

ISDN is a completely digital telecommunications system, it can carry all types of data on the same line in a fraction of the time it took for analog lines. Therefore, ISDN can accommodate applications such as video conferencing, desktop computer conferencing, and voice messaging systems, and it can accomplish these jobs more quickly and less expensively than separate systems.

These new developments in transmission capabilities are exciting news for technicians who work to make links work, but, ultimately, these delivery systems will be completely transparent to teachers and students. They will become the norm for obtaining information and instruction in the same way that people use the telephone to keep in touch with family and friends. In combination with readily available equipment like telephones, fax machines, VCRs, TV monitors, and cameras, these technologies affect how learners interact, what information resources they use, and how effectively the learning system functions.

Linking configurations. The part of a communications system that serves a given classroom or teacher will depend on the purpose of the link and the type of transmission technology. Some examples of typical workstations are described here to help teachers define the scope of activities and resources for linking to learn.

Microcomputer with modem, communications software, and telephone line. Perhaps the most common kind of communications setup in classrooms, this is also one of the least expensive means of linking students and teachers with resources they want. (See Figure 9.1.) A modem has the ability to change or *mod*ulate digital signals from a computer to form analog signals that can travel over telephone lines. It can also change the signals back or *dem*odulate them for access by the computer.

Modem buyers look at several kinds of features and decide how much they want to pay for each of them. These features include speed, internal/external connection, and sending options (e.g., fax capability). Speed depends on how fast the modem transmits data to and from the computer. It is measured by baud rate or bits per second (bps) and usually ranges from 2,400 up to 28,800 bps. More expensive modems are faster and have fax capabilities. Some users may prefer internal modem-to-computer connections, but either type of connection is acceptable.

Many modems come with communications software, but the user must purchase a program if one does not accompany the modem. Buyers evaluate communications software based on its emulation capabilities. The software makes the microcomputer work like a certain kind of terminal that can communicate with a mainframe system. Some typical terminal emulation standards are VT100, TTY, and TVI950. The communications software also specifies other information about the workstation, including communication parameters and file transfer protocols. To exchange information, two workstations must specify the same emulation modes, parameters, and protocols:

- **Communication parameters.** Transmissions must match baud rate, number of data bits transmitted at one time (usually 8), parity (whether an odd or even code or no code should be sent to indicate a transmission error), and number of stop bits (to indicate the end of data transmissions).

- **File transfer protocols.** These specifications govern how files will be transmitted and received (e.g., XMODEM, KERMIT, ASCII).

Audio-based systems. Some links communicate via sound (audio) only. Because they are limited to audio communications, these systems are not as common as other communication methods, but they meet certain educational needs inexpensively. Audio-only systems include one-way audio, audioconferencing (two-way audio), and audiographics.

- **One-way audio.** Audio programming may be delivered to individuals or classrooms via radio broadcasts or through standard telephone service. In this format, the teacher speaks to distant students, who cannot respond directly in real time (during the broadcast or phone transmission). Interaction can be built into the lesson through materials delivered prior to the instruction or facilitated on-site by a third party or classroom facilitator. A teacher may ask students to respond verbally (even though the teacher cannot hear the response) or in writing, and students monitor their own learning in other ways. This type of distance learning has been used extensively for student learning and teacher training in several developing countries, but its use in the United States has been very limited despite its extremely low cost (U.S. Congress, Office of Technology Assessment, 1989). Besides the low costs of delivery and duplication, other benefits of one-way audio programming include ease of use, low production costs, and its requirement for little specialized training.

- **Audioconferencing.** Audioconferencing or two-way audio allows two or more parties to interact in real time. Teacher and learner may converse over a simple telephone link, or a phone bridge may connect more than a dozen participants. Speaker phones at multiple sites can extend the numbers of people involved, but additional participants reduce the amount of

interaction possible in a given time frame. In today's instructional world, audioconferencing is frequently combined with other technologies (e.g., computers, fax machines, prerecorded videotapes, and texts).

- **Audiographics.** An advanced form of computer networking, audiographics augments computer interaction with real-time audio communication. Participants in an audiographics conference can communicate with one another via their computers and talk with one another at the same time, usually over the same telephone line. To do this, a special modem combines the computer and audio signals and sends them over one line. Audiographic systems can now capture, transmit, and annotate color visuals (U.S. Congress, Office of Technology Assessment, 1989). Audiographics systems provide their major benefits by allowing participants to add pictures, charts, and graphics to transmitted messages and by facilitating interaction, helping people to communicate and clarify complicated or detailed information. This supports new levels of collaboration and problem solving.

One-way video systems. Educational programming can be broadcast one way from one point (transmitter, uplink) to a single point or multiple points (receivers) via satellite, Instructional Television Fixed Service (ITFS), cable television, microwave, or public broadcasting systems. Broadcasts can be sent via open air or encrypted. Anyone with the necessary hardware and connections (as with a cable system) can receive open-air broadcasts, as can anyone with the coordinates to focus a satellite receiver dish. Encrypted programs require decryption devices. (Anyone who has ever watched pay-per-view television at home has probably rented such a device and installed it temporarily into the cable television system. (See Figure 9.2.)

Instructional programs can be broadcast in real time (live), or they can be delivered via broadcast for taping and showing at a later time (tape delay). Live broadcasts permit student-teacher interaction over telephones or special-response systems usually for a predetermined number of students. A later section will discuss this possibility more fully. A tape delay system can store one or more program segments for later teacher or learner control and use; this keeps delivery costs to a minimum while ensuring the timeliness of content, and it permits flexibility regarding both the time and place of instruction.

One-way video systems can accommodate interaction through several means involving both teacher and learner. Electronic mail and telephones can link teachers and students for discussion, problem solving, and sharing. Hands-on assignments can call for active discovery learning in the classroom. Both collaborative and competitive projects can require students to cooperate with other students and/or compete with classrooms in other school systems and states. Regular contests and on-air assessments can help to emphasize interactivity.

One-way video, two-way audio. Current technologies now permit completely interactive, real-time instructional course delivery. Programs of this type are not specifically intended for taping. (Although many participants do tape programs

Figure 9.2

Video programs are sent from a single point via a satellite uplink facility or television transmitter to a receiver (another single point or multiple points) and then distributed widely through a variety of broadcast and cable transmission systems.

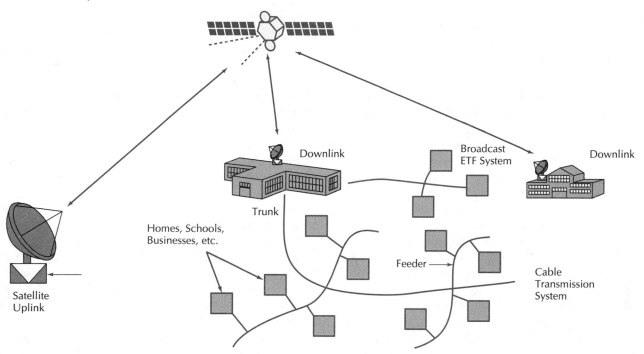

for study and review, stringent copyright restrictions govern what may be copied, for what purpose tapes may be used, and how long programs may be kept.) Instead, they are intended to function in the same way as more traditional teacher-to-student instruction. Although telecommunications technologies are employed, the general kind of interaction is much the same as that in any face-to-face educational situation: Teachers present material and ask questions of students; students take notes and ask questions of teachers. The advantages of interactive, video-based instruction come not necessarily from the instruction itself, but from the increased access to courses and expert resources that remote delivery allows.

Interactive two-way audio and video. Completely interactive two-way audio and video delivery systems allow distance learning to resemble classroom instruction most completely: Teachers can view students and their reactions, students can see other students, and interaction is much like that in a traditional classroom environment. (See Figure 9.3.) This kind of course may be produced in a classroom specifically designed for two-way transmission or in a studio. In either case, other technologies may supplement both teacher presentations and student interactions. Two-way remote learning systems are both the most complex and most expensive to manage, however, and these problems currently rule out widespread utilization within K-12 educational systems. The costs of video transmission services and hardware continue to drop, however,

while advances in existing copper-wire infrastructure and the rapid development of fiber optic systems combine to make this technology affordable and easy to use.

Figure 9.3

A completely interactive, two-way video and audio delivery system facilitates real-time interaction. Voice and video signals are transmitted using a single point-to-point system or through a hybrid system using broadcast and cable resources.

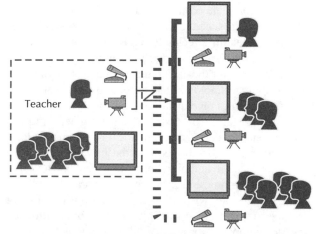

Source: Based on a diagram from Heinich, Molenda, Russell, and Smaldino, *Instructional Media and Technologies for Learning*, © 1996 by Merrill/Prentice Hall.

Implementation Issues for Linking Activities

A number of issues arise when schools implement linking activities. As with any innovation, time and experience will identify still more. Teachers can address some of these issues on their own, but others will require assistance from school, district, and even state administrators. Still, teachers should anticipate these complexities and their potential impact on the success of linking strategies.

Top-level policy/planning. State governments have come to play major roles in planning, supporting, and organizing distance learning activities because the responsibility for public education rests with state education departments and because government agencies disburse federal funds designated to promote national education goals and integration of educational technology. State-level involvement also helps educators to build coherent distance learning systems that meet the needs of all education stakeholders and, at the same time, uses state resources wisely. Only rarely have states even attempted to coordinate distance learning programs, delivery systems, and resources across pre-kindergarten-to-12, community college, and university levels. Despite the policy-making and resource-allocating role of the state governments, most decisions about actual utilization of distance learning technology and programs are made by local school administrators and subsequently supported or hindered by district school boards. Each of the issues identified below demand responses at both state and local levels.

Currently, significant barriers complicate projects with linking technologies in schools and institutions. Today's education, communication, and information policies and regulations were developed long before the major technological breakthroughs (in particular, technologies for managing digitally processed and transmitted information) that allow interactive learning to occur virtually anywhere in the industrialized world. These old arrangements must give way. For example, class schedules must become more flexible; a typical class period does not allow adequate time for students to complete projects using telecommunications or to participate in scheduled, interactive courses delivered via distance learning technology. To overcome obstacles like this, states, districts, and schools must make plans and develop policies for the use of telecommunications and distance learning technologies that accommodate the effects of those technologies on educational reform.

Future policies for distance learning will require involvement of federal, state, and local governments; local school districts and schools; and the private sector (See Inset 9.1.) As more and more institutions become involved in distance learning and produce a plethora of new programs, institutions will begin to move from postures of friendly cooperation toward unfriendly competition. In order for competing programs and institutions to survive and recover initial start-up costs, institutions will be forced to market their services aggressively. Additionally, as parents and students increasingly elect to learn outside traditional schools, resources that support traditional educational efforts will require different funding formulas and enabling policies. Fewer public schools and fewer teachers will be needed by reduced numbers of full-time or traditional (on-campus) students.

Teacher involvement and training. In order for any educational innovation to succeed, it must attract early involvement of classroom teachers. Teacher concerns about implementing new technologies must be determined and addressed or distance learning will never fulfill its early promise of vastly improving the educational system. The most successful strategies will involve teachers in early planning stages and then consistently seek meaningful teacher on input and participation. Classroom teachers may worry, for example, about being replaced by a "star" teacher on a television monitor. They may feel more confident when they understand that the role of the classroom teacher is even more critical in a system that employs cooperative learning strategies and attempts to individualize instruction to specific learner needs. Conversely, teachers who seek to participate actively in distance learning as "tele-teachers" must be allowed additional time to prepare to use the new technologies.

Educators need opportunities to learn how to operate and integrate the new technology tools, to implement radically different curricular approaches and associated classroom management strategies, and to become facilitators of learning for students. Yet, most districts spend less than a quarter of their computer budgets on training (Bruder, 1993). Moreover, professional development in technology applications to reform the learning environment has never kept pace with the purchase and distribution of equipment (Hawkins and Macmillan, 1993). Attention needs to shift to a new kind of sustained support for teaching professionals—support that combines learning about technology with instruction in how to realize new learning conditions through the new teaching practices.

States, districts, and schools must get involved in training teachers to use telecommunications, devoting at least as many resources to this effort as they have spent to train teachers to use other computer-based technologies. This seems especially important since the effort to become an "accomplished" technology-oriented teacher takes at least 5 or 6 years of sustained technology use (Sheingold and Hadley, 1990). Although technology can serve as a catalyst for change, teachers also need opportunities to reflect on their experiences, supportive school environments, and the freedom to experiment.

Technical limitations. Each of the technologies described in this chapter has certain advantages and limitations that must be analyzed carefully before selecting the linking method that will meet most needs. For example, teachers who employ one-way video, two-way audio programs worry about difficulty in establishing rapport with students

---

Inset 9.1
Recommendations of the United States Distance Learning Association

The United States Distance Learning Association (USDLA) is a nonprofit association founded in 1988 to promote the development and application of distance learning in education and training. USDLA has become the leading source of information and recommendations for government agencies, Congress, industry, and program developers. USDLA sponsored a National Policy Forum in 1991 to discuss needed changes in education and communication policy for the 21st century. It then developed a number of policy recommendations for federal, state, and local governments, which include the following:

1. Develop a vision for a national infrastructure recognizing the critical importance and interdependence of systemic reform and advanced telecommunications services.

2. Bring coherence to educational technology and distance learning funding and focus those resources on education restructuring projects. All future initiatives or policy should include distance learning as an option.

3. Develop national demonstration sites for educational technology and distance learning that disseminate research results, educational applications, and effective teaching strategies.

4. Provide incentives for teacher training institutions to restructure preservice and in-service programs recognizing the importance of communication and information technologies.

5. Provide incentives for regional and professional accreditation associations to recognize and encourage appropriate uses of distance learning technologies.

6. Ensure that financial aid programs recognize distance learning as a peer to traditional course delivery.

7. Address education use via distance learning technologies as an issue for special attention within copyright laws.

8. Provide incentives for states to remove barriers to distance learning based on teacher certification, textbook adoption, and accreditation practices.

9. Provide incentives for faculty who maximize resources and achieve quality instruction through use of appropriate educational technologies.

---

without visual interaction and classroom management at learner sites. Students become frustrated when they cannot get telephoned questions through to teachers attempting to interact with multiple classrooms of students. Still other issues require new policies. For example, state and local efforts to hold a school or even an individual teacher accountable for the progress of students are complicated when a portion of instruction is delivered by a teacher from another school, district, or state. Reciprocity agreements through which states accept certification of on-air teachers may not include uncertified but capable university or community college faculty as instructors for K-12 programs.

No professional requirements or widely varying standards specify in-field expertise necessary to support students in remote classrooms. Differences between different schools' responses to these and other issues only complicate matters further. Increasingly, audio interaction is being replaced by exchanges through computer networks, fax machines, and other computer-based applications like keypad response systems. As numbers of students enrolled per class continue to grow, audio systems will decline as focal points of interaction.

The technical limitations of one-way video, two-way audio programs also raise a concern with timeliness of materials. Like today's textbooks, packaged materials are difficult to change rapidly, so they may be out of date before new versions are available. Supporting materials may be misplaced or depleted. Facilitators and/or assessors may become inaccessible. Copyright protections may prohibit teachers or learners from adapting and/or copying materials. Costs for professionally produced programs and materials are high. Also, as happened with programs designed for filmstrips and 16mm film, the technology may change before the program itself is obsolete. The recent explosion of available programming and the current race to produce and deliver instructional courses through video confirm that the number and variety of educational alternatives to formal classroom instruction will continue to grow. In turn, these new programs offer continually expanded opportunities for learning to increasingly large numbers of students and other citizens.

Differentiated staffing. One reason for adding distance learning technologies to local classrooms is to import programs and resources that are otherwise unavailable. At the same time, both state and local rules may require staffing changes before distance learning programs can become effective in the classroom. For example, local school board policies may require the presence of a field teacher at class sessions, defeating the effort to import a remote teacher to remedy unavailability of local expertise. (This is an especially obvious challenge in foreign language programs where student expertise may surpass a local teacher's, yet local classroom facilitation is clearly needed.) In another instance, a distance learning class may have so many students enrolled that a single teacher simply cannot attend to all the details of managing the class, including responding fully to questions from remote sites, reviewing assignments, testing, and student recordkeeping. Excellent on-site teachers may not be especially effective on television, leading some to suggest using the most appealingly "packaged" teachers instead of teachers with excellent classroom rapport and proven, effective teaching skills. Solutions to these and other challenges will be varied, but the steady change of the teacher's role from "sage on stage" to "guide on side" is a consistent theme in all solutions.

Assessment. Implementation of distance learning in local classrooms complicates assessment in two basic ways. First, accountability for student achievement becomes split between the tele-teacher and the teacher/facilitator on-site. Second, educators must define an assessment system appropriate for the nontraditional learning that occurs, for example, when students collaborate via computers and fax to jointly solve a problem or produce a product. These two issues, accountability and alternative assessment, have multiple policy-based and practical solutions, and they are being addressed by federal, local, and state policy makers.

Throughout the United States, efforts are under way to shift decision making as well as accountability for student achievement to the lowest possible level. Funding is often inadequate or unequal, however, and teachers must contend with wide variation in student economic status and in the educational achievement and support of parents. This absence of homogeneous student populations often leads teachers to complain that the shift toward increased teacher responsibility is unfair and unwise, placing the accountability for learning on the teacher instead of the student. Both accountability and assessment of student achievement remain central to the full implementation of distance learning, and the paradigm shift from teaching to learning must resolve the problems.

Curriculum design and development for new and emerging standards. Many educators consider the technical design and development of the emerging telecommunications infrastructure less important than the design and development of quality instructional programs and services for teachers and students. National education goals, state standards and achievement criteria, and local school improvement plans indicate a universal need for significant changes in classroom instructional strategies and learning resources. At the same time, research in effective course and curriculum design has focused on overcoming the differences between distant and local classrooms. One study introduced the concept of "teletechniques," a set of components that teachers take for granted in face-to-face instruction, but that are not automatically found in distance education. Tele-teachers need to humanize the teaching experience (create rapport with students); they must encourage participation (ensure that students interact with other students and with teachers) and attend to message style (vary tone of voice and volume, using videos and other visual aides). They must also provide regular feedback (monitor student interest) (Portway and Lane, 1992).

Administrative support. While teacher ability is unquestionably important to the success of a computer network, administrative support for technology is critical. A 1992 study by Henry Jay Becker, a sociologist at the University of California at Irvine, suggests some factors that foster "exemplary computer-using teachers." The chief one involves the administrative support that such teachers enjoy within their schools, while teachers' inherent qualities have only secondary importance. If that is true, then by adopting appropriate policies, school administrators ought to be able to develop many more similarly skilled teachers (Leslie, 1993, p. 93).

Principals and other administrators make teachers skeptical when they advocate the use of technology in their offices and schools without using that technology themselves. Principals and administrators need not be experts; however, they do need to understand the different technologies and their good and bad features so they can be effective leaders and managers. This will not happen until principals share the same intensive training opportunities and peer support networks that teachers enjoy. Unfortunately, current training opportunities directed at preparing principals for the technological future fall far short of needs. To help district and school administrators implement new technologies, including delivery systems for distance learning, training must also address the leadership skills needed to plan for and manage change at the school level.

Interoperability. As they begin to use networked technologies, schools encounter a fundamental problem with the lack of uniform standards for interconnection. Current transmission technologies (including copper wire, fiber, cable, microwave, and satellite systems) vary considerably in availability and cost, and many current systems for receiving information are proprietary in nature, restricting reception to the same or technologically compatible equipment and systems. In contrast, computers have always communicated using a digital vocabulary, and eventually all electronic communication devices will use this same language. Once information is digitized, it can travel equally easily to a computer, satellite dish, television set, or telephone, and this dramatically raises the stakes in the race to digitize communication transmission technologies.

Standards for interoperability must cover more than a single standard for voice, video, and data delivery. For example, the December 1993 launch of AT&T's TELSTAR 401 transformed many satellite networks. Many of the nation's educational program providers opted to deliver their programs from purchased or leased transponders on this satellite, triggering a massive migration not only to the satellite location but also to its practices of digital compression and VSAT technologies. As a result, hundreds of schools employing Ku or C band receivers (different technology standards) were required to adapt their old satellite receiver dishes or purchase new ones to continue to receive old programs through the new compressed signal. The transition was further complicated by the unavailability of the hardware needed to adapt installed dishes. (At press time for this book, discussions were still underway between educators and major satellite equipment manufacturers to develop hardware that would adhere to an open standard; handle Ku, C, and digitally compressed signals; and be steerable at prices that would be affordable to the K-12 market.)

Obtaining required resources. Teachers who want to do modem-based linking activities complain most often about difficulties in obtaining telephone lines, computers, and modems in their classrooms. Over 75 percent of school districts participating in an annual study by a private educational research group reported that they do not use modems in conjunction with any distance learning programs. Within the remaining 25 percent of districts, less than one-third of the schools actually had modems (Quality Education Data, Inc., 1993). Even when students and teachers have access to networks and services, the cost of long distance calls can be prohibitive to many schools.

Schools wishing to participate in satellite-delivered distance learning must purchase a satellite dish. Prices at this writing range from $700 and up. Oates (1995) describes the myriad of factors schools must consider when buying a satellite dish. Schools must also purchase services with one or more distance learning providers. (See list of providers at the end of this chapter.)

Logistical problems. Successful classroom projects that involve links with other schools or classrooms call for elaborate coordination among members. Teachers must make sure that students have access to modems and telephone lines when they need them. Participants in a linking project must agree among themselves about a time frame in which to accomplish the activities. Working across different time zones adds another dimension to this problem; sometimes interaction cannot be in the form of real-time "chats," since one group of students may be in class while the other is getting a good night's sleep!

Planning a successful telecommunications-based curriculum. Online resources like the Internet can be a lot of fun to use and very seductive to both teachers and students. As appealing as these resources are, it seems an unwise use of school resources to go online for long periods of time just for fun. Every activity should have some identified purpose and link to established curriculum. Roblyer (1992) proposes two criteria to consider when designing curriculum for telecommunications. Activities should be (a) instructionally significant in terms of established state or district curriculum and (b) appropriate for the medium so they "take advantage of instructional features that only telecommunications can provide" (p. 17).

Ethical issues. People are spending increasing amounts of time communicating with each other and obtaining information online, and some problems with human behavior (and misbehavior) have already begun to emerge. The contents of messages and databases are not always monitored

in resources like the Internet, and they may contain language or information not appropriate for the public or for minors. Yet the Internet is also designed to make information easily obtainable. Some educational organizations have mechanisms and safeguards in place to keep their users from getting to these materials. In addition, a telecommunications etiquette for appropriate online behavior has also developed over the past few years. This code of behavior includes some important points:

- Begin a message with a salutation and identify yourself at the end.
- Avoid coarse, rough, or rude language.
- Avoid sarcasm, which unknown receivers may misinterpret.
- Acknowledge and return messages promptly.
- Do not forward a message without the permission of the original sender.

Equity and cultural issues. Dozier-Henry (1995) cited another group of issues that affect both usefulness and the limitations of linking as an instructional tool. She observed that, "As we look to technology to help meet the needs of students with diverse backgrounds and worldviews [these] kinds of equity issues seem important to consider" (p. 11):

- **Technology's built-in cultural bias.** Some cultures value face-to-face communications much more than machine-facilitated ones. Yet teachers may not recognize the validity of some students' need for a "low-tech" method of communication.
- **Access to information technology.** Dozier-Henry also questions whether all students will have equal access to information resources or whether these educational experiences will be reserved for a "deserving elite."
- **Technology's role in multicultural education.** Finally, she cautions against the view that telecommunications can or should accomplish all the goals of multicultural education. "There is a danger that multicultural education will be reduced to an appreciation of different food, heroes, and holidays" (p. 14).

Using Technology for Course Delivery

There are two general methods of delivering whole courses via telecommunications: video-based course delivery and interactive video instruction. The technical setups for these two strategies have already been described. This section will give some examples of how to undertake each type of instruction.

Video-based Course Delivery

Courses designed for live delivery typically employ one-way video, two-way audio systems. Courses are studio-produced to control the quality of both the audio and video signals and to allow the instructor to insert graphs, charts, video, and text to reinforce and supplement the presentation. The system can also present computer-generated and stored information, as well as segments of

Inset 9.2
An Overview of Linking Options for Educators

Linking Configuration Options

These are the physical linking configurations available to educators. Each makes possible several of the types of distance learning applications described at right:

Microcomputer with modem, communications software, and telephone lines (see Figure 9.1)

Audio-based systems:
- One-way audio
- Audioconferencing
- Audiographics

One-way video systems (see Figure 9.2)

One-way video, two-way audio

Interactive two-way audio and video (see Figure 9.3)

Distance Learning Options

These types of distance learning applications are available. (Note some overlap among the options; for example, online messaging can also be done on the Internet)

Distance learning options for course delivery:
- Video-based course delivery
- Interactive video instruction

Distance learning options to facilitate communications
- Audio messaging
- Classroom audiographics
- Teleconferencing
- Desktop videoconferencing
- Online messaging: e-mail, discussion groups, bulletin boards

Distance learning options to locate and integrate educational resources
- Networks/utilities (e. g., American Online)
- The Internet

---

other videotapes, photographs, and illustrations. Telephone lines provide student-teacher audio interaction. Also, a one-way video program can use real-time interaction through proprietary response systems (generally keypad configurations designed to tabulate student responses to teacher-generated questions). Fax machines can transmit questions and responses, and computer networks allow participants to share text and data.

One-way video, two-way audio programs are produced by numerous organizations and businesses, including state agencies, individual school systems, consortia of program developers, community colleges, universities, and both nonprofit and commercial enterprises. One organization, Distance Learning Associates, distributes programs for several developers and yearly delivers more than 5,000 hours of real-time, interactive programs for kindergarten through high school and adult education (including substantial teacher/staff development offerings) in eight states. A typical course would take a semester to complete, combine direct-instruction and constructivist strategies, share a standard assessment instrument or procedure, and substitute for the equivalent number of hours in a more traditional classroom environment (see Inset 9.2).

The following paragraphs describe some examples of courses delivered in this way.

The Integrated Science Curriculum. This video-based course developed at the University of Alabama in Tuscaloosa is distributed through the Satellite Education Resources Consortium (SERC). The video portion of the Integrated Science Curriculum for each grade level is delivered in one weekly broadcast and taped at schools in several states for use at times determined by teacher and school schedules. The three 20-minute video segments are supplemented with both teacher and student texts, an e-mail program for teachers, a science activity kit, sample tests and assignments, and a teacher stipend for consumable supplies needed for the science projects and experiments. Teacher participation follows an intensive, week-long training period that covers the course content, classroom management strategies, alternative assessments, and basic telecomputing skills needed to participate actively throughout the year-long course. Teachers maintain contact with other Integrated Science teachers and with program developers via electronic mail. Programs of this type are usually not time-sensitive and are intended to be used more than once. These courses or programs are generally intended to be completely self-contained, providing teacher and learner study guides, text, assessment instruments, work and study assignments, suggestions for further research, and other supporting ideas or materials.

The TQM Series. Another example of this type of course is a total quality management (TQM) series featuring W. Edwards Deming. The course is geared to the adult learner who would use the tapes on his or her own schedule to fulfill a personal decision or professional assignment to learn more about total quality management. Learners can proceed through the tapes at their own speed, complete and submit assignments, and assess progress in a multitude of ways. Since tapes can be viewed at home, or work or in a classroom situation, the learner can take courses at night or around work or family schedules. Teachers and learners can also augment learning through special educational or thematic video programs like those produced by National Geographic for airing on commercial or public broadcasting stations.

Interactive Video Instruction

Interactive video is an instructionally effective technology that allows more traditional student-teacher interaction than one-way audio, two-way video. The method is also expensive, however, and it requires teachers to modify both their teaching styles and curricula. In addition, it demands creative arrangements for student assessment and management. In a classroom completely devoted to interactive video, teachers and students can see and hear one another, allowing for questions as well as more comprehensive checking for understanding. Costs generally include hardware and software for video and audio capture capabilities (video cameras, lighting, microphones, etc.), transmission systems for the video and audio signals (fiber optics, cable, satellite, microwave), and signal reception equipment (e.g., televisions, monitors, speakers). Delivery of video and audio signals may also require hardware that converts signals to a different format for either transmission or reception. A fairly typical application of this technology would connect two geographically separate locations via the telephone or cable systems for real-time educational course delivery. The students must take the same class at the same time. The system uses a dedicated or leased connection to mimic traditional student-teacher interaction.

Using Technology to Facilitate Communications

Various telecommunications technologies that are not commonly used to deliver instruction can still meet a number of other communications needs that relate to learning. Some examples will illustrate the applications of these models.

Applications of Communications Links

Messaging. One-way audio systems that combine telephone and computer technology are currently being integrated into educational systems in many states as student-teacher and parent-teacher information sources. One example, originated by Jerold Bauch at Peabody College of Vanderbilt University, is the TransParent School Model. This model uses computer-based voice messaging to enhance teacher-parent interaction over the telephone.

Each teacher writes a short script and records a classroom message into a voice mail message box in a site-based computer. All parents can call and hear the message at any time from any touch-tone-based telephone. A school can also record a message to be sent automatically to any or all parents. When the model is fully implemented, parents can call one number and learn everything they need to know to help their children with home learning. They can also find information on meeting schedules, sports events, lunch menus, and special school-sponsored functions.

As of late 1993, this model was used in more than 400 schools in 27 states, leading to increases of up to 800 percent in school-parent involvement. Documented side effects include student grade improvements, increasingly positive parent attitudes toward schools, and even improved attendance rates (Bauch, 1993).

In another new use of an older technology, college instructors are giving tests, providing feedback, and assigning homework using computerized voice mail systems accessed by telephone from home or other remote locations. Future applications will expand voice mail systems to deliver audio messages over computer networks. Audioconferencing systems, which allow many telephone callers to participate in a shared conversation, are being used increasingly frequently for professional development activities, as well. For example, when introducing a new and potentially controversial health curriculum, school-level curriculum coordinators in one district "met" daily via telephone to discuss implementation issues and share individual concerns.

Classroom audiographics. Although classroom uses of audiographics are still limited compared to other forms of computer networking, some applications can be found. For example, in one training program for emergency medical technicians offered at several Florida sites, photographs of disaster scenes are shared with class participants, who must identify problems in procedures dramatized in the photos. Students share responses with all other sites and obtain feedback from both peers and the instructor.

Teleconferencing. Teleconferencing generally employs one-way video and two-way audio, although increasing

Figure 9.4 A Typical Teleconferencing System

From a single origination point, the system carries a broadcast to several downlink sites. Telephone systems establish audio links between remote sites and the program origination point.

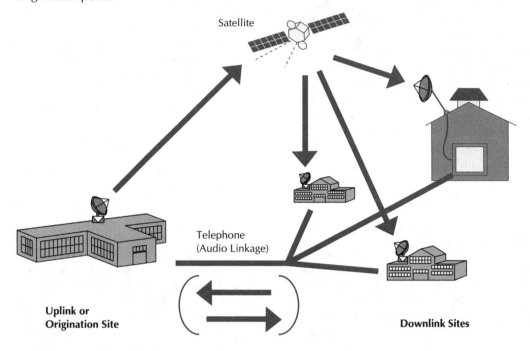

Satellite

Telephone
(Audio Linkage)

Uplink or
Origination Site

Downlink Sites

access to satellite uplink facilities allows more sites to become completely interactive. (See Figure 9.4.) Teleconferences are delivered live (in real time). The main difference between interactive, video-based instruction and teleconferencing centers on the content or subject of the transmission; one-time programs or series of programs on specific topics are standard fare for teleconferences. These sessions frequently serve staff development needs of professionals. For example, in the aftermath of Hurricane Andrew, the Florida Association of Independent Insurance Agents held a teleconference to explain changes to insurance rules and regulations in the state. In another example, North Carolina-based SERVE (SouthEast Regional Vision for Education) held a series of teleconferences in 1992 and 1993 on the integration of technology into classrooms. Each SERVE member state hosted a single program on a different application of technology.

The technology for teleconferencing is itself decades old, dating back to an AT&T demonstration during the 1964 World's Fair in New York. Yet, the method still seems not only new but also somewhat futuristic, perhaps because it has long been a staple in science fiction. Star Trek fans, for example, came to regard videoconferencing as a regular medium of communication between galaxies and lifeforms, making it both familiar and futuristic (Merwin, 1993). Current teleconferencing technologies enable teachers and program producers to integrate

new computer-based presentation technologies, as well as television production techniques for graphics and animation, into videoconferencing programs.

Limitations arise from the need to produce programs in a studio or at a site with its own satellite uplink capability or with access to this capability via a fiber optic or microwave connection. (The uplink facility transmits a signal up to a satellite, which rebroadcasts to desired receiver sites.) This adds costly studio and/or transmission expenses to satellite costs, which can vary considerably ($300 to $600 per hour for educational use in 1993). These costs make total production/transmission costs for teleconferencing somewhat high. In addition, participant response and involvement must be carefully planned and orchestrated so that questions and answers can come from and be heard by all participating sites. For obvious reasons, teleconferencing works best when the general flow of information is one-way; programs requiring high levels of response or interaction may be inappropriate for this medium.

In the future, teleconferencing will not be site dependent. Instead, users will be able to access it via phone and cable service sent directly into homes. They will interact through current phone systems, computers with modems, and fax machines.

Teleconferencing (also known as videoteleconferencing) and videoconferencing sound alike but are two very different technologies. Teleconferencing uses one-way video (e.g., one participant could broadcast to the others) with two-way interaction via the telephone among all participants. Videoconferencing is a point-to-point closed communications system connecting computers that are each equipped with video. Anyone with a satellite dish to pull down the signal can *see* a teleconference, but no one except the participants can see videoconferencing transmissions.

Desktop videoconferencing. The increasing availability of telephone switching technologies allow completely interactive communication between desktop computers. Computers adapted for complete two-way interaction, known as *personal interactive video learning terminals,* are beginning to be found in schools and classrooms across the country. In a typical desktop videoconferencing system, a video camera and microphone at each workstation or learning station allows the learner to be seen and heard by the person (teacher or learner) at the remote site. The system also includes a video monitor and speaker so that the remote person can be seen and heard (Chwelos, 1993). Signals are transmitted using modems and standard telephone lines. Teachers can use presentation technologies (e.g., LCD projection panels) connected to these video-capable computers to expand the number of persons who may observe the on-screen video. Learners may communicate directly with teachers, peers, and experts who use compatible systems.

Current applications of desktop videoconferencing are limited by several factors. First, the cost of transmitting video data over telephone lines is high because of the large amount of bandwidth those signals require. Second, current video transmissions employ analog signals, making them difficult to manipulate. Third, the absence of standards for interoperability limit interaction to others using the same, often proprietary, standards. Fourth, equipment costs interfere with wide-scale integration into schools. However, videoconferencing is on the increase, especially as equipment and line costs become cheaper and resources get easier to use (Jerram, 1995; Walsh and Reece, 1995; Barron and Orwig 1995).

E-mail, discussion groups, and bulletin boards. These three mechanisms are becoming as common as telephone calls. Electronic mail (e-mail) is the most common way to exchange personal, written messages between individuals or small groups. Discussion groups (called *Listservs* on the Internet) feature ongoing "conversations" via e-mail between groups of people who share common interests. If you belong to a Listserv, you send a single message which the system duplicates for everyone in the group. Bulletin boards resemble Listservs, but exchanges are more public; users publish announcements that will be seen by many people at once.

Using Technology to Locate and Integrate Educational Resources

Telecommunications technologies can transform schools into true learning centers with links that bring information resources throughout the world to learners of all ages. Teachers, the only professionals routinely denied phone service in their workplaces, can use networked technology to maintain regular contact with both colleagues and vast amounts of data. They also have access to the problem-solving expertise of both higher education and the larger industrial world, giving instruction a timeliness and relevance not possible in isolated classrooms. Students can access the world's information resources and collaborate on an international scale with others of similar interests and inclinations. Some examples of these activities are discussed in Inset 9.3

The impact of this access extends well beyond the ability to acquire the information and skills needed for economic survival, however. Today's networks and the world's steadily growing information highways endow individuals with the power to broadcast messages to large and diverse numbers of people. These links create intimacy among individuals who may never have met in person, and they give everyone the power of sharing information formerly held by only a privileged few (Cooke and Lehrer, 1993). Thus, the technologies of learning from a distance extend the role of learner to include active participation in the economic, social, and political action of the time.

Through the Internet, for example, some 30 million people in 40 countries (according to 1993 figures) are able to exchange information. Use is restricted only by an individual's access to a computer, a modem, and a password. While much of the communication on the Internet is claimed to result from serious needs for information, the Internet also allows people to build emotional bridges, helping diverse individuals see common interests and purposes not possible in any other widely available medium today. This occurs, in part, because distance learning technologies tend to blur social distinctions; race, gender, disabilities, physical appearances, even social status, lose their significance when all that matters is one's capacity for expression (Leslie, 1993, p. 90). The same technologies that allow for universal access to learning also foster a growing sense of community and relationship with people throughout the world. As a result, distance no longer impedes either learning or a sense of community that extends to individuals who may never meet face to face and to neighborhoods that are never visited.

Some Popular Online Information Sources for Education

Networks and utilities for educators. Anyone who has a computer, telephone line, and modem also has access to a world of information. Like finding one's way around a new city, however, locating specific people or materials in the maze of information can be confusing and

Inset 9.3
Example Telecommunications Projects

Example: Discovering Orlando—Alaska Style. Students in the Dillingham (Alaska) School District began their end-of-the-year trip to Orlando, Florida by establishing a computer link with peers in that city. Planning early for their April jaunt to Orlando, teachers and students at the William "Sonny" Nelson School in Ekwok, Alaska began communicating with other sixth through eighth grade students in Orlando using e-mail. Students continued to exchange information until the trip itself, visited briefly with the Orlando students, and continued their contact via e-mail after the trip. The travelers, primarily native Alaskans, and their new-found friends in Orlando learned about each others' states, customs, and cultures; developed expertise with telecommunications technology; were better prepared to learn from and enjoy their visit; and had a wonderful time, as well.

Example: Global Education Makes the World Smaller. WorldClassroom is a global telecommunications network for K-12 schools. It is available on a subscription basis and can be accessed through computer and modem. WorldClassroom links classes of students around the world to work together on structured activities in several areas:

- **Science.** Curriculum projects in life science and earth science encourage students to "do" science rather than merely

study about it. Students can, for example, collect and analyze water samples for pollutants.

- **Social studies.** Projects linking students from different countries allow comparisons on many levels and topics while expanding perceptions of other cultures and the global nature of today's world.

- **Language arts.** Students can use writing projects to share information, explore cultural diversity, and express differences and opinions. Guest speakers provide opportunities for students to interact with-well known authors and professionals in a variety of fields. A newsletter publishes student work for an international audience.

Standard WorldClassroom features include access to a variety of projects and activities online, access to all participating schools and classrooms worldwide, consultants, and ongoing communications. The program strives to provide a vehicle for students to become more globally aware and tolerant of different cultures. WorldClassroom is provided through Global Learning Corporation. For more information and subscription rates, the Internet address is global@glc.dallas.tx.us (or 1-800-866-4452).

---

time-consuming. Even someone who enjoys the challenge of exploring the new territory will probably need some help at various times. Companies are springing up around the globe to help people locate and use information. Dialog, CompuServ, Prodigy, and America Online are examples of information utilities that provide needed tools and training. Through accounts on one or more of these services, teachers and students can get access to some of the things they need to get started (e.g., e-mail addresses) and to some popular education databases like ERIC. Special networks designed for use in education (e.g., National Geographic Kids Network and the AT&T Learning Network) have also been set up to help teachers and students communicate among themselves.

An alternative may already be available, however, within a school, school district, or local higher education institution. Before investing in an account with an information utility or service provider, teachers may want to inquire locally to see if accounts are available to them through free or lower-cost sources.

The Internet. In the decade of the 1990s, technology began to play its most high-profile role in the history of the planet. Technology spread from technical labs and business offices into the living rooms of people around the world with unprecedented speed. The sudden rise of a multifaceted system called the *Internet* was typical of this groundswell of interest in and use of technology by people in all parts of society.

Educators have become interested in the Internet because everyone seemed to be on it or trying to get on it. It has become one of the most popular mechanisms available for communicating with others and locating information electronically. (See Inset 9.4.) The major benefit of the Internet is the comprehensive nature of its information and services. Once connected, educators and students can use the Internet to exchange messages and files with others anywhere in the world (see Figure 9.5 for a sample Web page). They can

Figure 9.5　White House Web Page

Inset 9.4
A Perspective on the Internet

How Did the Internet Originate? The earliest version of the Internet was developed during the 1970s by the U.S. Department of Defense (DOD) to encourage communications and exchange of information between researchers working on projects in about 30 locations. It was also seen as a way to guarantee that communications among these important defense sites would continue in the event of a worldwide catastrophe such as a nuclear attack. Since these projects were funded by the DOD's Advanced Research Projects Agency (ARPA), the network was originally called *ARPAnet*.

In the 1980s, just as desktop computers were becoming more and more common, funding from the National Science Foundation made possible a high-speed connection among university centers using the same kind of system as ARPAnet had used. By connecting networks together, universities could communicate and exchange information in the same way that DOD's projects had. However, these new connections had an additional benefit. A person accessing a university network from home or school could also get access to any site connected to that network. This connection began to be called a *gateway* to all networks, and what we call the *Internet* was born.

What Is the Internet? Networks connect computers to allow users to share resources and exchange information easily. The Internet has been called the "ultimate network" because it is a network of networks. It is a way for people in network sites all over the world to communicate with each other as though they were on the same local area network. The name means literally "between or among networks."

In order for people on different computers to communicate with each other, each must use a common technical communications system or protocol. Internet exchanges are possible because everyone uses a common communications system called an *Internet Protocol* or IP. Internet Protocol assigns a number designation to each Internet address. To navigate the Internet, you must have an account on a computer connected to it. Most users connect through high-powered computer systems running the Unix operating system. The combination of your account name (called a *user id*) and the Internet-registered name of the computer (called a *host name*) comprises your Internet address. For example, look at the components of the following Internet address:

darnell@freenet.fsu.edu

- The name *darnell* represents the user id.
- The @ symbol sets off the host name, in this case, *freenet.*
- The periods are essential parts of the address as well, because they separate key elements of the host name.
- Usually the institution or company, in this case *fsu,* follows the machine name.
- Then *edu* identifies the type of company or institution, in this case, an educational organization. Other types are org (organization), com (company), and gov (government).

Users in countries outside the United States append two-character codes to indicate their origins. For instance, *uk* identifies England and *au* is Australia. A company in Virginia known as INTERNIC (Internet Network Information Center) serves as a registry for Internet addresses, ensuring that no two Internet addresses or IP numbers are identical. Think of it as a database similar to that maintained by the Social Security Administration, which ensures that no two people have identical Social Security numbers.

To use the Internet in its native form requires familiarity with Unix and its commands. Fortunately, people have to know very little Unix these days to use the Internet effectively. Software interface tools like Mosaic, HGopher, and Eudora (called *clients* because they still work through host computers) have taken Internet use out of exclusive control of technical people and placed it in the hands of anyone with a personal computer. Programs like Netscape are designed to act as user-friendly, graphic interfaces between Unix and those who use the Internet. (Computer programs such as Telnet, FTP, World Wide Web (WWW), and Gopher were developed on Unix systems and still require some knowledge of Unix to use.)

What Can I Do on the Internet? An Internet user can perform five basic tasks: sending messages via electronic mail (e-mail), logging into remote systems with Telnet, sending and retrieving files by FTP, reading and posting information to Usenet, and browsing for information using Gopher and WWW.

- **E-mail.** Sending and receiving electronic messages (e-mail) is still the most popular use of the Internet. Communication between two individuals on the Internet is as simple as exchanging Internet addresses and using a mail program or client like Eudora to send and receive messages. Once using e-mail, people can exchange messages across the state or around the world at little or no cost. E-mail can be further used to participate in Listservs, which are discussion groups that communicate through computers set up specifically to distribute each member's message to others who register with the Listserv. For instance, a Listserv

Continued

known as Nettrain allows individuals who train people to use the Internet to share ideas and ask questions of others around the globe.

• **Using Telnet.** Telnet is a procedure that allows a user to log on remotely to a computer to use as they would at its location. One of the most common uses of Telnet is to log on to library systems and use their resources. Thousands of libraries offer information for exploration through Telnet.

• **FTP.** This can stand either for *File Transfer Program* or *File Transfer Protocol*. Sometimes you want to do more than just look at information; you want to get a whole report or maybe gain access to a neat application program like the Internet clients mentioned previously. FTP allows the user to log on to computers specifically set up to warehouse file data (known as *FTP sites*). Files such as application programs are stored on these systems as a service to other Internet users who want to get to them. *Get* is a command used in receiving files and *put* is a command for sending files to an FTP site. These commands are transparent to a user if they employ a client application like the Macintosh FTP client Fetch. Those who could cope with protocols like KERMIT and XMODEM used to dominate FTP. FTP allows users to transfer files to the machine they are currently using without bothering with modem protocols. Users can exchange files regardless of the program or language used to prepare them. What sorts of files are available for transfer (downloading)? Where do you get them? Just reading the newspaper these days or talking to colleagues can uncover dozens of FTP sites and files. A method for searching a great many FTP sites for files by name is called *Archie*. Archie is software designed to allow searches of a database index of all accessible FTP sites that it knows about. The user provides a program or file name and Archie searches the FTP sites to help find the file.

• **Usenet.** This has been referred to as "the world's newspaper." Its categories, referred to as *newsgroups,* divide the wealth of user-provided information into manageable topics. Some common newsgroups are rec.arts.sports and soc.culture. There are many more newsgroups, both logical and irrational (e.g., a cult favorite of 1994: alt.barney.die.die.die.die where people discuss topics related to killing the dinosaur character in a popular children's show). Because the information is user-provided and, in most cases, unmoderated, anything and everything is possible.

• **Web browsing.** Looking around for information on the Internet—either for a specific purpose or just to see what is out there—is called *browsing*. One browses with tools like Gopher and web browsers. Gopher uses a menu-based system to allow users to select information retrieval options. Web browers like Netscape and Mosaic allow users to move freely and effortlessly from one site to another through a text-based linking system known as *hypertext*. These systems make using the Internet as easy as point-and-click, because they include access to the other basic ways of using Internet.

What Can Educators Do on the Internet? As connections to the Internet become easier and as more resources of interest to educators become available, the Internet should become a major resource in linking for learning. In addition to a wealth of helpful information and trivia on a variety of topics, Dyrli (1994)

described several other items now available on the Internet that are particularly interesting to teachers:

• **Teacher discussion groups.** Some sites (e.g., KIDSPHERE at the University of Pittsburgh and EDNET at the University of Massachusetts) have set up online discussion groups in which teachers can exchange advice and information on a variety of topics of mutual interest.

• **Services for educators.** Some agencies have set up networks designed specifically to help teachers locate information on topics they are teaching (e.g., the University of Massachusetts/Amherst K-12 Information System, NASA Spacelink in Huntsville, Alabama).

Instructional uses for the Internet seem to be growing as fast as the Internet itself. A recent issue of the *Florida Technology in Education Quarterly* illustrates some of the ways it can be useful for teaching and learning:

• **Research projects.** Students locate information on the Net to answer specific questions on topics of interest and/or to develop summaries based on what they find. These activities can lead to individual or group classroom reports.

• **Co-development projects with other classes.** Students get together online to plan and produce joint products such as newsletters, magazines, travel booklets, collections of poems or stories, historical summaries based on family interviews, or exchanges of local cultures (e.g., food, local sayings, clothing, holidays).

• **Mentoring activities.** Students link up with experts who help them learn about topics and/or find resources for research and development projects.

• **Monitoring current events.** Many national and international events are reported via news services connected to the Internet before the news reaches media newsdesks. As part of social studies classes, some students can keep track of events in a certain region of the world or related to a chosen topic as they happen.

• **Penpal activities.** Students write to each other informally and/or as part of class writing assignments.

• **Teacher training opportunities.** Workshops, tutorials, and full courses are becoming available via the Internet.

In addition to Dyrli's (1994) article, a good resource for finding out more about Internet uses for education is through *Learning and Leading with Technology* (formerly *The Computing Teacher),* which has a monthly column called "Mining the Internet." This column has tips on late-breaking developments and products, as well as lesson ideas and reports of how educators are using the Internet successfully in their classroom activities. A special issue of Electronic Learning (1996) also has good resource lists and suggestions.

How Can Teachers Get Connected? Today, people can find two differnt types of gateways to the Internet. They can get access either through a direct connection already established by an organization such as a university or school district, or they can get accounts by subscribing to national, regional, or local services with their own connections:

Continued

- **Direct connection.** Many organizations with direct connections allow any of their members to use these connections, usually free of charge. For example, if a university has an Internet connection, anyone affiliated with the university (e.g., faculty, staff, students) may get an account that becomes, in effect, the person's address on the Internet. Although the members have the illusion that their connections are free, they are actually paid for by the organizations. People can also access the Internet via local Freenets (if they are available), but fewer software tools are available to make access easier. Also, use is often very slow because of the number of people trying to take advantage of it.

- **Service providers.** A variety of service providers have sprung up to sell accounts to those who do not have free access in some other way. Perhaps the best-known are international companies such as America Online, CompuServ, and Prodigy, but many regional and local companies also sell access to the Internet. Users of these accounts usually pay by the amount of time they spend connected. Service providers usually give users varied software tools in addition to user ids that act as addresses at which information may be sent and received.

New Issues for a New Era in Technology. Teachers, students, and citizens do not gain unprecedented freedom of access to information through the Internet without cost, even if they do not pay for access. Many educators and parents are frightened by the lack of limits placed on what goes on the Internet and what children can see if not strictly supervised. Elmer-DeWitt warns of the increase of both "cyberpunk" (1993) and "cyberporn" (1995). As quickly as these issues arose, software developers raced to develop "firewalls" or ways of blocking access to adult and X-rated Internet materials. One such program by a Vancouver company, Net Nanny, allows a parent to monitor everything going through the computer; it is programmed to shut down the connection if it encounters certain key phrases.

As use of the Internet increases, ethics in technology use becomes a topic of more-than-passing interest to nearly everyone from parents to publishers. Ethical use of information resources must be considered part of the fundamental guidance given to anyone who wishes to travel this new electronic pathway.

Source: Dyrli, O. (1994). Riding the Internet schoolbus: Places to go and things to do. *Technology and Learning, 15* (2), 32–40; Author. (1996, March/April). Networking and the Internet. *Electronic Learning 15*, 5, 1–13; Elmer-DeWitt, P. (1993, February 8). Cyberpunk! *Time, 141* (6), 58–65; Elmer-DeWitt, P. (1995, July 3). On a screen near you: Cyberporn. *Time, 146* (1), 38–45; Harris, J. (1994). Teaching teachers to use telecomputing tools. *The Computing Teacher, 22* (2), 60–63; Honey, M. (1994). NII roadblocks: Why do so few educators use the Internet? *Electronic Learning, 14* (2), 14–15; Krol, E. (1994). *The whole Internet user's guide and catalog.* Sebastopol, CA: O'Reilly and Associates; Monahan, B., and Dharm, M. (1995). The Internet for educators: A user's guide. *Educational Technology, 35* (1), 44–48; Truett, C. (1994). The Internet: What's in it for me? *The Computing Teacher, 22* (3), 66–68. Photo courtesy of Harry and Jennifer Buerkle.

also use it to locate information from virtually any place in the world. The person or site with the desired information need only be connected to the Internet and willing to provide an online listing of available resources.

Educational uses of the Internet are not yet as widespread as its popularity. Honey (1994) observes that the vast majority of educators still do not tap its potential. The same logistical problems that hamper use of any telecommunications capability (e.g., lack of training, time, and access to resources) also limit educators' use of the Internet. In addition, Honey noted, "The Internet was not designed with the K-12 community in mind. ... [T]eachers' enthusiasm over the potential of the Internet belies the reality of the resources currently available" (p. 14).

However, as educators can handle connections to the Internet more easily and as they can find more resources of interest to them, the Internet should become a major resource in linking for learning.

Types of Teaching and Learning Activities with Telecommunications

Some of the most exciting telecommunications or telecomputing applications call for student collaboration via technology to address real problems or communicate with people in other cultures throughout the world. In a series of articles written for *The Computing Teacher,* Judi Harris described three general types or models of educational telecomputing activities: interpersonal exchanges in which students communicate via technology with other students (or with teachers/experts), information collections which provide data and information on request, and student-oriented, cooperative problem-solving projects (Harris, 1994). Some specific ways to integrate these kinds of models include:

- **Electronic penpals.** This is the simplest instructional activity in which telecommunications plays a role. Each student is linked with a partner or penpal in a distant location to whom the student writes letters or diary-type entries. This kind of writing assignment has been shown to be very motivating to students (Cohen and Riel, 1989). Writing to communicate something to real people, rather than writing for teacher evaluations, encourages students to write more and with better grammar, spelling, and usage. This makes electronic correspondence an ideal activity for English and/or writing classes. Networking in this way also eliminates social bias regarding gender, race, age, and physical appearance. Without social and cultural cues to color interactions, two people who may never have communicated to one another in person are able to build positive relationships electronically. Though not a substitute for other, face-to-face activities in multicultural education, e-mail can provide an important way to begin building awareness of and appreciation for other cultures.

- **Individual and cooperative research projects.** Students can research a problem online working either by themselves or in groups gathering information from electronic and paper-based sources. These research activities usually culminate in a presentation to the class and subsequent discussion of the findings. For example, students may be asked to tap various online databases for articles and reports on contributions by the space program to people's lives and modern culture. They may supplement this information with online conversations with experts they locate on the Internet. When the research is completed, the class report might include actual examples of these contributions as well as summaries delivered via multimedia or presentation software.

- **Electronic mentoring.** Dyrli (1994) refers to subject matter experts who volunteer to work closely with students online as "electronic mentors" (p. 34). One source of aid in these activities, the Electronic Emissary Project at the University of Texas, links up classes across the country with mentors on topics ranging from Greek literature to life support in space (p. 35).

- **Parallel problem solving.** Through technological links, students in a number of different locations can work on similar problems (Harris, 1994). They solve the problem independently and then compare their methods and results or build a database with information gathered during the activity.

Students in one school collaborating on a joint air quality project were surprised when they discovered through comparisons that their classroom air quality failed to meet minimum public health standards!

- **Electronic field trips.** As Inset 9.2 explained, an electronic field trip in its simplest form fills classroom screens with visual images of a place considered to offer some educational value and to which students would not routinely be able to travel. Virtual trips are designed to explore unique locations around the world and, by involving learners at those sites, to share the experience with other learners at remote locations. (Inset 9.5 describes an example.) Trips may entail only video programs or they may include prepared curriculum guides, suggested preparation and follow-up activities, and discussion

---

Inset 9.5
The Jason Project

The Jason Project is the world's first venture to broadcast live color images from an actual site of scientific exploration. Through advanced satellite technology, underwater robotics, a two-way audio line, and television screens, viewers at 24 participating museums in the United States, Canada, England, and Bermuda can view the exploration site, interact with the ship's crew, and even steer a remotely operated robot vehicle. Participants at the museums share the experience with thousands of other viewers who "join" the expedition through local cable television, closed circuit television, microwave links, and other broadcasting technologies. (See Figure 9.6.)

Since 1989, the Jason Project has beamed back interactive scientific explorations from the Mediterranean Sea, the bottom of Lake Ontario, the Galapagos Islands, off the coast of Baja California, and Belize—a coastal Central American country roughly the size of Massachusetts. In 1993, the Jason Foundation and Mind Extension University teamed to create the Jason V Classroom Network, designed to prepare teachers and students for the expedition. Programs were aired as part of ME/U's network programming before, during, and after the Belize expedition, giving more teachers the opportunity to make the Jason Project an ongoing part of their classrooms.

The Jason Project has been called a "provocative vision of the future" because it reverses the dynamics of the traditional classroom; instead of transporting the teacher to the students, the students are brought to the teacher. This allows teachers and students to participate in activities that might not otherwise be possible or practical.

The Jason Project is inspired by the advanced technology for underwater exploration found in the ARGO and Jason systems developed by the Woods Hole Oceanographic Institute. The ARGO system is a series of television cameras and sonars that transmit both wide angle and close-up shots of the ocean bottom. It was used in the discovery of the *Titanic* in 1985. The Jason is a remotely operated vehicle capable of negotiating the ocean floor, collecting samples, and obtaining information useful to later human explorers. Together, ARGO and Jason represent a significant improvement in the rate and quality of deep underwater exploration.

Founded by Dr. Robert Ballard, the Jason Project was developed as a way to introduce students to the thrill of remote travel and scientific discovery. Both the curriculum and the expedition are designed to spark student interest in science, technology, and social studies. Through live participation via satellite-delivered images and two-way audio interaction, students get a taste of the varied and exciting fields that have direct impact on their lives. In addition, the local teachers and students who are involved have the opportunity to pose questions and suggest experiments that they can then see answered or carried out using the scientific method. This approach combines the power and reach of the media with the drama and experience of scientific exploration.

Each expedition is planned carefully over a period of months, and teachers and students prepare for the expedition by studying a special curriculum prepared by the National Science Teachers Association. Five 1-hour shows are broadcast to the receiving sites each day from Monday through Saturday during the 2-week expedition. During the expedition, students and teachers carefully monitor its progress, communicate through electronic mail, and participate by completing class activities and assignments. Although the Jason Project is aimed primarily at upper elementary and high school students, the general public also benefits from the Jason experience.

The expedition is sponsored by the Jason Foundation for Education, a partnership of private industry, scientific research facilities, educational organizations, and museums. Sponsors include the U.S. Navy, Electronic Data Systems (EDS) Corporation, Turner Broadcasting, The National Geographic Society, Cray Research, Inc., the National Science Foundation, the National Science Teachers Foundation, and the National Council for Social Studies.

The use of electronic field trips like the Jason Project to augment curriculum has grown considerably in the last few years, in part because of tremendous economic pressures on schools, and because of time, safety, and insurance issues associated with school-sponsored travel. Electronic field trips also enable hospitalized, home-bound, and incarcerated learners to participate in a meaningful way with peers in real learning situations and to observe the reactions and interactions of other learners.

questions to help correlate experiences with specific curriculum objectives. Learners interact with peers via telephone, computer, fax, and mail. Some typical examples of electronic field trips include a visit to the Great Wall of China, a walking tour of Washington, D.C., an exploration of an archeological dig, and a trip to a state capitol during a legislative session. Through electronic field trips, many students imagine themselves leaving their neighborhoods and cities; they experience in some fashion the excitement and wonder of new places and faces—and they learn from the experience.

• **Group development of products.** Teachers have developed many variations on this approach. For example, students may use e-mail to solicit and offer feedback on an evolving literary project, sometimes involving advice from professional authors. Students may work independently toward an agreed-upon goal, each student or group adding a portion of the final product. This is sometimes called *chain writing*. For example, two very different Missouri schools linked electronically to write and videotape a play. At one school, students developed a list of characters and a general plot outline and wrote Scenes 1 and 2. Students at the other school selected the topic for the play; developed the personalities, physical characteristics, and backgrounds of the characters; and wrote Scenes 3 and 4. Together the students developed the final story line and produced the play, exchanging the videotape portions that they produced independently.

• **Simulated activities.** In this type of problem-solving exercise, students participate in structured activities in which they carry out specific duties or responsibilities on which some aspect of the project outcome depends. One example is the Simulated Space Shuttle Program, in which different schools prepare for various missions that will take place during a specified "launch period." Success of the simulated mission depends on a series of smaller successes and appropriate student participation.

• **Social action projects.** Online communication gives students access to people in other countries that can support social problem solving. In this type of project, students are responsible for learning about and addressing important global social, economic, political, or environmental conditions. For example, students collaborating on a peace project write congressional representatives to voice concerns and present their viewpoints.

Sample lessons for each of these types of learning activities appear at the end of this chapter. Many other examples based on these categories of integration strategies are given in Armstrong's book *Telecommunications in the Classroom* (1995).

Getting started: Some ideas. Teachers can get involved in linking to learn in a variety of ways, ranging from taping regular television broadcasts for occasional classroom use to delivering instruction through completely interactive voice, video, and data transmission technologies. While many teachers approach innovative technologies cautiously, several applications of technology allow even these cautious or inexperienced teachers to

successfully integrate remote resources into their classrooms:

• Teachers can use school fax machines to pose student questions to politicians, scientists, business leaders, authors, etc. Questions and answers can generate discussion, develop topics for term papers or research, or build classroom databases on subjects of interest to students.

• Classes from different school districts, states, or countries can interact via letter, fax machine, computer and modem, or telephone conference call to conduct joint research projects, share differences in cultures, or simply exchange ideas and develop relationships.

• Teachers and students can visit local radio and television studios, cable television franchise operations, telephone service providers, police communications units, etc. to explore the latest transmission technologies and everyday applications.

• Teachers, parents, and local business people can create partnerships that center on some technology-based interaction. For example, a local telephone company may provide a videophone and long distance service for temporary use during a cooperative class project with a distant school. Parents can help fund a subscription to an online computer service like Classroom Prodigy.

• Many large companies like Wal-Mart and Sears are using remote technologies to train their employees. Teachers and students may visit and even explore using the facilities and equipment at these sites.

Figure 9.6 The Jason Project Transmission System

As Jason explores the ocean floor, it sends video images through fiber optic cable to its base ship. An uplink on the ship sends the audio and video signals to a satellite, which relays them to a receiving and broadcasting station in Atlanta for the addition of prerecorded sequences. The signal is then uplinked to one or more satellites and relayed again to receivers within each satellite's particular footprint, where students watch the whole process with only a few seconds' delay.

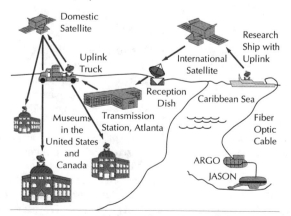

Source: U.S. Congress, Office of Technology Assessment's Linking for Learning: A New Course for Education (1989).

• On live radio talk shows, local leaders might accept student questions. Student representatives can be physically located at the station or in the school's administrative office near the telephone while the entire class listens to a program.

• Teachers with access to video equipment can help students produce their own video programs. In this way, absent classmates, parents, other classes, and schools can participate in some part of students' instruction, document the steps to a science experiment, or simply prepare a creative work on an assigned topic. Teachers can also videotape student assessments to share with parents and document student progress.

Opportunities are endless and seem limited only by teacher, parent, and student imagination. The real challenge for teachers seems to be taking the first step. Fortunately for all, radio, telephone, fax, and television (VCRs and videocameras) are all technologies that seem comfortable and reasonable as starting places for integrating linking technology.

Sample Linking-to-Learn Strategies

Strategy 1: A Simulated Activity
Developed by: Jamie Fagan, Christ Church Grammar School (Claremont, Perth, Western Australia, Australia)
Level: Grades 6–7
Purposes: To teach students about finding positions using a real situation, accessing information, geography, math, and lifestyles of people from different countries

Instructional Activities. Students were assigned a specific yacht participating in an annual race around the world. They were responsible for accessing computer databases of information giving latitude, longitude, elapsed time, and place. Each morning during the race, a student or group of students accessed a remote database and downloaded information on the assigned yacht. Students kept daily records of "their" yacht's progress from the beginning of the race until the boats arrived at a nearby harbor. Students ultimately presented each real skipper with a map of that yacht's progress and had the opportunity to discuss the race in person with some of the crew.

The activity incorporated several curriculum areas. Students learned how to use telecommunications technologies, how to locate positions using latitude and longitude, and how to calculate distance by comparing positions from one day to the next. In addition, students used the situation to develop writing assignments. Students devoted at least 30 minutes to 1 hour a day to this activity.

Problems to Anticipate. Students' interest in the project developed slowly until they demonstrated a yachts' movement and established some ownership in the individual yachts. Pictures of yacht and crew, news stories, and the increasing intensity of competition all helped to engage students. Real-time communication with the crews would have added another exciting dimension and allowed more culturally interesting interaction.

Strategy 2: Partners Online
Developed by: Jan Sims, Hickory Grove Elementary School (Bloomfield Hills, Michigan) and Bruce Simms, Della Lutes School (Waterford, Michigan)
Level: Grades 3–5
Purposes: Students become aware of and comfortable with technology integrated into the curriculum; students develop intergenerational relationships.

Instructional Activities. Students are paired with senior citizens for communication through telecommunications: messages are sent and received through local bulletin board systems. Students began the activity by becoming familiar with the computer keyboard and certain key functions. They then began working on questions to develop biographies which they subsequently shared with other students as well as their older partners. Instructors prepared the senior citizens and assisted them in logging on and off, entering passwords and user names, downloading messages, answering student questions, and sending and receiving files. Students used their questionnaires to interview the senior citizens to stimulate communication and obtain information needed to develop biographies. At the end of the activity, instructors arranged for students to meet their partners and present the biographies to them. Academic subjects were integrated into many parts of this activity. Students spent time on language arts, history, math, geography, and science skills. Intergenerational relationships developed through this activity were meaningful to both sets of partners. Teachers observed an increase in student's self-esteem, leadership, cooperative learning skills, and academic achievement.

Continued

Problems to Anticipate. The external partners needed much of the same support provided by the classroom student, making careful preparation and teacher accessibility an important part of the activity.

Strategy 3:	Group Development Activity/Information Exchange
Developed by:	Hillel Weintraub
Grade Level:	Grades 10–11
Purposes:	Students use telecommunications to explore the concept of humor from personal, cultural, and intercultural perspectives.

Instructional Activities. Within individual classes, students begin to search for things that they find personally humorous and discuss what and why they find particular situations, jokes, or people funny. In class and online with other schools/classes involved in the project, students then begin to look for cultural features of humor, discussing and sharing insights using e-mail as well as print materials prepared by the students. As the activity progresses, classes are encouraged to create products to send to other schools: original plays, interviews, original drawings, or jokes. Several final activities were suggested, but not planned in advance, giving students and teachers the opportunity to select their own culminating student, class, intraschool, or inter-school activities. This activity helped to deepen student understanding of humor and cultural differences in humor as well as to provide specific practice with computer interaction and collaborative project development. (Sharing information that is interesting to students is also an excellent international telecommunications activity.)

Problems to Anticipate. This activity deals with many kinds of materials, most of which cannot be sent online, but the conversations about the materials can be exchanged. For example, drawings and films need to be sent via regular mail while text or verbal humor can be sent online. In addition, some guidelines need to be established about what is appropriate humor within the context of school and community. (Of course, this provides yet another opportunity for lively debate!)

Strategy 4:	Electronic Penpals: Wired to the World
Developed by:	Patricia Weeg, Delmar and East Salisbury Elementary Schools (Maryland)
Level:	Grade 5
Purposes:	Students used the Maryland State Department of Education Network (METNET) and the Internet to compare customs from different countries and learn how similarly-aged children live, learn, and play.

Instructional Activities. Students compared customs in other lands by using their computers and modems to ask children in other countries how specific customs were alike and different. For example, in Peru, the tooth mouse rewarded Juan with "soles" for his lost tooth, enabling him to buy a "GI Joe" toy, which the children online in Iceland had never heard of before. In another activity, students translated their heights and weights using the metric system, figuring average heights and weights, determining who was taller or shorter, etc. Other students joined in an activity in which children at participating schools worldwide would stop once an hour for an entire waking day to write down what they were doing—learning about time zones, seasons, climate, bedtime rituals, television, attitudes toward homework, and so on.

Problems to Anticipate. This project was made possible because of the support of the METNET as a gateway to the Internet. Schools without this kind of support must identify resources for communications-related expenses and network services.

Strategy 5:	Social Action: The Trashy Project
Developed by:	Joyce Kutney, Orange Glen School (Escondido, California)
Level:	Grade 5–7
Purposes:	To gather and compare data regarding the collection and dispersion of trash to see if one region conserves better than others and to assess grade level differences and day of the week differences

Instructional Activities. Students collected information on the number of ounces of paper trash collected each day in their classroom. In addition to collecting data on the trash, students recorded the number in attendance each day to analyze the relative proportion of trash to the number of students. The data collected were then electronically transmitted to a central classroom where the information

Continued

was pooled for distribution. The students hypothesized about what the data would show and discussed local and collective data.

Strategy 6: Parallel Problem Solving: Integrated Science
Developed by: Star Bloom and Larry Rainey, University of Alabama, Integrated Science Project
Level: Teachers
Purposes: Students and teachers share science discovery projects via competition between classrooms in several states with results being shared online and via broadcast as part of the regularly scheduled Integrated Science curriculum segment for a specific grade level. Networked teachers discuss challenges, resources, and results while the activities are in progress.

Instructional Activities. Varied science projects are assigned via text and television broadcast and then tackled by individual students or groups of students. Teachers engage in discussion with other science teachers using the curriculum and with project developers to clarify activities, check results, and share data collected. Student projects are sometimes sent to project headquarters for exhibition or as part of regular competitions among schools. Teachers are also supported online with assessment materials and strategies, suggestions for further activities or alternate assignments, materials lists, and references for more information or research activity. Although students themselves are only beginning to use the telecommunications technologies available to the Integrated Science project to gather or share information, the teacher connectivity and ongoing online support make this project an exciting example of teacher collaboration and support using broadcast video and telecommunications technologies.

Source: Fagan, J. (1993). Telecomputing activity plan contest. *TIE News, 5* (2) (Winter), p. 14; Simms, J., and Simms, B. (1994). The electronic generation connection. *The Computing Teacher, 21* (7) (April), 9–11; Harris, J. (1994). Information collection activities. *The Computing Teacher, 21* (6) (March), 32–33; Making connections with telecommunications (1993). *Technology and Learning Magazine, 13* (8) (May/June), 33–34; Kelly, M. G., and Weiber, J. H. (1994). Telecommunications, data gathering, and problem solving. *The Computing Teacher, 21* (7) (April), 23–25.

Exercises

Exercise 9.1. Fill in the correct term for each of the following linking features or resources:

_____ 1. Communications device connected to a computer that allows one computer to send signals across telephone lines to another computer

_____ 2. A digital transmission system that carries all types of data on a single line in a fraction of the time it took on analog lines

_____ 3. The rate of speed at which data are transmitted to and from a computer

_____ 4. A two-way audio (telephone) link that could allow one teacher to talk with several students in different locations at the same time

_____ 5. A real-time, long distance video and audio conference among several participants in different locations

_____ 6. A network of networks that interconnects sites all over the world

_____ 7. Another term for communications at a distance, also known as *telecomputing*

_____ 8. The most common way to send messages to individuals electronically

_____ 9. A way to broadcast or post a message that will be accessible to a whole group of people

_____ 10. Linking students in distant locations so they can write to each other

Exercise 9.2. Identify the kind of linking option(s) or resource(s) appropriate for each of the following educational needs:

_____ 1. A group of remote, rural villages have a few students each who want to learn programming languages, but no teacher is available locally.

_____ 2. A teacher must visit, teach, and assess ten home-bound, physically disabled students. It is time-consuming and difficult to reach them all.

_____ **3.** Students would like to practice their German language skills by doing research at a German library or university and reporting on their findings in German, but travel costs for such activities are prohibitive.

_____ **4.** A classroom of students in Alabama would like to correspond with a classroom of students in the Newly Independent States (NIS) to learn about each other's countries and cultures. Snail mail is both slow and expensive.

_____ **5.** A writing teacher would like to motivate her students to write more by having them correspond with people they have never met in a country of interest to them.

References

Armstrong, S. (1995). *Telecommunications in the classroom.* Palo Alto, CA: Computer Learning Foundation and Eugene, OR: ISTE.

Barron, A., and Orwig, G. (1995). Digital video and the Internet: A powerful combination. *Journal of Instruction Delivery Systems, 9 (3),* 10–12.

Bauch, J. (1993). Telephones in classrooms: An unsung technology revolution. *Inventing Tomorrow's Schools, 3 (4),* 5–7.

Bruder, I. (1993). Technology in the USA: An educational perspective. *Electronic Learning, 13 (2),* 20–25; 27–28.

Bybee, D. (1996). Congress passes telecommunications act of 1996. *ISTE Update, 8 (5),* 1.

Bybee, D. L. (1993). *Comments on S. 1040 "Technology for Education Act of 1993."* Alexandria, VA: International Society for Technology in Education.

Chwelos, G. (1993). Telepresence: Technology for interactive distance education. *Education at a Distance, 6 (9),* J7–J10.

Clark, R. E. (1983). Reconsidering research on learning from media. *Review of Educational Research, 53 (4),* 445–459.

Cohen, M., and Riel, M. (1989). The effect of distant audiences on children's writing. *American Educational Research Journal, 26 (2),* 143–159.

Cooke, K., and Lehrer, D. (1993). The Internet: The whole world is talking. *The Nation, 257 (2),* 60–64.

Davidow, W. H., and Malone, M. S. (1993). *The virtual corporation.* New York: HarperCollins.

Dozier-Henry, O. (1995). Technology and cultural diversity: An uneasy alliance. *The Florida Technology in Education Quarterly, 7 (2),* 11–16.

Dyrli, O. (1994). Riding the Internet schoolbus: Places to go and things to do. *Technology and Learning, 15 (2),* 32–40.

Education Imperative (1992). *The Present Futures Report, 2 (9),* 1–8.

Fagan, J. (1993). Whitbread Yacht Race Plotting Project. *T.I.E. News, 5 (2),* 14.

Fox, J., Loutsch, K., and O'Brien, M. (1993). ISDN: Linking the information highway to the classroom. *Tech Trends, 38 (5),* 18–20.

Gallagher, W. (1993). *The power of place.* New York: Poseidon Press.

Gardner, H. (1991). *The unschooled mind: how children think and how schools should teach.* New York: Basic Books.

Garten, E. D., and Hedegaard, T. C. (1993). Computer conferencing technology and online education: New challenges for assessment and evaluation. *Education at a Distance, 7 (5),* J8–J12.

Gilder, G. (1989). *Microcosm.* New York: Simon & Schuster.

Gilder, G. (1992). *Life after television.* New York: W. W. Norton.

Gumbert, E. B., and Spring, J. H. (1974). *The superschool & the superstate: American education in the twentieth century, 1918–1970.* New York: John Wiley & Sons.

Hansell, K. (1992). Satellite education networks—growing and changing. *Via Satellite, 7 (11),* 32–34.

Harris, J. (1994a). Opportunities in work clothes: Online problem-solving project structures. *The Computing Teacher, 21 (7),* 52–55.

Harris, J. (1994b). Teaching teachers to use telecomputing tools. *The Computing Teacher, 22 (2),* 60–63.

Hawkins, J., and Macmillan, K. (1993). So what are teachers doing with this stuff? *Electronic Learning, 13 (2),* 26.

Herffernan-Cabrera, P. (1992). From there to there with K-12 distance learning. In P. S. Portway and C. L. Lane. *Technical guide to teleconferencing and distance learning.* San Ramon, CA: Applied Business teleCommunications.

Hezel Associates (1994). *Educational telecommunications: The state-by-state analysis 1994.* Syracuse, N.Y.: Author.

Holmberg, B. (1981). *Status and trends of distance education.* New York: Nichols.

Honey, M., and Henriquez, A. (1993). *Telecommunications and K-12 educators: Findings from a national survey.* New York: Center for Technology in Education, Bank Street College.

Honey, M. (1994). NII roadblocks: Why do so few educators use the Internet? *Electronic Learning, 14 (2),* 14–15.

Jerram, P. (1995). Videoconferencing gets in sync. *New Media, 5 (7),* 48, 50–55.

Johnston, J. (1987). *Electronic learning: From audiotape to videodisc.* Englewood Cliffs, NJ, Lawrence Erlbaum Associates.

Kearsley, G., Hunter, B., and Furlong, M. (1992). *We teach with technology.* Wilsonville, OR: Franklin, Beedle, and Associates, Inc.

Keegan, D. J. (1983). On defining distance education. In D. Sewert, D. Keegan, and B. Holmberg (Eds.). *Distance education—international perspectives.* New York: St. Martin's Press.

Kelly, M. G., and Wiebe, J. H. (1994). Telecommunications, data gathering, and problem solving. *The Computing Teacher, 21 (7),* 23–26.

Krol, E. (1994). *The whole Internet user's guide and catalog.* Sebastopol, CA: O'Reilly and Associates.

Leslie, J. (1993). Kids connecting. *Wired, 1* (5), 90–93.

Making connections with telecommunications (1993). *Technology & Learning, 13* (8), 33–34; 36.

Meier, D. W. (1992). Drop the false image of education's golden past. *Utne Reader, 61*(January/February), 80–82.

Merwin, A. (1993). Videoconferencing goes to work. *New Media, 3* (11), 59–64.

Monahan, B., and Dharm, M. (1995). The Internet for educators: A user's guide. *Educational Technology, 35* (1), 44–48.

Negroponte, N. (1995). Being digital. New York: Alfred A. Knopf.

Oates, R. (1995). Dishing it out: Buying a satellite receptor as technology begins to change. *Electronic Learning, 14* (4), 34–37.

Perelman, L. J. (1992). *School's out.* New York: Avon Books.

Perelman, L. J. (1993). Hyperlearning and the new economy. *Technos: Quarterly for Education and Technology, 2* (4), 8–11.

Portway, P. S., and Lane, C. (1992). *Technical guide to teleconferencing and distance learning.* San Ramon, CA: Applied Business teleCommunications.

Purcell, J. D. (1993). Designing new learning environments. Autodesk. *Supplement to Technological Horizons in Education Journal, 21,* 2–3.

Quality Education Data, Inc. (1993). *Technology in public schools 1992–93.* Denver, CO: Author.

Ringstaff, C. (1993). Trading places: when students become the experts. *Apple Education Review, 3.*

Roblyer, M. (1992). Electronic hands across the ocean: The Florida-England connection. *The Computing Teacher, 19* (5), 16–19.

Roblyer, M., and Darnell, L. (1995). What is the Internet and why are Florida teachers saying such wonderful things about it? *The Florida Technology in Education Quarterly, 7* (3), 14–27.

Russell, T. L. (1992). Television's indelible impact on distance education: what we should have learned from comparative research. *Research in Distance Education, 4* (4), 2–4.

Russell, T. L. (1993). *The "no significant difference" phenomenon as reported in research reports, summaries, and papers.* Raleigh: Office of Instructional Telecommunications, North Carolina State University.

Sauer, E. (1994). Creative collaboration online. *The Computing Teacher, 21* (7), 38–40.

Scholosser, C., and Anderson, M. (1994). *Distance education: Review of the literature.* Washington, DC: AECT Publications.

Schramm, W. (1977). *Big media, little media, tools and technologies for instruction.* Beverly Hills, CA: Sage Publications.

Sheingold, K., and Hadley, M. (1990). *Accomplished teachers: Integrating computers into classroom practice.* New York: Bank Street College.

St. Onge, J. (1992). Distance learning. *Curriculum Product News, 26* (9), 2–3.

Sewert, D. (1982). Individualizing support services. In J. Daniel, M. Stroud, and J. Thompson (Eds.). *Learning at a distance—A world perspective.* Edmonton, Alberta, Canada: Athabasca University, International Council for Correspondence Education.

Truett, C. (1994). The Internet: What's in it for me? *The Computing Teacher, 22* (3), 66–68.

U.S. Congress, Office of Technology Assessment, (1989). *Linking for learning: A new course for education.* Washington, DC: U.S. Government Printing Office, OTA-SET-430.

Wagner, E. D. (1988). Instructional design and development: Contingency management for distance education. Paper presented at the American Symposium on Research in Distance Education, July.

The Wall Street Journal Reports (1995, June 19). Technology. Special section.

Walsh, J., and Reese, B. (1995). Distance learning's growing reach. *T.H.E. Journal, 22* (11), 58–62.

Withrow, F. (1992). Distance learning: Star Schools. *Metropolitan Universities, 3* (1), 61–65.

Distance Learning/Telecommunications Associations

International Teleconferencing Association, 1650 Tysons Boulevard, Suite 200, McLean, VA, 22102

Public Broadcasting Service/Adult Learning Satellite Service, 1320 Braddock Place, Alexandria, VA 22314

United States Distance Learning Association, P. O. Box 5129, San Ramon, CA 94583

Periodicals Specific to Distance Learning/ Telecommunications

Classroom Connect, P.O. Box 10488, Lancaster, PA 17605-0488

ED Journal, Applied Business teleCommunications, P. O. Box 5106, San Ramon, CA 94583

Education at a Distance, Applied Business teleCommunications, P. O. Box 5106, San Ramon, CA 94583

The Heller Report: Internet Strategies for Education Markets, 1910 First Street, Suite 303, Highland Park, IL 60035-3146.

Inside the Internet, 9420 Bunsen Parkway, Suite 300, Louisville, KY 40220

Internet World, 20 Ketchum Street, Westport, CT 06880

MultiMedia Schools, 2809 Brandywine Street NW, Washington, DC 20008

Via Satellite, Phillips Business Information, Inc., 7811 Montrose Road, Potomac, MD 20854

Wired, 544 Second Street, San Francisco, CA 94107-1427

Distance Learning Companies and Networks and Example Offerings

Central Education Telecommunications Network, 2100 Crystal Drive, One Crystal Park, Suite 1100, Arlington, VA 22202

Distance Learning Associates, 190 West Washington Avenue, Pearl River, NY 10965

DLA Program Affiliates for 1995–1996:

The Academy at Ball State University

Fairfax Satellite Network

Prince William Network

Center for Agile Pennsylvania Educators Network

School Districts of Philadelphia Network

Massachusetts Corporation for Educational Telecommunications

South Florida Instructional Television Network with Barry University

Hispanic Informational Telecommunications Network

Satellite Communications for Learning, P. O. Box 619, McClelland, IA 51548-0619

Satellite Educational Resources Consortium, Inc., P. O. Box 50008, Columbia, SC 29250

Sea World Education Department, 1720 South Shores Road, San Diego, CA 92109

The Learning Channel, 7700 Wisconsin Avenue, Bethesda, MD 20814-3522

Ti-In Network, 1303 Marsh Lane, Carrollton, TX 75006

Turner Educational Services, 1 CNN Center, Atlanta, GA 30348-5366

Some Sources for Modems and Communications Software

Chameleon—NetManage Inc., 10725 North DeAnza Boulevard, Cupertino, CA 95014

Delrina Communications Suite—Delrina, 800-270-8031

Netcruiser—Netcom, 800-353-6600

Procomm Plus—Datastorm, P. O. Box 1471, Columbia, MO 65205

QModem—Mustang Software, Inc., P. O. Box 2264, Bakersfield, CA 93303

(Also see Appendix B for names and addresses of online services.)

Chapter 10

Emerging Technologies—Present Directions, Future Visions

This chapter will cover the following topics:

- Hardware and software components of artificial intelligence, virtual reality, and personal digital assistants

- Capabilities, applications, benefits, and limitations of these emerging computer technologies

Chapter Objectives

1. Identify terms and characteristics associated with each of the following emerging technologies:

   - Artificial intelligence (AI)
   - Virtual reality (VR)
   - Personal digital assistants (PDAs)

2. Identify appropriate applications of these technologies.

Introduction

Webster's Dictionary defines the future as "time that is to come." Yet many people tend to describe the future in terms of technologies likely to emerge and change people's lives in dramatic ways. Some of these futuristic technologies are often actually products or processes that have been around for some time in some form, but they seem to gain sudden importance and to influence and enhance many kinds of activities. Charp (1994) cites seven technologies to watch. These are products that are "coming out of the laboratory and into the hands of the educational community" (p. 8):

- **Wireless communications.** Connecting people with networks and voice communications over FM radio frequencies

- **The information highway.** This term lacks an agreed-upon definition, but it generally refers to "the capacity to interconnect . . . everything" (p. 8). Some people seem to include video-on-demand in this category as well as ready availability of data.

- **Asynchronous transfer mode (ATM) transmissions.** According to Charp, this may be the "pavement for the information highway" (p. 8). The latest communications technology, it can simultaneously carry voice, video, and data signals at speeds 50 times faster than other methods.

- **Integrated Services Digital Network (ISDN).** A network that uses existing telephone lines to provide integrated voice and data communications capability. (See Chapter 9 for more details.)

- **Multimedia applications.** Already in use in schools, multimedia software and systems will have an even greater impact on education in the future as publishers and developers learn more about exploiting its potential.

- **Personal digital assistants (PDAs).** Charp described these as "easy to carry, handheld devices that merge handwriting recognition, personal-organization tools, and state-of-the-art mobile communications in a compact package" (p. 8).

To this list of technology resources, Charp also adds *Total Quality Management* or TQM, which is a process, rather than a product. She claims TQM has had a positive impact on businesses, and that education is attempting to adapt TQM methods to achieve similar results.

Previous chapters of this book have already addressed many of the "futuristic applications" of technology Charp cites. This chapter discusses yet another of the applications on her list—personal digital assistants (PDAs)—and adds two more that may have a dramatic impact on integration of technology into curriculum: artificial intelligence and virtual reality. These applications may use some of the emerging technology developments described by Charp, but they are more closely linked to educational strategies than to communication systems such as the information highway. Although the technology resources described in this chapter strike many people as futuristic, they are, in fact, in current use in schools and classrooms throughout the country. Each promises significant contributions to educational technolo-

gy and to education itself. Integration methods for these technologies are still being developed, but this chapter will describe and illustrate some uses that already have emerged, and it will project some that may develop in the near future.

These technologies share a common thread: They are all designed to simulate something—intelligence, reality, or an assistant—to accomplish their purposes. They also combine hardware and software components. The specific components, in turn, determine the capabilities of each system. At the present time, AI, VR, and PDAs have seen limited use in education. This chapter will discuss benefits and limitations to show how each technology is used so that educators can determine which is the most appropriate for their needs.

Artificial intelligence is the oldest of the three technologies presented in this chapter, dating back to a 1956 conference at Dartmouth (Golob & Brus, 1990). AI has been around in some form for four decades, but because of its limited use and its continuing changes, it is still considered an emerging technology.

Most definitions of *artificial intelligence* (AI) include an expectation for a computer to perform in ways comparable to human behavior. The computer performance can be thinking, seeing, talking, understanding speech, or moving. Artificial intelligence is also described as a subdiscipline of computer science that uses combinations of hardware and software to simulate the functions of the human mind (Orwig and Baumbach, 1991–1992). Computers easily handle some functions, such as calculations, at speeds far faster than the human mind can attain. Other functions, such as recognizing and comprehending language, images, and touch, are much more difficult for computers to perform. The adage that applies to all computer applications—garbage in, garbage out—seems especially relevant for AI. Questions that fall even slightly outside a system's narrow expertise can completely skew the simulated thought process; as one author described it, "artificial intelligence quickly turns to artificial stupidity" (Walbridge, 1989). In addition to optional capabilities that can be programmed into an AI system, software components also vary, as do system applications. The following sections elaborate on the components, capabilities, applications, benefits, and limitations of AI.

Components of AI

The minimal components of an AI system are a standard computer and a software program. Additional hardware is added to provide (1) specialized input of sight, sound, or touch; (2) processing of this specialized input to achieve object recognition; and (3) output in the form of speech or movement. Every AI system has two software components: the knowledge base, or collection of information on a particular subject, and the inference engine, or computer program that makes sense of the knowledge base. The inference engine uses a logical routine defined by program statements to sort through the data in the knowledge base and identify patterns. The resulting pattern recognition is the artificially generated intelligence.

The knowledge base of an AI system is composed of facts. The inference engine is composed of software commands that process the facts and generate logical decisions. Both the knowledge base and the inference engine can be programmed in several computer languages. Different programming languages require different levels of expertise to use and different levels of control over the finished product. For example, a programmer can select a package that bypasses the need for certain programming skills, but this decision sacrifices some control of the program. Such a package allows a less experienced programmer to concentrate on the information in the knowledge base and the logic routines for evaluation. A more advanced programmer can choose to write all the software code and maintain complete control of the program, but at the cost of spending considerably more time developing the system. Both approaches to replicating human thought through computer software have yielded disappointing results to date. Performance has usually fallen short of both program objectives and developer expectations.

Capabilities of AI

An AI system can be programmed with many capabilities, from appearing to think like a human to "listening" and "speaking" like a human, as well. AI systems try to replicate certain aspects of the human thought process, recognize speech, and respond with a voice, among other complex aspects of human behavior. The capabilities programmed into a system both determine the system's usage and classify it in one of the subdisciplines of AI, such as fuzzy logic, neural networks, and expert systems.

The problem of duplicating a single sound in the English language illustrates some of the difficulties in programming a computer to recognize speech. The sound "tu" can be spelled *to, too,* or *two.* Since a computer cannot distinguish meaning based on sound alone, it must examine context. For example, if the "tu" precedes an adjective, then *too* is the most logical choice. However, if the "tu" precedes a noun, then either *to* or *two* are strong possibilities (e.g., to school, two students); the program would need to look for additional clues in the usage. Other confounding factors such as the accent or tone of voice of the speaker could also influence the ability of the computer to recognize speech. These problems hint at the complexity of a program capable of distinguishing similar sounds with different meanings while also accounting for all dissimilar sounds that could be variations of the same word. A computer with this powerful capability would, however, provide the user with a familiar and convenient tool.

Another human characteristic, subjectivity, further complicates programming a computer to display artificial intelligence. A computer's most basic operation stores a value of either 0 or 1 in a memory location, while humans often deal in shades of grey. A seemingly simple computer classification task can be complicated by the perspective of the human classifier. For example, a computer can easily classify a person as middle-aged, if the parameters for middle age are preset as older than 35 and younger than 55 (Anderson, 1993). But a child may classify someone 25 years old as middle-aged, and an 80-year-old may think 60 is middle-aged. The uncertain boundaries illustrate the subjective or "fuzzy" nature of the classification. In such a situation, the computer needs additional parameters to simulate the human classification process.

Fuzzy logic. An entire discipline of AI has been dedicated to simulating these fuzzy aspects of human thought. Fuzzy logic is not logic that is vague or uncertain, but a logic system designed to parallel the human perceptions and decision making by accounting for subjectivity and uncertainty. Rather than defining a completely new programming technique, fuzzy logic like all of AI, really just adds more programming. An inference engine with more program statements allows it to account for variable boundaries (such as the limits of middle age) in classification applications. The additional inferential functions make the program better able to simulate the subjectivity of humans. For example, adding program statements to account for the age of an individual who is classifying people as middle aged would enhance the capability of the program to simulate subjectivity.

At what point do additional programming and simulated subjectivity achieve artificial intelligence? An AI program is really only fancy programming. In addition, a program doesn't need to mimic "every nuance of human thought patterns" to be "intelligent" (Newquist, 1994, p. 51). The inventor of the modern form of fuzzy logic, Lotfi Zadeh, expressed a similar view about the disparity between computer systems and the capabilities of the human brain by posing the question, "Can we build a machine that can climb a tree like a monkey?" (Williams, 1993a, p. 20). Just as a machine does not need the capabilities of a monkey to climb a tree, a computer program that falls short of replicating human thought can still be intelligent (Dibbell, 1996).

Neural networks. A computer program needs more than fancy programming to qualify as the computer-generated form of intelligence called a *neural network.* A neural network program must also be able to adjust itself based on the difference between the actual result of a physical process and the output predicted by the program. This adjustment helps to minimize error in future trials, but requires training. If a neural network examines enough input and output data during the training phase, it can calculate adjustments in the program variables so that it can accept new input and generate accurate output values (Patton, Swan, and Arikara, 1993). In other words, a neural net can "learn" from its mistakes.

A testing phase after the training phase ensures that the output generated in response to "untrained" input is within tolerable error limits. If the predicted output differs too much from the actual value, then additional training is conducted. For example, a neural net used to classify people as middle-aged could learn to account for the age of the classifier during training. If the testing phase revealed a failure to adjust the limits of the age group sufficiently for each classifier, then additional training would be warranted.

Some types of neural nets are programmed with features that decrease the need for certain training. For example, after being trained to recognize one view of an object, programming can account for changes in the position, size, or orientation of the object. This eliminates the need to train the neural net for these additional capabilities (Neural-Network Object-Recognition program, 1993).

Expert systems. An expert system is a form of AI that computerizes human expertise (Walbridge, 1989). It attempts to duplicate the responses of an expert in a very narrow field of knowledge (Orwig and Baumbach, 1991–1992). Expert systems are not designed to accomplish some of the advanced input and output tasks of AI, such as manipulating robotic parts or understanding language. Instead, an expert system acts as an advisor and offers suggestions to assist the user's decision-making process, just as a human expert would. Like any computer program, an expert system is most effective in performing repetitive operations. In addition, it has the characteristic capability of evaluating information. An expert system in a media center, for example, might identify resources that match a specific student profile or supplement an instructional unit.

Some software packages help teachers to develop expert systems to capture their knowledge in a content area and their expertise in logically analyzing information. For example, a teacher could develop a system to serve as a nonhuman teacher's aide, freeing more time for less repetitive tasks. These expert systems could serve teachers as "software extensions of themselves" (Newquist, 1994). Teachers might also find expert systems to be helpful in content areas outside their primary areas of expertise. A student using such a system could become the resident classroom expert in that content area.

Current Applications of AI in Education

Current applications of artificial intelligence in education span a range of subjects and include systems that assist teachers as well as students. AI currently contributes to five kinds of educational strategies:

- **Help with problem solving.** One AI system was designed to help solve difficult mathematics problems (Mission accomplished, 1993b). This expert system has the capability to review a problem and display a solution set of equations. The user can then implement or override the program's solution.
- **Intelligent agents.** Salvadore (1995) described "intelligent agents" as software packages that do various jobs for students. For example, the programs may search online systems for "information tailored to the needs and interests of the user" (p. 14). They learn by observing the user's actions and noting pat-

terns of behavior (e.g., reading preferences); the programs then present articles based on revealed preferences. In the future, people may also program intelligent agents to search libraries and archives for specific information or to give suggestions on how to learn something or accomplish a desired task.

- **Personal tutoring systems.** One example of an intelligent tutoring system adapted for microcomputers is Smithtown (Raghavan and Katz, 1989). The Smithtown system uses an AI routine to tutor students in beginning economics courses. The program is geared to improve the user's problem-solving ability, and it provides discovery tools for self-directed experiments. The AI part of the program monitors the user's reasoning, tracks sources of errors, and offers advice on problem solving. The learning that occurs during use of the program is attributed to the discovery tools in the package. The AI programming provides coaching by an expert advisor prompted to respond or intervene based on preset criteria. The program allows instructors to vary the expert advice. In experiments conducted to assess its effectiveness, students using Smithtown achieved the same improvement in understanding as students taking an introductory course. Students taking the course and using Smithtown to supplement their instruction showed more improvement than the other groups. Schank's (1990) software case-studies are yet another example of this potentially powerful use.
- **Feedback on student performance.** Schools have used expert systems to integrate the accumulated knowledge of a subject area for self-testing. One expert system allows students to test their comprehension in a basic accounting course (Fogarty & Goldwater, 1993). The system is based on a compilation of test materials from various sources. Each test question was rewritten into a structure that allowed the program to substitute random numbers, permitting unlimited possible variations. The system offers options to display the correct answers to questions either immediately or after the student responds. The program also tracks student progress, offers unlimited privileges to retake exams, and generates output files on student behavior for instructors.
- **Developing critical-thinking skills.** Another expert system shell program, ES, helps high school students to develop their own expert systems (Dillon, 1994). The program's menu-driven functions allow students to input their own knowledge base facts and inference engine rules. After the facts and rules have been entered, students can test the system's output for various combinations of situations and then consult it as needed. As they generate the contents of the system, students develop critical-thinking skills. Some inferences depend on fairly straightforward logic, but more complex arrangements of facts into rules can tax the abilities of any student. For example, some inferences have only one condition and one conclusion. Many real-life problems that require the assistance of an expert, however, have numerous conditions and multiple solutions. The thought process involved in developing an expert system can provide an extremely productive student learning experience.

Potential Role of AI in Education

Potential applications of AI in education depend on two factors: integration of existing AI systems into the educational environment and development of knowledge bases, inference engines, and software packages that allow teachers to control AI system components. Teachers can integrate existing AI systems with relevant knowledge bases or inference engines into their classrooms without bothering about the development process. For example, one AI system to classify was developed as an option to statistical analysis. It has potential applications in both biology and statistics (Gertner & Guan, 1990). The contents of the knowledge base would make this system appropriate for a biology course and the logic of the inference engine would suit a statistics course.

An existing knowledge base can be reused in multiple AI applications. In the future, "knowledge-based systems will be assembled from components rather than built from scratch" (Neches, Fikes, Finin, Gruber, Patil, Senator, and Swartout, 1991). Students will learn by both designing and using the system. By adding to the knowledge base and modifying the computer code of the inference engine, students can gain additional learning opportunities. These opportunities would be useful both in the content area for which the program was developed and in courses studying human reasoning.

Benefits of AI

The flexibility for students to learn by both designing and using AI systems is one powerful benefit that few other learning materials can provide. Other benefits help to determine the specific type of AI application that best suits a given situation. All AI systems share some advantages, such as performing repetitive tasks rapidly and generating intelligent output that assists the user. Other advantages vary depending on the specific AI subdiscipline. For example, a fuzzy logic application provides benefits in subjective situations. A neural network has a great advantage over other computer programs in that it can modify its own processing to improve the accuracy of its predictions. Expert systems have the advantage of simulating certain contributions of a teacher or other knowledgeable professional, freeing the teacher to devote more time to activities that the computer cannot perform. Most classrooms would benefit from AI systems that could serve as advisors, help desks, or diagnostic tools.

Limitations of AI

Computer simulations of human behavior are limited by the array of hardware for inputting, processing, and outputting data and the processing routines in the software. Limitations result when a system lacks appropriate hardware to gather input data that would alter the thought process. For example, a teacher can react appropriately to a student who responds affirmatively, but hesitantly, to indicate understanding of the lesson material. An AI speech-recognition system would be unable to detect hesitancy in the student's voice and consequently would proceed as if the student confirmed unequivocal understanding.

Software limitations, like those based on hardware, revolve around the scope of the system components. A computer can be programmed to perform calculations like a

human does. To replicate effectively the entire thought process, however, the program must include all human computational processes. Simply identifying all of these computational processes can be a monumental task, but the resulting program would suffer if any step were omitted. For example, an expert system that provides advice on solving certain types of problems may not benefit younger students who do not understand the language of its output. Also, a teacher knows to target comments to match the level of the student. A software program without that process or ability could not adjust to students' levels.

The game of chess provides a clear example of trying to program a computer to account for human subjectivity. While humans are better chess players than computer programs today, chess champion Garry Kasparov found out in February of 1996 that this may not be the case much longer (Kasparov, 1996). Computers have trouble competing with humans because the software relies on programmed logic. However, continuing increases in processing speed allows computers to project more possible move combinations to counter this limitation.

Therefore, many limitations of AI do not result from inability of the hardware and software components to model human behavior. Instead, the limitations result from inability of programmers to account for all the possible variables in a certain situation. They encounter a trade-off in minimizing these limitations because the complexity and cost of a system eventually exceeds the value of its function. The classification procedure for middle age illustrates this trade-off. An AI system may be programmed to account for the age of the classifier, but without a means for inputting the classifier's age, this process will not make the necessary adjustment. The observer could input the age via keyboard, or a more elaborate system could more accurately simulate human behavior by determining the age of the classifier by voice, visual image, weight, or a combination of these factors. The ultimate use of the system in each situation should determine how to resolve the trade-off of complexity and cost versus function.

Virtual Reality (VR)

The term *virtual reality,* coined in 1974 (Helsel, 1992) refers to a relatively new technology also called *virtual environments, virtual worlds, artificial environments, artificial reality,* and *cyberspace* (Biocca, 1992a). At the most basic level, VR is a new way to present three-dimensional images (Marshall, 1991). In its most elaborate form, VR permits a complete simulation of reality or even the generation of a new reality. In an advanced version of VR depicted on the television show *Star Trek, The Next Generation,* a user of the ship's "Holodeck" is presented with images that cannot be differentiated from real objects (Pantelidis, 1993). The simulation is indistinguishable from reality in every human perception. Although beyond current capabilities (and possibly beyond future capabilities, as well), this example from the "final frontier" sets a standard for VR systems to achieve.

Virtual reality is "a computer-created sensory experience that so completely immerses participants that they barely can distinguish this 'virtual' experience from a real one" (Fritz, 1991, p. 45). The user is immersed in a "synthetic world, creating a sensory-based environment that interactively responds to and is controlled by the user's behavior" (Mission accomplished, 1993a, p. 16). VR can transcend time and space boundaries (Merrill, 1993) to create an environment which is "multi-dimensional" and "inclusive" (Winn and Bricken, 1992). A VR system provides a sense of inclusion, navigation, and manipulation (Helsel, 1992). These characteristics, which provide opportunities for immersion, exploration, and interactivity, are most commonly associated with the capabilities of VR.

The components of a VR system determine the degree of immersion, the scope of exploration, and the level of interactivity that the user experiences. The system's capabilities, in turn, determine its suitability for a specific application. The components, capabilities, and applications are covered first in this section on VR. Later discussions lay out the benefits and limitations, which depend to some degree on the system capabilities, although VR systems share some general benefits and limitations, as well. The concluding sections describe general information about the impact and issues of VR.

Components of VR

VR systems combine hardware devices that interface with computers and software that generates graphics within those computers (Merrill, 1993). A minimal VR system includes a head-mounted video display with a tracking device to detect head movements and a glove with sensors to detect hand movements (Helsel, 1992). These components relay information about head and hand movement to the computer, which updates a displayed image of the user's perspective.

Optical fibers in a typical data glove sense flexion and extension of the fingers (Hill, 1992). Some gloves have tiny metallic vibrators in the fingertips that tingle to approximate touch sensations to users' hands (Fritz, 1991). Since VR objects provide no weight or resistance, however, systems need improved physical sensations feedback. One system, for example, has integrated force feedback mechanisms into an exoskelton for the hand which is a more elaborate device than the data glove (Sensing and Force-Reflecting Exoskelton, 1993). The exoskelton allows users to control robot hands by generating feedback so they can handle objects gently without exerting excessive force. A prototype for a tactile glove with compressed air to provide sensory feedback about shape has also been developed (Rheingold, 1993).

Head-mounted displays have been introduced by several companies. Standard VR goggles present a slightly different image on a tiny screen in front of each eye to produce a 3D effect. At the same time, the goggles shield the user from any external sources of light to enhance the effect. Some displays include audio input and output and operate through radio signals (wireless communications). Some even have prescription lenses. Some newer and less expensive goggles are designed to produce 3D effects of program images on a computer monitor.

One option to VR goggles is a hood that fits over a computer screen and tapers to a binocular viewer that presents a 3D image to the user. The software sold with this commercially available product rotates and projects a pair of images of the same monitor view. The lenses in the viewer of the hood rotate the images to achieve a 3D effect. The optical hood offers a more practical alternative for the classroom at a fraction of the cost of goggles. In addition to the affordable price, the hood is attached to a monitor and, therefore, less likely to be damaged or lost at school.

Many other possible configurations combine more elaborate sensors for input of body movements, faster computers, and more sophisticated displays. VR technology encompasses an entire array of input and output hardware devices (Biocca, 1992b). Many devices besides data gloves can monitor body movements:

- Position trackers (mechanical, optical, magnetic, or acoustic sensors to monitor the locations of body parts)
- Exoskeletons (monitor the relative orientations of body parts)
- Treadmills (monitor foot movements with respect to the ground)
- Speech recognition hardware (monitors voice input)
- Muscle and brain electrical activity sensors (monitor performance of the autonomic nervous system)

In addition to head-mounted displays, output hardware may include:

- Spatial audio hardware (outputs sound that varies with the perceived distance from the source)
- Tactile feedback devices (simulate the sensation of touch)
- Force feedback devices (simulate resistive force)
- Full-motion display (displays movement of the entire body)

Additional components for VR systems continue to be developed. Some systems include a full-body data suit which covers the user with a "second skin" that relays tactile information to the wearer (Ferrington and Loge, 1992). A recently invented device allows paralyzed people to control VR systems by tongue movements (Rheingold, 1993). As system designers work to minimize the size and inconvenience of the VR hardware peripherals and optimize their simulation of reality, they will probably produce many generations of improvements on existing equipment and invent new input and output devices. Even the basic components that formulate a minimal system have changed. A type of virtual experience is now provided simply by running a software program using conventional computer hardware.

Future developments will probably result in systems that strive to present the most accurate representation of reality with the least expensive combination of components.

At this writing, few schools can afford to purchase all the components of a VR system. However, costs continue to fall and VR's potential in education is expected to justify its relatively high cost. The release of new products that are geared for educational settings or that teachers can easily modify will also help to curtail the overall expense. For example, a relatively inexpensive package ($3,500) released in 1991 allows users to develop VR systems for specific applications by writing minimum amounts of code (Mission accomplished, 1993a). More recent versions priced under $1,000 could allow integration into the classroom on a large scale.

Capabilities of VR

The capabilities of VR systems vary dramatically. A system composed of standard computer components might offer a low degree of immersion, provide an awkward form of navigation, and a minimal level of interactivity. The systems that most effectively simulate reality have many specialized components to monitor a user's movements and generate virtual images.

The components of a VR system determine the degree of immersion, the scope and sense of navigation, and the level of interaction that the user experiences. VR systems can be classified according to each of these capabilities so that a potential user can determine the most appropriate system for a given application and avoid wasting resources on an ineffective system for a specific purpose or a more expensive system than a task requires.

The immersion level of a virtual reality system indicates the degree to which the participant is involved. In a full-immersion system, the user typically wears headgear that generates the entire field of vision (Robertson, Card, and Mackinlay, 1993). The user then interacts with a three-dimensional environment instead of with the computer keyboard or mouse. In a nonimmersive VR system, the user interacts with a 3D environment by manipulating conventional computer peripherals such as a keyboard and mouse. A full-immersion system obviously gives a more realistic sensation, but a nonimmersive system requires no additional hardware, and therefore no additional expenses other than the cost of software.

Another classification system for immersion defines a scale partitioned into four levels that represent some of the more recent improvements in VR technology. At the first level, a nonstereographic desktop display shows images with perspective rendering, as exemplified by the standard PC flight simulator software package (Quinnell, 1993). The second level adds a stereo viewer (glasses or hood) for a true 3D effect. The third level requires hardware to track head movements and generates displays consistent with the user's view. The fourth level uses video images of the user to track movements, thereby producing the most immersive or realistic environment.

A more immersive environment creates a more realistic scope and sense of navigation. Navigating depends on both hardware and software. The standard computer interface devices such as a keyboard and mouse represent the least-natural end of the navigational spectrum. A full-body suit represents the most natural end. The range of movement allowed by the software combined with the speed and accuracy of updating an image also contribute to the naturalness of navigation. The navigational continuum spans from a restricted scope and limited sense of movement to an image that fully represents the user's movements.

The interactive capabilities of VR systems can be categorized in three levels of interaction that offer options for use with different student types (Ferrington and Loge, 1992). In Level I participation, users exist and interact in the virtual environment. The interactivity is limited to the options provided in the environment. Players of arcade simulation games experience this kind of interaction. In Level II, modification, the user can customize the environment with new objects, spaces, and activities. In Level III, construction, users have the knowledge and skills to build complete environments, beginning with the initial programming. Construction might produce a unique application using a VR software programming package.

Although improvement of one of the three capabilities (immersion, navigation, and interaction) does not necessarily mean that either of the others will improve, in most cases the capabilities affect one another. A system composed of components that yield a low immersion level will probably have minimal navigation, as well. A system that is designed to be high in any of the three factors will probably be highly rated in the other two.

Current Applications of VR in Education

Virtual reality is currently being integrated into the classroom in several ways, including tutorial applications, student project development, and student cooperative work. Some of the current applications of VR described here show the diverse uses of the technology and the wide array of content areas in which the technology can augment student learning.

Tutorial applications. One example of a tutorial application is a math program that offers students the opportunity to solve problems by exploration. In an effort to improve the students' classroom experiences, an elementary algebra curriculum is presented as a virtual world (Winn and Bricken, 1992). Spatial algebra represents algebraic concepts and procedures in a new way. Numbers are enclosed in cubes in space and the proximities of the cubes represent algebraic relationships. Users can manipulate the cubes in space like real objects.

In addition to the novel aspect of spatial algebra, this VR application also offers unique, programmable teaching enhancements. It is possible to program this virtual world to provide different types of guidance or support varied pedagogical strategies. For example, a system can be programmed to respond in a specified manner based on the ability of the student or the goals of the session. If a student performs a mathematically incorrect operation, the program could:

1. Automatically "float" the virtual objects back to the starting condition so that the user could view them being reset

2. Change an object to make the image mathematically correct

3. Allow the mistake to remain uncorrected so that the student could change it later

The system can also be programmed to take over some operations so the student can focus on other tasks.

Programming also permits students to take advantage of some control options. In a virtual physics lab developed to teach basic concepts, the user can control gravity, friction, and time within the virtual environment to conduct experiments in motion (Delaney, 1993). The user can slow time to observe fast motion or stop it to record data. In a networked system, the teacher could provide guidance as needed and monitor progress, as well.

Student project development. Students at a high school in England developed VR projects in foreign languages, arts awareness, and health and safety (Tait, 1993). In the health and safety project, students learned the importance of correct procedures in a dangerous environment. In a similar manner, a virtual office served as an environment for students to learn foreign languages. Contact with an object in the office caused its name to be displayed in English and in the foreign language. Several VR projects have been proposed by the San Diego City Schools for elementary, middle school, and high school students attending magnet schools (Kerney, 1993). The elementary school projects are designed to let students identify molecules, visit solar systems, manipulate math concepts, explore the digestive tract, build a world free of gravity, and interview virtual people. The middle school projects include developing science labs to study levers and safety with science equipment, designing VR games, and generating ancient cities. An example of a high school project is a virtual hospital where students handle medical emergencies.

Cooperative work. Other applications provide environments for students to work together. For example, virtual reality networks allow multiple users to share the same electronic space (Moore, 1993). Using a computer, CD-ROM, graphics software, and modem, a relatively inexpensive VR system was configured to transmit changes so that all users could see the same image at the same time. In another example of shared electronic space, children in three countries were networked via satellite to design cities in which they wanted to live (Stanfel, 1993). They explored the elements of their own environments and then defined "the parameters of habitable urban spaces." In this exercise, students investigated the interdependence of urban issues.

Potential Role of VR in Education

The contributions of VR in education depend on the effectiveness of new applications, access to VR technology both inside and outside the classroom, and educators' resolutions of implementation issues. Schools will not realize VR's full potential unless the new applications apply the VR technology in a way that supports effective learning strategies. The high cost of VR systems raises an important consideration about access to allow widespread use of the technology. Responses to these and other issues will determine the ultimate success and timing of VR in education.

Effective learning strategies. The integration of effective learning strategies into VR programs is essential. VR has the potential to "provide powerful educational experiences" but "technology does not, by itself, improve education" (Bricken, 1991, p. 184). Only through appropriate applications can technology become effective. Unfortunately, VR offers educators the option of substituting technological glitz for substance (Tinker, 1992). "Glitzy" technology can distract educators from less glamorous, but potentially more effective products. Thus, while the interactivity of technological systems is a primary attraction for educators, interaction alone does not promote learning. Instead, the integration of new information with the student's previously acquired knowledge base forms the basis for learning. Moreover, availability of technology does not guarantee integration. Observers do not expect difficulty achieving the necessary modifications in VR to facilitate learning, as VR is "consistent with successful instructional strategies: hands-on learning, group projects and discussions, field trips, simulations, and concept visualization" (Tinker, 1992, p. 26). Consider some examples:

• A NASA virtual wind tunnel simulates the airflow around aircraft (Schmitz, 1993). This permits students to both analyze the aerodynamic characteristics of an aircraft model and test constructed aircraft (Meyer & Dunn-Roberts, 1992). This wind tunnel could be modified to support visualization of concepts in physics courses.

• A VR application similar to that used by the U.S. Olympic bobsled team could provide hands-on learning in physical education. By connecting the steering gear of a bobsled to a computer and generating an image of a bobsled track, training time on actual tracks was decreased (Industry visualizes real uses for virtual worlds, 1993). Similar applications could be designed for learning or improving physical skills. Since VR requires the involvement of the user, a system could increase the level of participation and, consequently, the effectiveness of physical education.

• A strategy for industrial applications of VR could transfer naturally to an educational setting, possibly for a group pro-

VR headgear and data gloves

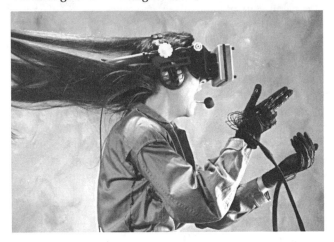

Source: Courtesy of NASA Ames Research Center

ject. VR goggles that superimpose an image of a finished product over real parts guide workers to improve the manufacturing process (Bylinsky, 1991). Adding graphics to an actual image generates an "augmented reality" (Drascic, 1993). If part of an actual image is obscured or not visible, computer-generated graphics could provide the necessary supplement to the view. Augmented reality could prompt students working on a group project after giving each student a chance to modify the image correctly.

VR system in use

Source: Courtesy of Victor Maxx Technologies, Inc.

Instructional tools could easily be incorporated into VR packages to add unique characteristics that would reinforce learning strategies. For example, navigational aids can be added to systems to "encourage students to follow appropriate paths for learning" (Ferrington and Loge, 1992, p. 19). Also, the novel and immersive nature of VR might facilitate collaborative participation by students and teachers.

In a particularly effective application to learning strategies, VR applications could address one of the most common complaints about educational systems: that school is not like the real world. Bricken (1991) has noted that "for the most part, our schools exist separately from the world that they teach about" (p. 178). VR can help promote needed change by simulating the presence of the student in the actual, physical world. This would be particularly useful in educational situations involving practical, work-related applications.

Access. Schools may gain access to VR systems more quickly if they can demonstrate that the technology is cost effective, the only way to perform certain activities, or provide protection for the safety of the student. Military training provides many examples of applications of VR to simulate costly maneuvers (Fritz, 1991). VR gives a less costly method for astronaut training than building a model of a space station. VR can simulate certain aspects of weightlessness. Simulations for repairing battle damage avoid the need to damage expensive equipment to teach soldiers how to repair it. As one example, the well-publicized repair of the Hubble telescope was tested using VR prior to launching the repair mission (Quinnell, 1993). A system developed by NASA scientists has been used to train surgeons in tendon transfer operations; this shows how VR allows students to perform activities when no other means are available (Bains, 1991). The surgeon is equipped with VR goggles and data gloves. Operating on a

VR head-mounted display and headgear

Source: Courtesy of Victor Maxx Technologies, Inc.

3D computer simulation of a human leg, the surgeon can reposition the muscle attachment on the bone. An additional 2D image displays the leg function based on the surgical changes. VR simulations of instructional applications in hostile environments would eliminate the danger to students or allow them to experience impossible situations. Such applications could include:

- Learning about nuclear reactors
- Exploring the ocean floor
- Navigating the surface of Mars (Pantelidis, 1993)

Networking is another possible means to increase access and stay within budget constraints. Pay-per-view access to VR over phone lines would eliminate the need for every user to purchase the most expensive equipment, making the technology available to a larger audience. In the future, networking to a central VR facility would minimize costs (Bricken, 1991).

To increase the benefit of VR to students, access to the technology could become available outside of the classroom. For example, students at a library could retrieve information as text, pictures, moving images, sound, and 3D objects. In a virtual library, a user could navigate in an "idea space" (Spring, 1993). The information in books and articles could be coded according to topic relevance, date, size, and other important characteristics. Each document could be represented as an image in the virtual environment to help a user find information with required characteristics.

Implementation issues. Numerous issues affect the implementation of VR systems. Naturally, teachers must address pedagogical issues. A question of morality also arises when people use VR. Resolution of these issues will influence the eventual design and ultimate effectiveness of the virtual environment for education. The generally less desirable and more commonly used alternative is to develop the technology and then hope to assimilate it into the learning process.

Applying VR to education presents a "backwards approach" (Roblyer, 1993). Instead of first developing "a more definitive vision of what we want teaching and learning to look like" and then determining how technology can make the vision become real, many educators attempt to adapt the technology to current educational objectives. The same concept applies to medicine, as doctors attempt to fit VR technology to the patient, rather than the patient to the technology (Nilan, Silverstein, and Lankes, 1993). Perhaps teachers could implement VR in education more prudently if they thought of students as patients.

The future locations of educational experiences have a crucial effect on planning and must be decided before allocating funds for technology (Roblyer, 1993). If the home will become the future educational environment, then educators must shift funds from salary and buildings to technology. However, if the school fulfills the function of gathering

people to communicate, then educators must integrate technology into the school environment. The bottom line is that "at some point we must begin the hard work of limiting our uses of technology to an achievable and meaningful vision of tomorrow's education" (Roblyer, 1993, p. 35).

Beyond establishing a game plan for using VR in education, other implementation issues concern the characteristics of the application. The nature of the image of reality, the accuracy of generated representations, and the control of the virtual environment all contribute to the uses of VR in an educational setting (Helsel, 1992). Each of the three characteristics can be explained in terms of a continuum. The nature of the reality in a VR system can be programmed as a computerized version of physical reality, a possible but not actual reality, or an alternative version of reality. Another continuum represents the naturalness of the user's interaction with the system. The most natural system would detect user actions without the need for any sensors on the user's body, as Star Trek's Holodeck does. A less natural system still allows typical body movements to signal the user's actions, but headgear and gloves are required to actually input those movements into the computer. The least natural system does not track the user's movements, but accepts input through a standard computer peripheral such as a mouse. The issue of control spans a continuum from complete user control (by student or teacher) to complete system control. Artificial intelligence capabilities could be programmed to allow for varying proportions of user and system control. Complete user control places no restrictions on how the user chooses to explore the virtual space, but this sacrifices any control over user behavior. At the other end of the spectrum, VR components maximize user constraints and minimize freedom of action.

Issues for special education. Special educators have a vested interest in equal access to new technologies (Woodward, 1992). Developing incentives for inventors to modify equipment for special populations can critically affect the availability of VR for certain groups. Use of VR by young children is even an issue. Lanier and Biocca (1992) ask at what age is it appropriate for children to use VR? Although the ages of 8 and 10 were only mentioned as possibilities, Lanier advocated establishing a definite age criterion. He acknowledged that it would probably be a controversial idea and stated that the criterion could be positively framed.

Potential impact of VR. Moral and ethical issues also concern many people (Ferrington and Loge, 1992). What type of behavior is appropriate within a virtual environment? One view proposes establishment of a code of ethics for VR before it is fully developed. An opposing view holds that "virtual reality is the one space in which individuals might be able to do anything they have never been able or allowed to do in physical life" (Ferrington and Loge, 1992,

p. 19). Some users will no doubt take the opportunity to misuse VR technology "as badly as books have been used" (Lanier and Biocca, 1992, p. 171). Some proponents of VR predict that users will eventually simulate "sex, drugs, rock and roll, and just about every other human activity" (Kantrowitz and Ramo, 1993, p. 43). Perhaps not surprisingly, some believe that underreaction to this new technology poses a greater risk than overreaction (Levy, 1992).

Finally, predictions about the magnitude of the impact of VR on education range from the cautious to the unrestrained. Some describe the assumption that high technology will improve the educational system as "a quantum leap in logic" (Roblyer, 1993, p. 35). Others see almost unlimited potential impact for VR. For example, Bylinski (1991) projected that VR could have as great an impact as writing. Educators' responses to these issues, combined with the steps taken to ensure integration of learning strategies and universal access, will determine if VR fulfills its potential in education.

Benefits of VR

The accolades for VR are numerous. Proponents claim that VR can improve the learning environment by offering a better presentation format for certain subject matter, by facilitating interaction with the computer, by providing student and teacher control, by making information more concrete and easier to process, and, most importantly, by increasing student involvement.

The most obvious improvement in the learning environment is the display that a VR system presents to the user. Even the minimal 3D display on a standard computer monitor is an improvement over two-dimensional displays. The 3D and real-time features of VR provide tremendous advantages over other presentation methods for certain information. For example, in the three dimensional and dynamic subject matter of anatomy (Merrill, 1993), VR presents material in a moving, stereographic manner that is not possible in two dimensions. Additional hardware components and the method of operation of a VR system can further improve a learning situation by facilitating interaction with the computer. Some VR systems bypass the interfaces of other computer-aided instruction systems (keyboard, mouse, touch pad) to reach the user's senses directly (Winn and Bricken, 1992). VR can then be operated by an intuitive, natural interaction of grasping and pointing to computer-generated objects.

For another improvement, both the student and the teacher can sometimes control a VR environment (Pantelidis, 1993; Sprague, 1996). Control provides the possibility for greater structure, more consistency, and more individualization in the learning environment. The teacher can tailor the instruction for a particular student. The student can also individualize a VR session through the inherent freedom of navigation that provides limitless ways to respond.

VR also makes information more concrete for children and adults by linking abstract ideas to physical objects (Fritz, 1991). The physical dimension increases the learning rate and retention time for tasks. For example, as an electron, a trainee could experience flow through an electronic circuit as a white-water raft ride.

VR can further improve the learning environment based on its original function of simplifying information processing (Bricken, 1991). With the option to explore new theories and a medium that is manageable by the student, VR adds a "what if?" potential to education similar to the potential that spreadsheets add to business.

The VR environment is conducive to student involvement. Participation is a key factor in learning, but students are often passive in a classroom (Bylinsky, 1991). It is almost impossible for a VR user to be passive (Pantelidis, 1993). "The motivation to learn hinges on interest, and most people find VR a very interesting experience" (Bricken, 1991, p. 180). The sensations from the virtual world are "pervasive and convincing" and cognitively and affectively immerse the student in the learning environment (Winn and Bricken, 1992). The idea of VR homework might offer an exciting way to involve students in school work.

As a result of the improved learning environment, several possible improvements in performance have been attributed to VR. The technology accentuates the user's vision, augments the ability to recognize patterns and make evaluations, and consequently, amplifies intelligence (Woodward, 1992). Using VR might actually improve reading and writing skills (Pantelidis, 1993).

Limitations of VR

At the present time, the primary drawback to VR is the cost of a system. Other limitations include the quality of the display, inadequate computer processing speed, and constraints of input and output devices. However, we are making continual progress on all of these limitations. Increasing performance of computers with accompanying price drops will assure a steady improvement in VR, regardless of developments in peripherals.

Some VR systems far exceed the budgets of most schools. Prices run as high as a quarter of a million dollars for some systems (Fritz, 1991). Cheaper systems run on computers with slower processors, but they compromise realism. The cost and complexity of the most expensive, full-immersion systems may limit applications of these superior units to situations that require their unique features (Pausch, 1993). For example, medical education could be enhanced by a system that allows students to probe a 3D CAT-scan.

The low quality of the display is another typical limitation of a VR system. VR headgear has poor resolution (Biocca, 1992a). Most VR systems have rather crude figures (Fritz, 1991). Educators must resolve a trade-off between very expensive displays and those with low resolution (Meyer and Dunn-Roberts, 1992).

The replication of real-time images in a VR system is generally inadequate. VR "is not quite viable" because the systems are not quite fast enough (Dutton, 1992). A screen may be redrawn or refreshed as infrequently as two times per second (Bains, 1991). Improvements are needed in speed, accuracy, and user mobility (Bricken, 1991). The speed of a VR system affects not only convenience but also health. At a display rate less than 60 frames per second, there is a "latency" or lag time between when the user's inner ear registers body movement and when the system updates the view. This delay can make the user suffer from motion sickness (Industry visualizes real uses for virtual worlds, 1993). The reports of simulation sickness are "persistent" and "very well documented" from research on military simulators (Biocca, 1992a). Adverse reactions to VR can occur immediately or as much as 18 hours after use (Greenfield, 1994). VR may even trigger a seizure disorder. The potential health disadvantages of VR are particularly important to consider in an educational setting.

Criticisms of the input devices focus on quality and expense. Input devices are cumbersome and expensive, and the tracking devices are limited (Schmitz, 1993). The tactual and force feedback of the output devices is limited (Meyer and Dunn-Roberts, 1992). The bulk and confines of the input and output devices inhibit user mobility (Bricken, 1991).

A final drawback of VR involves a suspected weakness in system effectiveness. Since VR systems are designed to offer a multitude of options, a student may focus on only what he/she experiences and, consequently, learn very little about the global aim of the package (Woodward, 1992).

Personal Digital Assistants (PDAs)

PDAs are by far the newest technology included in this chapter. Although they have been introduced to the market only in this decade, PDAs have attributes that may result in more widespread use than either AI or VR within a short time. In personal use, PDAs encounter less serious limitations than more complicated AI and VR systems, which usually depend on institutional policy.

The most basic problem with describing a personal digital assistant is that "there is no industry wide consensus on what constitutes a PDA" (Halfhill, 1993, p. 69). Some experts even claim that PDAs are misnamed, as "they do not offer you any assistance" (Newquist, 1994, p. 51). A PDA is a very portable computer that is generally equipped with an array of software applications such as an address book, an appointment book, a notepad, and other general-use programs. The hardware and software architecture of PDAs differs from that of standard personal computers. Options include fax capabilities and cellular phones. PDAs are compact, portable, battery operated, easy-to-use devices, usually with pen-operated input systems. PDAs are also characterized by what they lack: wires, keyboards, and high price tags. The designs of PDAs reflect the needs of their targeted users—people who do not currently use another kind of personal computer and may never buy one.

Components of PDAs

Although a PDA is similar in size to the smallest notebook and palmtop computers, it has unique hardware and software components. The display is usually sensitive to a pen stylus that allows input directly on the screen.

Personal Digital Assistant (PDA)

Source: Courtesy of Apple Computers, Inc. (Photographer, Frank Pryor).

The screen usually shows a limited selection of preprogrammed icons that access certain software functions, but the majority of the display provides an area for handwritten input. The PDA can convert handwriting to standard text by an available preprogrammed function or store the input as written. Beyond providing a unique but familiar input interface, the other major hardware component of PDAs is a means to transmit data via a wireless connection. Most PDAs provide for data transmission over fax systems, though they sometimes have the capability to send but not receive information.

Samis (1994) overviews the options available with various brands of PDAs. PDAs were generally not designed to run the same software packages that operate on desktop and notebook computers. The standard PDA software functions revolve around an input capability for messages, a display capability for scheduling, and an output transmission capability. Although the messaging and scheduling functions allow students to take notes and plan activities, most educational settings need to transmit data to students more than they need to receive transmissions from the students. Some PDAs have additional communications functions, with corresponding increases in price.

Although not completely unique to PDAs, credit-card-sized PCMCIA expansion slots are typically included to add electronic devices that expand, enhance, or replace internal capabilities. Many expansion capabilities that originally required standard-sized expansion slots within the computer are now available in the PCMCIA configuration.

Capabilities of PDAs

PDAs provide many typical computing capabilities that are standard to notebook, laptop, and desktop computers, but in easily transported packages. Because of their size limitations, PDAs have an architecture that does not accommodate most standard software packages. However, PDAs generally are not intended to provide these features. The user sacrifices the wide array of options offered by standard software for a smaller number of popular, easy-to-use options in a format that allows mobility.

Current Applications of PDAs in Education

Although interest in PDAs is growing rapidly, the number of applications in education is currently limited. In some current applications, teachers use PDAs to track student development. Students also use notebook computers as PDAs for word processing during class.

Teacher information management tool. The Florida Schoolyear 2000 initiative is a multiyear program designed to develop a model to enhance school production (Tycho, 1993). As part of that initiative, a teacher information manager called *Tycho* is specifically designed to meet the needs of teachers to record and retrieve information about student learning and transactions with students. The software also contains scheduling and curriculum information to indicate appropriate instruction for a student at a specified level. The system fills the need of the teacher to record more information than is currently practical. Tycho is initially being used by teachers of students 3 to 7 years old.

Student assessment tool. Learner profile software allows teachers to use PDAs to track student assessment in K–5 classes. The PDA offers the teacher a convenient way to assess repeatedly any student's demonstration of skill mastery. The software facilitates the assessment process by requiring only a touch entry by the student's name and outcome icon. The teacher can then rate the student's performance. The system also provides options for taking attendance and notes. PDA-based assessment software to support these activities is offered by companies such as Sunburst and Scholastic.

Student word processing. Although no examples of student word processing uses of PDAs were reported in the literature as of the time this book went to press, it is easy to imagine that such applications could become popular. Barrett (1994) reports that a teacher in Alaska used small "notebook computers" with high school, junior high school, and even elementary school students for word processing. The program provided every student in the class with a computer, leading to a noticeable increase in the amount of and enthusiasm about writing. The computing capabilities used by the students and described by the teacher as the optimal system are the functions of a typical PDA.

Potential Role of PDAs in Education

The mobile capabilities of PDAs are primary considerations in the future use of these devices in education. Conventional wisdom holds that the widespread use of PDAs will not be immediate. Yet predictions for 2000 assert that "wireless technology and PDAs will become so prevalent and important to personal computing, plugging in your computer will seem as impractical as driving a car attached to a cord" (Davis, 1994, p. 75). If that scenario truly develops, universal use of PDAs in the classroom could not be far behind.

Obviously, a PDA with an improved ability to translate handwriting would be an asset to students taking notes. Even better, the teacher could transmit class notes at the beginning of class that students could modify or enhance during class. If each student had a PDA with transmission capabilities, the logistics of attendance and test taking and correcting could be simplified as each one transmitted codes to signify his or her presence or answers to test questions. Using a program for a specific content area or even for a specific instructor, students could test themselves while traveling to school, eating lunch, or waiting for a class to start. PDA applications could improve the ability of students to gather information in the library and to utilize other resources. Students could begin outlining a project at school and quickly and simply transfer the file to a desktop computer at home to write a final draft. Creative teachers might also find ways to make homework more exciting using the PDA as the medium.

As a new product with a new architecture marketed by several companies in different versions, the role of the PDA in education hinges on how producers modify early models to suit the needs of educational users. These devices must offer capabilities to take notes and transmit information at least equal to the ease of the manual process. Other features that simplify the routine of students and teachers by saving time and/or work would help to determine the potential role for this technology in education.

Benefits of PDAs

The most obvious benefit of a PDA is to provide computing and communications in a package that is small and portable enough to be useful in an educational environment. Lighter than all but the smallest palmtop computers, a PDA also offers a stylus input device to make the interface more natural than the keyboard or mouse of a standard computer. Preinstalled software is designed to be consistent with features needed by the user in transit. Wireless communications capabilities are a decided advantage over typical desktop, wired configurations. Although current transmission capabilities may not be adequate for an educational environment, as the need or desire to communicate electronically with teachers, students, and resource materials develops, this may become the most important benefit of PDAs.

Limitations of PDAs

Although many have identified low price as a positive virtue of a PDA, the current prices of these units may seem high to educators. Current prices for PDAs start at about $500. Some are selling for thousands of dollars. Because of the price variation and problems with handwriting recognition, the technology has been labelled "still not ready for prime time" (Kantrowitz, 1993, p. 46). PDAs are not expected to become a mass market success until the price drops considerably. Fortunately, prices as low as $300 is the predicted threshold "for the market to take off" (Williams, 1993b, p. 18). When that happens, the devices could be an attractive option for educational uses.

Other limitations in PDAs have been cited. Insufficient memory and inadequate communications functions are limitations that prevent PDAs from fulfilling expectations (Allen, 1993). However, the PCMCIA slot provides a means for overcoming those limitations. As PCMCIA cards are developed to work with PDAs' operating systems, memory and communications limitations will be minimized or even eliminated. The PCMCIA slots also provide opportunities to enhance or replace limited software functions.

Summary

AI, VR, and PDAs are all emerging computer technologies that have both hardware and software components. Considerable instructional applications have been developed with AI and VR, although not many have been specifically designed for education. As the newest technology of the three, PDAs have just reached the market, but they could conceivably have more widespread use in education than either AI or VR within a short time. In probably the most important shared trait, AI, VR, and PDAs each have enormous potential for facilitating learning, both as supplemental tools and as equipment for students to develop projects. Each technology has some current form that can be incorporated into the learning environment, even if it is just for demonstration. As the costs decrease and the pre-programmed functions of each technology increase, assimilation into education will offer unlimited means for teachers to provide the most appropriate learning settings and activities for individual students.

Exercises

Exercise 10.1. Identify the correct emerging technology related to each term by noting *AI, VR,* or *PDA* beside each item:

_____ 1. Data glove

_____ 2. Handheld computer

_____ 3. Fuzzy logic

_____ 4. Head-mounted display

_____ 5. Expert system

_____ 6. Handwriting translator

_____ 7. Shared electronic space

_____ 8. Pen-operated input device

_____ 9. Neural network

_____ 10. Full-immersion system

Exercise 10.2. Identify the kind of emerging technology appropriate for each of the following educational needs by noting *AI, VR,* or *PDA* beside each item:

_____ 1. Architecture students need to be able to see a house "from the inside out" as it is being constructed. Ideally, they could visualize entering the house and walking around in it as it is built around them. They should be able to gather information by pointing to things and touching them.

_____ 2. A teacher writes notes about students' performance as he walks around the classroom looking at their work. He would like to have the information he writes automatically transferred into a computer file that he could later use in assessment.

_____ 3. Students can jot down outlines or make notes on an assigned topic wherever and whenever ideas occur to them and transfer the text to a computer at home or school to finish writing reports.

_____ 4. Students at different sites can collaborate on development projects in a shared electronic space.

_____ 5. An instructional system is designed to monitor each student's progress through a tutorial and give suggestions on more efficient learning strategies each student might use in the future.

References

Allen, D. (1993). Who will define the PDA? *Byte, 18* (10), 10.

Anderson, G. (1993). A guide to fuzzy logic. *Product Design and Development, 12,* 21–22.

Bains, S. (1991). Surgeons slice virtual leg. *New Scientist, 131,* 28.

Barrett, H. B. (1994). Notebook word processors in the classroom. *The Computing Teacher, 22* (2), 69–70.

Biocca, F. (1992a). Communication within virtual reality: Creating a space for research. *Journal of Communication, 42* (4), 5–22.

Biocca, F. (1992b). Virtual reality technology: a tutorial. *Journal of Communication, 42* (4), 23–72.

Bricken, M. (1991). Virtual reality learning environments: Potentials and challenges. *Computer Graphics, 25* (3), 178–184.

Charp, S. (1994). Editorial. *Technological Horizons in Education Journal, 22* (3), 8–10.

Coles, L. S. (1994). Computer chess: The drosophila of AI. *AI Expert, 9* (4), 25–31.

Davis, F. C. (1994). Windows in the wireless age. *Windows Sources, 2* (4), 75.

Delaney, B. (1993). Where virtual rubber meets the road. *AI Expert Virtual Reality Special Report,* 15–18.

Dibble, J. (1996). The race to build intelligent machines. *Time, 147, (13),* 56–58.

Dillon, R. (1994). Creating an expert system. *The Computing Teacher, 21* (7), 17–20.

Drascic, D. (1993, June). Stereoscopic vision and augmented reality. *Scientific Computing & Automation,* 31–34.

Dreyfus, H. L., and Dreyfus, S. E. (1984). Putting computers in their proper place: Analysis versus intuition in the classroom. *Teachers College Record, 85* (4), 578–601.

Dutton, G. (1992). Medicine gets closer to virtual reality. *IEEE Software, 9,* 108.

Ferrington, G., and Loge, R. (1992). Virtual reality: A new learning environment. *Computing Teacher, 19* (7), 16–19.

Fogarty, T. J., and Goldwater, P. M. (1993). An expert system for accounting education. *Technological Horizons in Education Journal, 21* (3), 89–91.

Fritz, M. (1991). The world of virtual reality. *Training, 28* (2), 45–47, 49–50.

Gertner, G., and Guan, B. T. (1990). Conceptual classification: An AI alternative to statistical procedures for classification problems. *AI Applications in Natural Resource Management, 4* (4), 25–32.

Golob, R., and Brus, E. (1990). *The almanac of science and technology.* Boston: Harcourt Brace Jovanovich.

Greenfield, R. P. (1994). Sick in cyberspace. *Newmedia, 4* (4), 27–28.

Halfhill, T. R. (1993). PDAs arrive but aren't quite here yet. *Byte, 18* (11), 66–86.

Helsel, S. (1992). Virtual reality and education. *Educational Technology, 32* (5), 38–42.

Hill, M. (1992). What's new in virtual reality? *Electronic Learning, 12* (2), 10.

Industry visualizes real uses for virtual worlds. (1993). *Machine Design, 65* (15), 12; 14.

Kantrowitz, B. (1993). This is your life. Maybe. *Newsweek, 122* (20), 45–46.

Kantrowitz, B., and Ramo, J. C. (1993). An interactive life. *Newsweek, 121* (22), 42–44.

Kasparov, G. (1996) The day I sensed a new kind of intelligence. *Time, 147* (13), 55.

Kerney, C. A. (1993). How soon will teachers use virtual reality in the classroom? A discussion of the proposed use of virtual reality in San Diego magnet schools. *Technology and Teacher Education Annual.* 178–181.

Lanier, J., and Biocca, F. (1992). An insider's view of the future of virtual reality. *Journal of Communication, 42* (4), 150–172.

Levy, M. R. (1992). Editor's note. *Journal of Communication, 42* (4), 3–4.

Marshall, D. (1991). A make-believe world. *Engineering, 231* (6), 15–16.

Merrill, J. R. (1993). Surgery on the cutting-edge. *Virtual Reality World, 1* (3 and 4), 34–38.

Meyer, C., and Dunn-Roberts, R. (1992). Virtual reality: A strategy for training in cross-cultural communication. *Educational Media International, 29* (3), 175–180.

Mission accomplished (1993a). *NASA Tech Briefs, 17* (7), 16, 18.

Mission accomplished (1993b). *NASA Tech Briefs, 17* (9), 14.

Moore, N. (1993). How to create a low-cost virtual reality network. *Educational Media International, 30* (1), 37–39.

Neches, R., Fikes, R., Finin, T., Gruber, T., Patil, R., Senator, T., and Swartout, W. R. (1991). Enabling technology for knowledge sharing. *AI Magazine, 12* (3), 36–56.

Neural-Network Object-Recognition Program (1993). *NASA Tech Briefs, 17* (9), 103–104.

Newquist, H. P. (1994). Random access AI. *AI Expert, 9* (4), 50–51.

Nilan, M. S., Silverstein, J. L., and Lankes, R. D. (1993). The VR technology agenda in medicine. *AI Expert Virtual Reality Special Report,* 33–37.

Orwig, G. W., and Baumbach, D. J. (1991–1992). Artificial intelligence. *SIGTC Connections, 8* (2), 11–12.

Pantelidis, V. S. (1993). Virtual reality in the classroom. *Educational Technology, 23* (4), 23–27.

Patton, A., Swan, D. M., and Arikara, M. (1993). Modeling car batteries with neural networks. *Machine Design, 65* (21), 133–134.

Pausch, R. (1993). Three views of virtual reality. *Computer, 26* (2), 79–80.

Quinnell, R. A. (1993). Software simplifies virtual-world design. *EDN, 38* (24), 47–54.

Raghavan, K. and Katz, A. (1989). Smithtown: An intelligent tutoring system. *Technological Horizons in Education Journal, 17* (1), 50–53.

Rheingold, H. (1993). Virtual reality, phase two. *AI Expert Virtual Reality Special Report,* 7–10.

Robertson, G. G., Card, S. K., and Mackinlay, J. D. (1993). Nonimmersive virtual reality. *Computer, 26* (2), 81, 83.

Roblyer, M. D. (1993). Technology in our time. *Educational Technology, 33* (2), 33–35.

Salvadore, R. (1995). What's new in artificial intelligence? *Electronic Learning, 14* (4), 14.

Samis, M. (1994). The digital apprentice. *Electronic Learning, 14* (2), 42–46.

Schank, R. (1990). Case-based teaching: Four experiences in educational software. *Interactive Learning Environments, 1* (4), 231–253.

Schmitz, B. (1993). Watching the trends. *Machine Design, 65* (15), 28–32.

Sensing and Force-Reflecting Exoskelton. (1993). *NASA Tech Briefs, 17* (10), 98–99.

Shapiro, S. C. (1990). *Encyclopedia of artificial intelligence.* New York: John Wiley & Sons.

Sprague, D. (1996) Virtual reality and pre-college education: Where are we today? *Learning and Leading with Technology, 23* (8), 10–12.

Spring, M. B. (1993). The virtual library. *Virtual Reality World, 1* (3 and 4), 53–66.

Stanfel, J. (1993). Virtual cities—a regional discovery project. *Educational Media International, 30* (1), 42–45.

Tait, A. (1993). Authoring virtual worlds on the desktop. *AI Expert Virtual Reality Special Report,* 11–13.

Tinker, R. (1992). Skip the glitz. *Electronic Learning Special Edition, 12* (1), 26.

Tycho (1994). The Schoolyear 2000 project. Tallahassee, FL: Florida State University, Center for Educational Technology (CET).

Walbridge, C. T. (1989). Genetic algorithms: What computers can learn from Darwin. *Technology Review, 92* (1), 47–53.

Williams, T. (1993a). Getting realistic about neural nets. *Computer Design, 32* (9), 20.

Williams, T. (1993b). PDAs comprise next wave of embedded systems. *Computer Design, 32* (11), 18.

Winn, W., and Bricken, W. (1992). Designing virtual worlds for use in mathematics education. *Educational Technology, 32* (12), 12–19.

Woodward, J. (1992). Virtual reality and its potential use in special education: Identifying emerging issues and trends in technology for special education. ERIC Document Reproduction No. ED 350 766.

Associations

American Association for Artificial Intelligence, 445 Burgess Drive, Menlo Park, CA 94025-3496

Educational Technology Center, Harvard Graduate School of Education, Nichols House, Appian Way, Cambridge, MA 02138

ERIC Clearinghouse on Information and Technology, Center for Science and Technology, 4th Floor, Room 194, Syracuse University, Syracuse, NY 13244-4100

International Neural Network Society, Suite 300, 1250 24th Street NW, Washington, DC 20037

International Society for Technology in Education, 1787 Agate Street, Eugene, OR 97403-1923

International Technology Education Association, 1914 Association Drive, Reston, VA 22091

U.S. Department of Education/Technology Resources Center, 555 New Jersey Avenue, NW, Washington, DC 20208

Periodicals

AI Applications, P.O. Box 3066, Moscow, ID 83843
AI Expert, 600 Harrison Street, San Francisco, CA 94107
AI Magazine, La Canada, CA
Byte, One Phoenix Mill Lane, Peterborough, NH 03458
Computer Design, 10 Tara Blvd., Nashua, NH 03062-2801
Digital Video Magazine, 80 Elm Street, Peterborough, NH 03458
Electronic Design, 1100 Superior Avenue, Cleveland, OH 44114-2543
EDN, 8773 South Ridgeline Blvd., Highlands Ranch, CO 80126-2329
Educational Media International, 120 Pentonville Road, London, N1 9JN, UK
Journal of Communication, 8140 Burnet Road, P. O. Box 9589, Austin, TX 78766
Machine Design, 1100 Superior Avenue, Cleveland, OH 44114-2543
New Directions for Teaching and Learning, 350 Sansome Street, San Francisco, CA 94104-1310
Newmedia, 901 Mariner's Island Blvd., Suite 365, San Mateo, CA 94404
Pen Computing, P. O. Box 640, Folsom, CA 95763-0640
Personal Engineering and Instrumentation News, Box 430, Rye, NH 03870
Scientific Computing & Automation, 301 Gibraltar Drive, Morris Plains, NJ 07950-0650
Virtual Reality Special Report, 600 Harrison Street, San Francisco, CA 94107
Virtual Reality World, 11 Ferry Lane West, Westport, CT 06880

Example PDAs

Boss—Casio, Inc., 800-962-2746
Envoy—Motorola, Inc., 708-576-1600
Magic Cap—General Magic, Mountain View, CA, 415-965-0400
Newton MessagePad—Apple Computer, Los Angeles, CA, 408-996-1010
Wizard—Sharp Electronics Corp., 800-237-4277
Zoomer—Tandy, 500 One Tandy Center, Fort Worth, TX 76102

Example AI Programs for Education

Acquire—Acquired Intelligence, Victoria BC, 604-479-8646
Cortex—Resolution Software, Nottingham, UK, 0602-206801
Derive—Soft Warehouse, Inc., Honolulu, HI, 808-734-5801

Example Virtual Reality Programs

Cyberspace Developer Kit—Autodesk, Inc., 111 McInnis Parkway, San Rafael, CA 94903
Provision—Division, Inc., 400 Seaport Court, Suite 101, Redwood City, CA 94063
Superscape—Dimension International, Zephyr One, Calleva Park, Aldermaston, Berkshire RG7 4QZ, UK
Virtus VR—Virtus Corp., 118 MacKenan Drive, Suite 250, Cary, NC 27511
VREAM—VREAM Inc., 2568 N. Clark Street #250, Chicago, IL 60614
WorldToolKit—Sense8, 100 Shoreline Highway, Suite 282, Mill Valley, CA 94941

Part IV

Integrating Technology into the Curriculum

The chapters in this part will help teachers learn:

to design instructional activities that integrate technology using either a single subject or an interdisciplinary approach.

255

Introduction

This part of the book addresses the use of technology in various subject areas. In an effort to model subject-area integration, the disciplines have been grouped in the following manner:

Chapter 11: Technology in Language Arts and Foreign Language Instruction

English and foreign language technology applications are grouped together in this chapter, since they are both language-related topics. Language arts in this chapter includes the communications skills (reading, listening, speaking), addressed primarily in the elementary grades, as well as English topics (writing and literature), which are the focus at secondary levels. Foreign language topics include the learning of foreign languages and English for Speakers of Other Languages (ESOL).

Chapter 12: Technology in Science and Mathematics Instruction

Science and mathematics topics are considered closely related and curriculum for them are often intertwined. This chapter looks at how technology applications help integrate the teaching of these topics, and how it addresses the special curriculum needs of each.

Chapter 13: Technology in Social Sciences Instruction

Technology applications covered in this chapter include those for history, social studies, civics, and geography.

Chapter 14: Technology in Music and Art Instruction

This chapter focuses on technology applications for the topics in arts instruction to which the majority of K-12 students are exposed: music (appreciation, theory and performance) and art (drawing, painting, and image production).

Chapter 15: Technology in Exceptional Student Education

This is the only chapter in this book identified by population rather than by topic. Technology applications for Exceptional Students (ESE) are addressed in Part IV because curriculum for "special students" has many unique characteristics. Needs of special students addressed in this chapter include those for learning disabled (LD, and Specific Learning Disabilities or SLD) and mentally handicapped (e.g., Educable Mentally Handicapped or EMH and Emotionally Handicapped or EH); physically disabled (e.g., hearing impaired, visually impaired, nonspeaking, wheelchair-bound); and gifted and talented students.

Trends in Technology Integration

Interdisciplinary and Subject-Area Emphasis: An Overview

This text focuses on how to integrate technology into the school curriculum. Although that sounds like a fairly straightforward process, it has been one of the major stumbling blocks to the effective use of technology in education. The discussion of integration strategies begins with a look at the general concept of curriculum structure.

The debate over interdisciplinary studies versus single-subject emphasis is not new in education. Plato thought that interdisciplinary instruction would allow a student to gain a broad understanding of complex ideas. Aristotle offered a different opinion: that pure knowledge of a discipline is necessary to gain true understanding. The debate over breadth versus depth has continued to this day. The traditional structure of the curriculum has usually divided it into discrete subject areas, teaching each in isolation. Curriculum designers saw this as the most efficient way to assure instruction is given in specific skills for a variety of content areas. In this century, many educators have sung the praises of interdisciplinary instruction. They cite two primary reasons for this approach: It reflects the true nature of objects and events in the world, and it reflects the way children think and learn.

Today, many point out reasons to shift the debate from depth *or* breadth to depth *and* breadth (Jacobs, 1990; Jacobs and Borland, 1986). People today live in a world with an ever-increasing body of knowledge, so educators must find a way to balance discipline-specific knowledge with insight into interconnections between the disciplines. Technology can help accomplish this complex goal in many ways.

Background on Subject-Area Emphasis: The Way Things Have Been

The climate in education today encourages an emphasis on interdisciplinary teaching, but this has not been an easy challenge to face. For many years, teachers have set a norm of delivering subject-specific instruction in a reductionist fashion. This reflects both the beliefs and the traditions of a large number of educators. The very structure of middle schools and high schools often gravitates around teaching a single subject at a time.

To a great extent, the integration of technology into the curriculum has followed a similar pattern. Software companies and developers have geared their products to traditional subject areas, and this method of integrating technology persists widely today.

Background on Interdisciplinary Emphasis: The Way Things May Be Going

The trend toward interdisciplinary instruction also appears likely to continue for a long time. It may, in fact, reflect a much larger societal trend. Many facets of society, from business to medicine, recognize that systems work much better when there are connections between components. If this pattern continues and is accepted as a long-term philosophy, the pressure on schools to teach subjects in an interdisciplinary fashion seems likely to escalate.

Evidence supports the presumption that teachers who become more comfortable with technology in their classrooms seem to take more flexible, even experimental, approaches to teaching (Sheingold and Hadley, 1990). This trend, coupled with a concerted effort by many software companies to offer products that model subject-area integration, portends an integrated environment for technology education.

Structure of Each Chapter

Issues and Problems Related to Technology Use in the Content Area

Chapters in this section provide information to support both single-subject and interdisciplinary integration strategies. Each chapter briefly describes current issues in a specific content area that affect the use of instructional technology. Most teachers ask not whether to use technology, but rather when, how, and for what to use it. Considering the dynamic nature of technological development, answers to these questions are likely to change. A broad consensus on the best ways to integrate technology seems elusive, but one of the biggest benefits of the infusion of technology into schools may come from the momentum that it has generated for reflection about teaching and learning.

Recommended Resources and Applications

Each chapter suggests general resources and applications for its subject area. Additionally, where applicable, the chapters recommend resources that are specific to directed instruction or the constructivist philosophy. Considering the plethora of software titles that are available, any attempt to provide comprehensive coverage of specific applications would be futile. Therefore, recommendations for specific titles emphasize general principles. The suggestions are designed to provide the reader with a feel for the educational benefits that instructional courseware promises to deliver.

Example Lesson Activities

Activities in Chapters 11 through 15 represent both resources and models for curriculum integration. Teachers are encouraged to use these activities and elaborate on them, customizing them to meet students' needs and match facility resources.

Example lesson activities: Interdisciplinary/thematic.

Some example lesson activities in this section of each chapter emphasize current trends in education (and in the larger society). They model a group-focused, integrative-thematic approach to infusing technology into the curriculum. They also stress higher-order thinking skills and workplace competencies. This approach and these types of skills respond to priorities identified by the Department of Labor's landmark study, the SCANS (Secretary's Commission on Achieving Necessary Skills) report (U.S. Department of Labor, 1992).

Example lesson activities: Single-subject emphasis.

In recognition of the emphasis in many schools on delivering instruction in a single subject at a time, Chapters 11 through 15 also include example lessons for single-subject curriculum. As with the interdisciplinary activities, these lessons also stress many SCANS competencies and/or basic skills.

Exercises: Designing Activities.

Exercises included in each chapter help the reader to develop a sense of proficiency in integrating technology into either a single-subject or integrated curriculum.

Preparing for Curriculum Integration

General Technical Skills: What Teachers Need

Stakenas, Tishkin, and Resnick (1992) suggest five areas of technological competence for K-12 classroom teachers:

- **Basic knowledge about computer technology.** Teachers must have a general grasp of how computers work and be able to use basic computer terminology. This gives them the vantage point from which to consider the social, economic, and ethical impact of computers on society.

- **Equipment operations skills.** Teachers must be able to perform standard computer operating procedures (e.g., formatting disks, loading and running programs, saving files, and printing documents) as well as troubleshooting for minor problems. Teachers should use a variety of input and output devices, including projection devices (LCD panels), interactive videodiscs, and modems as part of their instruction.

- **Productivity tools skills.** Teachers should use and teach word processing, database, spreadsheet, graphics, and desktop publishing software.

- **Instructional application skills.** Teachers must evaluate and use various types of specialized computer software (e.g., drill and practice, tutorial, simulation, and problem-solving packages) to accomplish specific educational objectives. From this base of experience, they can integrate appropriate applications of the computer in a variety of content areas using a variety of teaching/learning strategies, and they can design customized, multimedia learning activities.

- **Management application skills.** Teachers should use computers to manage and complete tasks such as recordkeeping, progress reports, report cards, attendance, worksheets, tests, letters to parents, and grade books. They should also understand electronic mail and local area networks (LANs).

This list of competencies should be viewed as an ideal to guide teacher development. It should not represent a list of essential skills that one must master before attempting technology integration.

General Technical Skills: What Students Need

Students of all ages are generally very willing to try new software applications, so teachers usually need not struggle to motivate students to use technology. Quality of use, however, can present a dilemma. Some students tend to play and experiment with a program without moving on to become proficient users. The teacher should stress to students that learning a computer program requires a substantial effort. Also, some students will become totally dependent on the teacher and ask for help with any problem. The teacher must balance the role of providing answers against the role of encouraging students to learn independently. The rate of change in the field of technology forces users to be able to learn as they work. Most software packages offer on-line help, and the teacher should encourage students to use this resource.

Preparation of the Classroom Environment

One of the biggest challenges that teachers face in integrating technology successfully is how to structure class time so that students use technology effectively and efficiently. Schools rarely have enough equipment to seat every student at a standalone computer or technology station. Teachers must plan for flexibility to make the best use of resources. The following table lists some guidelines for allocating limited technology resources:

Technology Available	Possible Arrangements
• One computer	Whole-class instruction with LCD panel or large monitor
• One computer	Whole-class instruction in cooperative groups with computer management of data (Tom Snyder's Decisions/Decisions software uses this model.)
• Limited computers	Classroom arranged in centers, some with computers, and students rotate through (This is common at the elementary level.)
• Limited computers	Cooperative group activity using computers as resources (CD-ROM encyclopedias or telecommunications stations support research in this way.)

Preparation of Equipment and Other Resources

Each addresses this question in its own way, but some general rules of thumb apply in all areas. When setting up for an activity, it is always best to test run an equipment setup before students arrive. In recent years, the reliability of equipment has improved considerably; however, small glitches invariably occur and often at the worse possible times. Many activities are disrupted by minor technical problems that do not require an expert to fix. It is well worth the teacher's time to become familiar with equipment troubleshooting procedures. Manuals often list such procedures. When equipment doesn't work, the cause is usually a small, easily corrected problem.

How to Design an Integration Strategy: Some Recommended Steps

- **Assess technology resources.** Determine what technology resources are available and for how long. Resources may be available at the school site or through a district media or technology center. Teachers often shy away from using technology in their teaching because they feel they lack reliable access to enough equipment. This is one of the primary obstacles to integrating technology into the classroom. The teacher can deal with this problem by viewing early implementation efforts as enrichment activities for both the students and the teacher.

- **Look for connections between what's available and what's needed.** Match curriculum needs to available resources. This is a good time to draw on resources such as colleagues with experience in teaching with technology. Lesson plans developed by others can also help.

- **Plan activities.** Use resources to plan lessons and activities. At this stage, the teacher decides whether to present single-subject or interdisciplinary activities. The chosen technology application may play a determining role. For example, an episode of the math videodisc program The Adventures of Jasper Woodbury teaches math within the context of physical science. It's natural at such a point to extend the study to science concepts. On the other hand, if a particular resource deals with only one content area, the teacher may decide to focus on that one subject. One of technology's great advantages for teachers is its flexibility in such roles.

- **Set up equipment.** This important and sometimes frustrating stage goes more smoothly if the teacher sets up everything before students arrive. Turn on all equipment beforehand and make sure that it is working properly. Student helpers can be a very good resource in dealing with equipment.

- **Try it!** A teacher may tell students that the class is trying something new and things may not work out as expected. This is a great opportunity to model risk-taking to the students.

- **Evaluate and revise.** Spend some time reflecting on classroom success and how to improve it. Think about alternative ways to set up equipment that might make things easier. It's also important to solicit feedback from students about how to improve activities. Teachers must understand that activities often do not work out as planned the first time. In the early stages of technology implementation, many teachers to feel that learning to use technology is more difficult than teaching without it. At these times, they need to persevere. Over time, their efforts will pay off, not only in better instruction for their students, but also in their own professional and personal development.

Practical Guidelines and Suggestions for Integration

The rest of this part introduction mentions a few general considerations for lessons that integrate technology. This discussion does not cover the subject comprehensively; instead, it describes a short list of important considerations.

Demonstrate Relevance

As they integrate technology into the curriculum, teachers should plan for a dynamic world and the future needs of their students. Students complain that they cannot see the connection between school and the real world. The SCANS report (1992, p. xvi) strongly recommends that educators teach "in context;" that is, students should learn content while solving realistic problems. Many technology applications, from multimedia programs to simulations, offer valuable resources to help teachers make real-world connections.

Strive to Prepare Students for the World in Which They Will Live

In developing activities and lessons for classroom use, the following list of workplace competencies and basic skills (SCANS, 1992) may serve as a useful reference for teachers. The report lists the workplace know-how that all workers of tomorrow will need in order to ensure solid performance.

Workplace competencies. Effective workers can productively use:

- **Resources.** They know how to allocate time, money, materials, space, and staff.
- **Interpersonal skills.** They can work on teams, teach others, serve customers, lead, negotiate, and work well with people from culturally diverse backgrounds.
- **Information.** They can acquire and evaluate data, organize and maintain files, interpret and communicate, and use computers to process information.
- **Systems.** They understand social, organizational, and technological systems; they can monitor and correct performance; they can also design or improve systems.
- **Technology.** They can select equipment and tools, apply technology to specific tasks, and maintain and troubleshoot equipment.

Foundation skills. Competent workers in the high-performance workplace need:

- **Basic skills.** Reading, writing, arithmetic and mathematics, speaking, and listening capabilities
- **Thinking skills.** The ability to learn, reason, think creatively, make decisions, and solve problems
- **Personal qualities.** Individual responsibility, self-esteem and self-management, sociability, and integrity

Utilize Community Resources

Technology offers tremendous potential to break down the walls between school and community. Students in business education classes at a high school in Florida fax their letters to local businesses to be proofread. The students get their letters faxed back the next day with suggestions for improvement. Some students have gone on to work with businesses that have seen the quality of their performance.

Educators will very likely apply the huge strides in distance learning and telecommunications in activities that will truly take advantage of the tremendous diversity of expertise in the community.

Plan for Technology's Effect on Teaching Practices

Integrating technology into teaching is no simple task, and teachers should expect the process to take time. Sheingold and Hadley (1990) conducted a nationwide survey of technology-proficient teachers. Those teachers reported working for up to 6 years to master computer-based teaching practices and approaches. As teachers become more comfortable with using technology in the classroom, researchers have observed shifts in teachers' behaviors and attitudes (Butzin, 1990; Butzin, Harris, and McEachem, 1991; Collins, 1991; Thomas and Knezek, 1991).

- **Teacher-centered to student-centered activities.** Teachers move from lecturing and directing to facilitating and coaching.
- **Whole-class to small-group instruction.** Teachers engage in fewer activities involving their entire classes at once. Instead, they promote activities involving small groups of students working independently.
- **Structured to exploratory instruction.** Teachers move away from rigid activities geared toward classroom control to open-ended activities and projects.
- **Competitive to cooperative activities.** Teachers involve students in fewer isolating, competitive activities and more activities that involve collaboration and cooperation.
- **Classroom to whole-world interactions.** Teachers expand the context of learning from issues and resources available in specific content areas to real-world problems that require interdisciplinary awareness and multiple resources.

While these guidelines certainly do not apply to every individual in exactly the same manner, they may help teachers to understand the ways in which teaching with technology may very well challenge them to reevaluate their basic beliefs about teaching and learning.

References

Butzin, S. M. (1990, March). Project CHILD: Not boring, but work that's fun and neat. *The Computing Teacher, 17* (6), 20–23.

Butzin, S. M., Harris, M., and McEachern, R. (1991, Spring). Project CHILD, a success story: Restructuring schools to tap technology's potential. *Florida Technology in Education Quarterly, 3 (*3), 33–44.

Collins, A. (1991). The role of computer technology in restructuring schools. *Phi Delta Kappan, 73 (*1), 28–36.

Jacobs, H. (1990). Interdisciplinary curriculum: Design and implementation. Alexandria, VA: ASCD.

Jacobs, H. and Borland, J. (1986). The interdisciplinary model: Theory and practice. *Gifted Child Quarterly, 30 (*4), 159–163.

Sheingold, K., and Hadley, M. (1990, September.) *Accomplished teachers: Integrating computers into classroom practice.* New York: Center for Technology in Education, Bank Street College of Education.

Stakenas, R. G.,Tishkin, D. P., and Resnick, M. M. (1992). *Best practices in developing teachers' knowledge and skills in using instructional technology.* Tallahassee, FL: Center for Policy Studies in Education, Florida State University.

Thomas, L. G., and Knezek, D. (1991, March). Facilitating restructured learning experiences with technology. *The Computing Teacher, 18 (*6), 49–53.

U.S. Department of Labor. (1992). *SCANS (The Secretary's Commission on Achieving Necessary Skills) report.* Washington, DC: U. S. Government Printing Office.

Chapter 11

Technology in Language Arts and Foreign Language Instruction

This chapter will cover the following topics:

- Current issues in language arts instruction

- Current issues in foreign language instruction

- Recommended resources and applications in language arts

- Recommended resources and applications in foreign language

- Sample lesson activities in language arts and foreign language instruction

Chapter Objectives

1. Identify current issues in language arts instruction and foreign language instruction that affect the selection and use of technology.

2. Describe the most popular uses for technology in the language arts and foreign language curricula.

3. Create instructional activities for language arts instruction and foreign language instruction that successfully integrate instructional technology.

"I should not like my writing to spare other people the trouble of thinking, but, if possible, to stimulate someone to thoughts of his own."
Ludwig Wittgenstein's *Philosophical Investigations* (1953).

Introduction

Communications skills (reading, listening, and speaking) are generally considered fundamental qualifications for basic literacy; most educators identify these skills as prerequisites to adequate performance in all other content areas. Curricula for the elementary to middle grades often group these skills together as "language arts" instruction, while later grades often cover them in combination or separately within English, literature, or writing courses. Many activities and technology resources are common to both levels and this chapter discusses them together, giving sample lesson activities for each level. In an effort to model integration of all subject areas, the chapter also addresses foreign language instruction. The crossover between language arts and foreign language instruction is unavoidable, especially as ESOL programs operate in many school systems.

Issues and Problems Related to Technology Use

Issues in Language Arts: Controversies and Trends

Background. The role of technology in teaching language arts has been shaped in large part by ongoing discussions and controversies about the best pedagogical strategies for this area. King and Vockell (1991) list and describe four instructional paradigms. The National Council of Teachers of English noted three of these in a 1980 statement of trends in the field (Mandel, 1980):

- **New Heritage model.** This perspective blends communications skills with an emphasis on literature as a means for learning about culture and sharing that knowledge with descendants.
- **Competencies model.** This model that assumes that "language arts skills can be broken down into components" (p. 6) and taught sequentially.
- **Process model.** This learner-generated (as opposed to teacher-generated) approach emphasizes each child's communication about his/her interaction with persons, places, and things.

King and Vockell (1991) added a fourth model that came into common use after NCTE's 1980 policy statement as a direct outgrowth of the process model:

- **Whole language model.** This principle focuses on meaning, rather than on language itself, in speech and literacy activities. It encourages children's self-expression in all forms.

Educators have developed technology resources to support instruction in language topics focused on each of these strategies; the tools teachers select and how they use the tools depend largely on their strategic preferences. Currently, the holistic or whole language strategy seems most widely used, although many educators still advocate combining instruction in subskills with holistic language processes. The following sections recommend resources that have general applications for all four models, and some that suit individual models.

Teachers should be prepared to address several issues prior to implementing any of the resources discussed in this chapter. Careful attention to some important guidelines will help to smooth efforts to incorporate technology resources into language arts and English instruction. When teachers neglect these concerns, promising resources can actually impede the instructional process.

Prerequisite technical skills. In language instruction perhaps more than any other content area, teachers must allow students enough time to become proficient in the use of equipment and software before assigning tasks on which they will be evaluated. Schwartz and Vockell (1988) have emphasized this need especially for word processing activities. They advise teachers to introduce word processing features patiently and gradually using readily available guides and assistants. Also, teachers should not expect that students' proficiency in one word processing package will automatically enable them to use another application. Students need time and practice to acquire a general facility that transfers to new resources.

Keyboarding issues. The controversy over whether or not to require keyboarding skills as prerequisite to any other computer use has inspired vigorous, ongoing debate that shows little sign of resolution. Whole language proponents in elementary schools seem to emphasize these skills less than business education teachers do, for example, but this is not always the case. This controversy affects teachers in language arts and English especially, since word processing software and other resources commonly used in these areas usually require a lot of typing. Teachers must decide whether to assign keyboarding practice based on knowledge of their students and the goals they are trying to achieve. If word processing is introduced in language arts activities at early grade levels, students usually acquire sufficient keyboarding skills by later grades to use any software. However, ten-finger typing is commonly considered a desirable skill to develop at some point in a student's education.

Issues in Foreign Languages: Controversies and Trends

Background. In recent years, educators have reevaluated methods of foreign language instruction. Traditionally, foreign language instruction followed the same basic pattern that dominated the teaching of English: the competencies model. Grammar was the backbone of this method, and students spent most of their class time learning the rules of grammar, memorizing vocabulary, or translating text. Over the past two decades, a new school of thought, commonly called *communicative language teaching,* has emerged. Although this trend has produced no clearly defined single model of communicative teaching (Whitley, 1993, p. 40), the term conveys "a broad ability to use language appropriately in natural situations." This translates to less classroom emphasis on rules of grammar and much more time spent speaking and listening to the second language, especially within the context of relevant dialogue.

Teaching culture. Many advocates of foreign language instruction promote the goal of teaching culture. Lambert (1967) argued that foreign language classrooms can realize their potential for intercultural growth only if students learn languages from the beginning in a setting that provides carefully developed cultural contexts.

Herron and Hanley (1992) found that children retained information about French culture after viewing a brief video with audio in the language related to but not identical to the content of a subsequent reading. This certainly could have implications for instruction in all foreign languages. A number of Level I videodisc programs try specifically to teach foreign culture (e.g., World Cultures and Youth Series, Mexico Vivo). Cultural instruction also needs to help students unlearn stereotypes acquired from popular media. The videodisc player provides a valuable resource for critically examining segments from movies that may promote negative cultural stereotyping.

Why should students study foreign languages? Foreign language educators think seriously about the issue of motivating students. In an attempt to convince students of the importance of foreign language courses, many highlight the critical importance of knowledge of foreign languages to the nation's role in the emerging global economy. However, others urge more idealistic motives; they stress the essential role of language skills in the quest for world peace and cooperation. Neither of these abstract goals are likely to have much influence on most students. To motivate students, foreign language teachers must provide intrinsically interesting, challenging, and rewarding instruction. Many believe that technology can make a major contribution to this effort. Telecomputing, videophones, and distance learning all facilitate much more immediate payoff for studying foreign languages as students from different countries learn to communicate in real time.

English for speakers of other languages (ESOL). Much of the change in foreign language instruction has built on innovations in ESOL and immersion programs. Salomone (1991) lists six successful techniques of immersion teachers that can transfer to all areas of language teaching: conducting housekeeping tasks in the second language, using nonverbal cues extensively, involving students as peer teachers, struc-

turing classroom routines, integrating various subject matter areas, and asking higher-order questions. These techniques can carry over to technology-oriented lessons as well.

Quality of computer software. Chun and Brandl (1992) contended that most of the foreign language software available today amount to nothing more than form-based drill and practice programs. Even those labeled *interactive* usually just provide feedback based on form and do not relate form to its meaning or communication goal. Computer-Assisted Language Learning (CALL) programs should offer a pedagogy centered on meaning that enables students to enter whole sentences as opposed to common one-word answers. In the future, speech recognition capabilities and other enhancements can support a major impact for instructional technology on foreign language instruction.

Distance learning. With the demand for foreign language instruction increasing and shortages of qualified teachers in many areas, distance education through satellites (as described in Chapter 9) may become extremely important. This method has provided a way for students to study some of the more exotic languages like Chinese, Japanese, and Russian despite local scarcity of qualified instructors. However, the technology is still expensive, and questions concerning its effectiveness in foreign language instruction still need answers.

Recommended Resources and Applications

Resources and Applications for Language Arts Instruction

Word processing. Generally viewed as the most versatile and powerful software for teaching language topics, a word processing package may have various features depending on its purpose and intended users. (See Chapter 5.) For example, word processing software designed for early elementary students may include draw features, easily imported graphics, and text-to-speech capabilities. (e.g., Kid Works II, Creative Writer, The Amazing Writing Machine). For upper elementary students, programs often include spell checkers and thesauruses; screen displays often look more mature in their appearance (e.g., The Writing Center, Children's Writing and Publishing Center). Middle school and high school students usually have no difficulty adapting to adult applications (e.g., Microsoft Works, ClarisWorks, WordPerfect Works). These applications offer word processors within integrated packages that enable users to capture information from databases and spreadsheets.

Desktop publishing. This kind of software usually features more powerful formatting capabilities than word processing software. It allowed users to develop documents such as newsletters and brochures that require more complicated layout and graphics capabilities. Desktop publishing functions usually differ from those of word processors in that they lay out and format the text and graphics of each page, rather than creating files of continuous text and then dividing them into pages; desktop publishing packages also tend to have more graphics and draw capabilities. Recently, much of the distinction between word processing software and desktop publishing software has disappeared with word processors including more desktop publishing features with each upgrade (e.g., WordPerfect, Microsoft Word, MacWrite Pro).

Many high school publishing classes use powerful DP programs (e.g., Pagemaker, QuarkXPress) to lay out yearbooks. These programs require large investments of time to master their intricacies, but they have tremendous capabilities. In a recent development, a powerful DP program, Microsoft Publisher, lets the user choose from a number of layout options, after which the program automatically sets up the layout. Elementary students use DP programs to develop big books, posters, banners, and newsletters (e.g., Toucan Press, Big Book Maker, Super Print).

Telecommunications. The capability of communicating via computer with other students has provided unexpectedly powerful support for language arts and English activities. Students frequently seem to be more motivated to write well (i.e., communicate more clearly and use better spelling and grammar) when they write for audiences at other sites (Cohen and Riel, 1989). Writing often takes the form of letters, but several teachers have designed successful telecommunications projects around creative writing themes such as developing and comparing reports, poems, and stories. The rapid growth in the quantity and quality of World Wide Web sites provides another major resource for educators and their students. This hypermedia component of the Internet offers students a wide variety of resources. Some examples include:

- **Publishing.** KidPub will upload students' work onto its home page. (For information, email KidPub@en-garde.com.) Thousands of people visit this site every week to read stories by students from all corners of the world. In addition, International Student Newswire (http://www.umass.edu/Special Programs/ISN/KidNews.html) allows students to write and present the news. In perhaps one of the most empowering uses of the WWW resource, Web66: A K-12 World Wide Web Project (http://web66.coled.umn.edu) provides students with online access to Web pages of schools around the world as well as information about creating a home page (see Figures 11.1 and 11.2 for sample home pages).

- **Research.** Additionally, numerous resources are designed to assist students in the research process: Kids Web (http://www.npac.syr.edu:80/textbook/kidsweb), Kids on Campus (http://wwwtc.cornell.edu/Kids on Campus/WWW Demo), and EdWeb (http://edweb.cndr.org:90).

Instructional software for language arts. In addition to word processing, which underlies many whole language and process writing activities, several other kinds of soft-

Figure 11.1   KidPub Home Page

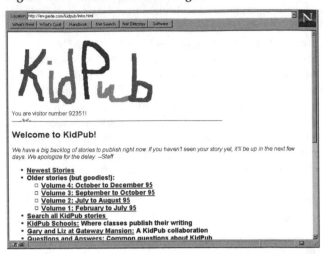

ware are available to support and encourage correct language and sentence structures in students' writing. According to Merrill, Hammons, Tolman, Christens, Vincents, and Reynolds (1992), software to support skill building within process writing includes spell checkers and grammar checkers (e.g., Grammatik), style analyzers, and other language analyzers. A new genre of more comprehensive language and reading programs are also seeing widespread use. IBM's Writing to Read program uses both CAI and word processing software in a complete early reading and writing program. Jostens Learning Corporation's Tapestry software generates a multifaceted story construction environment in which children learn about language and its creative uses.

Most software devoted to teaching reading subskills is based on the more traditional types of computer-assisted instruction (e.g., drill and practice, instructional games, tutorials). A variety of these kinds of software packages are available to teach prereading skills (e.g., Playroom,

Figure 11.2   Book Lures Home Page

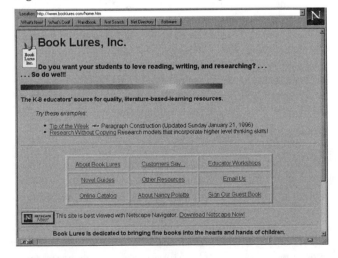

Muppet Word Book), vocabulary and other word skills (e.g., The Vocabulary Game, Crossword Magic) comprehension skills (e.g., Milliken's Comprehension Power), speed reading (e.g., Speed Reader II), and Cloze skills (e.g., M-ss-ng L-nks). Available packages can also analyze text readability levels to help teachers assign appropriate reading materials for students.

As with reading skills, grammar, usage, spelling, and punctuation skills are also addressed through traditional drill and practice, instructional game, and tutorial software. These packages frequently seem to induce students to spend more time practicing subskills on the computer than they would spend on pen-and-paper activities.

Several kinds of software can help teachers emphasize writing subskills. Creative aids to motivate students to write include Story Tree. Packages teach prewriting subskills such as outlining. Revision tools (e.g., RightWriter, Ghostwriter) motivate students to correct their work. Activity files for word processing packages provide exercises on various writing skills.

Level I and Level III videodiscs. Language arts teachers have used videodisc technology primarily in the areas of reading and literature. Some of these programs present the film versions of classic stories on disc and provide the user with bar codes to access scenes that support analysis of the story (e.g., *To Kill a Mockingbird, How the Elephant Got His Trunk)*. Other discs support only follow-up activities for specific books or sections. They offer supplemental material such as interviews with authors, illustrators, or critics (e.g., About Huckleberry Finn, Silver Burdett's Masterpiece Series). Recently, a number of major textbook publishers have begun marketing videodiscs as supplements to their reading series. These resources differ in their approaches but most offer comprehensive support materials in an effort to make the materials user friendly.

Level III reading and literature programs promote in-depth study of literature or genres. Many include the full text of their selections along with audio and video supplements. These programs typically combine widely varied resources, from glossaries to maps, which enable users to pursue individualized paths of study (e.g., Old Mother Witch, Literature Navigator Series). EduQuest's Illuminated Books and Manuscripts uses both videodisc and CD-ROM resources to help students understand and research important documents and literary works. It offers readers a unique variety of commentators with differing opinions on selected documents.

CD technology. With its tremendous storage and search capabilities, CD-ROM technology stores huge anthologies of work that give users access to phrases, quotes, or specific scenes, plus the ability to print or copy text (e.g., Shakespeare: The Complete Works). Both CD-ROM and CD-I technologies incorporate multimedia applications into reference materials (e.g., Compton's Multimedia Encyclopedia, A Visit to the Smithsonian). A number of

interactive books make the computer read to primary-grade children, and they offer teachers a variety of set-up options (e.g., Discus Books, Brøderbund's Living Books, Stories and More, Bravo Books).

Additional resources for directed instruction. A number of reading programs have been developed primarily for network implementation. These usually offer management systems that record scores and reading levels, and they help users isolate specific areas for needed improvement. Some of these programs, like Literature Series, focus on literature instructions, while others emphasize skills (e.g., Adult Education Reading). Most of these types of programs cover many reading levels. The publishers often market these packages as suitable resources for ESOL and adult remedial instruction.

The increasing supply of critical-thinking and problem-solving software programs represent another source of instructional material that is closely aligned with the area of language arts. Many of these programs cover skills that are essential for reading comprehension and writing (e.g., Thinkanalogy, Editor in Chief, Dr. DooRiddles Software).

Additional resources for constructivist teaching. The emergence of user-friendly multimedia authoring tools (e.g., HyperStudio, SuperLink, Multimedia Scrapbook) gives students capabilities to efficiently add sound, animation, and video to their writing. This multimedia approach to authoring has inspired many students to write after previously displaying little or no interest in the writing process. MicroWorlds Language is a language-focused authoring program that stresses integration of writing and art. It includes a number of projects that cover haiku, visual poetry, advertising, and cinquain. Perhaps its most unique feature, a music and sound center, makes it easy for students to set words to music.

Many great films on videodisc provide an excellent resource for teachers and students. Students can incorporate segments of video into activities that foster critical analysis of the film. The examination of video can also extend to instruction in media literacy, which many identify as a basic skill for the information age.

Resources and Applications for Foreign Language Instruction

Computer software. Most computer software programs for foreign language instruction provide grammar-based drill and practice (e.g., Spanish Grammar Series, Hagar the Polyglot, French Tutor). A number of programs also focus on vocabulary and translation (e.g., Berlitz Interpreter, French Vocabulary Beginners). Students can practice pronunciation skills by listening to spoken words and phrases and then repeating them. This type of software is commonly found on CD-ROM (e.g., Learn to Speak Series, Berlitz Think and Talk). In another approach, some programs use

Figure 11.3   From *Spanish Now!* by Transparent Language

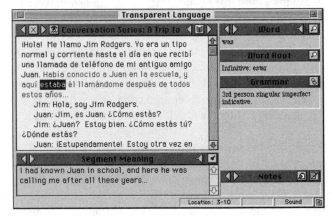

Source: Courtesy of Transparent Language, Inc.

Figure 11.4   Sample Foreign Language Page

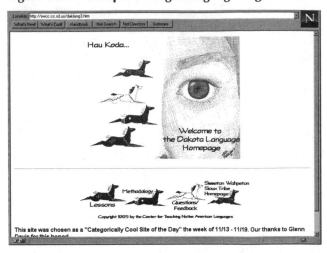

familiar games or the mystery genre to deliver instruction (e.g., French Micro Scrabble, Whodunit: French, German, Spanish). As a way of facilitating the writing process, Slavic Swiss Cyrillic Fonts and 3-D Keyboard provide easy ways to type a variety of foreign language characters. See Figure 11.3 for samples of Transparent Language software.

Telecommunications. Telecommunications technology has great potential as a CALL tool. Computer programs facilitate exchanges between students of different nations, although primarily in the English language (e.g., National Geographic's KIDSNET, World Classroom Network). Most efforts to facilitate second-language exchanges with students in other countries are limited in scope, but anticipated technological developments portend a bright future for synchronous discussions in other languages. LAN technology has shown potential to facilitate communication in foreign language instruction. As two or more students converse in real time on a network, reticent students are more likely to participate. The World Wide Web offers some sites that should prove valuable for instructors and students:

- **Human Languages Pages.** This kind of resource (http://www.willamette.edu/-tjones/Language-Page.html) offers a German-to-English and reverse dictionary along with Japanese lessons. See Figure 11.4 for a sample foreign language page.

- **Travel webs.** This site (http://www.travelweb.com) presents information about foreign customs and cultures for travelers.

Video. Several videodisc programs teach foreign languages within the context of native cultures (e.g., Espana Viva, A Vous la France!). These programs are typically divided into short (15-minute) segments tied to bar codes and student workbooks. Another use of videodisc technology in foreign language instruction, La Maree et Ses Secrets, tells a story of young people unwittingly involved in a dangerous mystery. Bar codes enable the user to quickly repeat segments of the program, thus making it easy to see and hear information as many times as needed.

Additional resources for directed instruction. Teachers can adapt many ESOL materials for foreign language instruction. Frequently subject specific, these programs provide both instructor and students convenient ways to experience subject integration. Many schools with ESOL populations have these programs on hand, giving teachers easy access to them.

Additional resources for constructivist teaching. The International Expanded Book Toolkit enables the user to create a hypertext, multimedia book in any of nine languages. In addition, students can use other multimedia authoring tools to create projects that might include recording of themselves reading in a second language, video clips from foreign films, or music performed in a second language. For younger children, Spanish Story Tailor is designed to integrate reading and writing in Spanish by inviting students to customize stories written by children's book authors. Finally, many videodiscs carry foreign language sound tracks, usually Spanish. These discs can be repurposed for the classroom. Windows on Science, for example, might support an activity that would integrate instruction in science and Spanish.

Sample Activities

Activity 1 (language arts or foreign language):	Designing and Producing a Brochure
Level:	Grades 5–12
Purpose of Activity:	The students will design and produce a brochure.

Instructional Activity:

Setting the Stage: The I. Care Plenty Scenario. World-renowned philanthropist, I. Care Plenty died recently and left a sum of $10 million to your group, the Environmental Protection Fund. Mr. Plenty's will stipulated that the money must fund construction of a wildlife refuge for endangered or near-endangered species. The refuge should, of course, provide a safe habitat for the animals and plants, but it should also serve as an educational center for the public. (Mr. Plenty believed that education is the only real, long-term solution to the threat of extinction to animals and plants.)

Your group has a great deal of latitude in how to spend the money, but the will does mandate a few guiding principles that you must follow. Above all, the Plenty family insists on a world-class facility in every respect that adheres to strict environmental standards. The refuge should serve as a model for others around the world and every aspect should exhibit the highest standard of quality. Finally, the overriding philosophy of the operation from its inception through the day-to-day functions should be driven by the motto "continuous quality improvement." Each person involved with the refuge should accept this motto as a daily commitment never to be satisfied with current achievements; they should always watch for ways to improve things and have fun doing it.

The Environmental Protection Fund has decided to put together an introductory brochure that will provide basic information about the planned refuge. It will both inform the community at large and build goodwill among this audience. A list of suggestions include:

- Give basic information (not too many details).
- Tell people how to get involved with the refuge.
- Include both a logo and a motto.
- Use graphics.
- Present scanned images of animals.
- Include the EPF's address and phone number.
- Look at what others have done. (Collect brochures and benchmark.)

Recommended Procedures

1. Students will need sufficient training on a desktop publishing program to feel comfortable taking on a project like a brochure.

2. Introduce the scenario to the class; this may range from passing out printed instructions to simply explaining it in your own words.

3. Thoroughly discuss the scenario. This crucial phase gives students an overview of the project.

4. In explaining the brochure assignment to the class, emphasize the need to avoid a quickie product and commit to invest hard work and cooperation. The tone for the assignment will be set at this point.

5. Lead the class through a concept mapping for the brochure project. (See the discussion of tools for brainstorming in Chapter 6.) This brainstorming activity encourages discussion of all aspects of the brochure. The teacher should help guide the discussion and add relevant information that students may not consider.

6. Discuss various team roles and responsibilities and how to divide up the work. Set a tentative timeline.

7. Divide the class into groups, each to create its own brochure. Stress cooperation and collaboration instead of adversarial competition.

8. Assign groups to meet and decide on a plan of action. The teacher should observe this process and intervene if the group gets off task.

Continued

9. As groups work on the project, the teacher should meet with them regularly to evaluate progress and offer assistance, including bringing in community resource people.

Suggestions for Teacher: This activity may have to be altered and/or scaled down considerably for the foreign language classroom. However, most of the project ideas are transferable.

Other Language Arts or Foreign Language Projects

- **Invitation.** Students create an invitation to a barbecue that EPF is sponsoring.
- **Signs.** Students make signs to indicate the species at the park with relevant information about the animals.
- **Presentation.** Students make a presentation to a community group showing some of the endangered species housed in the refuge. A videodisc player and bar codes would enhance the presentation.
- **Jingle/commercial.** A group might develop a public service commercial along with a jingle. They would need access to a video camera, audio CD, and cassette player.
- **Fund raising.** Students organize a fund-raising campaign for the refuge. This would encompass many language arts skills from advertising to letters to public officials requesting permits.
- **Big book for students.** Students produce a book that explains the plight of some of the endangered species. This book should be targeted for primary teachers to prepare their classes for field trips to the refuge. Programs like Toucan Press or Big Book Maker could be used.

Extensions to Other Subject Areas

- **Art.** Students create posters with thematic links to the plight of endangered species. The refuge would sell the posters to raise funds. A computer draw and graphics program could be used in this production.
- **Music.** Students produce a skit and perform in front of various groups to increase community awareness. As a component of the skit, they should incorporate music and sound effects using interactive CD players and bar codes.
- **Math.** The numerous possibilities may range from actually designing the park and creating a scale model to budget forecasting using a spreadsheet program. Technology could contribute substantially, as calculators and spreadsheets would make working with large figures painless for the students.
- **Social studies.** Students research the causes of species extinction over the years and compile results in a database. When they accumulate a sufficient number of entries, they can undertake a data analysis to look for underlying connections.
- **Science.** Students research the habitat requirements of various endangered species and choose those to support at a local refuge. Telecommunications resources could support the research and data could be organized using a database.
- **Foreign language.** When the refuge opens, public viewing areas will need informational/educational signs that inform visitors about the exhibits. These signs will need translations in French, German, Japanese, and Spanish. Students may also want to make audio tapes that provide information about park operations or the animals in a second language.

Activity 2 (language arts or foreign language): Talking It Up!
Level: Grades 3–12
Purpose of Activity: Students develop their observation and articulation skills.

Instructional Activity: In today's world, video applications are becoming more and more widespread in occupations that used to rely on text and still photos as the primary means of communicating. Fire marshals and insurance adjusters, for example, used to take still photographs of accident scenes and then write up extensive reports. Now, many visually record what they see on video cameras, simultaneously recording oral reports. This has resulted in higher-quality reports with much less paperwork.

In this activity, students working in pairs use video cameras to visually and orally record their own observations. They should be familiar with the basics of video camera operations and safety procedures. Together, the teacher and the students choose a scene, and the team makes a visual and oral

Continued

recording. The scene may be a room in school or even a bulletin board outside the classroom. High school students may want to document damage that is evident on some cars in the school parking lot.

The teacher may extend the lesson by playing the recording back to a group that turns away and does not look at the video image. They then hear it a second time as they watch the video, as well. They then give the reporter feedback on how well the audio matched up with expected visual images.

Suggestions for Teacher: Experiment with different ways of having partners work together; one might videotape while the other narrates, or they may take turns. Choosing and designing the scene could provide a lot of room for creativity. A pair could simulate a crime scene in which the person doing the recording watches for clues. In foreign language classes, this type of activity will provide an opportunity for students to practice and hear their pronunciation skills.

Activity 3 (language arts or foreign language):   Network Writing
Level: Grades 5–12
Purpose of Activity: Students converse online and respond to a high-interest issue. They must think and write almost simultaneously throughout the duration of this activity. This skill will become more important with the proliferation of computer networks.

Instructional Activity: This activity requires access to a telecommunications network with a chat feature. Many state education networks now have this feature. Some LAN software also facilitates chatting (e.g., GROUPwriter). This chat feature enables a number of people to carry on a written conversation in real time. The participating classrooms should prearrange a time and a topic for the discussion. Students should work in pairs, switching off between typing and advising. The discussion can be captured and then reviewed at a later date. A follow-up assignment could involve rewriting the dialogue of the conversation to make it easier to read. A student might also role play a position on an issue that differs from his/her own.

Suggestions for Teacher: Encourage students to prepare beforehand so that they have some knowledge of the subject. This type of activity can deteriorate into a nonsensical argument if the participants cannot draw on some knowledge. Give the students some say in what topics are discussed. Topics that are relevant to them will encourage more thought and debate. The teacher should also limit the length of the conversations; the correct time varies depending upon the level of the students. Although some networks allow a number of people to converse at the same time, it would be best to limit these chats to three sites at a time. Too many people trying to talk at the same time usually results in a weak dialogue.

Activity 4 (language arts):   Get-Well Coloring Book
Level: Grades K–5
Purpose of Activity: Students gain experience at writing, developing graphics, and publishing a book. They also gain the satisfaction of participating in a community project.

Instructional Activity: The students produce a coloring book for patients in the children's ward at a local hospital. To begin, the class should decide upon a theme for the book. This can vary; for example, they might choose a subject (animals), or a particular literary style (poetry). The students can work either individually or in pairs to develop their text and drawings. They can draw on the computer or on paper, then scan the results into a document. Each group can choose one of its drawings to add to the cover. Individual students would each color the cover of a book to be sent to the hospital.

Suggestions for Teacher: A desktop publishing program like PageMaker could be used to give the final products a polished look. If this software is not available, then the layout could be done by hand. Before sending them off to the hospital, the students could wrap the books in cellophane paper and include a small box of crayons.

Activity 5 (language arts):   Editing an Interview
Level: Grades 10–12
Purpose of Activity: Students will edit the transcript of an online interview to make it suitable for publication.

Instructional Activity: Online telecommunications services (e.g., America Online, Prodigy) frequently feature question-and-answer sessions with luminaries such as authors and business consultants. This lesson uses the text of

Continued

one of these interviews as the raw material for a student editing task. The students will take the text of the interview and edit it for publication. After completing the assignment, different students can compare their applications of the editing process, looking for ways that different editors diverged in interpreting the same interview.

Suggestions for Teacher: The script of the interview can be captured by either the teacher or a student who has the service at home. Discuss the interview process with students beforehand. Explain that many luminaries do not grant interviews because they are suspicious of the editing process. Discuss the ethical issues of taking something that somebody says out of context.

Activity 6: (foreign language) Cultural Awareness Project
Level: Grades 6–12
Purpose of Activity: The students will use a multimedia authoring tool to develop a cultural awareness program as a resource for teachers and students.

Instructional Activity: Numerous areas of the country are experiencing surges in registration of immigrant students, many from cultures familiar to neither the faculty or student population. In this activity, the class will develop a program that will enable teachers and students to quickly access information about different cultures. It should contain information about traditions, customs, government, religion, geography, food, and common phrases. When researching for this project, the students should remain alert for culturally sensitive information that could be particularly interesting to the users of the program. In other words, better information about a new student's culture reduces the likelihood that others will offend that person and start the relationship off on the wrong foot.

Suggestions for Teacher: Have the students work in teams, each team assigned a different country. The lesson should produce a viable product that people will actually use, so choose cultures that are pertinent to the geographic location. Use a wide variety of technologies; telecommunications and fax machines may help students reach out to valuable resources. Foreign language teachers may place more emphasis on the language aspect of the culture; for example, a program could offer the user help with the pronunciation of common words or phrases.

Additional Suggestions for Language Arts and Foreign Language Activities

The fax connection (language arts and foreign language). The fax machine is a wonderful tool for creating an audience for students' work. This activity requires the teacher to develop contacts in the community who are willing to be involved with the class's work. Upon completing a writing assignment, the student would fax it to a community resource who has agreed to edit or review the work and then fax it back to the student. This procedure could work in many ways: people at a local business could review business letters, advertising agency staff could review a brochure, a community person fluent in a certain language could review writing assignments in that language.

Concept mapping and brainstorming (language arts and foreign language). For this activity, the class would do a concept map on a particular subject such as the Russian culture. Certain software packages facilitate the brainstorming and concept mapping process (e.g.,

Inspiration, IdeaFisher). An LCD panel would facilitate this activity with the whole class. After the first concept map is completed, show the students a video on the given subject, then let them do another concept map. Compare and contrast the details and kinds of linkages of the two maps. This could also be done with small groups. The before and after concept maps could be used for portfolio assessment.

Multimedia greeting for a foreign student (foreign language). A multimedia authoring program is used to create a program that welcomes new ESOL students to the school and community. Students could use the sound feature of the program to record relevant information in the student's language. Photographs of points of interest could be included along with interesting facts about the history of the school, town, and state. The program could be used by the guidance counselor to familiarize the new ESOL student to his/her new home. ESOL teachers could provide a valuable resource for students who are working on this type of project.

Exercises

Exercise 11.1. Explain how each of the following current issues and trends in language arts instruction would affect your selection and use of technology:

- Teachers carry on continuing discussions about the most appropriate model for teaching English and language arts. What impact would these standards have on your selection and use of technology in these areas?

- Teachers of foreign languages also disagree about what skills their instruction should emphasize (e.g., communicative language versus grammar skills). What impact would this discussion have on your selection and use of technology in this field?

Exercise 11.2. Describe the applications of the following technology resources in both language arts and foreign language curricula:

- Word processing software
- Videodisc technology
- Video cameras
- Multimedia authoring software
- Desktop publishing software
- Telecommunications (e.g., Internet Web sites)
- Language arts software

Exercise 11.3. Create (a) individual instructional activities for language arts and foreign language and (b) instructional activities that integrate each of these disciplines with other curricula. Be prepared to teach the activity to a class. Each activity should meet the following criteria:

- Integrate one or more of the types of technology described in Exercise 11.2.

- Show how this activity could be adapted for large-group and small-group instruction.

- Describe classroom preparation that would have to occur before this activity could take place.

- Describe unique benefits you hope to derive from using technology resources in the lesson.

References

Armstrong, K., Yetter-Vassot, K. (1994). Transforming teaching through technology. *Foreign Language Annals, 27 (4),* 475–486.

Beauvois, M. H. (1992). Computer-assisted classroom discussion in the foreign language classroom: Conversation in slow motion. *Foreign Language Annals, 25 (5),* 455–463.

Birch, R. (1994). Every picture tells a story: The negotiation of meaning within an interactive social context at key stage 1. *English in Education, 28 (3),* 17–24.

Bristor, V. J. (1994). Linking the language arts and content areas through visual technology. *T.H.E. Journal, 22 (2),* 74–77.

Chun, D. M., and Brandl, K. K. (1992). Beyond form-based drill and practice: Meaning enhanced CALL on the Macintosh. *Foreign Language Annals, 25 (3),* 255–265.

Cohen, M., and Riel, M. (1989). The effect of distant audiences on students' writing. *American Educational Research Journal, 26 (2),* 67–72.

Daiute, C. (1992). Multimedia composing: Extending the resources of kindergarten to writers across the grades. *Language Arts, 69 (4),* 250–260.

Heaney, L. F. (1992). Children using language: Can computers help? *Gifted Education International, 8 (3),* 146–50.

Herron, C., and Handley, J. (1992). Using video to introduce children to a foreign culture. *Foreign Language Annals, 25 (5),* 419–425.

King, R., and Vockell, E. (1991). *The computer in the language arts curriculum.* Watsonville, CA: Mitchell-McGraw-Hill.

Lambert, W. E. (1967). *Children's views of foreign people.* New York: Appleton-Century-Crofts.

Madian, J. (1993). Using our gifts: I search, poetry, and technology. *Writing Notebook: Visions for Learning, 10 (3),* 38–39.

Mandel, B. (1980). *Three language arts curriculum models: Pre-kindergarten through college.* Urbana, IL: National Council of Teachers of English.

Marcus, S. (1993). Indiana writing project. *Writing Notebook: Visions for Learning, 10 (3),* 18.

Merrill, P., Hammons, K., Tolman, M., Christens, L., Vincent, B., and Reynolds, P. (1992). *Computers in education.* Boston: Allyn and Bacon.

Nourse, K. (1994). An enterprise technology project as part of German exchange. *Language Learning Journal, 9,* 26–27.

Pitkoff, E., and Roosen, E. (1994). New technology, new attitudes provide language instruction. *NASSP Bulletin, 78 (563),* 36–43.

Purcell-Gates, V. (1995). Research for the 21st century: A diversity of perspectives among researchers. *Language Arts, 72 (1),* 56–60.

Rose, D. H. (1994). The role of technology in language arts instruction. *Language Arts, 71 (4),* 290–94.

Salomone, A. M. (1991). Immersion teachers: What can we learn from them? *Foreign Language Annals, 24 (1),* 57–63.

Schwartz, E., and Vockell, E. (1988). *The computer in the English curriculum.* Santa Cruz, CA: Mitchell.

U.S. Department of Labor (1992). *SCANS (The Secretary's Commission on Achieving Necessary Skills) report.* Washington, DC: U. S. Government Printing Office.

Wepner, S. B. (1993). Technology and thematic units: An elementary example on Japan. *Reading Teacher, 46 (5),* 442–45.

Wepner, S. B. (1994). Saving endangered species: Using technology to teach thematically. *Computing Teacher, 22 (1),* 34–37.

Whitley, S. M. (1993). Communicative language teaching: An incomplete revolution. *Foreign Language Annals, 26 (2)*, 137–149.

Foreign Language Resources

Videodiscs
World Cultures and Youth Series—Coronet/MTI

Mexico Vivo, España Viva, A Vous la France! La Maree et Ses Secrets—Films Incorporated, available from EISI

Windows on Science/ Windows on Math—Optical Data Corp.

Software
Spanish Grammar Series—Intellectual Software, available from EISI

Hagar the Polyglot—Gessler, available from EISI

French Tutor—Queue, available from EISI

Berlitz Interpreter—Microlytics/Software Holdings, available from EISI

French Vocabulary Beginners—Microcomputer Workshops, available from EISI

Learn to Speak Series—HyperGlot Software, available from EISI

Berlitz Think and Talk—HyperGlot Software, available from EISI

French Micro Scrabble—Gessler, available from EISI

Whodunit: French, German, Spanish—Gessler, available from EISI

Slavic Swiss Cyrillic Fonts, 3-D Keyboard—Exceller Software Corp., available from EISI

National Geographic's KIDSNET—National Geographic

International Expanded Book Toolkit—Voyager Inc.

Spanish Story Tailor—Humanities Software

Language Arts

Word Processors for Children
Kid Works II—Davidson

The Writing Center, Children's Writing and Publishing Center—The Learning Company

Integrated Software Packages
Microsoft Works—Microsoft Corp.

ClarisWorks—Claris Corp.

WordPerfect Works—Corel, Inc.

Read-Listen-Speak Series—Transparent Language, Inc.

Desktop Publishing/Word Processing Programs
WordPerfect—WordPerfect Corp.

Microsoft Word, Microsoft Publisher—Microsoft Corp.

MacWrite Pro—Claris Corp.

Pagemaker—Adobe

QuarkXpress—Quark

Toucan Press— Pelican/Toucan

Big Book Maker—Pelican Toucan

Super Print—Scholastic Inc.

Other Software
Grammatik—WordPerfect Corp.

Writing to Read—EduQuest

Tapestry—Jostens

Playroom—Brøderbund

The Vocabulary Games—J & S Software, available from EISI

Crossword Magic—Mindscape Educational Software, available from EISI

Milliken's Comprehension Power—Milliken & I/CT

Speed Reader II—Davidson

Storybook Weaver—MECC

RightWriter—RightSoft/Que Software, Available from EISI

Literature Series—Sliwa/Queue, Available from EISI

Adult Education Reading—BLS,Inc, Available from EISI

Thinkanalogy, Editor in Chief, Dr. DooRiddles Software—Critical Thinking Press

Hypermedia Authoring (See Chapter 8)

CD-ROM and CD-I
Shakespeare—The Complete Works—CMC Research, available from EISI

Compton's Multimedia Encyclopedia—Compton's New Media

A Visit to the Smithsonian—National Geographic

Discus Books—Discus

Brøderbund's Living Books—Brøderbund

C. D.'s Story Time—Houghton Mifflin Co.

Videodiscs
To Kill a Mockingbird—Media Learning Systems/Image

How the Elephant Got His Trunk—Encyclopedia Brittanica

About Huckleberry Finn—Encyclopedia Brittanica

Masterpiece Series—Silver Burdett and Ginn

Old Mother Witch, Literature Navigator Series—BFA

Illuminated Books and Manuscripts—Eduqu

Chapter 12

Technology in Science and Mathematics Instruction

This chapter will cover the following topics:

- Current issues in mathematics and science instruction

- Recommended resources and applications in mathematics and science instruction

- Example activities for single-subject or integrated lessons in mathematics and science

Chapter Objectives

1. Identify current issues in mathematics instruction and science instruction that affect the selection and use of technology.

2. Describe the most popular uses for technology in mathematics and science curricula.

3. Create instructional activities for mathematics and science instruction that successfully integrate instructional technology.

"The challenge to designing a new math curriculum is to create one that is teachable. You cannot do that without technology."
Judah Schwartz, as quoted in Hill (1993)

"The world had changed in such a way that scientific literacy has become necessary for everyone, not just a privileged few; science education will have to change to make that possible."
James Rutherford and Andrew Ahlgren, as quoted in Bruder (1993)

Introduction

The surge of technological change is pushing the limits of humans' ability to adapt to it. At the forefront of this "knowledge age" are the disciplines of mathematics and science; they are closely intertwined with each other and with technology, in general. Indeed, citizens of tomorrow's world will not have the option of leaving mathematics and science for those who are perceived to have special gifts in these areas. People in all walks of life will need more proficiency in mathematics and science than most people have attained today. Education, of course, will be the key to preparing students for tomorrow's world. In this chapter, we will discuss the uses of technology in the mathematics/science curriculum and offer concrete suggestions on how to integrate instructional technology into these fields.

Issues and Problems Related to Technology Use

Issues in Mathematics: Controversies and Trends

Background. The driving force behind most changes in mathematics instruction today is the Curriculum and Evaluation Standards for School Mathematics, published by the National Council of Teachers of Mathematics in 1989. This comprehensive document calls for major revisions in the way that schools teach mathematics in order "to ensure that all students have an opportunity to become mathematically literate … and become informed citizens capable of understanding issues in a technological society" (NCTM, 1989, p. 4). The Standards for School Mathematics document identifies five broad goals designed to prepare American students for the demands of a global economy in the information age:

- **To value mathematics.** Students must appreciate and understand the value of mathematics in society. One of the math teacher's great challenges is to connect mathematics to the real world in ways that students appreciate.

- **To reason mathematically.** American culture often treats mathematics as a series of skills to memorize as opposed to a way of thinking. This antitheoretical bias, many believe, has its roots in the pragmatic nature of the American education system, which often stresses the product (the right answer) over the process.

- **To communicate mathematically.** Learning to speak, read, and write about mathematical topics is a natural extension of mathematical reasoning ability. The writing process is an ideal vehicle to help students clarify their own understanding, and it provides an excellent way to address various learning styles. Allowing students to work cooperatively with access to word processing software further extends these benefits.

- **To solve problems.** The ability to apply mathematical principles has gained paramount importance in an increasingly technological world. Students need to experience a variety of problems that differ in scope, difficulty, and context.

- **To develop confidence.** In the workplace of tomorrow, employees will need a mathematical fluency that relative few have developed today. The need for confidence in one's ability to apply mathematics has never been greater.

Textbook publishers and curriculum committees have begun to respond to the call for changes by redefining mathematics content to conform with the new NCTM standards. Today, more and more educational technology offerings are designed to meet NCTM standards.

Is technology necessary to meet NCTM standards? Vigorous debate still continues concerning the use of computers in mathematics instruction. Steen (1989) recognized a strong impact of computer technology on teaching in the K-12 classroom; however, computers do not become essential to the actual delivery of curriculum until the second year of college. Bruder (1993) suggested that educators should adopt a wait-and-see approach toward technology. She claimed that the graphing calculator has had a greater impact than the computer on instruction in higher mathematics. She stated, however, that since each student can have a calculator, that tool fits better with the prevalent model of instruction with students working alone.

Others have contended, however, that the computer should be part of mathematics instruction right from the beginning of school (Bitter, 1987; Schwartz, 1992; and Kaput, 1994). Computer advocates believe that the power and flexibility of the tool offers an opportunity for students to experience mathematics in too many ways to argue easily against its use. As Kaput (1994) sees it, "You can't really achieve what the standards suggest without technology" (p. 683).

Quality of software. The mathematics community must also contend with the quality of available software. Most mathematics programs released in the last 10 years are designed to teach basic skills; they lack a true problem-solving focus. Exceptions include some of the Tom Snyder products and the Adventures of Jasper Woodbury videodisc series, for example; but traditionally, mathematics has been one of the weakest curriculum areas for innovative product development.

Technology as a mathematics tool. The use of mathematics tools signified a major leap forward in software development. The Geometric Supposer, a program designed to let students create and experiment with the properties of various shapes and figures, was one of the earliest programs of its type. Programs like this empower users to explore and discover the subject in a risk-free environment. They enable students to see mathematics as an investigative, exploratory subject that has a beauty of its own. Appropriate tools help to ensure that mathematics instruction takes advantage of technological power.

The issues in summary. Observers seem to agree generally that the NCTM mathematics standards offer more than a set of goals. Educators must develop curriculum around them to improve mathematics instruction. No matter how intensively they integrate technology, its effectiveness depends on the strength of the overall curriculum.

Issues in Science: Controversies and Trends

Background. Today's rapidly changing world clearly demonstrates the importance of science in people's everyday lives. Science plays a part in a large percentage of the articles in a daily newspaper, ranging from health issues to crime scene investigations to environmental concerns.

This omnipresence of scientific information has led to a general realization of the importance of scientific literacy for all citizens. This represents a radical departure from previous notions that students should study science only if they really need it in their careers. A number of movements now seek to redefine scientific literacy, and therefore science education, in this country.

Standards. How instructional technology will contribute to this new emphasis on science is still ambiguous. The most ambitious effort to reform science instruction is the National Science Education Standards Project sponsored by the National Research Council (NRC), an agency of the National Academy of Sciences and the National Academy of Engineering. The National Science Education Standards were released in late 1995. The standards document states the following principles at the outset:

- All students should have the opportunity to attain higher levels of scientific literacy than they do currently.

- All students will learn science in the content standards.

- All students will develop an understanding of science that enables them to use their knowledge in personal, social, and historical contexts.

- Learning science is an active process.

- For all students to understand more science, less emphasis must be given to some science content and more resources must be devoted to science education.

These standards clearly favor an inquiry approach. According to the standards document, inquiry "involves making observations, posing questions, examining books and other sources of information to see what is already known, planning investigations, reviewing what is already

known in light of experimental evidence, proposing answers and explanations, and communicating the results" (NRC, 1993, p. 5).

Technology is mentioned directly in the following vein: "The use of tools and techniques ... will be guided by the questions asked and the investigations that students design. The use of computers for the collection, summary, and display of evidence is part of the standards. Students should be able to access, gather, store, retrieve, and organize data, selecting hardware and software designed for these purposes" (NRC, 1993, p. 31).

The standards do not state how curriculum should be organized, leaving that task to the states. They do ask teachers to help students become inquirers, but the standards do not pass judgment on how best to teach inquiry skills.

This inquiry-based approach is not popular with all teachers. Many teachers decry the standards' philosophy of depth over breadth and, instead, believe firmly that students need to know a wide variety of science terms and facts (Weiss, 1993).

Also, the new standards do not spell out clearly the role of instructional technology. But given the ambitious nature of the goals, a broader repertoire of teacher resources seems essential if schools are going to elevate students' science competency in the ways described. Industry will certainly be marketing a variety of technology-based products geared toward helping students attain the new standards.

Teacher empowerment. In addition, technology advocates view it as a means for enhancing the skills of science teachers. CD-ROMs can store vast numbers of quality lessons as in the Science Cap software. The tremendous search capabilities of this technology can quickly supply specific lessons suited to a teacher's needs. Computer networks offer great promise for promoting teacher collegiality. For collaboration within the school building or across national borders, networks offer vast potential.

Assessment. Technology seems a natural tool for alternative assessment, especially as both mathematics and science curricula are moving toward more process-oriented assessment methods. The tremendous evolution in computer power could realistically deliver practical systems for performance-based assessment. For example, the National Science Teachers Association is developing a product using CD-I technology that is designed to alleviate labor-intensive tasks such as recording and analyzing student performance.

The issues in summary. As with mathematics, many observers of science education are skeptical about jumping too quickly into technology. Publication of science standards will clarify the debate over technology's role and focus attention on the issue. In the meantime, schools are certainly purchasing extensive instructional technology resources for science, and many teachers seem eager to integrate those resources into their teaching.

Recommended Resources and Applications

Resources and Applications for Mathematics Instruction

Calculators. Discussions of instructional technology often overlook these familiar tools, but this mistake ignores their tremendous, largely untapped potential. The NCTM has endorsed the use of calculators in all classrooms and views them as tools for exploration and problem solving. In high schools, the graphing calculator is probably the most prevalent technology used in mathematics instruction today. These devices add a helpful visual dimension to abstract concepts. Many types of calculators are on the market today, some manufactured specifically for the education market.

Video. Video technology has enhanced mathematics instruction in two principal ways: adding a visual element to direct instruction and placing mathematics in the context of a story with real-life implications.

Directed-instruction videodisc programs offer both Level I and Level III lessons in basic mathematics and algebra instruction. The narrated lessons use graphics, animation, and real-life examples to enhance content delivery (e.g., ModuMathematics).

Video displays in other types of programs help to add relevancy to the teaching of mathematics. For younger students, video resources usually present cartoon characters and engaging stories (e.g., The Adventures of Fizz and Martina). For upper elementary and middle school students, videos embed mathematics in problem-solving scenarios using actors, often in the same age range as the students (e.g., The Adventures of Jasper Woodbury, The Perfect Pizza Caper). As of this writing, the Optical Data Corporation just released a videodisc-based mathematics curriculum called Windows on Math. This program for K-5 students includes a major hands-on component. Their random search capabilities of videodisc-based programs (e.g., The Adventures of Jasper Woodbury) offer much more interactivity than those that use the videotape format.

Computer software. Computer software offers an invaluable resource for students and teachers in mathematics/science instruction. Tremendous strides have brought vast improvement over early software, which basically emulated accounting worksheets. Many newer programs offer interactive instruction to teach students to think mathematically, scientifically, logically, and analytically and to apply that understanding to problem solving.

Elementary-level applications develop mathematical and logical thinking through challenging visual/spatial tasks (e.g., Building Perspective, Divide and Conquer, Probability Constructor, Measurement in Motion). Many programs develop higher-level mathematical thinking by making abstract ideas in geometry, physics, calculus and probability theory more concrete. Technology has also enhanced the opportunities of math students for hands-on learning. Programs are available for all grade levels that let users manipulate, resize, and measure objects within exploratory environments (e.g., Hands-On Mathematics: I, II, III; GeoExplorer; Geometer's Sketchpad).

The Cruncher is a spreadsheet program specifically designed for elementary students. This program adds sounds and clip art to make the spreadsheet metaphor more interactive and appealing to elementary age children.

Child Using Math Software

Internet resources for mathematics. A wide range of resources on the Net appeal to those studying math. Students can find current stock market quotes from Wall Street News (http://Wall-Street-News.com/forecasts/index.html). MathMagic offers problem-solving projects in which teams of students communicate about math (http://forum.swarthmore.edu/mathmagic/what.html). Math Answers, the math "swat team," will answer K-12 math questions (mail dr.math@forum.swarthmore.edu, see Figure 12.1). A geometry website (http://www.geom.umn.edu/apps/gallery.html) offers activities in visualizing and experimenting with ideas in two, three, and higher dimensions.

Additional resources for directed instruction. Through the use of a computer and an LCD panel, certain software programs offer electronic chalkboards (e.g., Mathematics Exploration Toolkit, The Mathematics Teacher's Workstation). This type of program enables the teacher or student to manipulate symbols and solve advanced problems quickly in full view of the class. A number of comprehensive software packages are designed to teach specific mathematics skills (e.g., Mathematics Sequences, Skills Bank II-Mathematics). Schools usually offer these programs over networks, but they can also present the software at a standalone workstation. The software usually includes a management program that allows the teacher to individualize instruction.

Figure 12.1   Ask Dr. Math Home Page

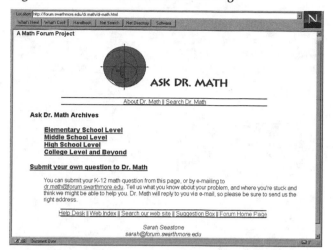

Science represents an attempt to develop a total science curriculum on videodisc. Some school districts have adopted this direct instruction program in lieu of textbooks. More commonly, videodiscs pertain to specific areas of science (e.g., Weather: Air in Action Series, Garbage: The Movie—An Environmental Crisis). As in other areas of instructional technology, passing time brings improvements in the quality of the available programs. Recent releases of Level I videodisc programs are geared toward developing critical-thinking skills in science (e.g., Science Sleuths, The Great Solar System Rescue, Innovations).

Level III programs sometimes offer huge visual databases along with premade lessons and presentation-tool components (e.g., BioSci II, Interactive NOVA, Physics of Sports). Computer software can combine with videodiscs to run simulations enhanced with high-quality visuals (e.g., The Great Ocean Rescue).

CD-ROM resources. Many CD-ROM titles are now available for the science classroom, usually with a number of multimedia features. It is easy to understand how pictures, sound, video, and animation can enhance science learning. The variety of CD-ROM titles for science run the gamut from games to simulations to tutorials. One of the most innovative applications is the Science Sleuth series by Videodiscovery. In transferring the episodes from videodisc to CD-ROM, the developers added an amazing amount of interactive challenges for the user.

Microcomputer-based laboratories. Commonly referred to as *probeware*, these resources enable students to perform a variety of hands-on scientific experiments that help build skills in data collection and analysis. Typically, an MBL package includes probes and software to measure, record, and graph heat, sound, motion, pH, and pulse rate. High school programs (e.g., Personal Science Laboratory, MI-1000 MultiFunction Interface) tend to perform more powerful and complex tasks than elementary and middle school versions (e.g., Science Toolkit Plus, Temperature Experiments).

Internet resources. The connection between science and the Net has generated a tremendous number of sites that can serve as resources to young scientists. Here are a few examples:

- Boston Museum of Science (http://www.mos.org/)
- Grand Canyon National Park (http://www.kbt.com.gc)
- Plasma Physics Laboratory at Los Alamos, New Mexico (http://www.pppl.gov/)
- European Space Agency (http://www.esrin.esa.it/)
- The Eisenhower National Clearinghouse for Mathematics and Science Education (http://wwwenc.org) fast becoming one of the most valuable Internet resources for these topics. See Figures 12.1, 12.2, and 12.3 for web resources.

Additional resources for constructivist teaching. The database and spreadsheet components of integrated software packages (e.g., ClarisWorks, Microsoft Works) provide excellent tools for manipulating and analyzing mathematical data. These programs also offer supplemental materials designed to help the teacher integrate technology (e.g., ClarisWorks in the Classroom, Microsoft Works for Teachers). CD-ROM technology offers students and teachers access to huge resources of statistical data that can be used for mathematical exploration (e.g., The Multimedia World FactBook, The CIA World Factbook). In an especially innovative approach to teaching mathematics, students may develop projects that are not always perceived as mathematical, but that require mathematical thinking and reasoning (e.g., MicroWorlds Mathematics Links).

Resources and Applications for Science Instruction

Computer software. One of the most common types of software for science instruction is the simulation. Simulations are common for environmental science classes (e.g., Ecology Treks, Balance of the Planet, SimEarth). Simulations have also had a large impact in life science; programs that simulate animal dissection have become very popular (e.g., Biology Labs, Operation Frog, Visifrog, The Lab: Experiments in Biology).

Science teachers also use computer courseware quite extensively for directed instruction. A number of programs are designed for single users (e.g., The Science of Living Things, Science in Your World). Others make efforts to connect science with other content areas, mathematics in particular (e.g., CHEMiCALC, Chemistry of Achievement I: Mathematics of Chemistry, Water Budget).

Videodiscs: Level I and Level III. In science, probably more than any other subject area, video resources have affected the instructional process. A multitude of Level I videodisc programs support science instruction at all grade levels. In one of the most ambitious efforts, Windows on

Additional resources for directed instruction. To make full use of the tremendous quantity of science videodiscs,

Figure 12.2   European Space Agency Home Page

Figure 12.3   Grand Canyon National Park Home Page

the teacher can save videodisc presentations for reuse. A number of utilities offer capabilities to edit video segments and then save them as either bar codes or computer files (e.g., Multimedia Report Generator, CBE MultiMedia Sequencer). Some programs can even overlay text on the monitor screen when the teacher wants to label something on a video frame (e.g., The Videodisc Toolkit, MediaMax).

Additional resources for constructivist teaching.
Through electronic links with other classrooms, students

can collect research data and work collaboratively with others to analyze it. Programs like National Geographic's KIDSNET support structured activities of this kind, and students can interact informally over LANs or WANs. CD-ROM technology offers resources for students to use in developing projects. The Database of Images, for example, provides a huge collection of images that users can import into multimedia reports. Multimedia authoring programs (e.g., HyperStudio, Linkway Live, MediaText, HyperCard) help students to develop multimedia projects with text, audio, video, and animation.

Example Activities

Activity 1 (mathematics or science):   Developing a Persuasive Presentation
Level: Middle or high school
Purpose of Activity: Students will develop a presentation designed to persuade an audience to accept their points of view.

Instructional Activities: **Setting the Stage: The Ilova Textbooka Scenario.** In an effort to improve mathematics and science achievement in your school district, the school board created the position of mathematics/science quality supervisor and hired a recent Russian immigrant, Dr. Ilova Textbooka. She has an excellent background in basic mathematics and science, and has worked in the Russian educational system for years. Dr. Textbooka can best be described as a traditionalist. She believes that the teacher should remain the focal point in the classroom, delivering instruction in content primarily through lectures and demonstrations.

This philosophy has created a problem for many stakeholders in the school district. They have recently become aware of the merits of educational technology and its probable future importance. Dr. Textbooka is in the process of determining how to allocate next year's budget, and she has agreed to evaluate new instructional technologies with an open mind. In order to gain input for her decisions, she has asked schools to prepare presentations that state the case for technology in instruction. Your school welcomes the opportunity to demonstrate the merits of educational technology to Dr. Textbooka, but it now needs to come up with a strategy for the presentation. The administrators, teachers, and parents at your school believe that the best way to influence Dr. Textbooka is to have students demonstrate to her the power and effectiveness of instructional technology.

Continued

Background. Each group must present a technology-centered program to the new mathematics/science quality supervisor to convince her to purchase it for your school. The team must develop a persuasive presentation/demonstration that will convince Dr. Textbooka that its technology program will be a valuable addition to the mathematics/science department of your school.

Software requirements. This will depend on available resources. Other possible alternatives include borrowing programs from other schools or temporarily obtaining preview programs from vendors.

Recommended Procedures

1. Introduce the scenario to the class, perhaps by passing out copies of paper documents or simply explaining it in a verbal presentation.

2. Discuss the scenario in detail. At this important phase, the teacher must present the lesson in a way that motivates students. Students like the idea of taping their presentations and actually broadcasting them on local access cable television.

3. Explain the presentation procedure to the class. The teacher has great latitude at this point in determining the scope of the project.

4. Lead the class through a brainstorming session for the activity. Ideas will have to remain general since students have not chosen their programs yet. This activity should get students thinking about the task ahead. Brainstorming software like IdeaFisher may be useful, but it is not essential.

5. Discuss various roles and responsibilities and how to divide up the work. Set a tentative timeline rather than leaving the planning stage open ended. Students will need considerable guidance at this point as their organizational skills may not be very well-developed. Software packages like In Control might facilitate project planning.

6. Divide the class into groups of no more than four. Each group will develop its own presentation using a different instructional technology program in either mathematics or science.

7. Groups meet to plan their work. The teacher should observe the process and intervene if the class gets off task.

8. Groups work on the project. Stress cooperation and collaboration between group members instead of adversarial competition.

9. Students deliver presentations. The demonstration could accommodate many possible variations. The class could play out the scenario, casting the teacher or some other adult as Dr. Textbooka. Retired mathematics and science teachers from the community might be brought in to comprise a panel to judge presentations. Administrators from the district office could be another source. Get student input for this decision. Encourage the audience to give constructive feedback to the presenting teams.

Activity 2 (mathematics and science):	Health-care Education Project
Level:	Elementary or middle school
Purpose of Activity:	Students will develop an educational program on the circulatory system.
Instructional Activity:	**Setting the Stage: The StudentPro Scenario.** The President's Health Care Task Force has decided that reform of the nation's health-care system should include an important educational component designed to get the American people to practice a wellness approach toward health care. The task force believes that if people, particularly children, take more proactive steps to promote their own physical well-being, then a long-term solution to exorbitant health-care costs will be easier to achieve.

The educational consulting firm StudentPro (Student Productions) has contracted with the class to develop a student-generated informational/educational program on the circulatory system. The program should include a large display component to appear in museums, hospitals, and malls around the country to educate children about the human circulatory system and its care. StudentPro has granted the class a great deal of leeway to decide how to reach children. However, the company does insist that all of the products give medically accurate information and follow proper spelling and grammar principles.

The classmembers must now educate themselves on the circulatory system and researchers' suggestions for people to take care of their bodies, in general, and their circulatory systems, in particular.

Continued

After researching basic facts, students will brainstorm ideas on how to reach young people and convince them to care for their bodies and prevent common problems.

Suggested Preliminary Activities

- **Inductive activity.** Any number of activities from activity books could help to grab the students' interest and demonstrate the subject area. A microcomputer-based lab (MBL) pulse rate test might be one activity.

- **Videodisc lesson.** A disc like National Geographic's The Incredible Human Machine can help to create interest and inform students. Rather than just playing the disc, use it interactively in order to provoke thought and questions. Display the timer or frame counter while playing the disc and encourage students to record the addresses of segments that they might want to view again.

- **Guest speaker.** Invite speakers from a museum, hospital, university, American Heart Association, etc. Check with the local branch of the American Association of Retired People for a speakers' bureau. See if speakers would be willing to serve as consultants as students develop products. The question-and-answer session could be recorded on audiotape or videotape for future use in multimedia projects.

- **Field trip.** The nature of the trip depends on local resources, but it could include visits to museums, universities, hospitals, health clubs or wellness centers where students could take an aerobics class. Have students videotape the trip for possible use in future projects.

Suggested Related Projects

- **Multimedia/hypermedia display.** Using authoring software, create an educational program that will inform the user about some aspect of the circulatory system.

- **Brochure.** Create an educational brochure that will inform the reader about the causes and preventions of heart disease. Use a desktop publishing program and include student-made graphics.

- **Bar-code activity.** Develop a bar-code activity to access selections from a videodisc on the circulatory system.

- **Song/Jingle.** Create a song or jingle that relates to the theme. Students could record it and include it in a multimedia presentation. Use a word processing package for writing and editing. Bar codes could be used to edit music from an audio CD. Some software enables students to actually write music (e.g., MicroWorlds Project Builder).

- **Poem.** Students could write a poem, record their reading, and incorporate it into a multimedia project. They could also publish the poem in a newsletter, greeting card, or sign.

- **Commercial.** Produce and videotape a public-service commercial to air on the school TV station.

- **Big book.** Develop a book resource for primary classrooms. A primary class could create this product itself with the teacher's guidance.

- **Poster.** With a draw program, students could create an educational poster. A primary-grade teacher might use an LCD panel and draw program to draw a poster in front of the class with student input. Then students could color it.

- **Slogan/banner.** Develop a motto that promotes the lesson's theme, then make a banner using an appropriate program.

- **Activity for elementary classroom.** Students in upper grades could develop activities to be taught to primary students, writing up plans on word processing software.

- **Newsletter.** Design a news resource for the general public with articles and research summaries on the circulatory system.

- **Script for skit.** Write and produce a skit that educates the audience about health problems. Perform at a PTA meeting.

- **Database on foods and fat content.** Bring in labels from foods at home and enter nutritional data into a database.

- **Refrigerator chart with fat content of foods.** Use draw program or clip art to make an attractive and humorous chart to hang on refrigerators.

- **Newspaper cartoon.** Create an educational cartoon. Older students could incorporate satire or sarcasm. A draw program or scanner would facilitate publication.

Continued

- **Survey for families.** Design and conduct a survey to collect data to evaluate dietary practices.

- **Chart to record family blood pressure.** Design and produce a chart to record family blood pressure. Students would gather data with either home or drugstore-type blood pressure devices. A database or draw program could be used for this project.

- **Create slideshow.** Use a presentation program to create a slideshow as part of a presentation to other classes or as a standalone display.

- **Videotapes of interviews with experts.** Invite guest speakers to visit class. A team could tape the speaker and make the tape available to the school library.

- **TV special.** Produce a program for the school TV network or a local cable access channel. Groups would handle different aspects of production.

- **Telecommunications survey.** Using a telecommunications network, send out a health survey to a number of classes around the state. Students produce the survey, retrieve the data, and analyze it using spreadsheet and database software.

- **Faxed conference with a cardiologist.** This is a good way to check the medical accuracy of products.

- **Spreadsheet activity.** Create a spreadsheet that would enable the user to enter in a type of food along with the total calories per serving and the grams of fat. A formula should calculate the percentage of calories in the food attributed to fat.

- **Script for CD encyclopedia.** Write directions on how to locate pertinent sections of a CD-ROM encyclopedia. This could be part of a display.

Activity 3 (mathematics):	Designing Your Classroom
Level:	Grades 3–8
Purpose of Activity:	Students enhance their spatial acuity by designing a workspace.

Instructional Activity:

In this activity, students work in small groups (two or three per group) using draw software to design alternative furniture layouts for their classroom. They should be experienced with the drawing tools of the chosen software.

Students must resolve the following problem: A classroom has 29 student desks, one teacher desk, and 3 work tables. (This information can be adapted to match actual class data.) Each team must create five possible arrangements for the furniture and list the advantages and disadvantages of each arrangement. Students draw a typical student desk table. They then duplicate both (using the software's copy and paste feature) enough times to match the number they need. Using the mouse, they can move the objects around the screen to try different arrangements. As students complete each design, they can save it and then print it at a later time. The artists must list the pros and cons of each arrangement either in a typed supplement or labels on each layout.

In possible extensions for this activity, the groups might present their favorite layouts to the class and let the whole class vote on one favorite. The teacher may then decide to let the students arrange the room in that fashion. Students in upper grades might use a CAD program with scaling features. The teacher may ask more of the group by requiring them to draw objects to scale.

Suggestions for Teacher:

Students may need help conceiving constructive pros and cons of particular layouts. The teacher may need to intervene and explain things like traffic patterns, accessibility to storage areas, etc. A copy of team layouts and documentation may be placed in student portfolios. For younger students, the teacher may want to create the objects and store them in a file to help discuss as a class the pros and cons of designs.

Activity 4: (science):	How Are Things Alike/How Are They Different?
Level:	Grades K–5
Purpose of Activity:	Students enhance their observation skills.

Instructional Activity:

This activity is designed to develop students' observation skills and help them recognize similarities and differences. The teacher will need to prepare by looking at various still frames from a CAV videodisc that relates to an area of science (animals, for example). Look for sets of two frames that students can compare based on similarities and differences. The age of the students will affect the choice of frames.

Continued

Show the class first one frame and then the other and ask them to look for things that are the same and things that are different. Flipping back and forth between the frames a number of times gives the class time to observe. This activity is designed to be open-ended, so be flexible in accepting answers if students can justify their ideas. The videodisc player lends itself to a number of activities like this. For example, show a short video segment, then ask students to recall everything they saw in the clip and describe it to partners who have not see the video segment.

To extend the lesson, the class might develop some categories from the observation data. This activity could also be done at a classroom center with teacher directions.

Suggestions for Teacher: It is much easier to manage this activity if the selected frames are accessed by bar codes. A bar-code program could be used to make codes for specific frames. This practice enables the teacher to concentrate on the content of the lesson rather than operating the remote control.

Additional Ideas for Mathematics/Science Activities

Stock market simulation (mathematics). Students simulate investing in the stock market. A spreadsheet program provides an excellent way for students to track advances and losses. Many spreadsheet programs also have built-in charting features that could be used to make attractive bulletin board displays. Telecommunications services also deliver current stock price quotes.

What's in litter? (mathematics and science). Students collect litter from the school grounds and then categorize it by type (paper, glass, metal, etc.). They enter data into a spreadsheet to calculate percentages and averages. They could then compare this data with that collected from students' neighborhoods or that received from other schools via telecommunications. Look for clues to indicate why certain types or amounts of litter show up (e.g., proximity to convenience store or fast-food restaurant, income level of community, dogs running loose). This model can be used for a variety of activities; an analysis of junk mail is a good way to lead up to Earth Week.

Script for mathematics lesson (mathematics). A teacher could repurpose a videodisc-based mathematics program for an innovative lesson. The teacher plays a segment of the program that shows how to solve a particular algorithm, perhaps long division. After the students are comfortable

with the process, the teacher turns off the audio track and assigns teams to write the scrip for the narrator. Students will need access to a player and the correct segment of the disc. They will need to use the remote control to apply some special features like Still/Step and slow motion. When they finish, the groups can simultaneously read their scripts and play the video. As an alternative to reading the script, have a student teach the lesson without audio from the disc.

Project AIMS materials (mathematics and science). Project AIMS has developed a series of activities that engage students in collecting scientific data and then analyzing it via mathematics. Many of these activities invite students to apply database and spreadsheet technology. The series covers grades K-8.

School drivers (mathematics and science). Survey the entire junior and senior classes at a high school to determine the number of automobile accidents in which each student driver has been involved. Create a database with the information. The survey may include questions involving the make, year, and color of car and seatbelt use. Data analysis can then look for correlations between things like the age of vehicle and accidents. Students could compare data with that of the general population or of another school with which classes telecommunicate. Students could perform more complex data searches using the Boolean search feature of a database program.

Exercises

Exercise 12.1. Explain how each of the following current issues and trends in mathematics and science instruction affects the selection and use of technology:

• NCTM and NRC have set new curriculum standards and goals for their respective fields. What impact would these standards have on your selection and use of technology?

• Several controversies cloud the choices of technology applications in math and science. How would each controversy affect your selection and use of technology resources?

• Software for mathematics generally has been criticized for poor quality. Does this criticism still apply? How would this affect your selection and use of technology resources?

Exercise 12.2. Describe the applications in mathematics and science curricula of the following technology resources:

Mathematics:

a. spreadsheet and database software

b. instructional software

c. calculator

d. videodisc

e. word processing software

f. telecommunications (e.g., Internet sites)

Science:

a. videodisc

b. instructional software

c. MBL

d. desktop publishing

e. LCD panel

f. CD-ROM

Exercise 12.3. Create (a) individual instructional activities for mathematics and science and (b) instructional activi-ties that integrate mathematics and science skills. Be pre-pared to teach the activity to a class. Each of these activities should meet the following criteria:

- Integrate one or more of the types of technology described in Exercise 12.2.

- Show how to adapt the activity for large-group and small-group instruction.

- Describe the preparation that the activity would require.

- Describe the benefits you would hope to derive from using technology resources in the lesson.

Exercise 12.4. Research ways in which people use tech-nology for math or science in the workplace. Study the question either formally or informally (perhaps by talking with neighbors). Write a brief summary of your findings and be prepared to discuss them.

Exercise 12.5. With a partner or two, brainstorm ways that technology can enhance student learning in both math and science. Be prepared to share your thoughts with the class.

References

Allen, D. (1995). Teaching with technology: Software that's right for you. *Teaching PreK-8, 25* (8), 14–17.

American Association for the Advancement of Science (1989). *Science for All Americans: A Project 2061 Report on Literacy Goals in Science, Mathematics, and Technology.* Washington, DC: American Association for the Advancement of Science.

Barba, R. H. (1991). Technology in the science classroom. In S. K. Majumdar et al. (Eds.). *Science education in the United States: Issues, crisis, and priorities.* Easton, PA: Pennsylvania Academy of Science.

Becker, J. P. (1993). Current trends in mathematics education in the United States, with reference to computers and calculators. *Hiroshima Journal of Mathematics Education, 1,* 37–50.

Ben-Chaim, D. (1994). Empowerment of elementary school teachers to implement science curriculum reforms. *School Science and Mathematics, 94* (7), 356–366.

Bitter, G. G. (1987). Educational technology and the future of mathematics education. *School Science and Mathematics, 87,* 454–465.

Bitter, G. G. (1989). Teaching mathematics with technology: Finding number patterns. *Arithmetic Teacher, 37* (4), 52–54.

Bruder, I. (1993). Redefining science. *Electronic Learning, 12* (6), 20–29.

Hill, M. (1993). Mathematics reform: No technology, no chance. *Electronic Learning, 12* (7), 24–32.

Hoyles, C. (1994). Learning mathematics in groups with computers: Reflections on a research study. *British Educational Research Journal, 20* (4), 465–483.

Kaput, J. J. (1994). Technology in mathematics education research: The first 25 years. *Journal for Research in Mathematics Education, 25* (6), 667–684.

Lehman, J. R. (1994). Technology use in the teaching of mathematics and science in elementary schools. *School Science and Mathematics, 94* (4), 194–202.

National Council of Teachers of Mathematics. (1989). *Curriculum and evaluation standards for school mathematics.* Reston, VA: National Council of Teachers of Mathematics.

National Research Council. (1995). *National science education standards.* Washington, DC: National Academy Press.

Rice, M. (1995). Issues surrounding the integration of technology into the K-12 classroom: Notes from the field. *Interpersonal Computing and Technology Journal, 3* (1), 67–81.

Ritz, J. M. (1995). Lunar exploration resources in technology. *Technology Teacher, 54* (7), 15–22.

Robinson, M. (1994). Using e-mail and the Internet in science teaching. *Journal of Information Technology for Teacher Education, 3* (2), 229–238.

Schwartz, J. L. (1992). Of Tinkertoys, technology, and the educational encounter. *Technos, 1* (2), 15–18.

Skovsmose, O. (1994). Toward a critical mathematics education. *Educational Studies in Mathematics, 27* (1), 35–37.

Steen, L. A. (1989). Teaching mathematics for tomorrow's world. *Educational Leadership, 7* (1), 18–22.

U.S. Office of Technology Assessment (1988). *Power on: New tools for teaching and learning.* Washington, DC: U.S. Government Printing Office.

Weiss, I. (1993). *A profile of science and mathematics education in the United States.* Chapel Hill, NC: Horizon Research.

Willis, S. (1995, Summer). Reinventing science education. *ASCD Curriculum Update, 37,* 2–7.

Wilson, T. F., and Utecht, G. (1995). The Internet at Eagan High School. *T.H.E. Journal, 22* (9), 75–79.

Zech, L. (1994). Power on! Bringing geometry into the classroom with videodisc technology. *Mathematics Teaching in the Middle School, 1* (3), 228–233.

Resources

Math Software

Adventures in Flight—Sanctuary Woods
Counting on Frank—Creative Wonders
Money and Time Workshop—Scott Foresman
Math Keys—MECC
Geometry—Brøderbund
Physics and Calculus, Probability Theory—Cross
Hands-On Math: I, II, III—Ventura Educational Systems
GeoExplorer—Scott Foresman
Mathematics Exploration Toolkit—Sterling Swift Software
The Mathematics Teacher's Workstation—Sterling Swift Software
Math Sequences—Milliken
Skills Bank II-Mathematics—Skills Bank Corp.
ClarisWorks in the Classroom—Claris
Microsoft Works for Teachers—Microsoft Corp.
MicroWorlds Math Links—LCSI

Math Videodiscs

The Adventures of Jasper Woodbury—Optical Data Corp.
The Adventures of Fizz and Martina—Tom Snyder Inc.
The Perfect Pizza Caper—HRM Video
Windows on Math—Optical Data Corp.

Science CD-ROM Resources

The Multimedia World FactBook—Quanta
The CIA World Factbook—Quanta

Science Software

Science Cap—Demco
Ecology Treks—Magic Quest/EarthQuest
Balance of the Planet—Brøderbund
SimEarth—Maxis
Biology Labs—Cross
Operation Frog—Scholastic Inc.
Visifrog—Ventura Educational Systems
The Lab: Experiments in Biology—Mindplay
The Science of Living Things—Victoria Learning Systems
Science in Your World—Macmillan/McGraw-Hill
CHEMiCALC—Chemical Concepts Corporation
Chemistry of Achievement I: Mathematics of Chemistry—Microcomputer Workshops
Water Budget—EME
Physical Science Laboratory—Focus Media
Science Toolkit Plus—Brøderbund
Temperature Experiments—Hartley
Multimedia Report Generator—Laser Learning Tech
CBE MultiMedia Sequencer—Emerging Technology Consultants
The Videodisc Toolkit—Voyager
MediaMax—Videodiscovery
Science Sleuths—Videodiscovery
Database of Images—Digital Imaging

Networks

KIDSNET—National Geographic

Science Videodiscs

Windows on Science—Optical Data Corp.
Weather: Air in Action Series—AIMS Media
Garbage: The Movie—An Environmental Crisis—HRM Video
Science Sleuths—Videodiscovery
The Great Solar System Rescue—Tom Snyder Inc.
BioSci II—Videodiscovery
Interactive NOVA: Animal Pathfinders, The Miracle of Life, Race to Save the Planet—Scholastic Inc.
Physics of Sports—Videodiscovery
The Great Ocean Rescue—Tom Snyder Inc.

Chapter 13

Technology in Social Science Instruction

This chapter will cover the following topics:

- Current issues in social sciences instruction

- Recommended resources and applications in social science instruction

- Example activities for single-subject or integrated lessons in the social sciences

Chapter Objectives

1. Identify current issues in social science instruction that affect the selection and use of technology.

2. Describe the most popular uses for technology in social science curricula.

3. Create instructional activities for social science instruction that successfully integrate instructional technology.

"The words of this song are stained with our blood,
Within them are sorrow and grief,
Yet your camp song will carry beyond these barbed wires,
To a distant place unknown to you."
From the Piesn Obozowa, *Camp Song* written by Zbigniew Kocsanowicz at the Falkansee Concentration Camp in Poland (April, 1945). On the U.S. Holocaust Memorial Museum web site.

Introduction

The term *social sciences* encompasses the disciplines of history, social studies, geography, civics, and economics. Perhaps no other content area has so clearly delineated its components as this segment of the curriculum. When educational trends point toward quality over quantity, and educators strive to integrate subject matter, these partitions have major implications for the social sciences.

Technology will very likely drive significant changes in the content and methodology of social sciences instruction. It will enable students to interact with the subject matter in a way that builds a greater understanding of world politics and geography. This chapter will discuss the rationale for this optimistic view.

Issues and Problems Related to Technology Use

The Focus of Social Sciences Instruction

Around what central theme should teachers organize the social sciences curriculum? That is the major unresolved dispute in this area (Shaver, 1992; Parker, 1991; and Evans, 1992). Consensus in the field seems to target citizenship education as the central purpose, but Shaver (1992) saw largely superficial agreement. The true dispute focuses on how to teach citizenship education. The predominant notion advocates a survey of history and social sciences to give students the necessary background to make informed decisions. Many others (Engle, 1989; Nelson, 1993; Saxe, 1990), however, advocate social studies education to teach certain issues. They believe that students learn information and concepts most effectively as they confront issues.

Reforming the Social Sciences

In recent years, three reform proposals have drawn a great deal of attention in the social studies community: California's History-Social Science Framework, the Bradley Commission's Building a History Curriculum, and the National Commission on Social Studies in the Schools Charting a Course: Social Studies for the 21st Century. All of these reform proposals stress the importance of teaching history and geography throughout the curriculum, even in the primary grades.

These recent proposals also agree that effective social sciences instruction requires limits on the quantity of essential content that all students should learn. Rather than superficially covering numerous topics, reformers urge schools to provide each student access to core subject matter.

According to Parker (1991), all of these reform proposals contradict themselves in a critical way. They articulate goals of developing critical understanding in history, geography, and participatory democracy. In reality, the planned curriculum exposes students to huge quantities of geography and narrative history instead of attaining the stated goals. A focus on quantity pervades these reform proposals as a result of the century-old battle for dominance among the university disciplines and its imperfect resolution in the integrated school subject known as *social studies* (Saxe, 1990).

Standards for Social Studies Instruction

By the end of 1995, specialists in the disciplines of history, geography, civics, and social studies intended to develop a set of national standards. If funding is found, economics should have standards in the near future. These sets of standards seem likely to differ in significant ways.

History. Projected history standards would focus on content in U.S. and world history. A favored chronology would divide U.S. history divided into 10 historical periods. Content standards guide instruction for each period and five themes have been identified that run through the period.

Social studies. On the other hand, social studies standards do not treat history in a chronological manner. Instead, they identify ten major themes: culture; time, continuity and change; individuals, groups, and institutions; power, authority, and governance; science, technology, and society; individual development and identity; production, distribution, and consumption; global connections; civic ideals and practices; and people, places, and the environment. The social studies standards also strongly emphasize assessment.

Civics. This topic area will draw its standards from the voluminous 1991 curriculum guide, *Civitas*. The curriculum's 22 exit standards for 12th grade students are categorized around five "organizing questions:"

- What is government and what should it do?
- What are the foundations of the American political system?
- What are the roles of the citizen in the American political system?
- What is the relationship of American politics and government to world affairs?
- How does the government established by the U.S. Constitution embody the principles and purposes of American democracy?

Geography. Geography specialists have identified 18 standards divided into three grade-level blocks: K–4, 5–8, 9–12. These standards are explained primarily in performance terms. The field stresses the interaction of geographic literacy and higher-order thinking skills.

Parker (1994) expresses concern that these individual standards will serve to only reinforce the tendency toward "separateness" in the social sciences at a time when school reformers demand integration. At this time, only the social studies standards project seems actively interested in integration, although the other areas recognize the need to integrate. In the future, the groups that represent the separate entities under the social sciences umbrella should work with their peers to decide on a set of core standards that integrate principles from all of the social sciences.

New World Order

Rapid political change has had a number of implications for social studies instruction. First, information is becoming obsolete virtually overnight. Map publishers today are hard pressed to keep up with almost daily changes. The flexibility advantage of digital technology over printed media will make technologies like telecommunications and CD-ROM extremely relevant in the delivery of social studies instruction.

Multiculturalism

Demographic data suggest a major shift in the ethnic composition of the U.S. population. In the future, people of color will comprise a much larger segment of the citizenry. The Eurocentric view of both American and world history appears to be losing emphasis. Instead, school curriculum will focus more on the contributions of a variety of African, Asian, and Latin cultures. As the world economy becomes more closely interwoven, more and more people will work with others from a variety of cultures. Schools will have to provide students with skills to function in a culturally diverse environment. Technology, through the use of telecommunications and video resources, will support students' experiences and interactions with a variety of cultures.

Recommended Resources and Applications for the Social Sciences

Computer Software

Traditionally, facts have driven social science instruction more than any other content area. The nature of the subject matter, with its historic dates and geographic names, has certainly contributed to this methodology. Since computers organize and present such information quite effectively, the abundance of computer software for social sciences instruction should not surprise anyone. The software falls into several broad categories.

Simulations. These tools contribute to both the traditional model of social sciences instruction and the issues-oriented approach. Simulation is a powerful method today, and future increases in computer power promise to deliver still more elaborate computer simulations that include virtual

reality and an almost unlimited number of "what if" scenarios to explore (Kay, 1991). Many simulation programs lend themselves to group work and engage students in higher-order thinking activities. These programs are often accompanied by written materials integrated into the software's routine (e.g., Tom Snyder's Decisions, Decisions series, Simpolicon). Some simulations propel the user back in time (e.g., Pilgrim Quest, Railroad Tycoon), while others place them in contemporary settings (e.g., SimCity, Mastering the Market Economy, How a Bill Becomes a Law).

Instructional games. Sometimes only a fine line separates programs that teachers might call *games* and those they might call *simulations*. Generally speaking, simulations present more complex model-building activities geared toward developing thinking and cooperative-learning skills. Games, on the other hand, are usually directed more toward mastering factual information. Many are built around competition, either between the user and the computer (e.g., Operation U.S. Presidents) or against other users (e.g., Knowledge Master Basics: Social Studies). Another kind of program presents information to users in a game format and then quizzes them on the content (e.g., Women of Influence, State Smart).

Map utilities. Through their graphics and animation capabilities, computers can help students to examine cartography in a way that is not possible without technology. Map View 3D is a drawing tool that enables the user to draw a contour map and, with a click of the mouse, see a three-dimensional view of the terrain. Other types of mapping software offer large collections of premade maps along with drawing tools to add features (e.g., David Smith's Mapping the World by Heart, BaseMap). Geopuzzlers represents another approach: presenting geographic jigsaw puzzles. The user chooses a difficulty level and customizes the puzzle in many other ways. Practically any drawing utility can help teachers develop mapping skills. ClarisWorks for Teachers includes a lesson that demonstrates how students can gain mapping experience via the draw and paint functions of the program.

Databases. Database programs enable students to sort through enormous quantities of data with relative ease. They have begun to make the traditional report obsolete. With the rapid growth of automated systems to retrieve factual information, teachers are now realizing the need to focus more on how to synthesize and utilize data. Database programs contribute extensively to the study of geography, where they help students navigate through huge amounts of geographic data. Programs like Social Studies Toolkits and the Research Assistant Series go beyond just providing data; they also help students to identify important information and guide them in effective uses of information. Many database programs let students input their own data and then generate activities for the students based on that data (e.g., School Works, Global Studies with Country

Level III Video Social Studies Lesson

Databases). In another valuable use of database technology, Timeliner and Time Patterns Tool Kit provide vast collections of historical data and then help users search for trends.

Videodiscs: Level I and Level III

Video resources effectively capture much of the information of social sciences in real-time recordings or as dramatizations. A number of videodisc series contribute to history instruction. The American Lifestyle Series offers biographies on important Americans from Andrew Jackson to Booker T. Washington and Pearl Buck. Large collections of videodiscs also teach about other cultures. Some, like World Cultures and Youth, teach within the context of stories, while others use a documentary approach (e.g., Asia: An Introduction, South America—People and Culture). The Vital Links: Multimedia History system is a middle school/high school product that incorporates a variety of media into the study of U.S. history, including videodisc, CD-ROM, videotapes, audio tapes, and presentation software. A large selection of videodiscs provide actual film footage of important events of the 20th century (e.g., Eyes on the Prize—The Civil Rights Years, Roosevelt and the Fireside Chats). A number of programs dramatize events to within their historical contexts (e.g., American Documents: The Lincoln-Douglas Debates, Had You Lived Then). Other video materials provide valuable support for social science instruction, despite a lack of teacher support (e.g., National Geographic Specials, NOVA). The quality and affordability of these programs give them valuable advantages.

Although not as plentiful as the Level I videodiscs, social studies teachers can use a number of Level III programs. GTV by National Geographic uses engaging multimedia presentations to teach American history and geography at the elementary level. ABC Interactive offers titles dealing with AIDS, tobacco, war, and communism. These types of programs offer extensive text-based information to support their video components. Most also include presentation tools that enable the teacher or students to fashion customized presentations.

CD-ROMs

The multimedia potential and tremendous storage and search capabilities of this technology will help it flourish in the social sciences arena. CD-ROMs make research much more practical than book-printed methods. As more and more classrooms get CD-ROM equipment, teachers and students can have tremendous information resources available literally at their fingertips. In the study of geography, CD-ROM atlases offer a wide variety of resources like maps, statistical data, music, video clips, and details of customs (e.g., World Vista, The Importance of Place, Great Cities of the World). Engaging programs for both world history and U.S. history provide a wealth of information through interactive multimedia displays (e.g., Rome and the Celts, The Middle East Diary, Time Traveler). Publishers also use this technology to store large collections of documents on CD that can offer ready access to researchers. These text-based programs enable users to search and print valuable information (e.g., U. S. History, Terrorist Group Profiles).

One big advantage of CD-ROM is its capacity for storing audio. Some social studies programs take advantage of this strength by including large selections of quotes or speeches (e.g., Black American History—Slavery to Civil Rights). Audio CDs, which can play on CD-ROM players, are another potential resource for education (e.g., Great Speeches of the 20th Century).

Telecommunications

Distant communications links seem poised to dramatically increase the amount of interplay, and hopefully understanding, among the cultures of the world. This great potential will very probably reach fruition soon, and its impact will carry over to education. A number of telecommunications projects have electronically linked students from different corners of the world; however, due primarily to limited resources at schools, applications of overseas telecommunications have not yet reached the mainstream in education. Nevertheless, most commercial online services (e.g., America Online, Prodigy) now offer access to the Internet, so the future will bring a dramatic increase in telecommunications between students of the world. Indeed, much of this activity is likely to take place in the home rather than at school. These links could certainly lead to a variety of new types of homework assignments.

The World Wide Web also offers a wide variety of social studies resources (see Figures 13.1 and 13.2). These offer students access to a voluminous amount of timely information (Dyrli,1995):

- **Library of Congress.** This site (http://lcweb.loc.gov) details exhibits and events, congressional information, historical collections, and online indexes.

- **Smithsonian Institution.** The Smithsonian (http://www.si.edu) catalogs the people places, activities, exhibits, and products of "America's Treasure House of Learning."

Figure 13.1 Sample Social Studies Web Page

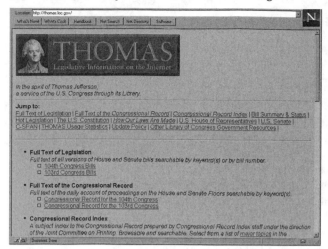

Figure 13.2 Sample Social Studies Web Page

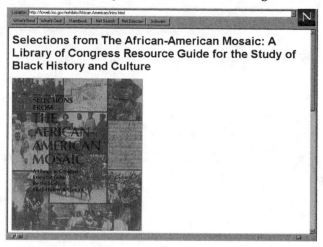

- **Thomas Legislative Information on the Internet.** This Web server (http://thomas.loc.gov) presents a searchable Constitution of the United States, full text of the Congressional Record, pending legislation, and e-mail addresses for members of Congress.

- **White House.** This site (http://www.whitehouse.gov) covers tours and multimedia information about the executive branch and the first family.

- **United States Holocaust Memorial Museum.** This site (http://www.ushmm.org/) is a document and photo archive to support study of the Holocaust.

- **Tiger Map Service Experimental Browser.** This Web server (http://wings.buffalo.edu/geogw/) allows one to search for any U.S. name place.

- **Xerox PARC Map Viewer.** The Xerox map viewer (http://pubweb.parc.xerox.com/map/) lets one click on a location anywhere in the world and get a map of the area.

Additional Resources for Directed Instruction

A number of tutorial programs present the content of social sciences material to students in a rather straightforward manner (e.g., European History, History of Western

Civilization, U.S. Government). Teachers can use these types of programs as the primary vehicles for instruction or to supplement lecture and textbook methodologies. In another directed-instruction approach, publishers develop programs for the specific purpose of helping users perform well on tests (e.g., A+ U.S. History, American History Review Bank).

Additional Resources for Constructivist Teaching

Students can use any of the multimedia/hypermedia authoring tools (e.g., SuperLink, HyperStudio, HyperCard, MicroWorlds Project Builder) to develop social studies projects. With so many visual and audio resources available, these tools provide tremendous benefits for students as they develop products. Any videodisc can be repurposed, or used to meet specific needs of the user other than those envisioned by the publisher. For example, a section from a science disc may support instruction in a social sciences concept. In another example of repurposing, a student might present a scene from the movie *Lawrence of Arabia* to demonstrate a geographic concept about deserts.

Example Activities

Activity 1:	Training for Cultural Awareness
Level:	Grades 9–12
Purpose of Activity:	To familiarize students with the intricacies of other cultures through the development of a project
Instructional Activity:	**Setting the Stage: The Training Departments Dilemma.** Over the past 2 years your company, NUTECH, has experienced a surge in overseas business. This has resulted in a tremendous increase in the amount of foreign travel for NUTECH employees, who have conducted business in locales where they knew very little about the local cultures. The employees report that they believe that this has put

Continued

them at a distinct disadvantage, and they would like to get training on how to relate more effectively to indigenous populations when traveling abroad.

Your training team has been assigned the task of putting together a multicultural center that will provide the employees with an easy-to-access compilation of resources. At a team meeting, a brainstorming session identified several ideas for potentially valuable resources.

Ideas for Multicultural Resource Center

- **Brochure.** The brochure should include a brief description of the history of the country along with relevant geographic and cultural data. It would be helpful to stress any cultural difference that visitors should recognize; for example, in Thailand it is considered very rude to sit with one leg crossed over the other with a foot pointing at another person. Students should use a desktop publishing or word processing program to develop the brochure. Graphics should enhance the layout.

- **Videotape.** Create a videotape that provides useful information to a traveler in a specific country. The information should enable that person to function more effectively in the local culture by stressing customs, values, and historical perspectives.

- **Bar-coded videodisc presentation.** Develop a bar-code-driven program that accesses relevant segments of a videodisc that pertain to a specific culture. Students should use a bar-code generator and a word processor for the project materials, which they should mount on tagboard and laminate.

- **Multimedia display.** Create a multimedia display that provides suggestions for travelers in a particular country. Video segments would enhance the program, either imported into the program or through interaction with a videodisc player.

- **Database.** Develop a database of resources for each country to which employees might travel. This should include magazine and newspaper articles, videos, books, etc. Students could access much of this information via the Internet.

Suggestions for Teacher: For this activity, students will need to choose a country on which to focus. They must understand the purpose of the product—to provide a resource for someone who needs help functioning in another country's culture. To develop a quality product, students will need to strive to truly understand the culture of the chosen country. Encourage students to use telecommunications resources to locate information. The Internet may offer them an opportunity to actually converse with citizens of the chosen country. They may also have access to foreign nationals living nearby.

Activity 2: Database of Inventions
Level: Grades 4–8
Purpose of Activity: Students design and build a database to observe, compare, and sort numerous inventions,

Instructional Activities: The students develop a database to add to the school media center as a resource for students studying about inventors. Working in small groups or as individuals, students begin by deciding what fields to include in the database. The teacher may want to brainstorm this question with the whole class and decide upon a number of required fields to which students may add others of their choosing. When they have completed the database, the students can create a scavenger hunt game in which players use the sort/find/report features of the database. For example, they might answer questions like, "find the name of an inventor who was alive in the 1920s, was English, whose last name begins with *T*, and who invented something between 1947 and 1956."

Suggestions for Teacher: Some possible database fields are invention, year invented, inventor, dates of inventor's birth and death, inventor's country, description, and information source. When students complete the project, they may want to exchange scavenger hunts and see who finishes first. The projects could also be shared or exchanged with classes in other schools or even other countries.

Activity 3: Attributes of Explorers
Level: Grades 6–12
Purpose of Activity: The teacher and students will compare and contrast the characteristics of various modern-day explorers by examining video accounts of their exploits.

Instructional Activity: The teacher shows various segments from the videodisc National Geographic Explorers. This disc, like many National Geographic specials, is divided into a number of thematic segments. In this activity, the

Continued

teacher shows a segment and then leads a discussion concerning the attributes of the explorers. As attributes are identified, a student volunteer should list them on the chalkboard. As a follow-up to this part of the activity, the students work in pairs and write an account of a fictional explorer. Encourage them to use the list of attributes as a resource.

Suggestions for Teacher: This activity, although geared to a specific videodisc, is designed to be used as a model. It involves using video from the National Geographic Explorers videodisc to compare and contrast the attributes of modern day explorers, based on the way they are portrayed in the video accounts of their explorations. In this model, the video provides fodder for a class dialogue concerning which attributes explorers seem to share. It may be used as a standalone activity or as a lead-in to an indepth study of explorers. It's important that the teacher play an active role in keeping discussions going. It will also require that the teacher preview the disc and note beginning and ending points of segments.

**Activity 4 (geography/
physical science):**   Acid Rain: Let's Act Now!
Level: Grades 5–8
Purpose of Activity: To design and develop an environmental project dealing with the issue of acid rain

Instructional Activity: This activity can employ the Acid Rain component of the National Geographic's KIDSNET curriculum package. The KIDSNET program includes activities and telecommunications software. The activity should not be limited, however, to participants in the National Geographic network. Classes that have access to telecommunications equipment may want to link up with classes located in other geographic areas for electronic exchanges of acid rain data. These projects are intended to follow up a curriculum unit on acid rain, and they presume that the students understand the scientific and political aspects of the acid rain issue.

Groups of two to three students select projects from those listed below, or they develop their own ideas or adaptations (contingent on teacher approval). Each group will need to compare results with at least one other group (preferably in a distant geographic location) working on the same project. During the development of the projects, the partners will collaborate via telecommunications.

Suggested Projects

- **Research report.** The Sierra Club has heard of your involvement in the study of acid rain. They have commissioned you to write a detailed report concerning the causes and dangers of acid rain as well as a prognosis for the future. They would like you to present this paper at their national convention. Use a word processor to compile the report and create overhead transparencies for the presentation.

- **Contrasting viewpoint.** The American Forest Association, an industry trade group, has hired you to write a report that discusses its side of the acid rain issue. The association maintains that the danger of acid rain has been greatly exaggerated, and it cites plenty of research to back up the claim. Utilize technology in the same manner as for the Sierra Club report.

- **Newspaper editorial.** You are concerned about the apparent apathy of your community regarding acid rain. Compose a letter to the editor stating your concern and explaining why more people need to work to reduce acid rain levels.

- **Contrasting viewpoint.** You are concerned about a perceived overreaction to the alleged acid rain problem. Write a letter to the editor that refutes the contention about an acid rain problem.

- **Multimedia presentation.** Ranger Rick Online magazine wants to make students aware of the problem of acid rain. The editors have asked your group to create a poem or a short story to inform young people of the dangers of acid rain. Incorporate the poem or story into a multimedia project.

- **Contrasting viewpoint.** Jobs for America, Inc., a progrowth political action committee, would like your team to develop a multimedia presentation to display on major telecommunications networks. The presentation should convince viewers that the issue of acid rain is overstated and resulting overregulation of business costs people their jobs.

- **Television commercial.** Your advertising company, LET-US-PUSH-IT, has been commissioned to design a nationwide campaign to "Save the Forests." Your team needs to design a 30-second television commercial (complete with visuals) to inform the general public about the threat to woodlands from acid rain.

Continued

- **Contrasting viewpoint.** LET-US-PUSH-IT has been hired to develop a national campaign to balance the strong effort by environmental groups to toughen emissions standards on automobiles and power plants. The campaign should convince people that the acid rain problem is grossly overstated.

Suggestions for Teacher: It will be important to fully prepare students to understand another point of view. Particularly with environmental issues, students often have one-sided opinions, and they may resist even looking at the opposite position. When developing these projects, encourage students to fully utilize technological resources. Online telecommunications services may be good sources of information. For example, a sample search under the entry *acid rain* on the World Wide Web yields a number of valuable resources. Some of these include acid rain laboratory activities, EcoNet acid rain resources, scientific data from precipitation monitors, and a list of references on the subject.

Activity 5: Exploring Geography
Level: Primary grades
Purpose of Activity: To expose students to a number of geographical concepts and broaden their geographic knowledge base

Instructional Activity: The teacher uses the videodisc player to illustrate various geographic concepts like mountains, beaches, snow, tornadoes, and hurricanes. The teacher should ask questions and have students walk up to the monitor and point out things in the picture or video. If the narration on the disc exceeds the students' comprehension level, the teacher can shut off the audio at the videodisc player and provide the narration.

Suggestions for Teacher: It is helpful to do this activity before the students read or listen to reading. Evidence suggests improved reading comprehension when students see a segment of video related to what they will be reading. Many videodiscs could contribute to this lesson. Teachers can utilize video that may not have been intended for instructional purposes (or "repurpose" the video resource). Movies or National Geographic specials prove unexpectedly valuable. With a CAV disc, the teacher can enhance the lesson with the slow-motion capability of the videodisc player.

Activity 6: Social Studies Kiosk
Level: Grades 3–12
Purpose of Activity: Students will design a bar-code driven guide that helps the user retrieve information from a social studies related videodisc.

Instructional Activity: Students will develop an interactive social studies display for the school media center. It should occupy a visitor to the media center for 10 to 15 minutes. The display will require a videodisc player with bar-code reader, a relevant videodisc, and the student-developed guide. The guide should help the user to quickly and easily access some important social studies fact or concept. The group will need to decide which videodisc to use, what outcomes to achieve, and what grade levels to target. With these guidelines determined, they will need to spend time previewing the disc and selecting clips or slides for which to make bar codes. A word processor with graphics capabilities can help them produce the guide. This could be a booklet or even a poster with bar codes attached.

Suggestions for Teacher: Make sure that students understand the need to do much more than just select the video segments. The written guide is central to the project. Students must ensure that their text and video selections support their intended outcomes. The kiosk can be decorated to match the theme of the program. The class might complete this lesson as a school improvement project, changing the program every month.

Activity 7 (economics): Provisioning for a Journey
Level: Grades 5–8
Purpose of Activity: Students gain experience at allocating resources and maintaining budgets.

Instructional Activity: This activity can extend any number of social studies simulations of journeys (Voyage of the Mimi II, Oregon Trail). The students allocate a set sum of money to finance their group's needs for a simulated trip or expedition. The students work in groups using resources that they have helped to collect (catalogs, newspaper ads, etc.) that provide price data. They track expenditures under a set conservative budget based on relevant information like the duration of the trip and number of people. They must

Continued

| Suggestions for Teacher: | figure out what to buy and how to allocate the budget. When they complete the activity, each group will deliver a presentation of its budget.

This activity corresponds well to the SCANS workplace competency dealing with allocation of resources. Groups should work with calculators. The activity also presents an excellent opportunity to integrate spreadsheet software. The teacher should work with the whole class to develop a spreadsheet template that performs the following calculations: unit price × quantity, subtotal, subtotal × tax, subtotal × shipping, total, amount left in budget. Groups should review "what if" questions that spreadsheets can answer. Groups could use their spreadsheets as in their budget presentations. |
|---|---|

Additional Suggestions for Social Sciences Instruction

Social studies family day (social sciences, language arts, art). The school could sponsor a thematic social studies fair. Booths could integrate technology in a number of ways, including making posters and banners, publishing a program of events and displays, presenting student-made multimedia displays, demonstrating social studies resources on the Internet at a telecommunications center, showcasing social studies software, etc. Community involvement could include presenters who are native to another culture, or businesses with international transactions like travel agents. Peace Corps recruiters may also be willing to participate.

Music of the world (geography, music). Using the CD-ROM Musical Instruments, demonstrate to the class the tremendous variety of instruments. Be sure to stress the country or region of the world where each one originated, linking geography with music. Students of all ages will enjoy hearing the sounds of different instruments. This could be an ongoing activity that the class does once or twice a month for 15 minutes or so.

Trading numbers (social studies, math). Students gather geographic, historic, and economic data for their community. They then enter the data into a spreadsheet and use graphing tools to create a series of graphs. They can also exchange data with students in other communities. If possible, they should communicate over electronic links. Depending on the level of the students, they can undertake further analysis of the data or create a bulletin board using graphs from other classes.

Public service project (social studies, math, language arts). The class works with residents of a nursing home to develop a yearbook that includes a brief biography of each resident. The students can collect data through personal interviews. The book could focus on a number of important historical events that occurred during the lifetime of the elderly people. The publication could feature pictures of the subjects. If feasible, some of the nursing home residents could contribute to the production process. Sell advertising space to local businesses to finance the printing of the book.

Public access TV show. Students produce a television show to air on either the school's closed-circuit system or possibly a local public access channel. It could deal with any number of social studies issues. Once students choose their issue, they select representatives for both sides.

Bar codes for teachers. Customized bar codes can make a particular videodisc valuable for a wide variety of learning experiences. Often, though, the teacher lacks time to make the bar codes. In this activity, the teacher gives students certain learning objectives that he/she needs to teach and the students search through social studies discs for relevant video frames or segments. After the teacher has previewed the selections, the students generate and print the bar codes. This activity can engage older children in developing resources for children functioning at lower academic levels. One alternative placement school in Florida has students make bar codes for the primary classrooms in nearby elementary schools. The developers can then go and watch the teachers or students use the tools that they have produced.

Exercises

Exercise 13.1. Explain how each of the following current issues and trends in social sciences instruction affects the selection and use of technology:

- The focus of classroom instruction inspires an ongoing controversy in social sciences education. Several curriculum reform proposals seek to address this issue. What impact would these controversies have on your selection and use of technology in this field?

- New standards have been proposed for history, social studies, civics, and geography instruction. What impact would these standards have on your selection and use of technology for these areas?

Exercise 13.2. Describe the applications of the following technology resources in social sciences curricula:

- Word processing software
- Videodisc technology

- Video cameras
- Multimedia authoring programs
- Desktop publishing software
- Telecommunications (e.g., Internet Web sites)
- Social studies coursesoftware

Exercise 13.2. Create (a) individual instructional activities for one of the social sciences and (b) instructional activities that integrate social sciences with other curricula. Be pre-

pared to teach the activity to a class. Each of these activities should meet the following criteria:

- Integrate one or more of the types of technology described in Exercise 13.2.
- Show how to adapt each activity for large-group and small-group instruction.
- Describe classroom preparation necessary for this activity.
- Describe unique benefits of technology resources for the lesson.

References

Brady, R .H. (1994). An overview of computer integration into social studies. *Social Education, 58* (5), 312–314.

Brooks, D. L. (1994). Technology as basic to history-social science: It's long overdue. *Educational Technology, 34* (7), 19–20.

Carroll, T. (1995). Carmen Sandiego: Crime can pay when it comes to learning. *Social Education, 59* (3), 165–169.

Dyrli, E. (1995). Surfing the World Wide Web to education hot spots. *Technology and Learning, 16* (2), 44–51.

Eisner, E. W. (1991). Art, music, and literature within social studies. In J. P. Shaver (Ed.). *Handbook of Research on Social Studies Teaching and Learning.* New York: Macmillan.

Engle, S. H. (1989). Proposals for a typical issue-centered curriculum. *The Social Studies, 80* (5) 187–191.

Evans, R. W. (1992). Introduction: What do we mean by issues-centered social studies education? *The Social Studies, 83* (2), 93–94.

Kay, A. C. (1991). Computers, networks, and education. *Scientific American, 265* (3), 138–148.

Lombard, R. (1995). Children, technology, and social studies. *Social Studies and the Young Learner, 7* (3), 19–21.

Massialas, B. G. (1992). The "new social studies"—Retrospect and prospect. *The Social Studies, 83* (3), 120–124.

Nelson, M. (1993). Hip, hype, hope: social studies reform for the 1990s. *International Journal of Social Education, 8* (1), 50–58.

Parker, W. C. (1991). Trends: Social studies—the newest reform proposals. *Educational Leadership, 48* (3), 85.

Parker, W. C. (1994). The standards are coming. *Educational Leadership, 51* (5), 84–85.

Peterson, G. A. (1994). Geography and technology in the classroom. *NASSP Bulletin, 78* (564), 25–29.

Saxe, D. (1990). A plea for rapprochement in social studies reformation. *Social Education, 54* (6), 351–352.

Semrau, P. (1995). Social studies lessons integrating technology. *Social Studies and the Young Learner, 7* (3), 1–4.

Shaver, J. P. (1992). Rationales for issues-centered social studies education. *The Social Studies, 83* (3), 95–99.

U.S. Department of Labor (1992). *(The SCANS Secretary's Commission on Achieving Necessary Skills) report.* Washington, DC: U.S. Government Printing Office.

White, C. (1995). Two CD-ROM products for social studies classrooms. *Social Education, 59* (4), 198–202.

Wilson, E. K., and Marsh, G. E. (1995). Social studies and the Internet revolution. *Social Education, 59* (4), 203–207

Resources

Social Studies Software

Decisions, Decisions—Tom Snyder Inc.

Simpolicon—Cross Cultural Software

Pilgrim Quest—Decision Development Corp.

Railroad Tycoon—Microprose

SimCity—Maxis

Mastering the Market Economy—Interactive Market Technologies

How a Bill Becomes a Law—Intellectual Software

Operation U.S. Presidents—Tanager

Knowledge Master Basics: Social Studies—Academic Hallmarks

Women of Influence—Wintergreen Software

State Smart—Platypus Software

Map View 3D—Philosoft

David Smith's Mapping the World by Heart—Tom Snyder Inc.

BaseMap—GeoPoint

GeoPuzzlers—Platypus Software

ClarisWorks for Teachers—Claris Corp.

Social Studies Toolkits—Tom Snyder Inc.

Research Assistant series—Focus Media

School Works: Social Studies—K-12 MicroMedia Publishing

Global Studies with Country Databases—TEC

Timeliner—Tom Snyder Inc.

Time Patterns Tool Kit:—Tom Snyder.

History of Western Civilization Queue

U.S. Government—EPC

A + U.S. History—American Education Corp.

American History Review Bank—TEC

European History—TEC

Videodisc Programs

The American Lifestyle Series—AIMS Media

World Cultures and Youth—Coronet/MTI

Asia: An Introduction—BFA

South America—People and Culture—Windows on Social Studies—Optical Data Corp.

Vital Links—Davidson

Eyes on the Prize—The Civil Rights Years—Image Entertainment

Roosevelt and the Fireside Chats—Coronet/MTI

American Documents: The Lincoln-Douglas Debates—Coronet/MTI

Had You Lived Then—AIMS Media

National Geographic Specials—National Geographic
GTV—National Geographic
ABC Interactive—ABC News Interactive

CD-ROM Resources
World Vista—Applied Optical
Great Cities of the World—InterOptica
Rome and the Celts—Queue
The Middle East Diary—Quanta

Time Traveler—Orange Cherry New Media
Terrorist Group Profiles—Quanta
Black American History—Slavery to Civil Rights—Queue
Great Speeches of the 20th Century—Rhino Records

Telecommunications Services
America Online—See Telecommunications chapter
E-World
Prodigy—See Telecommunications chapter

Chapter 14

Technology in Music and Art Instruction

This chapter will cover the following topics:

- Current issues in music and art instruction

- Recommended resources and applications in music and art instruction

- Example activities for single-subject or integrated lessons in music and art

Chapter Objectives

1. Identify the effects of current issues in music and art instruction on the selection and use of technology.

2. Describe the most popular uses for technology in music and art curricula.

3. Create instructional activities for music and art instruction that successfully integrate instructional technology.

"... [T]he language of the new technology ... images, sound, movement, voice, drama, and text ... is also the language of the arts. Arts teachers need to build on this connection by helping students exploit the expressive potential of electronic media and by educating them to become active, critical consumers of its message."
R. Robinson and C. Roland, from *Technology in Arts Education* (1994)

Introduction

Arts educators have often resisted pressure to use computers and other instructional technologies. They have frequently complained of a contradiction in blending impersonal machines with traditionally humanistic endeavors. In reality, however, technology has always played a part in the arts. Over the centuries, technology has provided tools, materials, and processes that aided artists' creative expression. In more recent times, the phonograph in music and the camera in the visual arts have changed people's definitions of *art*. Therefore, integration of computers and other forms of electronic technology represents the next logical step in the evolution of the arts.

Issues and Problems Related to Technology Use

Issues Common to Music and Art Instruction: Controversies and Trends

Why teach music and the arts? A combination of the back-to-the-basics movement in the early 1980s and budget constraints in recent years have led many American schools to systematically dismantle arts programs (Fowler, 1989). The National Center for Educational Statistics estimates that almost half of all American schools have no full-time art teachers on staff.

The arts have been easy prey for legislators and administrators who believe they are less important than other subjects to a school's curriculum. Many assume that schools can either ignore arts instruction or cover it with the traditional classroom teacher. Many who consider art classes as educational "frills" do not comprehend tremendous social shifts. In fact, technological advances compel schools to place more weight than ever on teaching the arts. Arts instruction will help students develop the values and sensibilities that will enable them to function as healthy citizens in an increasingly artificial, high-tech environment (Robinson and Roland, 1994).

A background in the arts helps to foster uniquely human qualities like willingness to take risks and challenge convention, self-discipline, self-assessment, and a commitment to developing individual creative talents.

Standards. New national standards for education in the arts were released in the spring of 1994. They emerged

from the most important national effort in years to define the role of the arts in education. Proponents feel that these standards will strongly influence the ongoing debate over philosophy and methodology in arts instruction. They also hope that the standards will lift the arts into a more prominent position in debates over the content of a basic education for every student.

Some prominent features of the new standards include:

- An emphasis on long-term, broadly defined, sequential programs of study rather than merely occasional exposure or access. For each art form, standards are organized according to three interdependent areas of competence: creating and performing, perceiving and analyzing, and understanding cultural and historical contexts. Local considerations affect the weight accorded to each area. The standards set guidelines for grades K through 4, grades 5 through 8, and grades 9 through 12.

- There are two types of standards. Content standards specify what students need to know and to accomplish. For example, one standard for visual arts defines a capability for "Selecting and using visual arts media, techniques, and processes to effectively communicate ideas" (Willis, 1994, p. 2). Achievement standards specify the experiences that students should successfully complete during grades 4, 8, and 12. For the quoted content standard, for example, a related achievement standard for the end of 4th grade ensures that students "are introduced to, become familiar with, and are able to safely use different art materials such as pencils, paint, clay, fibers, paper, and wood" (Willis, 1994, p. 3).

- One set of universal achievement standards for grades K through 4 and grades 5 through 8. The achievement standards for grades 9 through 12 are divided into two levels: proficient and advanced. All students, the standards document suggests, should reach the proficient level in at least one of the arts. Students who have elected specialized arts courses may reach standards for the advancement achievement level.

These arts standards do not offer any panacea. Proponents will have to sell their beliefs to an educational establishment that does not always see the connection between arts instruction and students' roles as productive citizens.

The arts in the information age. Many educators and members of the community question the need for instructional technology in the arts curriculum. Even some proponents of technology applications in other disciplines balk at investments in technology for the arts. Robinson and Roland (1994) offer four reasons to link the goals of a school arts program with rapidly developing instructional technologies:

1. By integrating new technologies into the arts curriculum, instructors expose students to new and exciting modes of artistic expression. All media have a place in the curriculum if they enable students to achieve desired instructional outcomes. New technologies warrant special attention because they constitute entirely new genres of art that may alter people's paradigms about art.

2. The new technological culture requires today's students to develop a whole new set of literacies that go far

beyond the term *computer literacy*. Arts instruction provides many unique opportunities for students to hone analytical skills to critically evaluate the flood of messages that fill a technologically saturated environment. The communicative language of the new technologies—sound, animation, music, drama, video, graphics, text, and voice—is also the language of the arts. Thus, arts teachers are particularly well-positioned to help students develop skills as both critical producers and critical consumers of electronic media.

3. In the workplace of tomorrow, workers will often have to generate creative solutions to problems. An arts program that develops students' potential for innovation in the areas of music, animation, graphics, multimedia, desktop publishing, and other emerging technologies will enable those students to compete in tomorrow's global business environment.

4. The arts counterbalance the massive infusion of technological change that society is experiencing. Technology can be very seductive, and people need to keep in sight unique human abilities. Citizens of tomorrow's world will need coping skills that enable them to keep their aesthetic sensibilities in the face of breathtaking technological advances. Arts education will help develop and maintain these skills.

Integration with other content areas. One of the most popular current ideas for adding more arts instruction to the curriculum seeks to weave it into other content areas. This path leads to some pitfalls, though. Content area teachers may try to integrate the arts simply by having their students listen to some music or draw pictures that relate to their studies. This shallow approach does little to broaden the students' understanding of the arts. Additionally, it may prove to inhibit the movement to elevate art education to the status of a major curriculum area by reinforcing the notion that regular classroom teachers can easily cover the arts, as well. Collaboration between arts and content area teachers may resolve this dilemma.

Arts teachers can work more closely with classroom teachers in order to bridge the gap between their fields. To ensure quality integration, the content area teachers will need help from the arts specialists to decide just how to present arts education. Often, the arts teachers will also help classes acquire needed materials. This synergistic relationship promises to enhance the quality of classroom instruction and bring the arts teacher into the mainstream of the curriculum.

Issues in Art Instruction: Controversies and Trends

Academic versus studio. Many visual arts educators advocate redefining the field to give it a larger role in the school reform movement. They want to rethink the recipients of art education and how it is taught. In many school districts, students receive an hour or so of art instruction per week. The lessons often focus on producing some art products, leaving little time for curriculum that introduces students to other aspects of art, such as art history, aesthetic principles, or criticism. At the end of elementary or middle

school, schools direct students with obvious talent toward elective courses that continue the focus on producing art. Critics of this approach argue that it reaches a relatively small number of students, and even then it narrowly defines art education. From a political perspective, this approach gives art educators a very weak power base and subjects the discipline to the constant scrutiny of budget cutters.

But how can art instruction give all students strong art backgrounds that go beyond just producing art? Recently much debate has centered on a philosophy of art instruction called Discipline-Based Art Education (DBAE). Proponents of DBAE want to give students broad and rich experiences with works of art in four ways:

- Making art (art production)
- Responding to and making judgments about properties and qualities in visual forms (art criticism)
- Acquiring knowledge about the contributions of artists and art to culture and society (art history)
- Understanding how people justify judgments about art objects (aesthetics)

Art production, art history, art criticism, and aesthetics are the foundation *disciplines* that make up DBAE. Critics of the DBAE approach worry that studio training within art instruction will suffer and that students' creativity will be stifled. Some contend that proponents of DBAE are more interested in producing a population of museum patrons than in developing individual artistic expression. Others believe that DBAE's emphasis on academics will discourage students who may not perform well in the traditional classroom, but find the process of creating art fun and motivating.

DBAE seems to have influenced the development of the visual arts standards. Proponents believe that if art will gain acceptance as a mainstream subject only by broadening instruction to include more than just the students who draw or paint well. As they see it, DBAE offers just such a broad framework.

Issues in Music Instruction: Controversies and Trends

Theory or performance? The community of music educators defines a broad spectrum of opinions on how to teach music. At one end of the spectrum, some believe that music theory is a basic competence that students need to truly understand and appreciate music. At the other extreme, some believe that students learn by doing and students should spend most of their instructional time performing music. The new national music standards offer a balanced interpretation of this issue.

Another issue in middle schools, and especially in high schools, concerns band competitions. Many music educators criticize perceived overemphasis on competitions. This replaces a great deal of pressure on the student and teacher for a perfect performance of a particular musical selection. Therefore, the student spends inordinate amounts of time practicing a few pieces over and over instead of broadening his or her musical expertise. On the other hand, supporters of band competitions note that this area of arts instruction in schools often inspires strong support in the community.

Recommended Resources and Applications

Resources and Applications for Music Instruction

Hardware. Music teachers typically use several types of hardware resources:

- **MIDI.** The Musical Instrument Digital Interface (MIDI) standard enables computers, electronic musical instruments, and software to communicate. MIDI-compatible electronic keyboards are equipped with ports through which they connect via cables to other MIDI devices. The standard governs translation of analog musical sound into a digital format that computers can process. MIDI technology gives users a special ability to create, edit, and recreate music. It can be compared to the piano roll that allows a player piano to replay a tune. Holes punched in the roll trigger a mechanical device that plays a piano note. MIDI does what the old piano roll does but in a digital format. It assigns a number to every aspect of musical sound including pitch, length, and instrument. When the MIDI device receives these numbers from a computer, it plays the appropriate musical sound. Because MIDI is an industry standard, equipment and software from various manufacturers work together. Some of the MIDI interfaces that are available are Passport MIDI Pro Interface for the Apple II, Roland MPU-IPC for IBM compatibles, and Sentech Mini MAC for the Macintosh.

- **Synthesizers.** Electronic musical instruments constitute the cornerstone of music technology. Piano-like keyboards replicate the sounds of both acoustic instruments, like the guitar, and original electronic sounds such as television sound effects. Synthesizers can also split tones to produce more than one sound at a time. In effect, this provides the user with his or her own orchestra. This shows how technology offers capabilities that traditional methods cannot replicate.

- **MIDI networks.** It is now possible to network MIDI devices in much the same manner as computers are networked. In such a network, students work at MIDI keyboards instead of computers. These keyboards are then connected to a single, teacher-controlled computer. This system lets the teacher monitor student progress and musical achievement in a very efficient manner. The student gets immediate feedback, which promotes independent development. Research has indicated that a MIDI network increases student achievement (Moore, 1992).

Software and media. Music instruction also relies on several types of software and media resources:

- **Sequencers.** Software in this category (e.g., Performer 4.2) enables the user to record, edit, and play back music. A composer can save short sequences and then easily manipulate them and paste them together to form larger works. In many ways, the sequencer brings to music composition what word processing software brought to the writing process. After saving work as a MIDI file, the user can easily transport the music and revise it virtually at will. This provides tremendous

Music Editor and Synthesizer

flexibility for musical composition and performance on any hardware or software configuration. In addition, programs like Master Tracks Pro 5 can synchronize music to film, video, multimedia presentations or multitrack audio tape.

- **Musical notation software.** Musicians use this type of software (e.g., Finale 3.1, Composer's Mosaic, MusicTime) to print scores for music they create using synthesizers and sequencers. It is a particularly helpful tool for students working on creative and composition skills. As with creative writing, seeing the results of one's effort in print seems to drive the creative process (Moore, 1992). This software also facilitates portfolio assessment. Programs like MusicPrinter Plus combine sequencing and music notation software, enabling users to edit any size score, lead sheet, or choral style.

- **Multimedia resources.** In multimedia applications, students combine music with video or animation. They can generate soundtracks with sequencers and synthesizers or record audio from other sources using microphones. Students can also alter sounds through cut-and-paste techniques. In addition, they can create special effects using capabilities such as playing sounds backward and changing their pitch or length characteristics. Programs that support this work include HyperCard, LinkWay Live, HyperStudio, Media Text, and MicroWorlds Project Builder.

- **Instructional software.** Music teachers employ a number of different types of instructional software. Play It by Ear and Practica Musica provide music students with ear training. Other programs, like PianoWorks, help students improve their music and performance skills through guided exercises. This type of

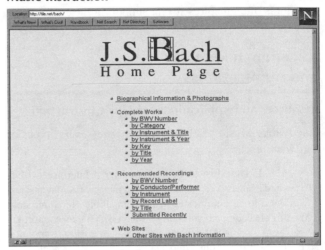

Figure 14.1 Sample Home Page for Music Instruction

program records the student's performance and begins a new session at the previous ending point. Another type of application teaches various aspects of music like rhythm, melody, and form (e.g., MetroGnomes' Music, Melodious Dictator).

* **Videodiscs: Level I and Level III.** A number of Level I videodiscs provide recorded presentations of music performances (Domingo Live from Miami, Beethoven & Mozart, Imagine the Sound). Educators like these resources for their high-quality video and audio outputs along with the capability of randomly accessing segments of music. They include no software or teacher's guides, so successful applications depend upon the teacher or students repurposing them. These discs also have the advantage of low prices. Instruments of the Symphony Orchestra and Bachdisc are examples of Level I videodiscs produced for educators along with teacher support materials. Level III programs include The University of Delaware Videodisc Music Series and Mozart: The Dissonant Quartet. These programs provide recorded concert music along with analyzed scores, scholarly commentary, and slides of cultural and historical highlights (see J.S. Bach home page in Figure 14.1).

* **CD-ROM Technology.** Programs like The Magic Flute contain complete musical performances, on-screen commentary and annotation, and extra audio tracks to help students better understand the opera or other piece of music. Some CD-ROMS strive to help students better understand the roots of music by providing extensive historical data (Composer Quest, So I've Heard, Vol. 1: Bach and Before). Another group of CD-ROMs presents the sounds and history of musical instruments of the world (e.g., Microsoft's Musical Instruments). The interactive formats of these resources offer a variety of search options.

Internet resources. The vast resources of the Internet include music information such as:

* **Music on the Web** (http://.art.net/Links/musicref.html)

* **The Piano Page** (http://www.prairienet.org/arts/ptg/home page.html)

* **Music Library** (http://www.musicindiana.edu/)

Additional resources for directed instruction. The CD Companion Series guides the user through a number of famous pieces of music. This type of program often offers a game or games that test the user's proficiency level. A number of computer-assisted instruction programs deliver music instruction directly to the student (e.g., Alfred Music Achievement Series, Diatonic Chords).

Additional resources for constructivist teaching. Many multimedia authoring tools (e.g., HyperStudio, LinkWay Live) enable the user to access segments of a CD audio program through a CD-ROM drive. This application allows students, for example, to play a segment of music to accompany a poem or story that they have written. In another use of audio CDs, a user can play selections or entire pieces on a videodisc player using bar codes. Collections of copyright free music (e.g., Clip-Sounds, FracTunes, Grooves) give the user selections to incorporate into multimedia presentations. These music databases are usually available on CD-ROM.

Resources and Applications for Art Instruction

Hardware. The most common type of hardware resource in art instruction is image digitizing equipment. Graphic scanners are computer peripherals that transfer print materials into digital images on the computer. A scanner can transfer any image, photograph, line drawing, or text into a graphics file in a cost-effective and efficient manner. An artist can also capture an image from a video source (camcorder or VCR) using digitizing software like Computer Eyes. Finally, a digital camera (e.g., ZapShot) records images directly to disk. This equipment provides the user with an extremely flexible system for developing digital images. Students can then manipulate digital images in a number of creative ways. Indeed, the ability to digitize still images and video has opened up a whole new genre of art.

Software and media. Several types of software and media resources commonly aid art instruction:

- **Graphics software.** Computer programs create two major categories of graphics: object-oriented or draw images and bitmapped or paint images. Art students may think they can create an image using graphics tools in software packages or by importing graphics via scanners or clip art collections without knowing which category of image they have made. However, a basic understanding of the differences helps them to modify their images. A bitmapped image resembles a puzzle made from tiny tiles. Each tile is called a *pixel*, short for *picture element*. A bitmapped image is both filled and outlined with pixels so modifying it requires the artist to edit pixels. Scaling or resizing bitmapped graphics can be difficult, since pixels do not adjust mathematically. Consequently, resized images often display coarse or jagged details. Paint programs offer some tools not available in draw programs, including the eraser, paint bucket, marquee, and lasso tools. Popular paint programs are MacPaint, Canvas, KidPix, and SuperPaint 3.0.

Object-oriented graphics programs (draw tools) construct computer images from lines, shapes, and text stored as sets of mathematical instructions rather than collections of pixels. Object-oriented images are generally smaller than bitmapped images, so they occupy less storage space. All drawing programs include object-oriented features which are particularly useful in producing technical illustrations, diagrams and charts, maps and graphic decorations, and other, more mechanical images. An object-oriented program like MacDraw II and FreeHand 3.1 stores the user's menu choices and mouse movements as a list of drawing instructions. The program can then resize, rotate, and refill the drawn objects with no loss of quality. The images are not tied to a particular resolution, so they print with greater accuracy than bitmapped images on any type of printer.

To decide whether to use a painting program or a drawing program, artists must consider first the type of art work they want to produce. Painting programs create and edit images with great detail using effects like shading and stylized brushstrokes. Only paint programs can alter digitized images, which are always bitmapped. Drawing programs are useful for producing illustrations, posters, or architectural-type layouts. Technical drawings, which require capabilities for rotating and resizing also suit drawing applications. An art student should work with both types of tools to understand how and when to use each.

Print Graphics Software

A Graphics Program

- **Animation software.** Art educators can choose among a number of options for animation software. Low-end programs do simple cell-type animation (e.g., Flip Book). More advanced programs (e.g., FantaVision) offer features like "tweening," which enables the user to change one frame to the next by moving through as many as 64 intermediate positions in between. Other programs are specifically geared toward cartoon production (e.g., The Animation Studio, Cartoonin'). They allow artists to add music and sound. Today's more powerful programs speed up the process by providing two-dimensional animation (e.g., Animation Works).

- **Graphics and image manipulation software.** An art studio would not be complete without a program that enables students to creatively manipulate and edit digitized photos. High-end programs (e.g., Adobe Photoshop 2.5) provide hundreds of options and special effects for altering images. Morphing software (e.g., Morph 1, Morph 2.0) enables the user to smoothly transform from one shape to another. This technique offers tremendous potential for artistic expression.

- **Videodisc: Level I and Level III.** Level I and Level III titles guide users through art experiences using a variety of approaches. Some study specific artists (e.g., Van Gogh Revisited, Michelangelo Self-Portrait); others simulate journeys through famous museums (e.g., The Louvre, National Gallery of Art). The high quality of videodisc images along with the technology's quick access gives both students and teachers a powerful and flexible teaching and learning tool.

Internet resources. A gopher site (gopher unix5.nysed. gov) provides useful resources for art education. This site gives links to thousands of animations, clip art, and

Figure 14.2   Home Page for Art Instruction

other images. See Figure 14.2 for an additional art education resource.

Additional resources for directed instruction. In a videodisc, Draw and Color Funny Doodles with Uncle Fred, the teacher (Uncle Fred) leads children through pre-cise, patient steps to create drawings. Video footage of animals and other objects highlight the drawings. Another software package, Draw to Learn, uses the computer to teach pencil-and-paper drawing skills. This type of program always leaves the student in control; the computer cannot run ahead or lag behind. Some programs (e.g., BrushStrokes) come with complete step-by-step instructions on computer paint techniques along with formal instruction in color theory and composition. In another use of videodisc technology, some programs (e.g., National Gallery of Art Tour Guide, With Open Eyes) enable the instructor to easily create customized slide presentations.

Additional resources for constructivist teaching. CD-ROM technology offers students and teachers large quantities of copyright-free clip art images that they can incorporate into projects (e.g., Fresh Arte, ClipArt Plus, Graphics at Your Fingertips). Software programs like Video Jam, an open-ended creativity tool, enable students to create their own animated music videos. Copyright-free video clips and stills are also available (e.g., Visual Almanac) to augment student multimedia productions.

Sample Activities

Activity 1 (art and economics):	Designing an Art Calendar
Level:	Grades 5–12
Purpose of Activity:	Students use computer-generated art to produce a school calendar to sell as a fundraiser.
Instructional Activity:	Discuss with students the idea of using art that they create on the computer in a calendar for sale in the community. Explain that it will take a lot of work to complete a serious attempt to develop an aesthetically pleasing product. Each piece of art should have a title and the artist(s) should write a brief description of the piece. The students may use any available resources as long as they incorporate computer technology in the product. Students can develop the actual calendar part of the project with a calendar-making program. Relevant dates for the school should be noted. This team-focused activity will require extensive collaboration among students. They will have to make a number of production decisions ranging from choosing pieces of art to setting a price for the calendar.
Suggestions for Teacher:	Students conceive a strong sense of ownership if they choose the theme for their art. The teacher may want to break the class into subgroups that decide on a theme for each month. Depending upon printing resources, the calendar could be limited to black-and-white drawings. The project could simply develop the calendar without selling it, or the PTA may give or lend funds to cover printing costs, thus allowing students to produce a high-quality product that the school could sell all over the community.
Activity 2 (art and geography):	Designing a Postcard
Level:	Grades 4–8
Purpose of Activity:	Students use computer art tools to create a postcard that depicts a geographic location in another county.
Instructional Activity:	Explain to students that they will create a postcard to promote visits to a geographic location of their choice. Begin by showing the students some sample promotional postcards from local stores, tourist

Continued

areas, museums, etc. The students will need to research their chosen areas to understand the locations they will promote. National Geographic magazine would be a great resource for this research. Encourage students to sketch out the postcard before they attempt to draw it with the computer program. When students complete the artwork, they can print it out and mount it on index cards.

Suggestions for Teacher: Telecommunications services and the Internet are allowing students to arrange pen-pal relationships with others in foreign countries. If this is feasible, it would motivate students to know that they would send the postcards to a class in another country (perhaps even the location they designed the postcards to promote).

Activity 3 (art): Developing Drawing Skills
Level: Grades 2–5
Purpose of Activity: This series of activities is designed to familiarize students with the power of drawing tools and provide them with stimulating activities that build confidence in their image-making abilities.

Instructional Activity: The activity consists of four different lessons, each of which would take a class period (30 to 45 minutes) in a computer lab setting.

1. **Visual memory approach.** Students work in pairs at each computer. The teacher lists a variety of themes from which students choose. These themes should enable them to generate images from memory of familiar, everyday surroundings (e.g., home, room at home, favorite store, ballpark).

2. **Stimulated imaginative approach.** The students begin this activity with a picture from a graphics file that falls into a certain theme (e.g., whales). The picture should have all textures in the internal shapes erased. Students begin using the computer tools (pencils, patterns, erasers) to add texture to the drawing. The teacher may want to guide the students in this process by using overhead transparencies. Next, the students should use drawing tools to generate their own fantastic sea creatures.

3. **Observation drawing approach.** Students use pencil and paper to draw an object chosen by the teacher. When they complete this task, they go to the computer and transfer their images using draw and paint tools.

4. **Scanned imagery approach.** In this activity, the students scan an image that they have created using traditional media. They then load the file into a computer draw/paint program to enhance and further develop the image. In a twist on this activity, students could do two enhancements of the same image, one using only draw tools and the other using only paint tools. They could also do one enhancement in color and another in black and white.

Suggestions for Teacher: A teacher with younger students will need to have the images scanned into the program prior to the activity. Through all of the activities, students must recognize the drawing process as the focus of the lesson.

Activity 4 (music): Creating New Sounds for Music
Level: Grades 4–12
Purpose of Activity: Students use current technologies to create new timbres for musical sounds by layering and manipulating sounds on a synthesizer.

Instructional Activity:
1. Students listen to a variety of sounds made by a synthesizer.
2. In small groups or individually, students choose two or three timbres (depending on the capability of the synthesizer) and then combine them (e.g. French horn, trumpet). They also combine two or three that they feel do not go together (e.g., trombone and snare drum).
3. Students then write melodies and play them with combinations of timbres that do and do not go together. The class listens to the melodies and compares the effects.
4. While the class listens to each group's melodies, have the class judge which timbres go together and which do not.
5. Combine different groups' melodies and sounds so that students hear new possibilities for musical pieces based on their examples.

Suggestions for Teacher: As a prerequisite for this activity, students need basic knowledge of music terms such as *tone*, *color*, and *timbre*. In addition, students will need a multitimbral synthesizer capable of playing two

Continued

timbres at the same time. Students might tape their sounds and compositions for inclusion in their portfolios.

Activity 5 (music and writing):	Repurposing Music
Level:	Grades 4–12
Purpose of Activity:	Students combine music with oral reading to enhance the dramatic or comic effect of written material.

Instructional Activity: Many newer videodisc players can play audio CDs. Bar'n'Coder 3.0 for the Macintosh enables the user to create bar codes that access segments of CDs. This activity combines bar-code technology and word processing graphics capability to create a story with musical enhancement.

Activity 6 (music):	Conductor in Action
Level:	Grades 6–12
Purpose of Activity:	Students closely watch the actions of a conductor while simultaneously listening to the music.

Instructional Activity: A number of videodiscs feature classical music performances by some of the world's finest orchestras. The video images often focus a great deal of attention on the conductor. The random access flexibility of the videodisc player allows the teacher and the students to study the actions of the conductor in detail. This activity is particularly helpful for band directors, but it can also benefit general music appreciation classes.

Suggestions for Teacher: Bar codes could help the teacher repeat segments with relative ease. For a large group, a video projection system or LCD panel may increase the effectiveness of the image.

Activity 7 (music and social studies):	Instruments of the World
Level:	Grades 4–8
Purpose of Activity:	Students will observe the rich diversity of musical instruments that have evolved in different cultures.

Instructional Activity: Students to use Microsoft's CD-ROM Musical Instruments to develop a presentation for the class. Working in pairs, students select three musical instruments from the CD-ROM. The teams research the development of the instruments using the CD and other resources. Upon completion of their research, they present their instruments to the class using the CD for visuals and sound and discussing differences and similarities between the instruments. Older children should also include information concerning the cultures that originated the instruments.

Suggestions for Teacher: The teacher will need to help groups discern the musical similarities and differences between the instruments. Students will also need help to find connections between cultures and the types of instruments that evolved. The activity should not focus too much on finding a single, right answer; it should emphasize the process of hypothesizing about connections between music and culture.

Activity 8 (art and music):	Technology and the Arts Expo
Level:	Grades 9–12
Purpose of Activity:	Students organize and stage a fine arts festival that exhibits the role of modern technology in the arts. As a secondary purpose, parents become aware of the new world of arts and technology.

Instructional Activity: Students can create any number of technology-based exhibits for an arts festival. Some choices might include:

- **Famous works of art.** Students develop a multimedia display using a visual arts videodisc (e.g., National Gallery of Art, The Louvre). The display should help the viewer gain a better understanding of art. Programs such as HyperCard, HyperStudio, LinkWay Live, or MediaText could contribute to this activity. A similar project for music could employ either a videodisc or an audio CD played through a CD-ROM drive.

- **Student presentations.** This activity uses Level I videodiscs with bar codes. Students select a particular visual arts or music concept to express. The team then finds selected slides or musical

Continued

segments that help make this point. They generate bar codes for the relevant materials and use them to display elements of the presentation.

- **Demonstrations.** As a component of the festival, students could conduct ongoing demonstrations of visual arts and music technology. Music technology demonstrations could showcase MIDI-compatible synthesizers, sequencers, assorted software, multimedia programs, and videodiscs. Visual arts demonstrations could display paint and draw programs, videodiscs and slideshow software, animation software, scanners, and digital cameras. Provide an opportunity for the audience to try the different technologies.

Suggestions for Teacher: Creative work does not have to reinvent the wheel. Locate others who have staged similar shows, and use some of their experiences as benchmarks. Computer bulletin boards often have conferences for teachers; post a message asking for suggestions. Invite prominent community and business people to attend the festival. They may provide support for your program in the future. Include artists from the community as a part of the festival; architects could demonstrate CAD technology; companies specializing in multimedia presentations could show their tools; local musicians may expose listeners to music technology.

Activity 9 (art): Self-portrait
Developed by: Mark Hampton, Sabin Junior High School (Colorado Springs, Colorado)
Level: Grades 6–12
Purpose of Activity: Students create self-portraits collage that emphasizes artistic uses of values to achieve realistic images.

Instructional Activity: Begin the lesson with a discussion of self-portraits. Show examples of famous self-portraits by artist, such as van Gogh, Rembrandt, and Picasso. Discuss the concept of value and how faces and images can be defined by different values instead of by outlining parts with lines. Explain the collage art form and assign students to create a self-portrait collage that uses values to achieve a realistic image.

The assignment begins with a digital image of each student. Optional methods include scanning a photograph of the student into the computer using either a flatbed or handheld scanner. A digital camera photograph, or a camcorder along with Computer Eyes could also capture the student's image.

Next, import each image into a desktop publishing program (e.g., PageMaker) and enlarge to 7 by 10 inches. After completing the document, print out each student's portrait on a separate piece of paper. Individual students then outline areas of value in their portraits. The resulting line drawings need enough detail to make them recognizable as portraits. Students then transfer outlines to a piece of stiff poster board using carbon paper. Finally, the work on the collage begins.

Students should search through old newspapers and magazines and, referring to the original computer printout, tear out different values and textures. They need to find different textures and values to fill in the various shapes on the line drawing. Tear the different values into small pieces and paste them within the appropriate areas. Challenge the students to identify one another's completed self-portraits.

Suggestions for Teacher: To help students discern the areas of value on the digitized photograph, suggest that they squint their eyes. This will often help observers see different values more clearly.

Activity 10 (music and reading): Karaoke Sing Along
Level: Grade 2–5
Purpose of Activity: Students develop a positive attitude toward singing, develop musical skills, and improve reading skills.

Instructional Activity: Using a karaoke style videodisc that displays the words to songs on the screen, the teacher guides the students through singing the songs. The videodisc offers flexibility to access exact segments of a song for which students need extra work. The teacher might also make a bar code for a particular song that plays the audio track but not the video images. This encourages students in memorization. A bar code can also specify video images without the audio track. In a live performance, the TV monitor could face the chorus to act as a TelePrompTer.

Suggestions for Teacher: This activity could employ a number of karaoke type videodiscs. Walt Disney, Inc. has a number of titles available that show the animated Disney characters singing songs. A large selection of discs also feature more contemporary songs. These discs are accompanied by music videos, some of which have content that may not be appropriate for schools. Many of the karaoke discs have two audio tracks, one that plays the music and background vocals and one that plays only the music.

Exercises

Exercise 14.1. Explain how each of the following current issues and trends in music and art instruction affects the selection and use of technology.

- New curriculum standards and goals govern music and art instruction. Describe the impact of these standards on your selection and use of technology in these fields.

- A controversy continues about the role technology should play in music and art instruction. How would this controversy affect your selection and use of technology resources?

- Ongoing controversies surround the kinds of activities that music and art instruction should emphasize (e.g., academic versus studio, theory versus performance). How would these controversies affect your selection and use of technology resources?

Exercise 14.2. Describe the applications in music and art curricula of the following technology resources:

Music:
- Music notation software
- Videodisc technology
- Video cameras
- Sequencers
- CD-ROM programs
- Synthesizers

Art:
- Draw program
- Videodisc technology
- Video cameras
- Animation software
- Digital cameras
- Scanners

Exercise 14.3. Create (a) individual instructional activities for music and art and (b) instructional activities that integrate music and art with each other and with other curricula. Be prepared to teach the activity to a class. Each of these activities should meet the following criteria:

- Integrate one or more types of technology described in Exercise 14.2.

- Show how this activity could be adapted for large-group and small-group instruction.

- Describe classroom preparation that this activity would require.

- Describe unique benefits you hope to derive from using technology resources in the lesson.

References

Bell, B., and Vecchione, B. (1993). Computational musicology. *Computers and the Humanities, 27* (1), 1–5.

Berg, B., and Turner, D. (1993). MTV unleashed: Sixth graders create music videos based on works of art. *TechTrends, 38* (3), 28–31.

Bristor, V. J., and Drake, S. V. (1994). Linking the language arts and content areas through visual technology. *T.H.E. Journal, 22* (2), 74-77.

Brouch, V. (1994). Navigating the arts in an electronic sea. *NASSP Bulletin, 78* (561), 43–49.

Chia, J., and Duthie, B. (1993). Primary children and computer-based art work: Their strategies and context. *Art Education, 46* (6), 23–26, 35–41.

Dunnigan, P. (1993). The computer in instrumental music. *Music Educators Journal, 80* (1), 32–37, 61.

Fowler, C. (1989). The arts are essential to education. *Educational Leadership, 47* (3), 60–63.

Fuller, F., Jr. (1994). The arts for whose children? A challenge to educators. *NASSP Bulletin, 78* (561), 1–6.

Gouzouasis, P. (1994) . Multimedia constructions of children: An exploratory study. *Journal of Computing in Childhood Education, 5* (3–4), 273–284.

Hicks, J. M. (1993). Technology and aesthetic education: A crucial synthesis. *Art Education, 46* (6), 42–47.

Kersten, F. (1993). A/V alternatives for interesting homework. *Music Educators Journal, 79* (5), 33–35.

Madeja, S. S. (1993). The age of the electronic image: The effect on art education. *Art Education, 46* (6), 8–14.

Matthews, J., and Jessel, J. (1993). Very young children use electronic paint: A study of the beginnings of drawing with traditional media and computer paintbox. *Visual Arts Research, 19* (1), 47–62.

Michael, J. A. (1991). Art education: Nurture or nature: Where's the pendulum now? *Art Education, 44* (4), 16–23.

Moore, B. (1992). Music, technology, and an evolving curriculum. *NASSP Bulletin, 76* (544), 42–46.

Nolan, E. (1994). Creativity with instant feedback. *Technology, Teaching, Music, 2* (3), 36–37, 55.

Robinson, R., and Roland, C. (1994). *Technology and arts education.* Tallahassee, FL: Florida Department of Education.

Roland, C. (1990). Our love affair with new technology: Is the honeymoon over? *Art Education, 43* (3), 54–60.

Turner, D. (1994). Creating music videos with works of art. *Teaching Pre K–8, 24* (6), 55–57.

U.S. Department of Labor (1992). The SCANS (Secretary's Commission on Achieving Necessary Skills) report. Washington, DC: U.S. Government Printing Office.

Wagner, M. J. (1988). Technology: A musical explosion. *Music Educators Journal, 75* (2), 30–33.

Wagner, M. J., and Brick, J. S., (1993). Using karaoke in the classroom. *Music Educators Journal, 79* (7), 44–46.

Webster, P. R. (1990). Creative thinking, arts, and music education. *Design for Arts in Education, 91* (5), 35–41.

Music Resources

MIDI Interfaces
Passport MIDI Pro Interface—National Educational Music
Roland MPU-IPC—Electronic Courseware
Sentech Mini MAC—Electronic Courseware

Sequencers
Performer 4.2—Mark of the Unicorn
Master Tracks Pro—National Educational Music
Musical Notation Software
Finale 3.1—Coda Music
Composer's Mosaic—Mark of the Unicorn
MusicTime—Coda Music
MusicPrinter Plus—Temporal Acuity

Related Resources
HyperCard—Apple
LinkWay Live—EduQuest
HyperStudio—Roger Wagner I
Media Text—Sunburst
MicroWorlds Project Builder—LESI

Instructional Software
Play It by Ear—Practica Musica
PianoWorks—Temporal Acuity
MetroGnomes' Music—The Learning Co.
Melodious Dictator—Temporal Acuity

Videodiscs
Domingo Live from Miami—Ztek
Beethoven & Mozart—Voyager
Imagine the Sound—Voyager/S.M.S. Optical
Instruments of the Symphony Orchestra and Bachdisc—
    Britannica Software
The University of Delaware Videodisc Music Series—
    Videodiscovery
Mozart: The Dissonant Quartet—Voyager

CD-ROM Resources
Magic Flute—Ztek
Composer Quest—Compton's New Media
So I've Heard Vol. 1: Bach and Before—Ztek
Musical Instruments—Microsoft
CD Companion Series—Voyager
Alfred Music Achievement Series—Alfred Publishing Company

Diatonic Chords—Temporal Acuity
Clip-Sounds—Monarch Software
FracTunes—Quanta
Grooves—Media Design

Art Resources

Digitizing Equipment
Computer Eyes—Digital Vision
Zapshot—Canon

Graphics Software
MacPaint—Claris
Canvas 3.5—Deneba Software
KidPix—Brøderbund
SuperPaint 3.0—Adobe Systems
MacDraw II—Claris
FreeHand 3.1—Adobe Systems

Animation Software
Flip Book—Intellimation
FantaVision—Wild Duck
The Animation Studio—Walt Disney/Buena Vista
Cartoonin'—Remarkable Software
Animation Works—Claris

Graphics and Image Manipulation Software
Adobe Photoshop 2.5—Adobe Systems
Morph 2.0—Gryphon/ACS

Level I and Level III Videodisc
Van Gogh Revisited—Voyager
Michelangelo Self-portrait—Voyager
The Louvre—Voyager
National Gallery of Art—Voyager
Draw and Color Funny Doodles with Uncle Fred—Fred
    Lasswell, Inc.
National Gallery of Art Tour Guide—Laser Learning Tech
With Open Eyes—Voyager
Visual Almanac—Voyager

Instructional Software
Draw to Learn—Draw to Learn Associates
BrushStrokes—Claris

CD-ROM Resources
Fresh Arte—Quanta
ClipArt Plus—Triad Venture
Graphics at Your Fingertips—Vicki Legu

Chapter 15

Technology in Exceptional Student Education

This chapter will cover the following topics:

- Current issues in exceptional student education

- Recommended resources and applications for exceptional student education

- Example activities for single-subject and integrated lessons in exceptional student education

Chapter Objectives

1. Identify current issues in exceptional student education that affect the selection and use of technology.

2. Describe the most popular uses of technology for various groups of exceptional students.

3. Create instructional activities for exceptional student education that successfully integrate instructional technology.

"... Jeff sat down in front of the computer and began typing questions about the Miami Dolphins football players.... [F]or the first time, he was able to communicate his thoughts, desires, questions, and feelings. And Jeff smiled. And his father cried. Jeff was 35 years old when he began writing."
Janeen Clinton (1993)

Introduction

As Chapters 11 through 14 have shown, content areas such as science and mathematics are no longer as separate from each other as they once were. In the same way, the once-distant lines between instruction for regular and "special" students are now becoming increasingly blurred. Whitworth (1993) observed that now more than ever before, many children fall into the "gray area between regular and special education" (p. 133). Several factors contribute to this ambiguity. Some experts in the subject called *special education* or *exceptional student education* (ESE) are also beginning to recognize more common learning needs between "special students" and "regular students" than disparate needs. Hanley (1993) observed that it is often difficult to differentiate the needs of special students from those of other students. He asserted that "for a very large number of 'special education' students, the regular classroom is a critically important educational setting" (p. 167). Consequently, he assigned many of the same benefits—and limitations—to technology for all students, regardless of the groups in which school systems place them. Finally, legislation shows a definite trend toward making sure that all students receive the same educational opportunities regardless of their disabilities. Whitworth (1993) identified additional responsibility on non-ESE educators. "'Shared responsibility' (means that) ... teaching and preparing students with disabilities to enter the adult world is the responsibility of everyone, not just special educators" (p. 132).

As the curricula of regular and special-needs students merge, it is not surprising that the ESE literature describes many of the same technology resources and integration activities that have already been described in previous chapters for the content areas. Some technology resources have been designed especially to meet unique needs of students with certain disabilities (e.g., Braille printers, word prediction software). Also, schools use some types of resources and applications in ways that fairly specifically meet the needs of special students (e.g., telecommunications, drill software). Still, schools seem to use most of the same products with students identified as having special needs as they use with "regular" students, and they use those products in very similar ways. This chapter will describe technology applications to meet the regular and special learning needs of learning disabled and mentally handicapped, physically disabled, and gifted students.

Issues and Concerns Related to Technology Use

Issues in Exceptional Student Education

The impact of federal legislation. Several federal laws enacted in recent years have addressed the need to provide education to special needs students. These laws have dramatically changed the requirements placed on schools to educate these students and the need to increase technology applications for them.

- **The Education for All Handicapped Children Act of 1975 (Public Law 94–142).** This landmark legislation sought to "ensure that all disabled children have a free and appropriate public education, to ensure that the rights of disabled children and their parents are protected, and to assist state and local agencies in providing this education" (Clinton, 1992, p. 307).
- **The Technology-Related Assistance for Individuals with Disabilities Act of 1988 (Public Law 100–407).** This law speaks directly to the need for schools to provide handicapped students and their families with access to technology resources. It provided funds to help states obtain and deliver assistive technology service. Lewis (1993) noted that, as of 1991, approximately half of the states had been awarded such funding.
- **The Individuals with Disabilities Act (IDEA) of 1990 (Public Law 99–457).** This amendment to Public Law 94–142 included specific references to the need to provide assistive/adaptive technologies (and related services) for the disabled.
- **The Americans with Disabilities Act (ADA) of 1990 (Public Law 101–336).** This law amounted to a "civil rights bill intended to eliminate discrimination against persons with disabilities" (Lewis, 1993, p. 12). Although not directed specifically toward education, it placed many requirements on public schools to provide access to the same services for both handicapped and nonhandicapped students and employees.

Together, these laws have increased the responsibility of schools to provide equal access to resources and activities for both disabled and nonhandicapped students. Schools have not always agreed with parents on how to interpret legal requirements, and lawsuits have resulted. No one doubts, however, that meeting these special needs, especially technology needs, has resulted in substantial additional costs to education systems.

Controversies related to mainstreaming and inclusion. Federal laws have driven a trend in schools to place students with special needs in regular classrooms whenever possible. Schools seem to see little choice, despite frequent objections from both teachers who are not skilled in dealing with special needs students and parents of other students. This is definitely the largest and most far-reaching issue in the field today, and it will have consequences for curriculum development and implementation methods for the forseeable future. Schools are attempting to comply with legal requirements for equal access to educational opportunities in two ways:

- **Mainstreaming.** This strategy, which grew out of schools' efforts to comply with PL 94–142, places students diagnosed with disabilities in regular classes whenever possible. However, it retains the classification of special education students and leaves the primary responsibility for planning for and tracking their progress to special education teachers.

- **Inclusion.** The premise underlying inclusion differs slightly from that of mainstreaming. All students attend regular classes unless the school can show a compelling reason they should not. Friend and Cook (1993) claimed that this approach is increasingly common.

Educators seem to perceive technology as a double-edged sword. It can make the mainstreaming and inclusion processes more feasible, but it can also create extra demands on school systems to make sure that *all* technology opportunities are accessible to *all* students.

Transition needs. In spite of the tremendous legal push in recent years toward assisting students with special needs and the obvious dedication of educators in the field, ESE programs for mildly handicapped or learning disabled students have not shown much success in graduating employable people. IDEA (PL 99–457) inspired one strategy in which students identified with disabilities had to have Transition Plans (TPs) as well as Individual Educational Prescriptions (IEPs) to show how their educational experiences help to prepare them for the world of work. Technology resources can help with the TP/IEP requirement in at least two key ways. First, software products can assist with paperwork and recordkeeping related to TPs and IEPs. Second, adaptive devices such as voice recognition systems have helped people with disabilities to demonstrate marketable skills.

Access issues based on ADA. Previous laws implied the need to provide equal access to resources, but PL 101–336, commonly known as ADA, gave legal teeth to this requirement. As one implication of ADA, schools must be careful not to create any barriers for any children (ESE students or others) as they configure technology systems. For example, if a school purchases an Integrated Learning System, it may have to provide adaptive devices such as special switches and voice recognition software to assure that all physically handicapped students at the grade level can use the instruction.

Costs versus potential benefits. Technology can certainly make life-changing differences for people with special needs, especially for those with physical disabilities. Some technology systems can allow people to communicate and move around on their own for the first time, giving them a level of freedom and self-determination they would not otherwise have. Lewis (1993) and Male (1994) described in detail the benefits of technology for educating people with special needs. However, since these technologies serve only limited markets, they are sometimes among the most expensive educational technology resources. The federal

laws described earlier have allocated substantial amounts of funds for purchasing and using these technologies. As Lewis (1993) noted, however, the speed of technology changes makes it difficult and expensive for schools to keep up with new developments. Hardware/software compatibility problems seem even greater for adaptive devices and special needs software than for other products. Lewis cautioned that, although technology can make many kinds of contributions to education for special needs students, "costs must be weighed against potential benefits" (p. 10). Schools that already face budget problems must often decide whether or not to purchase expensive new systems that will benefit only a few students.

Frustrations with technology. Educators who attempt to apply technology solutions to instructional problems of physically disabled students report three kinds of problems (Clinton, 1995). First, parents and/or those who provide support often have unrealistic expectations that technology will cure handicapping conditions. "Assistive technology has been sold as 'the great equalizer' and, in some cases, has lived up to the billing. But individuals with severe physical and/or cognitive disabilities cannot be normalized by using technology tools…. [However, students] have opportunities to maximize their abilities with less stress and frustration than … with conventional tools" (p. i). ESE teachers and students alike feel frustrated when hardware or software requires extensive initial training. Clinton cautioned that assistive technology should offer plug-and-play solutions when possible so students can direct their energies to learning. Finally, despite hopes of teachers and students and some amazing recent advances in assistive technologies, Clinton noted "some handicapping conditions for which present technology offers no satisfactory solutions" (p. i). There is hope that future advancements in technology will lessen or eliminate these frustrations.

The Role of Technology: Two Trends

The divergent characteristics of directed-instruction and constructivist teaching strategies are reflected vividly in curriculum for exceptional students. The literature for ESE reports numerous very traditional applications of technology to teach basic life skills and basic academic skills (e.g., drill and practice, tutorial). But other goals and activities are clearly constructivist: developing problem-solving skills, teaching students to work together in cooperative groups, helping them visualize concepts and gain experience with unfamiliar topics, and making them feel empowered as learners. Yet these differences seem to spark less controversy in ESE than in other areas of education. Both strategies and aims are seen as necessary and appropriate depending on students' needs and abilities (Whitworth, 1993; and Cates, 1993).

Although constructivist strategies have seemed well-suited to the special needs of gifted students (Morgan, 1993), successful constructivist activities are also being reported with other ESE groups. For example, Hasselbring (1994) described video-based instructional strategies to benefit hearing-impaired students, although the original work was actually done with learning disabled (LD) students. These methods are designed to help students build mental models in order to grasp abstract concepts that may be unusually difficult to understand for those who lack verbal skills. These methods also help to motivate students by anchoring instruction in familiar and interesting situations. Miller (1993) and Woodard (1992) see virtual reality as a potentially powerful way of giving physically impaired students more meaningful learning experiences by providing an environment in which they can explore and discover concepts free of constraints.

This move toward less formal, less structured strategies for special needs students may result from the research finding that skill-by-skill instruction frequently has not resulted in levels of competency needed to make students employable after graduation. Since instruction in isolated skills has also been criticized as unmotivating and unproductive for other students, the trend may reinforce Hanley's (1993) observation that technology applications can help those categorized as "special needs students" in the same ways that they can help all students.

Recommended Resources and Applications: Some General Trends

ESE encompasses several very different student populations that range from the learning disabled and physically disabled (e.g., hearing impaired or visually impaired students) to gifted students. The widely varying needs of these groups demand dramatically different technology resources and applications. However, some general benefits of technology seem common to most or all types of ESE students:

• **Improved motivation and self-concept.** Technology tools have demonstrated a well-recognized ability to motivate ESE students, get them to spend more time on instructional tasks, and generally improve their self-confidence. Several characteristics promote this ability, other than the simple temporary novelty of working with interesting gadgets and colorful software. First, the patience and privacy of a computer-based learning environment, which appeal to many non-ESE students, provide especially important protection for those who have failed often in traditional settings. For a learning disabled or physically disabled student, even the completion of a simple drill and practice assignment becomes an accomplishment and a source of pride. Second, students who have experienced little control over their bodies and their minds find powerful motivation in the ability to control their learning environments and undertake activities completely on their own. Finally, many ESE students achieve new respect from their peers and their families by demonstrating proficiency with computer hardware and software. Lewis (1993) claimed that computer use can improve students' own feeling of self-worth and make others perceive their capabilities (see Figure 15.1).

Figure 15.1   Child Using *My School* Software

Source: Courtesy of Laureate Learning Systems

• **Increased opportunities to use communication and interaction skills.** For students who have had problems expressing themselves clearly or who have been unable to communicate at all, technology can fulfill its promise as "the great equalizer." Holzberg (1994, 1995) and Lewis (1993) reported some dramatic successes in which ESE students used word processing software MIDI synthesizers to express themselves in words or music. This increased ability not only improves students' self-concepts, but it also allows them to interact more equally with other students. Holzberg noted that "Traditionally, kids with behavioral and emotional problems rarely have a chance to experience such cooperative activities as working together to write a song or a research report" (1994, p. 20). She quoted one school principal as saying that, "Technology helps students with disabilities uncover and release an intellectual potential that has been buried under layers of frustration and emotional conflict" (p. 20).

Recommended Resources and Applications for Learning Disabled and Mentally Handicapped Students

Many students with learning disabilities, behavioral problems, and mental handicaps find that reading skill deficits represent their main obstacles to school achievement. Thus, most technology tools and applications for these students have focused on remedies for reading problems. In addition to this area of emphasis, a variety of other technology applications serve this population.

Traditional applications (tutorials, drill and practice, games). As Higgins and Boone (1993) observed, over the past 10 years technology applications for children diagnosed with reading disabilities have centered on tutorial and drill and practice software to build fluency in basic skills. The greatest concentration works on decoding and vocabulary skills. But Spence and Hively (1993) criticized that students need fluency not only in reading skills, but also in other basic skill areas such as writing and arithmetic. "You can't read

Figure 15.2 Talking Word Processors Screen

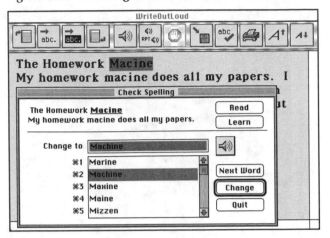

Source: Courtesy of Don Johnson, Inc.

Figure 15.3 Word Prediction Software Screen

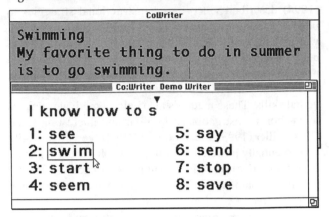

Source: Courtesy of Don Johnson, Inc.

meaningfully when you are spending most of your energy trying to decode words.... You can't write effectively when you are worrying about how to form letters or how to spell.... In [math] exploration and problem solving ... solutions come more readily to the fluent mind" (p. 15). These authors found that computer-based drill and practice has proven the ideal means of providing needed practice for LD students. Many examples also illustrate uses of game-type software to motivate students with learning handicaps to stay on task longer and focus more on skills they need to acquire.

Word processing. After drill and practice, the second most common application for children with LD and behavioral problems seems to be word processing. Holzberg (1994) reports that word processing has helped students with a variety of disabilities and emotional problems to make great strides in improving their written language skills. "Children who may be incapable of writing an essay, paragraph, sentence, or even a word on paper ... find they can write with a word processor" (p. 19). Talking word processors such as Write Out Loud or Intellitalk can be very useful with these students (see Figure 15.2).

Other technology applications. Although drill programs and word processing applications have been the most common technology uses for this population, some indications suggest growth in more open-ended, constructivist technology applications for these children as much as for the population at large (Higgins and Boone, 1993; Whitworth, 1993; Holzberg, 1994).

• Teachers involve mildly handicapped students in both regular and special classrooms in cooperative group activities (e.g., writing, multimedia development).

• Graphics and drawing software are gaining wide popularity.

• Holzberg described a videodisc-based antivictimization training program directed at mentally impaired teenagers, as well as several multimedia applications to teach communications skills.

• Higgins and Boone (1993) reviewed several successful projects that have used videodisc and hypermedia technology in reading programs for learning disabled students.

• Cognitive organizers like Idea Fisher and Inspiration (described in Chapter 6) are often helpful with LD students.

• Steele and Raab (1995) describe an expert system to identify appropriate teaching strategies for students with learning disabilities.

• Finally, LD students frequently find help in word prediction software. Although originally designed to allow physically disabled students to construct sentences on a computer screen as Figure 15.3 illustrates, these prompting systems have helped LD students to build their language skills.

Recommended Resources and Applications for Physically Disabled Students

Educators who have worked with physically disabled students often praise technology eloquently and passionately as a means of freeing these students to develop their potential. Clinton (1993) explained, "There is really no question that computers and related technologies are providing ... unique possibilities for the physically disabled. There are thousands of [such students] who have spent a part of their lives trapped inside dysfunctional bodies. The quality of their lives will remain dramatically diminished until teachers provide the technical tools that give them avenues for communication and environmental control" (p. 70). In the highest-profile uses of technology for physically disabled students, adaptive devices let them communicate and/or use the computer for word processing or speech (Clinton, 1995). However, other successful uses abound.

Input devices to compensate for various disabilities. Clinton (1992, 1993) observed that many students with physical disabilities such as cerebral palsy cannot use traditional input devices like the mouse and keyboard. Several alternative devices allow these students to use the computer: a wide range of switches (see Figure 15.4); touch screens; touch tablets; customized, alternative expanded

Figure 15.4 Switch Used in Lieu of Keyboard

Source: Courtesy of Don Johnson, Inc.

keyboards (Figure 15.5); optical pointers; voice-controlled devices; and word prediction software systems.

Output devices for visually impaired students. Fitterman (1993) reviewed output devices that visually impaired students can use to overcome their disability. Software or special hardware such as closed circuit television (CCTV) can enlarge computer images and text; speech output devices can "tell" what a program does; printers can produce large print or Braille; finally, tactile output devices that scan a page and translate the text into a vibrating, tactile display that a trained person can "read."

Augmentation technologies for nonspeaking students. Clinton (1992) described resources for nonspeaking persons to use to communicate with the computer or with others via the computer. These systems, known as *augmentation aids*, combine input and output devices to serve as voices for nonspeaking students. Giordano and Stuart (1994) described uses of these augmentation systems to enhance reading skills of nonspeaking students.

Figure 15.5 Alternative Keyboard

Source: Courtesy of Don Johnson, Inc.

Figure 15.6 Special Education Web Page

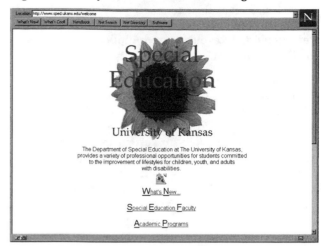

Captioned video and other technologies for deaf learners. Captioned video simply provides subtitles for television and other video presentations so that hearing impaired people can read what others hear. Hairston (1994) reported on a variety of projects that have used captioned video successfully in instructional programs for the deaf and hearing impaired. Hairston also noted that speech recognition systems have brought great benefits to those with hearing deficits as well as to visually impaired people.

Telecommunications applications for physically disabled students. Telecommunications technologies offer several unique advantages for physically disabled students. Telecommunication devices for the deaf (TDD) allow many hearing impaired people to communicate over the telephone (Hairston, 1994). While this is not a significant benefit to instruction, it does allow students and teachers to communicate over distances when necessary. Coombs (1993) described present and potential uses of distance learning and telecommunications programs for "physically challenged" students. Telecommunications can equalize opportunities for students with many kinds of disabilities (especially hearing impairments), since students at the other end may not even know that a caller is disabled. Rozik-Rosen (1993) reports on technology that brings increased access to educational opportunities for homebound students. See Figure 15.6 for a sample special education web page.

Virtual reality. Finally, the future holds the promise of even greater freedom for physically disabled students through new technologies. Woodward (1992) and Miller (1993) describe almost limitless capabilities that virtual reality could provide for those who cannot walk or control their movements. Learning in "cyberspace" could allow complete freedom of life-like movement in simulated environments. Students with no mobility at all can simulate walking around a VR room or driving a car or piloting an airplane in a VR simulator.

Recommended Resources and Applications for Gifted Students

At the other end of the ESE spectrum, special problems and challenges also confront students who are exceptionally talented and/or intelligent. Teachers often report that the most gifted students can be the most difficult to motivate and the least likely to sustain interest in a topic. Morgan (1993) describes several technology applications that can help.

Productivity tools. Many students quickly grow impatient with mundane tasks such as writing and looking up information as part of all research, problem-solving, and production work. Resources such as word processing systems, statistical and graphing packages, and searchable online and CD-ROM databases can make these tasks more interesting, as well as easier to accomplish.

Telecommunications. Gifted students find motivation in open-ended resources that give them freedom to explore as well. Morgan (1993) wrote that, "Telecommunications activities offer the opportunity to expand the learning environment beyond the confines of the classroom" and put students in touch with an unlimited number of people and information resources. The Internet is a "virtual playground" for those who like to discover new things and communicate with new people. He noted that e-mail helps teachers to keep in touch with students and provide the individual attention and feedback they often need to keep on task.

Multimedia production and presentation activities. The more constructivist aspects of multimedia production and presentation projects seem especially appealing to gifted students. In addition to providing motivational goals that focus their attention and abilities, these activities allow gifted students to work cooperatively with others—a skill area in which they are frequently lacking.

Robotics and other emerging applications. The literature has reported a variety of uses of new and emerging technologies with gifted students. Smith (1994) described a project with robotics that focused on teaching creative thinking and problem solving. Mann (1994) reviewed instructional activities involving image processing and expert systems. The novelty of these new resources appeals to gifted students, as does the opportunity they provide for exploration and discovery.

Example Activities

Activity 1:	Let's Write It Right!
Developed by:	Lucie Zaugg, Newberry Middle School (Gainesville, FL) (Reported in *Florida Technology in Education Quarterly*, Winter 1991, p. 75)
Level:	Specific learning disabilities (SLD) and emotionally handicapped (EH) middle school students.
Purpose of Activity:	Many SLD and EH students labor to produce illegible handwriting. Word processing allows these students to focus their energies on expression rather than production. Better legibility also results in better capitalization and punctuation. The software is also very motivational. Students love selecting fonts and pictures for their stories, as well as the act of storing and retrieving them.
Instructional Activity:	Begin by showing students examples of good work that has already been produced on word processing software such as the Children's Writing and Publishing Center (CWPC). Introduce the goal (e.g., writing three stories during the semester) and offer a reward (e.g., students who contribute three stories to the class book will get a copy of it). Begin by developing sentence writing skills using resources such as Kansas Learning Systems' Sentence Writing materials. (This sentence writing practice continues throughout the semester.) Also, review keyboarding skills as necessary. Demonstrate the word processor's capabilities for fonts, graphics, and print options. Have students practice using these options as a class; have them contribute to a description of the school or class as you (or a capable student) select fonts and enter text at the front of the room. Students work on their stories and sentence skills throughout the semester. Make sure each student gets a chance to do some actual keyboard work with the word processor. At the end of the semester, proofread all stories, edit as necessary, and combine them into one book. Make copies for all students.
Activity 2:	Welcome to My House
Developed by:	Based on a lesson designed by Razia Pullen and Beth Sanders (Broward County, FL) (Reported in a 1994 workshop handout)
Level:	Autistic and LD elementary-age children

Continued

Purpose of Activity:	Students with autism and severe learning disabilities have unique styles of learning. They acquire concepts best through repetitive, concrete activities. Highly visual, interactive software provides a natural medium for this kind of learning. The activity develops increased attention to tasks and frequency and quality of communications to teachers and among students.
Instructional Activity:	Set up several learning centers around the classroom, each with a "My House" theme. At one center, My House software (Laureate) gives students practice in identifying objects around the home and naming their functions. Another center has drawing software (e.g., Kid Pix) to let students draw their own pictures of objects around the house. Use word processing software (e.g., Kid Works 2 by Davidson) to write brief stories about students' homes. Let students work in pairs or small groups to combine pictures and stories. The class can continue this theme and expand upon it throughout the semester with related topics (e.g., moving day, cleaning time).
Activity 3:	Story Time
Level:	Physically disabled students in grades 6 through 8 with varying ability levels
Developed by:	Mary Dunbar (Topeka, KS) (Reported in *The Computer Learning Foundation's Special Education Lesson Plans*, 1989, p. 10)
Purpose of Activity:	Students learn about sentence structure and cause–effect relationships by creating a fun, animated story. For a group of students with varying levels of disabilities, the computer acts as the equalizer, allowing each to interact and contribute.
Instructional Activity:	The teacher must set up a color monitor, speech synthesizer, and programmable, touch-sensitive keyboard. The story maker software must be compatible with the speech synthesizer. The teacher also creates the following keyboard overlays:

- **Overlay for Story Level 1.** Divide the keyboard into two sides, one picture on each side. When the child selects a picture, the story appears on the screen already typed.
- **Overlay for Story Level 2.** Present four pictures, each representing a sentence. The child selects one sentence at a time to create a story. The child can also sequence the pictures.
- **Overlay for Story Level 3.** Pictures represent two subjects, two verbs, and two objects. Students select pictures to create sentences. Software automatically writes the words on the screen, along with the appropriate picture and any necessary articles or prepositions.
- **Overlay for Story Level 4.** Pictures represent eight subjects, eight verbs, and eight objects. Students select pictures to create sentences as in Level 3.
- **Overlay for Story Level 5.** Words appear instead of pictures for articles and prepositions; pictures again represent subjects, verbs, and objects. Students select words from the overlay in the order desired for the sentence. Words appear on screen as if the student had typed them letter by letter.
- **Overlay for Story Level 6.** Only words appear. Students select each word for their sentence by pressing the word.

Have students progress through the levels, as their ability dictates to create stories. The words and sentences should be spoken through the speech synthesizer and animated on the screen by the software.

Activity 4:	Survival Vocabulary
Developed by:	Lynette Wright (Topeka, KS) (Reported in *The Computer Learning Foundation's Special Education Lesson Plans*, 1989, p. 33)
Level:	Physically disabled students in Grades 9 through 12
Purpose of Activity:	Disabled students must be able to communicate information about themselves and the communities where they live. The computer gives them practice with vocabulary words in an assisted environment so they automatically know the words, allowing them to communicate more effectively without assistance.
Instructional Activity:	Use a computer, speech synthesizer, and programmable, touch-sensitive keyboard. Create keyboard overlays with the survival vocabulary and one for each student with personal information. Also create flash cards for practice independent of the computer. Pretest students on their knowledge of the vocabulary. Students practice the vocabulary by using the software, which poses questions. When the student presses a word from the keyboard, the speech synthesizer says the word. Take students on field trips

Continued

into the community to allow them to use and generalize the vocabulary. Use the keyboard overlays with students' personal information in the same way. Keep a chart of each student's progress on the individual target vocabulary and keep each student's chart near the desk.

Activity 5:	Geotour USA
Developed by:	Based on a lesson designed by Kathy Bradley, Belleview Middle School (Escambia County, FL) and Michelle Dodds, Holley Navarre School (Santa Rose County, FL) (Reported in *Florida Technology in Education Quarterly*, Winter 1991, p. 97)
Level:	Gifted students in Grades 4 through 6
Purpose of Activity:	This activity was designed to familiarize students with the research and cooperative skills involved in creating a group product centered around geography content. It also helped them memorize geography facts such as the states and capitals without the usual tedium involved in such activities.

Instructional Activity:

Begin by discussing the various states that students have visited on vacations and trips; point out each place on a map and put a colored pin there. Discuss the five regions of the United States. Study one region at a time, beginning each session by providing some historical background. Ask students for any personal knowledge they might have on the region. Have students proceed through the following stations in small groups (changing materials in the centers to reflect the currently studied region):

- **Learning Center 1: Research.** Offer computerized atlases, books, magazine articles, and blank regional maps along with word processing software. Students use the computer and text resources to complete the blank maps and answer questions about the region. They may also pick one thing about the region that interests them and write a brief description about it on the word processor.

- **Learning Center 2: Practice.** Students practice naming states and capitals in the region using appropriate drill software.

- **Learning Center 3: Development.** Using manipulative materials or authoring software (e.g., HyperCard), students make a trivia game of their own to allow players to review and practice important information about the region.

- **Learning Center 4: Development.** Use previously developed student games to review information about the region.

Finally, have each group fill in a blank map of each region, showing states, capitals, and other information in their own creative way.

Activity 6:	Student Weather Broadcast
Developed by:	Based on a lesson plan designed by Joseph Gatti, Powell Middle School (Brooksville, FL) (Reported in *Florida Technology in Education Quarterly*, Spring 1993, p. 63)
Level:	Gifted students in junior high or high school
Purpose of Activity:	Telecommunications technology combines with video production resources to deliver an exciting way of learning about how weather information is gathered and disseminated. Students learn about the weather by simulating the job activities of a professional television weather forecaster.

Instructional Activity:

Set up a portion of the classroom to look like a modern TV newsroom. Each class should have its own tape. The teacher begins by demonstrating a sample weather broadcast and how it was developed. Be sure to teach students the prerequisite skills: how to use a computer and software to download images and information from an online weather information service and how to use a camcorder. If necessary, guide students through the steps to assemble a sample broadcast. Then divide students into groups and allow them to choose their roles: e.g., writer, anchorperson, camera operator, cue cards, sound.

Groups download current weather information and satellite images. Then they design and film their broadcast based on this information and the local weather. If possible, they "broadcast" it to the school or other classes.

Exercises

Exercise 15.1. Explain how each of the following current ESE issues and trends affects the selection and use of technology.

• Recent federal legislation related to ESE and individuals with disabilities

• Trends toward mainstreaming and inclusion for all students, regardless of disabilities

• Traditional emphasis on directed-instruction models and new emphasis on constructivist instructional models for various types of ESE students

Exercise 15.2. Describe the kinds of uses being made in exceptional student education curricula of the following technology resources:

For learning disabled students and students with other mental/emotional handicaps:

a. Traditional applications (e.g., drill, tutorial, game)

b. Word processing

c. Videodisc and multimedia applications

For physically challenged students (hearing impaired, visually impaired, wheelchair-bound):

a. Adaptive input devices

b. Adaptive output devices

c. Augmentation technologies

d. Captioned video

e. Telecommunications applications

f. Virtual reality

For gifted students:

a. Productivity tools

b. Telecommunications

c. Multimedia and presentation activities

d. Robotics and expert systems

Exercise 15.3. Identify a specific ESE population and level. Create instructional activities that integrate technology appropriately for that population and level. Each of these activities should meet the following criteria:

• Integrate one or more types of technology described in Exercise 15.2.

• Show how to adapt this activity for large-group and small-group instruction.

• Describe the required preparation for this activity.

• Describe the benefits you would hope to derive from using technology resources in the lesson.

Exercise 15.4. Identify a physically disabled student in your local community and learn what role, if any, technology has played in the student's learning activities. Could you recommend some other technology options for the student?

Exercise 15.5. Locate a regular, mainstream class that includes students with various disabilities in its regular activities. Find out what role, if any, technology plays in meeting the special needs of these students. Can you recommend some other technology options that could help meet these students' needs?

Exercise 15.6. Visit a resource class or organization (e.g., a magnet school) designed to serve the needs of gifted and talented students. Review how these students are using technology. Can you recommend some additional technology resources or activities to help meet these students' needs?

References

Alliance for Technology Access. (1994). *Computer resources for people with disabilities.* Alameda, CA: Hunter House.

Bradley, K., and Dodds, M. (1991). Geotour USA. *The Florida Technology in Education Quarterly, 3* (2), 97–98.

Cates, W. (1993). Instructional technology: The design debate. *The Clearinghouse, 66* (3), 133–134.

Clinton, J. (1995, March 2). Taming the technology. Materials distributed at the Florida Assistive Technology Impact Conference, Orlando, FL.

Clinton, J. (1993). Why use technology to teach disabled students? Why ask why? *The Florida Technology in Education Quarterly, 5* (4), 64–79.

Clinton, J. (1992). Technology for the disabled. In G. Bitter (Ed.) *Macmillan encyclopedia of computers.* New York: Macmillan.

Coombs, N. (1993). Global empowerment of impaired learners: Data networks will transcend both physical distance and physical disabilities. *Educational Media International, 30* (1), 23–25.

Dunbar, M. (1989). Story Machine. In *The Computer Learning Foundation's Special Education Lesson Plans.* Palo Alto, CA: The Computer Learning Foundation.

Ellsworth, N., and Hedley, C. (1993). What's new in technology? Integrating technology: Current directions. *Reading and Writing Quarterly, 9* (4), 377–380.

Fitterman, J. (1993, May). Present vision—Future vision: Technology for visually impaired students. (ERIC Document Reproduction No. ED 363 321).

Foster, K., Erickson, G., Foster, D., Brinkman, D., and Torgeson, J. (1994). Computer-administered instruction in phonological awareness: Evaluation of the DaisyQuest program. *Journal of Research and Development in Education, 27* (2), 126–137.

Friend, M., and Cook, L. (1993). Inclusion. *Instructor, 103* (4), 52–56.

Gatti, J. (1993). Student weather broadcast. *The Florida Technology in Education Quarterly, 5* (3), 63–64.

Giordano, G., and Stuart, S. (1994). Pictorial literacy activities for children with disabilities. *Day Care and Early Education, 21* (3), 44–46.

Hairston, E. (1994). Educational media technology for hearing-impaired persons. *American Annals of the Deaf, Special Issue, 139*, 1–11.

Hanley, T. (1993). The future has been a disappointment: A response to Woodward and Noell's article on software development in special education. *Journal of Special Education Technology, 12* (2), 164–172.

Hasselbring, T. (1994). Using media for developing mental models and anchoring instruction. *American Annals of the Deaf, Special Issue, 139*, 36–44.

Higgins, K., and Boone, R. (1993). Technology as a tutor, tools, and agent for reading. *Journal of Special Education Technology, 12* (1), 28–37.

Holzberg, C. (1994). Technology in special education. *Technology and Learning, 14* (7), 18–21.

Holzberg, C. (1995). Technology in special education. *Technology and Learning, 15* (5), 18–23.

Lewis, R. (1993). *Special education technologies: Classroom applications*. Pacific Grove, CA: Brooks/Cole.

Male, M. (1994). *Technology for inclusion: Meeting the special needs of all students*. Boston: Allyn and Bacon.

Mann, C. (1994). New technologies and gifted education. *Roeper Review, 16* (3), 172–176.

Miller, E. (1993). Special experiences for exceptional students: Integrating virtual reality into special education classrooms. ERIC Document Reproduction No. ED 363 321.

Morgan, T. (1993). Technology: An essential tool for gifted and talented education. *Journal for the Education of the Gifted, 16* (4), 358–371.

Pullen, R., and Sanders, B. (1994). Thematic units: My house. Handout from a Broward County, FL workshop.

Rozik-Rosen, A. (1993). Special needs, special answers: The story of TLALIM at the service of sick children. *Educational Media International, 31* (1), 36–41.

Smith, R. (1994). Robotic challenges: Robots bring new life to gifted classes. *Gifted Child Today Magazine, 17* (2), 36–38.

Spence, I., and Hively, W. (1993). What makes Chris practice? *Educational Technology, 33* (10), 15–20.

Steele, J., and Raab, M. (1995). FERRET: An expert system for identifying teaching strategies for students with learning disabilities. *Tech Trends, 40* (3), 13–16.

Whitworth, J. (1993). What's new in special education: An overview. *The Clearinghouse, 66* (3), 132–133.

Woodward, J. (1992). Virtual reality and its potential use in special education: Identifying emerging issues and trends in technology in special education. ERIC Document Reproduction No. ED 350 766.

Zaugg, L. (1991). Let's write it right! *The Florida Technology in Education Quarterly, 3* (2), 75–77.

Resources

Networks and Databases for People with Special Needs

ABLEDATA. A national database of assistive technology information. You can call, write or fax to request a database search; access the database via modem by calling the bulletin board service; or get a copy on disk or CD-ROM. Service hours are 8:00 a.m. to 6:00 P.M. EST, Monday through Friday. Fact sheets with information on a device or a type of device are also available. ABLEDATA is available from The National Rehabilitation Information Center.

CO-NET (The Cooperative Database Distribution Network for Assistive Technology). A CD-ROM with the cooperative Electronic Library. This disk includes:

* ABLEDATA database

* REHABDATA database of publications and reports

* Cooperative Service Directories database of services, service providers, and organizations

* Publications, Media and Materials database of publications and articles related to disability

SeniorNet on America Online. Network for senior citizens

WIDNet on Delphi. A network through the World Institute on Disability

Electronic Bulletin Boards with Information for People with Disabilities

ADAnet—Nework of the Disability Law Foundation (205) 854-0698

Black Bag BBS—General medical information (302) 731-1998

Body Dharma—Health-related conferences (510) 836-4717

BrailleBank BBS—For visually impaired (215) 244-9937

Deaf Comm—Association of Late Deafened Adults (312) 262-6173

DDConnection— Developmental Disabilities Connection (817) 277-6989

DEN (Disability Electronic Network)—Focuses on disabilities, autism (201) 342-3273

Denver Deaf Net BBS—(303) 989-9245

4Sights Network—Greater Detroit Society for the Blind (312) 272-7111

Department of Justice—Disability Civil Rights Division (202) 514-6193

HandicapNews—Also has lists of other boards (203) 337-1607

Handiline—Network with focus on general disabilities (703) 818-2660

Hearing Aid BBS—(305) 653-2589

HEX BBS—Disability issues, lists of BBS (301) 593-7357

Information 90 BBS—Adaptive devices for computers (215) 411-2237

National Federation of the Blind—(301) 752-5011

Parents of the Visually Impaired—(209) 825-8537

Project Enable—Information on rehabilitation (304) 766-7842

Seattle Hearing—Of interest to those with hearing impairments (206) 526-2744

SpecialNeeds—Focuses on social issues and barriers (219) 659-0112

SYNAPSE—General disability information (202) 543-9176

The Alliance for Technology Access (1994) recommends locating other free bulletin board systems with areas of interest to people with disabilities by going online with some of the numbers above and asking the system operator or looking in free, computer-oriented newspapers.

**Hardware and Software Resources for
People with Disabilities**

Alternate keyboards—Don Johnson, Inc.; Exceptional Computing, Inc.; Intellitools; Sunburst; TASH, Inc., Zygo Industries

Arm and wrist supports—ErgoFlex Systems; KLAI Enterprises

Braille displays—American Thermoform Corporation; HumanWare, Inc.; TeleSensory

Braille translators and embossers—American Thermoform Corporation; Blazie Engineering, Inc.; Enabling Technologies Co.; TeleSensory

Closed circuit TVs–HumanWare, Inc.; Seeing Technologies, Inc.; TeleSensory

Electronic pointing devices—Ability Research; Inocomp; Madenta Communications; Prentike-Romich; Words+, Inc.

Interface devices—AbleNet, Inc.; BEST; Brown & Company; Consultants for Communication Technology; Don Johnson, Inc.; TASH, Inc.; Words+, Inc.

Joysticks—KY Enterprises; McIntyre Computer Systems; Penny and Giles; Prentke Romich Co.; TASH, Inc.

Keyboard accessories (Keyguards, moisture guards, alternate labels)—Don Johnson, Inc.; Intellitools; Prentke Romich Co.; TASH, Inc.; Toys for Special Children

Menu customizing programs—Apple Computer Co; Edmark Company; Microsoft Corporation; Symantec

Pointing and typing aids (sticks or wands for those who cannot hit keys with fingers)—Crestwood Co.; Extensions for Independence; Maddak; North Coast Medical

Reading comprehension programs—Advanced Ideas, Inc.; CompuTeach; Continental Press; Creative Learning, Inc.; Davidson and Associates; Don Johnson, Inc.; Great Wave Software; Hartley Courseware, Inc.; IBM Special Needs Systems; Laureate Learning Systems; The Learning Company; MECC; Milliken Publishing Co.; Optimum Resources, Inc.; Scholastic; SVE; Sunburst; Tom Snyder, Inc.

Screen enlargement software—Apple Computer, Inc.; Arctic Technologies; Berkeley Systems; HumanWare, Inc.; Microsystems Software, Inc.; TeleSensory

Screen reader software for the blind—American Printing House for the Blind, Inc.; Berkeley Systems, Inc.; Biolink Research and Development; GW Micro; IBM Special Needs Systems; OMS Development; TeleSensory

Speech synthesizers—American Printing House for the Blind, Inc.; AICOM, Inc.; Digital Equipment Corporation; Echo Speech Corporation; HumanWare, Inc.; TeleSensory

Switches (with software)—Creative Switch Industries; Don Johnson, Inc.; Luminald, Inc.; Prentke Romich Co.; TASH, Inc.; Toys for Special Children; Zygo Industries

Text telephones (TT) for the hearing impaired—AT&T Accessible Communications Products; IBM Special Needs Systems; KRI Communications; TeleSensory

Touch screens—Edmark, Inc.; Information Strategies, Inc.; Microtouch Systems

Trackballs—APT, Inc; CoStar Corporation; Kensington Microwave Ltd.; Logitech; Mouse Systems, Inc.; Penny and Giles Computer Products Ltd.

Voice recognition systems—Apple Computer Co.; Articulate Systems; Dragon Systems; IBM Special Needs Systems; Kurzweil Applied Intelligence; Microsoft Corporation

Word prediction programs—Don Johnson, Inc.; Innovative Designs; Microsystems Software; Pointer Systems; Prentke Romich Co.; Words+, Inc.

Word processors with speech and large print—Davidson and Associates; Don Johnson, Inc.; Hartley Courseware, Inc.; IBM Special Needs Systems; Intellitools; SkiSoft Publishing Co.

Writing aids and writing skill software—Blissymbols Communications International; Creative Learning, Inc.; Davidson and Associates; Don Johnson, Inc.; Hartley Courseware, Inc.; Humanities Software; IBM Special Needs Systems; Scholastic Inc.; Teacher Support Software; Tom Snyder, Inc.; William K. Bradford Publishers

Appendix A

Getting Started with Microcomputer Systems

This appendix will cover the following topics:

- How to operate a computer system with a mouse and keyboard

- Definitions and descriptions of parts of a desktop for the Macintosh and Windows operating systems

- How to perform basic operations with a desktop and disks and with folders, programs, and files in either a Macintosh or Windows system

- A description of common problems teachers are likely to encounter with Macintosh and DOS systems and how to remedy them, including troubleshooting checklists

Appendix Objectives

1. Complete fundamental operations with a microcomputer, including the following:

   - Use proper procedures to start (boot) up and shut down the system.
   - Use proper procedures to operate a keyboard, mouse, disk drive, and hard disk.
   - Complete operations with windows (e.g., opening, closing, sizing, viewing contents in various ways).
   - Insert and eject disks.
   - Prepare disks to receive programs (format, label, name, and rename disks).
   - Manage and organize a hard disk.
   - Copy files from one disk to another.
   - Copy an entire disk.
   - Transfer files from one folder or subdirectory on a disk to another.

2. Tell whether an icon represents a program, file, or folder/subdirectory.

3. Identify the correct procedures to accomplish desired tasks with equipment, software, and disks.

4. Complete basic troubleshooting procedures to identify the nature of a problem and decide whether to try to resolve the problem or call a technician.

Introduction

Where to Begin, How to Keep Up

Many people feel overwhelmed by the "information explosion," with all its associated tools and techniques. This can be true for beginners as well as for those who use technology in their daily lives. Some simply fear technology, or they fear failing, appearing stupid, or seeming slow to catch on in front of others. Others are intimidated by the reputation of the computer as a powerful "brain" or of the reputed complexity of a particular resource; they are afraid that something may prove too complicated for them to master. Still others are afraid that, in their ignorance, they may damage the equipment.

A great many people just feel inundated by the daily flood of new terms to learn, new concepts to master, new resources (or new versions of old resources) to track. They see so much to learn and so little time to learn it all. This feeling causes some people to procrastinate on beginning to learn a new method or software; it causes some not to begin at all!

This appendix is provided to help those just getting started, as well as those who have been using technology for some time but feel that their conceptual grounding is a little shaky. It provides a recommended sequence of concepts and skills for all microcomputer users to address. A final section gives some common problems to avoid with microcomputer systems and some troubleshooting procedures and checklists. This appendix is a primer for the novice user, but it also a good summary for the skilled "cookbook user," who has been doing a lot of things without knowing exactly why. It may not completely eliminate that overwhelmed feeling, but it can provide some building blocks that can help. The information in this chapter should be a helpful reference both for beginners and those who teach them.

Equipment Platforms

Educators use several "platforms" or types of microcomputer equipment today. Platforms are defined by the type of operating system that runs them. Recent surveys show that most schools have a combination of equipment with the Microsoft Disk Operating System (MS-DOS), Macintosh, and Apple platforms. (The MS-DOS platform is associated with IBM and IBM-compatible brands.) It seems best for teachers to learn operation skills first on newer equipment, rather than on platforms that schools are phasing out such as the Apple. New equipment purchases will primarily consist of various models of Macintoshes and MS-DOS equipment that use Microsoft's Windows or the Windows 95 operating system. Thus, the information in this chapter focuses primarily on these systems. For those who are learning on other types of machines, some of the concepts and information should still prove helpful, but they must locate learning sequences and exercises in manuals written specifically for these machines.

Macintoshes all have an operating system that displays a "desktop," a graphic depiction of the options available to a user and the information stored on disks in the system. Since the Apple Computer Company was the first to exploit the concept of a desktop, early MS-DOS systems did not feature this friendly software that made it easier for nontechnical people to use computers. The software that displays the desktop and lets people interact with it is called a *graphic user interface* (GUI-pronounced "gooey"). Then the Microsoft Company developed its own GUI, Windows, which gave MS-DOS machines their own kind of desktop. Although the first versions of Windows were really applications programs, more recent versions (Windows 95) are actually operating systems in themselves. The Macintosh desktop and the desktop generated by Windows are very similar in appearance, but they have many operational differences. This chapter will introduce some general procedures and troubleshooting steps for each of these popular systems.

Suggested Learning Sequence for Operating Macintosh and Systems with Windows or Windows 95

Every microcomputer user should know how to complete the following set of activities, regardless of the specific kind of microcomputer. The order of the list suggests an optimal sequence for learning these activities. System manuals that accompany newly purchased computers give specific information about how to complete these tasks.

General Activities

___ 1. Learn about your system:
- How many disk drives?
- What size hard disk?
- How much RAM?
- What version of the operating system?
- What size/kind of monitor?
- What kind of printer?
- What other peripherals?

___ 2. Locate all the ON/OFF buttons and turn on all the devices of the system.

On a Macintosh System

___ 1. Learn about the keys on your keyboard, especially:
- Spacebar
- ESC, Tab, Caps Lock, Shift, Control, Delete
- Arrow keys: up, down, left, right
- Open-Apple, Option
- Return (or enter)
- Optional: Numbered function keys

___ 2. Identify the parts of the desktop:
- Pull-down menus: Apple, Edit, View, Label, and Special
- Icons: See icons for disk drives and trash

___ 3. Learn how to use a mouse (point, click, double-click, drag) as you learn about windows:

- Open a window by double-clicking.
- Make a window inactive by clicking outside it and active by clicking inside it once.
- Identify the other parts of a window: close box, zoom box, arrow, scroll bars/boxes, title bar (handle), information bar.
- Size a window using the zoom and dragging methods.
- Scroll the display in a window using scroll arrows and scroll boxes.
- Move a window around on the screen by its handle.
- Place two or more open windows beside each other so you can see them both at the same time.
- Practice using command key sequences to complete operations with windows.
- Drag icons around within windows.
- Close a window by clicking the close box.

___ 4. Use pull-down menus. (Learn other pull-down menus later under specific programs.)

- View-Select various ways to view contents of a window. (See the arrow by the selection.)
- Special-Eject a disk, restart, and shut down.
- Edit-Highlight all the icons in a given window at once (Select All).

___ 5. Within windows, identify icons for programs, files, and folders.

___ 6. Carry out operations with folders:

- Create and name a new folder.
- Rename a folder.
- Place a folder inside another folder.

___ 7. Insert a blank disk into the drive.

- Initialize/format it.
- Learn how to rename a disk.
- Prepare a label for the disk and place the label on it.

___ 8. Learn how to move folders from the hard drive to the new disk:

- Highlight an icon for a folder.
- Highlight two or three icons at once. (Use the shift+click method.)
- Drag icons of desired folders from one disk to the other.

___ 9. Learn how to remove files and folders from a disk.

___ 10. Learn how to copy a file or folder from one disk to another.

___ 11. Learn how to copy an entire disk.

___ 12. Learn how to remove a disk from the disk drive.

___ 13. Learn how to use some system options.

- Locate a desired file on the hard drive.
- Adjust the mouse and sound controls.
- Change the background color/pattern on the desktop.

___ 14. Learn how to use some system accessories.

- Calculator

- Notepad
- Alarm clock
- Keycaps
- Clipboard/scrapbook

___ 15. Shut down the computer system and turn off the devices in the system.

In other sessions, learn how to organize a disk/hard drive and do troubleshooting procedures. See pages 327–329.)

On an MS-DOS System with Windows

___ 1. Learn about the keys on your keyboard, especially:

- Spacebar
- ESC, Tab, Caps lock, Shift, Control, Delete
- Numbered function keys
- Arrow keys: up, down, left, right
- Alt, Backspace, Enter

___ 2. Identify the parts of the desktop:

- Pull-down menus: File, Options, Window, Help
- Icons: See the icons for disks, the Program Manager, and the File Manager.

___ 3. Learn how to use a mouse (point, click, double-click, drag) as you learn about windows:

- Open a window.
- Make a window active by clicking inside it and inactive by clicking outside it.
- Identify the other parts of a window: title bar, minimize, maximize, and restore buttons; scroll bars and boxes; and control menu box.
- Minimize and maximize a window. Size it using the dragging method.
- Restore a window.
- Scroll items in a window using scroll arrows and scroll boxes.
- Move a window around on the screen by its title bar.
- Place two or more open windows beside each other so you can see them both at the same time. (Use tile and cascade.)
- Drag icons around within the Program Manager window.
- Practice using command key sequences to carry out operations with windows.

___ 4. Identify icons for programs, files, and groups.

___ 5. Use pull-down menus in the Program Manager: File, Options, Window, Help.

- File-Arrange icons
- Options
- Window-Tile and cascade
- Help-Look at the Help Options

___ 6. Look at program groups:

- Move programs and/or files around in groups.
- Delete programs and/or files from a group.

___ 7. Use disks:
- Initialize/format a disk.
- Learn how to rename a disk.
- Remove a disk from the disk drive.
- Prepare a label for the disk and place the label on it.

___ 8. Use pull-down menus in the Program Manager: File, Options, Window, Help.
- File-Arrange icons
- Disk
- Tree-Expand a branch
- View
- Options
- Window
- Help

___ 9. Identify the icon for the File Manager and complete operations with folders/subdirectories:
- Identify the main or root directory.
- Create a subdirectory.
- Rename a subdirectory.
- Place a subdirectory inside another subdirectory.

___ 10. Learn how to move files or folders from the hard drive to a new disk:
- Highlight an icon for a folder.
- Highlight two or three icons at once. (Use the shift+click method.)
- Highlight all the icons in a given window at once (Select All).
- Drag icons of desired folders from one disk to the other.
- Remove files and folders/subdirectories from a disk.

___ 11. Learn how to copy a file or folder from one disk to another.

___ 12. Learn how to copy an entire disk.

___ 13. Learn how to use some system options.
- Activate the task list to see what programs are currently running.
- Locate a desired file on the hard drive.
- Adjust the mouse, keyboard, and sound controls.
- Change the background color/pattern on the desktop.

___ 14. Learn how to use some of the system accessories in the Main Window:
- Calculator
- Notepad
- Clock
- Calendar
- Paintbrush

___ 15. Shut down the computer system and turn off the devices.

In other sessions, learn how to carry out troubleshooting procedures (see page 329).

Pointers on Using Microcomputer Equipment

A computer system's hardware is a machine, or more precisely, a series of machines connected together by wires and cables. As with any machine, a computer system works best if it is well-maintained and if the person using it knows and follows the correct procedures. For recommendations on maintenance and security, see Chapter 2. Assuming the computer is working properly, the best place to begin is by examining the parts of the system:

- Locate the input devices (usually the keyboard and mouse), the output devices (usually the monitor and printer), and the computer cabinet itself.
- Locate the on/off switches for each of the devices. Sometimes, a switch is located on the keyboard as well as on the computer casing.
- Find out about the capabilities of the system either by asking an instructor or by using the procedures in the computer manual to locate the following information:
- How much RAM does it have?
- How many disk drives?
- Any hard disk drive? If so, what is its capacity?
- Is there a CD-ROM drive?
  How large is the monitor? Color or monochrome?
- What kind of printer does the system have (e.g., dot matrix or laser)? How fast will it print? Is the printer shared among more than one computer?
- Is the computer connected to or part of a network? What kind of network is it?
- Does the system include a modem or other optional equipment such as a microphone and/or speakers?

Using the Keyboard

Keyboards are the most commonly used input devices for computer systems. Macintosh and IBM/IBM-compatible systems feature varied keyboard configurations, but certain keys are becoming increasingly standard. (The checklists noted some important, frequently used keys.)

Using a Mouse

The mouse, like the keyboard, is becoming a standard input device on most microcomputers. It lets users point to things on the screen in a way that communicates information to the computer. Pointing to things on the screen with a finger cannot tell the computer what the user wants to do. The mouse controls an arrow pointer to indicate objects (e.g., icons and menus) on the screen. This procedure tells the computer to select an item or display its contents. Many people find it very difficult at first to learn the motor skills required to use a mouse. For example, many beginners have difficulty learning that they can pick up the mouse off the desk or mousepad and move it without moving the arrow on the screen; the arrow moves only if the ball of the mouse moves, usually while touching the surface of the desk or mousepad. Practice is the only answer to resolving this kind of problem.

The four operations a user does with a mouse are:

- **Pointing.** This is the process of moving the arrow to a desired location on the screen. Usually, this is the first step in accomplishing one of the following operations.

- **Clicking.** Depending on the software, the user may be able to select an option on the screen or complete another desired step simply by pressing the mouse button. This action is called *clicking* the mouse on an object.

- **Double-clicking.** Some software calls for completing a step by clicking the mouse button twice in rapid succession. This action is called *double-clicking*. The choice between single clicking and double-clicking depends on what the software requires.

- **Dragging.** Holding down a mouse button while moving a mouse on the mousepad or desk is called *dragging*. Objects on the screen can be moved around by clicking on them and dragging them. Dragging is also the way to look at options on menus.

Procedures for Starting Up and Shutting Down a Microcomputer

When a user turns on a computer, electrical current begins running through the computer's circuitry and the system is said to *boot up*. As the system boots up, the operating system is read into the computer's RAM. When booting up is completed on Macintosh computers and on MS-DOS computer systems with Windows and with Windows 95, the screen displays the desktop. Users should remember three warnings about starting up and shutting down any computer system:

- **Don't turn off the computer until you finish using it.** Turning the system off and on is hard on its mechanical parts (e.g., switches, power supply, motors). Therefore, do not turn off the computer until you finish using it. Some beginners erroneously turn off the system automatically when they make errors. Try to correct an error without turning off the system, if you can.

- **Don't turn the system off and on quickly.** Again, this action is hard on the system's devices. It won't hurt anything the first time you do it, but repeated on/off sequences will eventually break down the devices.

- **Make sure to turn on all devices in the system.** All the devices may be attached to one power switch, but sometimes a user may have to press separate power switches for devices such as the monitor.

Specific steps to start up and shut down a Macintosh system. If it is working properly, the Macintosh automatically shows the desktop when it boots up. If a Macintosh screen shows a picture of a computer with a frown or an *X* on it instead of the desktop, something is wrong with the system. (See troubleshooting procedures in this chapter.) When shutting down a Macintosh, don't turn off the computer's power switch before selecting the Shut Down option from the Special menu. This step makes sure all

programs and files are properly stored in memory before the system loses power. Depending on the problem, selecting Shut Down may not be possible, but follow this recommended shutdown procedure if at all possible.

Specific steps to start up and shut down an IBM/IBM compatible system. When an MS-DOS system starts up, it usually first goes through a memory check, a procedure in which the operating system software stored in ROM checks each memory chip for proper functions. Then the system checks to make sure required equipment like the keyboard is connected and working correctly. (If something is not working properly, the screen will display an error message that directs the user to instructions in the system manual to determine what is wrong.) The system then looks for the rest of the operating system. It looks first on a disk in Drive A; if it does not find the correct file, the system will look on the hard drive. After the operating system is loaded in an MS-DOS system with Windows, the screen will show one of three things, depending on how the system was originally set up by the person who installed it:

- **The Windows desktop.** If the system has been configured to load the Windows software automatically when the system boots up, the Windows desktop will appear.

- **A menu.** Sometimes the lab manager or someone else in charge of the systems will either design or obtain a program that displays a menu of all the programs available on the system. Some programs work well under Windows and some do not. Therefore, Windows may be only one of the programs available on the menu.

- **C:\>.** This is called a *C prompt*. The user types a command (usually either *WIN* or *WINDOWS*) that calls the Windows software into memory.

- A Windows 95 system automatically shows the desktop.

When shutting down an MS-DOS system, a user first quits the Windows Program Manager and then exits the Windows program itself prior to turning off the machine.

Using a Desktop

The ingenious concept of a desktop originated at Xerox Corporation. However, it came into widespread use for the first time in 1984 as part of a system christened the Macintosh, which represented the Apple Computer Company's plan to build on the success of its popular Apple II platform. The desktop's graphic user interface (GUI) makes it easier for nontechnical people to use microcomputer systems. Macintosh, MS-DOS systems with Windows, and those with Windows 95 have similar-looking desktops, but they also feature many differences. Compare the components of the two desktops shown in Figures A.1 and A2.

Common basic elements include:

- **Icons.** Each desktop shows pictures of a number of small objects. These are called *icons*. An icon is a picture that symbolizes an object or concept. For example, a disk icon sym-

Figure A.1   Macintosh

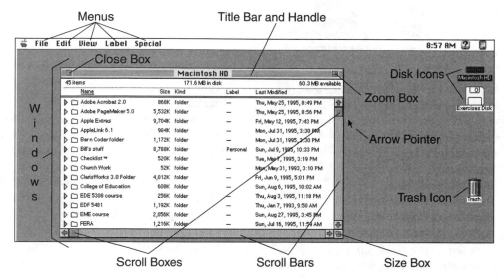

Menus   Title Bar and Handle

Close Box

Disk Icons

Zoom Box

Arrow Pointer

Windows

Trash Icon

Scroll Boxes   Scroll Bars   Size Box

bolizes the disk itself; the trash can icon on the Macintosh desktop symbolizes a place to discard or throw away things (e.g., files, folders). In both systems, users move icons by dragging them around on the desktop. (Example icons appear in following text.)

• **Windows.** These rectangular boxes display information such as the contents of a disk or a folder on the screen. An open window shows a list of the items in the disk or folder along with some information about each one. However, in Windows and Windows 95 systems, a window also displays the programs available on the system.

Using a Macintosh Desktop

A Macintosh displays only one main screen—the desktop—when the user turns on the computer. The Macintosh desktop includes windows, menus, and icons.

Macintosh windows. When the user opens a window, it lists the contents of a disk on the desktop. Windows can have varied sizes set when the user adjusts a zoom box or size box. They can also move around on the screen so the user can see more than one window at one time. If the window is too small to show all of its contents at one time, the user can scroll up and down the list to view hidden information.

Macintosh menus. The computer displays a number of built-in choices in the form of a series of lists or menus. To select a menu choice, point and click on the menu name and drag down the pointer to the desired option. A Macintosh user can also select many menu options by pressing the command key (⌘) and another letter. For example, to select the File-Open option, press ⌘-O. A Macintosh desktop has six menus: Apple (⌘), File, Edit, View, and Special. Dimmed menu items, like some of those shown in Figure A.1, are not available to select.

Macintosh icons: Programs, files, and folders. Icons that represent programs, files, and folders on the Macintosh desktop look very similar. Beginners may tend to think of them as essentially the same thing with different names. However, the icons serve very different functions on the desktop and within the computer system:

• Program icons represent software packages that fulfill various functions required by the system or its users. Chapter 2 discussed applications software and systems software. The system may include a program called *After Dark*®, which is designed to show a picture or graphic display as a screen saver. A word processing package is also an applications program.

Program icons often feature more elaborate designs than file icons. Example program icons are shown below:

Microsoft Works 3.0

SuperPaint 2.0

HyperCard 2.0

- Files store information created by programs or for use with those programs. For example, a word processing document created in WordPerfect becomes a file when stored on disk. Sometimes, a set of files accompanies a program. For example, After Dark® calls on a set of files to display different screen images. Example file icons are shown here:

Vandivort article

Diagram

Art Ideas

- Unlike programs and files, folders do not symbolize software, products of software, or data. They exist only as images on the screen and places in computer memory. Folders extend the concept of a computer screen as the top of a desk. Folder icons represent groups of programs and files that the user wants to locate together, just as file folders on a real desk gather and organize papers. Folder icons represent visual storage devices in the same way that real file folders are physical storage devices. Without folders, the desktop would have to show all programs and files in one continuous list. It would take a long time to search through this list to find and open a required program or file. Like disk icons, folder icons may be opened and their contents displayed in windows. Example folder icons are shown here:

Subjects

Letters

Art work

Using a Windows Desktop

If Windows were to duplicate the Macintosh desktop exactly, it would violate a copyright held by the Apple Computer Company. However, the main difference between the Macintosh operating system and versions of Windows prior to Windows 95 is that the latter are all applications software designed to act like systems software rather than real systems software. In Windows, everything on the desktop is part of a window. The other elements that are part of the Macintosh desktop—menus and icons—also appear in a Windows desktop, but always as parts of the two main windows in the system: the Program Manager and the File Manager.

The Program Manager window. Figure A.2 shows the first screen (or window) to come up from the desktop, the

Figure A.2   Program Manager Window with Program Groups

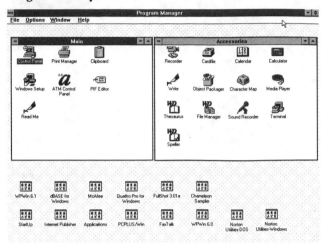

Program Manager. This window shows an icon for each program group stored on the hard disk. Program groups are collections of programs grouped under a logical heading. For example, installation of Quattro Pro places several programs on the hard disk. Each program has its own icon, but all are grouped together under a single program group icon in the Program Manager window. If the user double-clicks on the program group icon, a window appears showing all the programs in that program group. In this way, a program group icon acts as an organizer, like a folder icon in a Macintosh system.

The File Manager window. One of the programs listed in the Program Manager window, the File Manager, provides a way of organizing files on a desktop. When the File Manager is opened, it shows lists of files stored on the drives. But instead of calling these lists *folders*, Windows calls them *directories* and *subdirectories*. While the Macintosh allows the user to create folders (and folders within folders) to organize disks and the hard drive, Windows lets the user do this through the File Manager.

Application windows. Users manipulate windows in a Microsoft Windows display in the same way as in a Macintosh system. For example, double-clicking opens a window and clicking on the close box closes it. The user can make a window larger or smaller and scroll to see hidden contents. The screen can show more than one window at a time.

Menus. Five pull-down menus are available in the Program Manager: Control (represented by a small box), File, Options, Window, and Help. The File Manager offers these same menus and two more: Disk and Tree.

Using Disks

Types of Disk Resources

The three most common types of removable disks (those that the user can insert and take out of disk drives) are discussed at length in Chapter 1.

Inserting and Ejecting Disks on Macintosh and IBM Systems

Floppy 5 1/4-inch disks are always inserted into a computer's disk drive with the shiny side up and the end opposite the label inserted first. Microdisks (3 1/2-inch diskettes) are always inserted into the disk drive metal side first and label side up. Users can remove (eject) these disks from a Macintosh computer in two ways. An Eject Disk command on the Special menu (or the equivalent command key combination) returns the disk, or the user may drag the disk icon to the trash can icon. After the Eject Disk command, the operating system usually keeps an image of the disk in RAM until it is physically ejected. This enables the system to copy one disk to another with only one external disk drive. On IBM systems, the procedure to get a disk out of a disk drive is simpler: Press the manual eject button.

Preparing Disks to Receive Programs

To store programs and files on disk, the user must first prepare the disk by placing a small program on it that links it to the computer's operating system. In effect, this program is part of the operating system. The process of placing this program on the disk is usually called *initializing* the disk in Macintosh systems and *formatting* in MS-DOS systems. In either system, it is done with a few simple commands from the desktop.

Initializing/formatting and naming. Each disk is formatted or initialized only once, since this process begins by erasing the disk. As part of this process, the user names the disk, usually in a way that indicates the information stored on it. For example, if a teacher wants to store only letters to parents and students on a disk, the disk might be called *Letters*. A disk to store all handouts and tests for all classes in a given year might be called *Class Materials-1996*. The default name for a disk is *untitled* for a Macintosh and *unnamed* for a Windows disk.

Labeling. The name of a disk appears with the disk's icon on the desktop. Its label is a paper stuck to the outside of the disk. The label should be typed or written first and then affixed to the disk. For ease of use, the label should contain two pieces of information: the name of the person or place (e.g., the lab) to which it belongs and its purpose.

If a teacher is sending a disk to another person or an organization (e.g., an article submitted to a journal for publication), the label might carry other information. For example, the label might need to specify the word processing program used to prepare the file. The recipient of the disk usually specifies the kind of information to put on the label.

Care and Feeding of Removable Disks

Since computer users depend on disks to keep important programs and files, they should follow carefully maintenance routines. Disks are sensitive to extreme heat and cold, and direct sunlight can warp them. They also should stay away from sources of magnetism (e.g., TVs), which can erase stored information. Floppy disks should always be stored with protective envelopes or sleeves, but microdisks need not be stored in the plastic envelopes in which they are shipped. All disks should be stored in a disk storage box or carrier to guard against accidental loss or damage.

Although disks are fairly sturdy storage media, they can become damaged as a result of age, misuse, computer problems, or flaws in component materials. This risk should lead users to keep a backup disk for every data disk. Losing files when disks are lost or defective is a major problem. A later section of this chapter covers procedures for copying programs and files to a backup disk and for copying the entire contents of one disk to another.

If a disk does fail or become damaged and no backup is available, special programs can help users to retrieve some or all of the files. This process is not always successful, but it is always worth a try if the disk is not so badly damaged that it will not fit in the disk drive. The section headed Disk Utility Software lists some of these programs in the Resources section at the end of this chapter.

Disk Care: A Summary Checklist

___ 1. Treat disks gently; do not bend them, place them between books or other objects, or jam them quickly into disk drives.

___ 2. Keep disks out of direct sunlight and at comfortable temperatures: not too cold (below 40°) or too hot (above 100°).

___ 3. Keep away from sources of magnetism: TV sets and any machines with electric motors.

___ 4. Do not touch any shiny surfaces beneath the plastic covering or allow dirt on these surfaces.

___ 5. Store disks correctly. Place floppy disks in protective sleeves/envelopes, and keep all disks in disk boxes or carrying cases.

___ 6. Back up disks. Make copies of all disks with programs or files stored on them.

Managing the Hard Disk Drive

Frequently used programs and files are usually stored on a hard disk drive. This drive takes less time to boot up a program or file than a floppy disk or microdisk. It also saves a changed file in less time. Users keep less essential programs and files on removable disks (e.g., microdisks). Although

Figure A.3   Example Macintosh Hard Drive Organized with Folders

Name	Size	Kind	Last Modified	
📁 Apple File Exchange Folder	--	folder	Sun, May 5, 1991	2:29 PM
📁 AppleLink 6.1	--	folder	Sun, Jun 26, 1994	10:37 AM
📁 Art work	--	folder	Sun, Sep 11, 1994	11:53 AM
📁 Church Work	--	folder	Mon, May 31, 1993	3:10 PM
📁 College of Education	--	folder	Sun, Jun 19, 1994	11:26 AM
📁 EDF 5481	--	folder	Thu, Jan 7, 1993	9:50 AM
📁 EME course	--	folder	Sun, Sep 4, 1994	1:51 PM
📁 FTEQ	--	folder	Sun, Sep 11, 1994	12:11 PM
📁 Graphics applications	--	folder	Sun, Sep 11, 1994	12:13 PM
📄 Howard article.Fall 1994	9K	document	Sun, Aug 21, 1994	3:45 PM
📁 HyperCard	--	folder	Sun, Sep 11, 1994	11:52 AM
📁 Letters	--	folder	Sun, Sep 11, 1994	11:53 AM
📁 Microsoft Works 3.0 Fol...	--	folder	Sun, Sep 11, 1994	11:42 AM
📁 Personal files	--	folder	Wed, Aug 24, 1994	8:51 PM
📁 Preservice Project	--	folder	Sun, Sep 11, 1994	12:13 PM
📁 QUICKEN	--	folder	Wed, Jun 9, 1993	8:48 PM
📁 Radius Software	--	folder	Fri, May 3, 1991	6:09 PM
📁 Ready ,Set,Go! Folder	--	folder	Sat, Sep 10, 1994	2:48 PM
📁 Red Ryder	--	folder	Sat, Sep 10, 1994	9:11 AM
📁 System Folder	--	folder	Sun, Sep 11, 1994	11:52 AM
📁 Technology Text	--	folder	Sat, Sep 10, 1994	2:47 PM
📁 Virus folder	--	folder	Tue, May 17, 1994	8:42 AM
📁 WordPerfect 2.1	--	folder	Sun, Sep 4, 1994	1:51 PM
📁 ● PICT file pictures	--	folder	Wed, Aug 17, 1994	9:48 AM

hard disks in each new generation of microcomputers have larger and larger capacities, they rarely provide enough space to store all of a user's programs and files. Each user must decide how best to use hard drive space. A program usually remains on the hard drive if it is used every week or is so large that it takes a long time to reinstall on the drive. Files also remain on the hard drive if they are used every week or may serve as templates or models from which the user can derive other files. As with removable disks, a user should always keep backup copies of information stored on a hard drive. Depending on the size of the hard disk, backups can take dozens of disks; however, if the information is important, the insurance that backups provide is worth the time, trouble, and expense.

Managing Macintosh hard disk drives. As discussed earlier, folders help a user organize programs and files on the Macintosh hard drive. Figure A.3 shows an example of a hard drive organized with folders. Each of these folders may have other folders inside it, as well as copies of programs and files. For example, the Letters folder may have two other folders: a Parent Letters folder containing word processing files for letters a teacher sends home and a Faculty Memos folder containing word processing files for correspondence with other teachers and administrators. To get to a particular parent letter, a user would have to open the Letters folder and then open the Parent Letters folder. (See Figure A.4.)

Managing hard disk drives in Windows. The Windows File Manager organizes files on the hard drive, as discussed earlier. Files are stored under directories and subdirectories symbolized by icons that look very much like real folders.

Figure A.4   Folders Subsumed under Other Folders

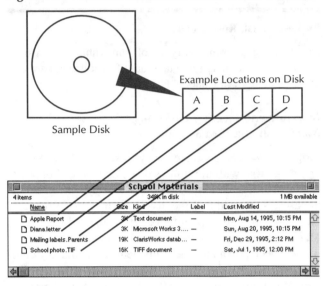

Saving Files: Computer Storage and Memory Concepts

Computer users often struggle with some of the most important concepts related to using disks. It is important to develop a working knowledge of some of those concepts:

• **Files created with applications programs are stored on a disk** *by name.* The name designates a specific physical location on the disk. The user asks the computer to show a file by indicating the name of the file. The user must remember the name under which the system stored the file to retrieve it from its location on the disk. (Otherwise, the user would have to load and look at each file that might be the right one!) The only way to save two different versions of the same file (e.g., two copies of a word processing letter that differ only in the address) is to save the versions under two different names.

• **Files must be saved** *each time* **the user changes them.** What is on the screen when the user saves it to disk takes the place of what was there before under the same name.

• **In order to load a file, a user must first load a copy of the application program that created it.** A file created on one program (e.g., Microsoft Works, Lotus 1-2-3) will usually not work with another program. (Exceptions occur when a file is stored as a text or ASCII file or when a program is designed to read files created with other software. See the chapter on word processing for details.) If the application is on a Macintosh hard drive or disk along with the file, the operation of loading the file automatically loads the application program first. MS-DOS machines can also be set up to load applications automatically when the user loads a file.

Copying Files

Computer users often need to share files. If you are working with someone else on an article or report, you might give that person a copy of your word processing document to make additions or changes. You might have developed a handout or a spreadsheet template that works well for you and want to share it with a colleague. You would do this by copying the file from your disk to the other person's disk. This process requires a hard drive plus one other disk drive or two disk drives, but it is easier if you have a hard drive and two disk drives. Files can be copied from one disk to another without opening the programs that created them. The user drags a file's icon from one place to another. To store or save a file on a given disk on a Macintosh desktop, the user clicks on the file icon, drags it over to the disk icon (or to the window for the disk), and releases the mouse. On a Windows desktop, the user selects the Copy option under the File Manager menu. Both of these procedures are designed to transfer a file from one place to another, but neither removes the file from its original location.

Copying Disks

A user might need to copy an entire disk either to create a backup or to share the whole disk with a colleague.

In a Macintosh system. To copy from disk to disk in a Macintosh system, follow the directions that apply to your setup:

- **With two floppy or microdisk drives.** Drag the icon of the disk to be copied to the disk that will receive the copy. When the icon for the second disk darkens, release the mouse and copying will proceed.

- **With a hard drive and one other disk drive.** Copy the entire contents of the disk to be copied on the hard drive temporarily, then copy from there to the second disk.

- **With only one drive.** This can be done, but it takes longer. This operation involves ejecting the first disk with the Eject Disk command, then dragging the icon of the first disk to the second one and following displayed directions to insert first one disk and then the other until copying is completed.

Copying the entire contents of a hard drive onto removable disks is a little more involved and a lot more time-consuming. The user drags folders one at a time from the hard drive to disk icons. However, removable disks have limited storage capacity, and the entire contents of a folder may not fit on one disk. In that case, the folder will have to be divided between two or more disks, and disks will have to be named and labeled according to their contents (e.g., Graphics Files 1, Graphics Files 2).

In a Windows system. In Windows, disks are copied by selecting the Disk menu within the File Manager. One of the options under this menu, Copy Disk, activates procedures for making copies of disks.

Basic Troubleshooting Concepts

Most teachers have little interest in becoming technical experts with computers any more than the typical driver

wants to become an automotive expert. However, as you learn more about the causes of simple problems and steps to correct them, you become less dependent on someone else to do things for you. One of the greatest challenges for technology-using teachers is determining whether they can correct problems themselves or call in a technician. The information in Table A.1 is intended to help educators identify some of the most common problems and categorize them. It is by no means an exhaustive list. Those who wish to know more than the most common problems should consult their system manuals or take computer maintenance and troubleshooting workshops.

General Troubleshooting Tips

Manuals from Apple Computer, Inc. (1995) and Microsoft Corporation (1993) provide a great deal of useful information on troubleshooting and maintenance. These manuals provide important general tips and suggestions for dealing with problems:

- **Restarting.** Many problems can be resolved simply by clearing the computer's memory. This is true for peripheral devices such as printers, as well. This should not be the first option you try, but it sometimes can work when nothing else does. Save all work first, if you can, and either select Restart from the screen or turn off the computer, wait 10 seconds, and turn it on again. To restart another device (e.g., a printer), turn it off and be sure to wait 10 seconds to let the motor power down completely before turning it on again.

- **Check copies of your programs and system.** Errors in the copy of the program stored in RAM can cause many kinds of problems. Sometimes these errors are introduced by a random power fluctuation or simply by the user doing a sequence of activities the programmers did not anticipate. Sometimes errors affect the original program disk, and the disk has to be replaced. Occasionally, the version of a program is incompatible with the version of the operating system on a machine (especially if the program has just been moved to a new machine). Check with the software company to see if you need an upgrade. Reinstalling the system on a startup disk can also correct many kinds of problems.

- **Take your time.** To assist the technician you will call for stubborn problems, try to note the sequence of steps that occurred before the problem happened. Write down any error messages before you turn off the system. If any other experienced system users are around when the error occurs, ask them if they have had a similar problem. They may be able to help you before you turn off the system.

- **Rebuild the Macintosh desktop.** Rebuilding the desktop is a process that helps the system organize and keep track of items on the startup disk (e.g., a hard disk drive). Do this at least once a year by holding the Option and Command keys down while starting up the computer.

- **Make frequent backups of programs and all work.** Anticipate problems. Always make at least two copies of everything you produce. Make backup copies of programs.

Table A.1 Common Technology-Oriented Problems and Solutions	
Problems with Equipment **What the user observes:**	**Possible causes of problem and what to do:**
• You turn on the device and nothing happens.	• Check to see that the device is plugged into the electrical outlet or power strip. • The power strip is bad, circuit breaker is tripped, or the outlet has no electricity. (Try plugging in another electrical device on the outlet to see if it works.) • Make sure that both ends of the power cable are tightly connected. • If none of these is the problem, the component may be defective. Consult a technician.
• You turn on the computer and it seems to be starting up, but the screen is dark.	• The monitor may be on a separate switch from the computer; turn on all switches. • Make sure both ends of the monitor cable are tightly connected. • Turn up the monitor's brightness control. • If none of these is the problem, the monitor or video card may be defective. Consult a technician.
• The mouse will not move the arrow pointer.	• Make sure both ends of the mouse cable are connected. • The mouse may need cleaning. Remove it and clean it with alcohol or a special mouse-cleaning solution. • The mouse ball may be missing.
• The mouse is "jumpy." That is, it is difficult to control the movement of the arrow pointer.	• The mouse setting on the system control panel may be too fast. Check it and adjust the speed, if necessary. • The mouse needs cleaning or has become defective with age. Clean it and retest the movement. If it is still hard to control, replace the mouse.
• You turn on the system. The disk drive runs and the screen is bright, but the system does not load; the desktop never appears. However, the following displays may appear: -On a Mac: A sad face or computer with an *X* -On a DOS: A warning: "nonsystem disk"	• Either the disk drive is not working correctly or the system software is not stored on the disk or is not working properly. Try turning off the system, waiting 10 seconds, and turning it on again, or try using another disk that you know contains the system. If that doesn't work, call a technician.
• The keys don't seem to be typing what their labels indicate.	• Someone may have started a macro, either accidentally or intentionally. (Macros are programs designed to run when a certain sequence of keys is pressed.) Check to see what macros are stored and delete them, if necessary.
• When you try to load a program or file, you get a screen message that the computer does not have enough memory to do the operation.	• The system may actually lack sufficient RAM to do the operation. Check the memory required for the program against the RAM in the computer. • The system may have enough RAM, but it is allocated inadequately. Check the memory requirements of the program against the RAM allocated for the program. (Select Get Info from the File menu.) Increase the allocation to the required amount, if necessary and if possible. • In Macintoshes, the RAM cache may be inadequately allocated. The RAM cache is a memory amount set aside for the system's use in running all programs to speed hard drive performance. If the RAM cache is turned on and two or more programs are running at the same time, the out-of-memory error may occur. In this situation, it may be best to turn off the RAM cache. If you want to run only one program and have a version of System 6, try turning off the Multifinder. This will release some memory. • If none of these is the problem, one or more memory chips may be defective or the system may have another kind of memory allocation problem. Consult a technician.

Table A.1 *Continued*

Problems with Software: General

What the user observes:	Possible causes of the problem and what to do:
• The program is behaving erratically. It may quit suddenly or display unexpected things on the screen.	• Use a virus detection program to check for a virus on the disk or hard drive. If not, the copy of the program in memory may have become faulty due to a power fluctuation or other electrical problem. Try reloading the program or restarting the system.
• The system suddenly "hangs," that is, the pointer stops on the screen in the middle of what you are doing, and the mouse will not move it anymore.	• Check the mouse first, using the equipment troubleshooting steps given previously in this section. If not caused by a mouse connection, this problem is usually due to a power surge throughout the system or a problem with the program you are running. The system must usually be restarted. If the problem happens before you had a chance to save a file, changes to that file are lost.

Problems with Software: Macintosh

• When you try to load a file, you get an error message saying the program is missing or busy.	• The program under which the file was created may not be on the disk or the hard drive.
• A system bomb icon appears on the screen.	• If this happens before you had a chance to save a file, changes to that file are lost. If the dialog box provides a Restart option, click on it to restart the system. If this doesn't work, turn off the computer, wait 10 seconds, and turn it on again. If the system will not boot, try using a different startup disk. If this works, try reinstalling the system software on the disk that has the problem. Make sure that two different versions of the system software are not stored on the disk and that no old versions of desk accessories are incompatible with the system. If you are not sure how to ascertain this, consult a technician.

Problems with Saving and Disks

What the user observes:	Possible causes of the problem and what to do:
• You cannot find the program or file you thought should be on the disk.	• Try looking in other folders/directories on the same disk. You may have placed the program or file accidentally in another folder.
	• If you are sure about the name under which the file was saved, try the Find File option on one of the desktop menus. Enter the name and let the system look for the file. It will tell you if a file anywhere on the disk matches that name.
	• You may have named the file something other than what you remember. Try loading and looking at other files that may have the document you developed but under a different name.
	• You may have saved the program or file to another disk. The most common error is saving an item to the hard disk drive instead of the floppy disk.
• You give the proper commands to save a file, but the screen message says the disk is locked.	• Eject the disk and make sure the locking tab is in the unlocked (down) position. Check over the disk to make sure it is undamaged. If it appears to be unlocked and undamaged, reinsert the disk and check the information box (Get Info from the File Menu) to make sure it has not been locked from there. If the box has an *X* in it, click it to uncheck the box. If you conclude the disk is not really locked but you are still getting the message, try another disk. If you still get the same message, the fault may be in the disk drive. Consult a technician.
• You give the proper commands to save a file and it appears to save, but the disk will not accept any data.	• Either the disk or the disk drive is faulty. Try another disk. If it works, the first disk is bad; throw it away. If the same thing happens with the second disk, it is probably the disk drive. Consult a technician.

(continued)

Table A.1 *Continued*	
• When you try to load a disk in the disk drive, the screen display says it is unreadable and needs to be initialized. You know it worked previously because you stored files on it.	• Follow this sequence: First, try ejecting the disk and reinserting it. If this doesn't work, try another disk in the disk drive. If it doesn't work either, the drive may be defective; use the disk on another computer and drive. (Call a technician to repair the defective drive.) If another disk does work on the drive, be sure to try the first disk on another drive anyway. There may be a difference in disk drive speed between the drive on which you created the disk and the one on which you are now trying to load it. If the disk you are trying to use will not work on either drive (and you are sure it was created on the same operating system: Macintosh or MS-DOS), it has been damaged; use a backup copy of the disk. Try a disk repair program on the bad disk, or simply make another copy from the good copy. Do not reuse the bad disk; throw it away.
• When you try to eject a disk on a Macintosh system, it will not come out.	• If the disk does not move at all from the internal drive when you give the command, hold down the Command and Shift keys and press the number 1 key. If the disk does not move from the external drive, hold down the Command and Shift keys and press the number 2 key.
	• If the disk seems to start to move but the mechanism appears to be jammed, shut down the system and disconnect the power cable. Locate the small hole to the right of the disk drive opening, and push a small, sharp instrument such as a straightened paper clip into it. This lifts the mechanism up to eject the disk unless the metal plate on the disk has become lodged in the drive or unless a label has partly come off. In the latter case, you may need to consult a technician to remove the disk.

Problems with Printers and Printing

What the user observes:	Possible causes of the problem and what to do:
• You give commands to print something, but nothing happens.	• Check to see that the printer is plugged into the electrical outlet or power strip and turn the switch on. On a dot matrix printer, check to see that the printer is online. (The Online button should be lit up.) For a laser printer, make sure it is warmed up and has a toner cartridge inside.
	• If multiple computers share the printer, make sure the system you are printing from has been identified either from the network or from the switch box that connects the computer to the printer.
• You get an error message saying the printer cannot be located.	• On a Mac, look at the Chooser option under the Apple menu. Make sure the printer you want is selected and turned on. Its icon or name should appear on the right-hand side of the chooser box. If its icon is not there, the driver for that printer may be missing. Look at your computer manual to learn how to install a driver.
• Your printer is connected to a network and the printer displays an error message after it prints your document.	• The printer may have gotten conflicting instructions. Check your version of the printer driver to make sure it is the same one that others on your network are using. If not, see your manual or a technician to learn how to install the correct driver.
	• The cable connecting the printer to the computer may be bad or improperly connected.
	• If the printer is connected to more than one computer (on a switch box) or on a network, another computer may be chosen instead of the one on which you are working.
• When you try to print a document, you either get an error message or the print dialog box closes.	• Make sure you have a version of the printer driver that is up-to-date and compatible with the computer's operating system.

Exercises

Exercise A.1. Basic Usage Concepts. Read the following scenarios and describe what has happened and/or what the user should do:

1. On one disk you have a folder named *Worksheets* that holds three word processing files. On the hard drive you have another folder also named *Worksheets*, but this one has no files in it. Your intention is to copy the folder with files over onto the one without files. Instead, you accidentally do the opposite. You copy the folder without files onto the one with files. Where are the three word processing files now?

2. You load a word processing file from your disk into the computer. Suddenly, the maintenance workers turn off the electricity in the building, and all the computers go off. What effect does this have on the disk copy of your document?

3. You have a number of word processing files on your disk: letters, course materials, and a variety of other things. The list in the window for the disk is so long that it is taking you an increasingly long time to scroll down and locate the name of a given file. If you want to keep all these files on your hard drive, what do you probably need to do with the files to organize them and make them more efficient to use?

4. You have a Macintosh disk with a lot of graphics files for handouts; one of your colleagues has requested a copy of the disk. Your system has a hard disk drive and one internal microdisk drive.

   • What procedure would you follow on a Macintosh to make a copy for your colleague?

   • What procedure would you follow on a Windows or Windows 95 system to make a copy for your colleague?

Exercise A.2. Troubleshooting. Read each scenario, describe the likely problem, and tell how to correct it.

1. You create a database on your Macintosh microcomputer at home using a program called Microsoft Works. However, when you bring the disk with the file to the Macintosh lab at school and try to load it, it will not work. It gives you a message "Application busy or missing." What is the most likely reason the file won't load?

2. You create a word processing file on your IBM microcomputer at home, but when you try to load the disk into the Macintosh at school, the computer says the disk is unreadable. What is the problem, and what can you do?

3. When you try to save a file to a Macintosh disk, you get a message saying the disk is locked. You know you didn't lock it, and it was not locked the last time you used it. What can you do?

4. As you work on your computer system, every once in a while you get a message that says, "The phantom strikes again." Then the program shuts down before you can save your work. This message is not supposed to appear in your software. What is the most likely source of the problem and what should you do?

5. One of the people working in the microcomputer lab cannot get her disk out of the disk drive. When she finally does, the metal plate for the disk has become lodged in the drive and she cannot move it. One of your colleagues suggests inserting metal tweezers into the drive to try to dislodge the plate. What do you think you should do?

6. In your school lab, you start up a Macintosh microcomputer from the hard disk, then insert a microdisk in the internal disk drive. You boot the word processing application program from the hard drive and prepare a letter to a parent. After the letter is completed, you name the file *Parent letter*, save it, and shut down the system. The next day, you come back in the lab and put your disk in a microcomputer, but the file is not in your Letters folder.

   • What may have happened?

   • What steps should you follow to ascertain the problem?

7. You insert into a disk drive a disk on which you have several important files. You get a message that says the disk is unreadable and asks if you want to initialize it. What do you do?

References

Apple Computer, Inc. (1993)
Macintosh Reference Manual. Cupertino, CA: Author.
Microsoft MS-DOS6 Operating System Plus Enhanced Tools.
   Redmond, WA: Author. Microsoft Corporation (1993).

Recommended System Utility Software

Troubleshooting and Security/Antivirus Utility Software
(for disk repair and virus prevention/detection/removal)
Dr. Soloman's Anti-Virus Toolkit—for MS-DOS (S&S)
Norton Utilities— versions for Macintosh, MS-DOS, and
   Windows/Windows 95 (Symantec)
SAM—versions for Macintosh, MS-DOS, and
   Windows/Windows 95 (Symantec)
Virex with Speedscan—versions for Macintosh, MS-DOS, and
   Windows/Windows 95 (DataWatch, Inc.)
VirusScan for Windows (McAlfee Associates)

Disk Security Utility Software
(to protect hard drive and/or network from unauthorized use)
DiskGuard (ASD Software)
FileGuard—for Macintosh (Kent-Marsh, Ltd.)
Full Armor (Micah Development)
Norton Disklock (Symantec)
VEIL—for MS-DOS (TECSEC)

Conversion Utility Software
Conversion Plus for Windows (Dataviz, Inc.)
MacAccess (Syncronys Softcorp)
Mac-in-DOS for Windows (Pacific Micro)

Disk Management Utility Software
(manages files, optimizes disk space, and/or allows maximum
   storage)
Disk Express (Alsoft)
Norton DiskDoubler (Symantec)
Norton Navigator (Symantec)
RAM Doubler (Connectix)
StuffIt Deluxe (Aladdin Software)

System/File Management Utility Software
(pinpoints system startup problems)
Conflict Catcher (Casady and Greene)

General System Enhancements
(make various operations more efficient)
Aladdin Desktop Tools (Aladdin Software)
UnInstaller—for Windows (MicroHelp)

Appendix B

Educational Technology Resources

Sources of Educational Technology Products
Company Names and Addresses

ABC News Interactive, Inc.
7 W. 66th Street
New York, NY 10023
(800) 524–2481
(212) 456–2000

Activision, Inc.
11601 Wilshire Blvd.
Los Angeles, CA 90025
(310) 473–9200

AccuLab Products Group-Precision
Tech, Inc.
50 Maple Street
Norwood, NJ 07648
(201) 767–1600

Addison-Wesley
2725 Sand Hill Road
Menlo Park, CA 94025
(800) 552–2259

Adobe Systems Inc.
1585 Charleston Rd.
P.O. Box 7900
Mountain View, CA 94039
(800) 833–6687
(415) 961–4400

Advanced Ideas Inc.
591 Redwood Highway, Suite 2325
Mill Valley, CA 94941
(415) 388–2430

Aladdin Systems
165 Westridge Drive
Watsonville, CA 95076-4150
(800) 761–6200
(408) 761–6200

Aldus Corp.
411 First Ave. S.
Seattle, WA 98104
(206) 622–5500

Alfred Pub. Co.
16380 Roscoe Blvd.
P.O. Box 10003
Van Nuys, CA 91410

Allegant Technologies, Inc.
6496 Weathers Place
San Diego, CA 92121
(619) 587–0500

AIMS Media
9710 DeSoto Ave.
Chatsworth, CA 91311
(800) 367–2467
(818) 773–4300

Apple Computer
One Infinite Loop
Cupertino, CA 95014
(800) 767–2775
(408) 996–1010

Applied Optical Media Corp.
1450 Boot Rd., Bldg. 400
West Chester, PA 19380
(800) 321–7259

Asymetrix
110 110th Ave. NE #700
Bellevue, WA 98004
(800) 448–6543
(206) 462–0501

Aurbach & Associates
8233 Tuland Ave.
St. Louis, MO 63132
(314) 726–5933

AutoDesk, Inc.
111 McInnis Pkwy.
San Rafael, CA 94903
(415) 507–5000

AutoDessys
2011 Riverside Drive
Columbus, OH 43221
(614) 488–9777

Baudville Computer Products, Inc.
5380 52nd St. SE
Grand Rapids, MI 49512
(616) 698–0888

Bentley Systems
690 Pennsylvania Drive
Exton, PA 19341
(800) 778–4274
(610) 458–5000

Big Top Productions L.P.
548 Fourth St.
San Francisco, CA 94107
(415) 978–5363

BLOC Development Corp.
800 Douglas Entrance
Coral Gables, FL 33134
(305) 567–9931

Bookmaker Corp.
2470 El Camino Real, #108
Palo Alto, CA 94306
(415) 354–8160

Borland International
P.O. Box 660001
Scotts Valley, CA 95067
(800) 437–4329
(408) 431–1000

Britannica Software Inc. (See also: Compton's New Media)
345 Fourth St.
San Francisco, CA 94107

Brøderbund Software
500 Redwood Blvd.
Novato, CA 94948
(800) 521–6263
(415) 382–4400

CADKey, Inc.
4 Griffin Road North
Windsor, CT 06095-1511
(203) 298–8888

Canon U. S. A., Inc.
P.O. Box 1000
Jamesburg, NJ 08831
(908) 521–7000

CD-i Association of North America
11111 Santa Monica Blvd.
Los Angeles, CA 90025
(310) 444–6613

CD-ROM Directory
Pemberton Press
462 Danbury Rd.
Wilton, CT 06897-2126
(800) 248–8466

CE Software Inc.
1854 Fuller Rd., Box 65580
West Des Moines, IA 50265
(800) 523–7638

Chickadee Software, Inc.
RR2 Box 79W
Center Harbor, NH 03226
(603) 253–4600

Churchill Media
6901 Woodley Ave.
Van Nuys, CA 91406
(800) 334–7830
(818) 778–1978

Claris Clear Choice
5210 Patrick Henry Dr.
Santa Clara, CA 95052
(800) 3–CLARIS
(408) 727–8025

Coda Music Technology Inc.
6210 Bury Dr.
Eden Prairie, MN 55346
(612) 937–6911

Computer Teaching Corp.
1713 S. State St.
Champaign, IL 61820
(217) 352–6363

Compton's New Media
2320 Camino Vida Roble
Carlsbad, CA 92009
(800) 862–2206
(619) 929–2500

Computer Curriculum Corporation
1287 Lawrence Station Rd.
Sunnyvale, CA 94088–3711
(408) 745–6270

Conduit, University of Iowa
100 Oakdale Campus M306 OH
Iowa City, IA 52242
(800) 365–9774
(319) 335–4100

Connectix
2600 Campus Drive
San Mateo, CA 94403
(800) 950–5880

Continental Press, Inc.
P.O. Box 1063
St. Cloud, MN 56302
(612) 251–5875

Corel Systems Corp.
1600 Carling Ave.
Ottawa, Ontario K1Z 8R2, Canada
(800) 554–1635
(613) 728–8200

Coronet/MTI: Now called CoroNet Systems
5150 El Camino Real, #C22
Los Altos, CA 94022
(415) 960–3255

Cricket Software: Now called Computer Associates International, Inc.
1 Computer Associates Plaza
Islandia, NY 11788
(516) 342–5224

Cross Educational Software
P.O. Box 1536
Ruston, LA 71270
(800) 768-1969

Curriculum Associates, Inc.
5 Esquire Rd.
North Billerica, MA 01862
(508) 667–8000

CUE Softswap
Box 271704
Concord, CA 94527-1704
(415) 685–7265

Davidson and Associates, Inc.
19840 Pioneer Avenue
Torrance, CA 90503
(800) 545–7677
(310) 793–0600

D. C. Heath and Company
2700 North Richardt Avenue
Indianapolis, IN 46219
(800) 334–3284

DeLorme Pub. Co., Inc.
P.O. Box 298
Freeport, ME 04032
(207) 865–1234

Demco, Inc.
P.O. Box 7488
Madison, WI 53707
(608) 241–1201

Deneba Systems Inc.
7400 SW 87th Avenue
Miami, FL 33173
(305) 596–5644

Developmental Learning Materials (DLM)
Box 4000
Allen, TX 75002
(214) 248–6300

Didatech Software
4250 Dawson St., Suite 200
Burnaly, BCVSC 4B1, Canada
(800) 665–0667

Digital Imaging Association
10153 York Rd., Suite 107
Hunt Valley, MD 21030
(800) 989–5353

Digital Impact
6506 South Lewis Ave., Suite 250
Tulsa, OK 74136
(800) 775–4232

Digital Vision
270 Bridge St.
Dedham, MA 02026
(617) 329–5400

Discovery Communications
7700 Wisconsin Ave.
Bethesda, MD 20814
(301) 986–1999

Discus Knowledge Research, Inc.
90 Sheppard Ave. East, 7th floor
Toronto, Ontario M2N 3A1, Canada
(800) 567–4321

Diskovery Educational Systems
1860 Old Okeechobee Rd., Suite 105
West Palm Beach, FL 33409
(407) 683-8410

DK Multimedia Publishing, Inc.
95 Madison Ave.
New York, NY 10016
(212) 213–4800

Dorling Kinderly Publishing, Inc.
DK Multimedia
95 Madison Ave.
New York, NY 10016
(800) 225–3362
(212) 213–4800

Earthquest
125 University Ave.
Palo Alto, CA 94301
(415) 321–5838

Eastman Kodak Company
Electronic Photography Division
343 State St.
Rochester, NY 14650
(800) 242–2424
(716) 724–4000

Educational Activities, Inc.
1937 Grand Ave.
Baldwin, NY 11510
(516) 223–4666

Educational Resources
1550 Executive Dr.
Elgin, IL 60123
(800) 624–2926
(847) 888–8300

Educational Software Institute (EISI)
4213 South 94th St.
Omaha, NE 68127
(800) 955–5570

EDUCORP Computer Services
7434 Trade St.
San Diego, CA 92121
(800) 843–9497
(619) 536–9999

Edmark Corp
6727 185th Ave. NE
Redmond, WA 98052
(800) 362–2890
(206) 556–8400

EduQuest/IBM
P.O. Box 2150
Atlanta, GA 30327
(800) IBM–4EDU
(404) 238–3100

Egghead Software
P.O. Box 7004
Issaquah, WA 98027
(800) 726–3446
(206) 391–0800

Electronic Arts
1450 Fashion Island Blvd.
San Mateo, CA 94404
(415) 571–7171

Electronic Courseware Systems
1210 Lancaster Dr.
Champaign, IL 61821
(217) 359–7099

EME
P.O. Box 2805
Danbury, CT 06813
(800) 848–2050

Emerging Technology Consultants, Inc.
P.O. Box 120444
St. Paul, MN 55112
(612) 639–3973

Encyclopedia Britannica
Educational Corp.
310 South Michigan Ave.
Chicago, IL 60604
(800) 554–9862
(312) 347–7000

Exsym
301 North Harrison St., Bldg. B, Suite 435
Princeton, NJ 08540
(609) 737–2312

Facts on File, Inc.
460 Park Ave. S.
New York, NY 10016-7382
(212) 967–8800

Focus Media
839 Stewart Ave.
Garden City, NY 11530
(800) 645–8989

Fred Lasswell Inc.
1111 N. Westshore Blvd., Suite 604
Tampa, FL 33607
(813) 289–4486

Gessler Publishing Co., Inc.
55 W. 13th St.
New York, NY 10011
(212) 627–0099

Gold Disk, Inc.
3350 Scott Blvd., Bldg. 14
Santa Clara, CA 95054
(408) 982–0200

Graphisoft U.S., Inc.
400 Oyster Point Blvd., #429
South San Francisco, CA 94080
(415) 737–8665

Great Wave Software
5353 Scotts Valley Drive
Scotts Valley, CA 95066
(408) 438–1990

Grolier Electronic Publishing
Sherman Turnpike
Danbury, CT 06816
(800) 365–5590
(203) 797–3500

Group Technologies Corp.
10901 Malcom McKinley Dr.
Tampa, FL 33612
(813) 972–6426

Gryphon Software Corp.
7220 Trade St., Suite 120
San Diego, CA 92121
(619) 536–8815

Harmonic Vision, Inc.
906 University Place
Evanston, IL 60201
(800) 644–4994
(847) 467–2395

Harper/Collins Interactive
10 E. 53rd St.
New York, NY 10022
(212) 207–7000

Hart, Inc.
320 New Stock Rd.
Asheville, NC 28804

Hartley Courseware Inc.
3451 Dunckel Rd., #200
Lansing, MI 48911
(517) 394–8500

Harvard Associates
10 Holworthy St.
Cambridge, MA 02138
(617) 492–0660

Houghton Mifflin Co.
222 Berkley St.
Boston, MA 02116
(617) 251–5000

HRM Video
175 Tompkins Ave.
Pleasantville, NY 10570
(800) 431–2050

Humanities Software
Box 950
408 Columbia, Suite 222
Hood River, OR 97031
(800) 245–6737
(503) 386–6777

Human Relations Media
175 Tompkins Ave.
Pleasantville, NY 10570
(800) 431–2050
(914) 769–7496

Ideafisher Systems Inc.
2222 Martin #110
Irvine, CA 92715
(714) 474–8111

Inspiration Software, Inc.
2920 S.W. Dolph Ct.
Portland, OR 97219
(503) 245–9011

Intelligent Software
9609 Cypress Ave.
Munster, IN 46321
(800) 521–4518

Intel Corp.
P.O. Box 58119
Santa Clara, CA 95052
(800) 538–3373
(408) 765–8080

Intellimation
130 Cremona Dr.
Box 1922
Santa Barbara, CA 93117
(800) 346–8355
(800) 443–6633
(805) 968–2291

Interactive Image Technologies
111 Peter St., Suite 801
Toronto, ON M5V-2H1, Canada
(800) 263–5552

ISTE (International Society for Technology
in Education)
1878 Agate St.
Eugene, OR 97403-1923
(800) 336–5191
(503) 346–4414

Jostens Learning Corp.
9920 Pacific Heights Blvd.
San Diego, CA 92121
(619) 587–0087

K-12 MicroMedia Publishing
16 McKee Dr.
Mahwah, NJ 07430
(800) 292–1997

Key Curriculum Press, Inc.
2512 Martin Luther King Jr. Way
Box 2304
Berkeley, CA 94702
(510) 548–2304

Knowledge Adventure
4502 Dyer St.
La Crescenta, CA 91214
(800) 542–4240
(818) 542–4200

Laser Learning
3324 Pennsylvania Ave.
Charleston, WV 25320
(800) 70–LASER

Laureate Learning Systems Inc.
110 E. Spring St.
Winooski, VT 05404-1837
(802) 655–4755

LCSI (Logo Computer Systems Inc.)
P.O. Box 162
Highgate Springs, VT 05460
(800) 321–LOGO

The Learning Company
6493 Kaiser Dr.
Fremont, CA 94555
(800) 852–2255
(510) 792–2101

Learning in Motion
500 Seabright Ave., Suite 105
Santa Cruz, CA 95062
(800) 560–5670

Learning Services Inc.
P.O. Box 10636
Eugene, OR 97440
(800) 877–9378

LEGO Dacta
555 Taylor Rd.
P.O. Box 1600
Enfield, CT 06083-1600
(203) 749–2291

Light Source: Now called Computer
Images Inc.
17 E. Sir Francis Drake
Larkspur, CA 94939
(415) 461–8000

LOGAL Software
125 Cambridge Park Dr.
Cambridge, MA 02140
(800) 564–2587
(617) 491–4440

Lotus Development Corp.
55 Cambridge Pkwy.
Cambridge, MA 02142
(800) 343–5414
(617) 577–8500

Lumivision Corp.
877 Federal Blvd.
Denver, CO 80204
(800) 776–5864
(303) 860–0400

MacPlay
17922 Fitch Ave.
Irvine, CA 92714
(714) 553–3530

Macromedia
600 Townsend Street, #408
San Francisco, CA 94103
(415) 252–2000

Manhattan Graphics
250 East Hartsdale Avenue
Hartsdale, NY 10530
(914) 725–2048

Mark of the Unicorn
222 Third St.
Cambridge, MA 02142
(617) 576–2760

Maxis, Inc.
2 Theatre Square
Orinda, CA 94563
(800) 33–MAXIS
(510) 254–9700

McDougal Littell/Houghton Mifflin
P.O. Box 1667
Evanston, IL 60204
(847) 869–2300

Minnesota Educational Computing Corp.
(MECC)
6160 Summit Dr. N.
Minneapolis, MN 55430
(800) 685–6322
(612) 569–1500

Media for the Arts
Budek Films and Slides
P.O. Box 1011
Newport, RI 02840
(401) 846–6580

MiBAC Music Software
P.O. Box 468
Northfield, MN 55057
(507) 645–5851

Medio Multimedia, Inc.
2703 152nd Ave. NE
Redmond, WA 98052
(800) 788–3866
(206) 867–5500

Microsoft Corporation
One Microsoft Way
Redmond, WA 98052
(800) 426–9400
(206) 882–8080

Microsystems Software, Inc.
600 Worcester Rd.
Framingham, MA 01701
(508) 879–9000

Milliken Publishing Company
1100 Research Blvd.
St. Louis, MO 63122
(314) 991–4220

MindPlay
160 W. Fort Lowell
Tuscon, AZ 85705
(800) 221–7911
(602) 888–1800

Mindscape, Inc.
476 Dieno Dr.
Wheeling, IL 60090
(708) 983–3550

Misty City Software
10921 129th Place NE
Kirkland, WA 98033
(206) 828–3107

Monarch Software
P.O. Box 147
Hasum, WA 98623
(800) 647–7997

Multimedia Nuts and Bolts
115 Bloomingdale Lane
Woodbridge, Ontario L4L 6X8, Canada

Musicware, Inc.
8654 154th Ave. NE
Redmond, WA 98052
(800) 99–PIANO
(206) 881–9797

National Geographic Educational Services
P.O. Box 98019
Washington, DC 20090-8019
(800) 368–2728

National Instruments Corp.
6504 Bridge Point Pkwy.
Austin, TX 78730
(521) 338–9119

NCS Education, Inc.
112 Kings Hwy.
Landing, NJ 07850
(201) 770–0800

New Castle Communications, Inc.
229 King St.
Chappaqua, NY 10514
(800) 723–1263

New Media Schoolhouse
P.O. Box 390
Pound Ridge, NY 10576
(800) 672–6002
(914) 764–4104

Novell, Inc.
1555 North Technology Way
Orem, Utah 84057-2399
(801) 429–7000

NTERGAID
60 Commerce Pkwy.
Millford, CT 06460
(800) 254–9737

Opcode Systems, Inc.
3950 Fabian Way, #100
Palo Alto, CA 94303
(415) 856–3333

Open World Interactive
330 E. 70th St.
New York, NY 10021
(800) 570–4190

Optical Data Corporation
30 Technology Drive
Warren, NJ 07059
(800) 524–2481
(908) 668–0022

Optimum Resources, Inc.
P.O. Box 23317
Hilton Head Island, SC 29925
(800) 327–1473
(803) 689–0022

Orange Cherry New Media/
Schoolhouse, Inc.
P.O. Box 390
Pound Ridge, NY 10576
(914) 764–4104

Oxford University Press
198 Madison Ave.
New York, NY 10016
(212) 679–7300

Panasonic Corporation
Two Panasonic Way
Secaucus, NJ 07094
(800) 524–0864
(201) 348–7000

Paradigm Software, Inc.
P.O. Box 668
Concordia, MO 64020

Paramount Publishing
1230 Avenue of the Americas
New York, NY 10020
(212) 698–7000

Passport Design, Inc.
100 Stone Pine Rd.
Half Moon Bay, CA 94019
(800) 443–3210
(415) 726–0280

PBS Video
1320 Braddock Place
Alexandria, VA 22314
(800) 424–7963

PC Globe, Inc.
500 Redwood Blvd.
Novato, CA 94948
(415) 382–4400

Penguin Books USA, Inc.
375 Hudson St.
New York, NY 10014
(212) 366–2260

Philips Interactive Media of America
11111 Santa Monica Blvd.
Los Angeles, CA 90025
(800) 223–4432
(310) 444–6600

Pierian Springs Software
5200 SW Macadam Ave., Suite 570
Portland, OR 97201
(800) 472–8578
(503) 222–2044

Pioneer New Media Technologies, Inc.
2265 East 220th St.
Long Beach, CA 90810
(800) LASER–ON
(310) 952–2111

Powerup Software
2929 Campus Drive
San Mateo, CA 90003
(414) 345–5900

Prentice-Hall (See Paramount)

Prentke Romich Company
1022 Heyl Rd.
Wooster, OH 44691
(216) 262–1984

Quanta Press
1212 5th St. SE
Minneapolis, MN 55414
(612) 379–3956

Quanta Technology, Inc.
1632 Enterprise Pkwy.
Twinsburg, OH 44087
(216) 425–7880

Quark, Inc.
1800 Grant St.
Denver, CO 80203
(800) 476–4575
(303) 894–8888

Que Software
201 W. 103rd St.
Indianapolis, IN 46290
(317) 581–3500

Queue
338 Commerce Dr.
Fairfield, CT 06430
(800) 232–2224

Random House School Division
400 Hahn Rd.
Westminster, MD 21157

Realtime Learning Systems
2700 Connecticut Ave. NW
Washington, DC 20008
(800) 832–2472

Roger Wagner Publishers
1050 Pioneer Way, #P
El Cajon, CA 92020
(800) 421–6526
(619) 442–0522

Sanctuary Woods Multimedia Corp.
1825 S. Grant St., #410
San Mateo, CA 94402
(415) 286–6110

SAS, Inc.
P.O. Box 21990
San Antonio, TX 78221
(210) 921–7300

Scholastic, Inc.
730 Broadway
New York, NY 10003
(212) 505–3000

Science for Kids
9950 Concord Church Rd.
Lewisville, NC 27023
(919) 945–9000

Scott-Foresman (See HarperCollins)

Sensible Software
335 E. Big Beaver, Suite 207
Troy, MI 48083
(313) 528–1950

Sierra On-Line, Inc.
3380 146th Pl. SE, #300
Bellevue, WA 98007
(206) 649–9800

Silicon Beach Software, Inc.
9770 Carroll Center Rd., Suite J
San Diego, CA 92126
(619) 695–6956

Smartek Software, Inc.
2223 Avienda De La Playa, #208
La Jolla, CA 92037
(619) 456–5064

Softeast Corp.
Knox Trail Office Bldg.
2352 Main St.
Concord, MA 01742
(508) 897–3172

Softkey International, Inc.
201 Broadway
Cambridge, MA 02139
(800) 227–5609
(494) 612–1200

Software for Kids
9950 Concord Church Rd.
Lewisville, NC 27023
(919) 945–9000

Software Toolworks, Inc.
60 Leveroni Ct.
Novato, CA 94949
(415) 883–3000

Software Publishing Corp.
P.O. Box 54983
Santa Clara, CA 95056
(408) 986–8000

Sony Electronics, Inc.
1 Sony Drive
Park Ridge, NJ 07656
(800) 472–7669
(201) 930–1000

Spinnaker Software Corp.
201 Broadway
Cambridge, MA 02139
(800) 323–8088
(617) 494–1200

SPSS Inc.
444 N. Michigan Ave.
Chicago, IL 60611
(312) 329–2400

Sunburst Communications
101 Castleton St.
Pleasantville, NY 10570
(800) 321–7511
(914) 747–3310

Strata
2 West St. George Blvd.
St. George, Utah 84770
(800) STRATA3D

SVE, Inc.
6677 N. Northwest Hwy.
Chicago, IL 60631
(312) 775–9550

Swifte International, Ltd.
P.O. Box 219
Rockland, DE 19732
(302) 234–1740
(800) 722–7202

Symantec Corp.
10201 Torre Ave.
Cupertino, CA 95014
(800) 441–7234
(408) 253–9600

Tanager Software
1933 Davis St., Suite 208
San Leandro, CA 94577
(800) 841–2020

Teacher Support Software, Inc.
1035 N.W. 57th St.
Gainesville, FL 32605
(904) 332–6404

Techware Corp.
Box 151085
Altamonte Springs, FL 32715
(407) 695–9000

Temporal Acuity Products
300 120th Ave. NE, Building #1
Bellevue, WA 98005
(800) 426–2673

Terrapin Software Inc.
10 Holworthy St.
Cambridge, MA 02138
(800) 972–8200

Testmaker
912 Kingsley Dr.
Colorado Springs, CO 80909

Texas Instruments Inc.
6620 Chase Oaks Blvd., M58501
Plano, TX 75023
(800) 842–2737
(214) 575–5729

3M Learning Software
Software Media & CD ROM Services
3M Center Bldg. 220-2W-04
P.O. Box 33220
St. Paul, MN 55133-3220
(800) 219–9022

Third Wave Technologies
4544 S. Lamar Blvd., #100-D
Austin, TX 78745
(512) 892–4070

Timeworks International, Inc.
625 Academy Dr.
Northbrook, IL 60025
(708) 559–1300

Tom Snyder Productions, Inc.
80 Coolidge Hill Rd.
Watertown, MA 02172
(800) 342–0236
(617) 926–6000

Triad Venture Inc.
P.O. Box 12201
Hauppauge, NY 11788
(516) 732–3771

Transparent Language, Inc.
22 Proctor Hill Road
Hollis, NH 03049-0575
(800) 752–1767

Turner Educational Services, Inc.
P.O. Box 105366
Atlanta, GA 30348
(404) 827–2252

Ventura Educational Systems
P.O. Box 425
Grover Beach, CA 94383
(805) 473–7383

Vernier Software
8565 S.W. Beaverton Hillsdale Hwy.
Portland, OR 97225
(503) 297–5317

Virtual Reality Development System
(VREAM)
2568 N. Clark St., #250
Chicago, IL 60614

Voyager Company
578 Broadway, #406
New York, NY 10012
(212) 435–5199

Videodiscovery, Inc.
1700 Westlake Ave. N., Suite 600
Seattle, WA 98109
(800) 548–3472
(206) 285–5400

Video Fusion, Inc.: Now Video
Freedom, Inc.
5414 Oberlin Dr.
San Diego, CA 92121
(619) 658–9030

Virtual Entertainment
200 Highland Ave.
Needham, MA 02194
(617) 449–7567

Walt Disney Computer Software
500 S. Buena Vista St.
Burbank, CA 91521
(818) 567–5360

Warner New Media
3500 Olive Ave.
Burbank, CA 91505
(818) 955-9999

Weekly Reader, Software
(See also Optimum Resources)
10 Station Place
Norfolk, CT 06058
(203) 452–5553

Wild Duck Software
979 Golf Course Dr., Suite 256
Rohnet Park, CA 94928
(707) 586–0728

William K. Bradford Publishing
310 School St.
Acton, MA 01720
(800) 421–2009
(508) 263–6996

Wisconsin Foundation for Vocational,
Technical and Adult Education, Inc.
2564 Branch St.
Middleton, WI 53562-9965
(608) 831–6313

World Book, Inc.
101 N.W. Point Blvd.
Elk Grove Village, IL 60007
(847) 290–5300

Xerox Corp.
P.O. Box 1600
Stamford, CT 06904
(203) 968–3000

Zedcor, Inc.
E. Speedway Blvd., #22
Tucson, AZ 85712
(800) 482–4567

ZSoft Corp.
450 Franklin Rd., Suite 100
Marietta, GA 30067
(404) 428–0008

Ztek Corporation
P.O. Box 1055
Louisville, KY 40577-1768
(800) 247–1603

Integrated Learning Systems (ILS)

Computer Curriculum Corporation
Box 3711
Sunnyvale, CA 94088-3711

Computer Networking Specialists, Inc.
Rte. 1, Box 286-C
Walla Walla, WA 99362
(800) 372–3277
(509) 529–3070

Control Data Corporation
PLATO Educational Services
8800 Oueen Ave. S.
Bloomington, MN 55431
(800) 328–1109

DEGEM Ltd.
Two Park Ave.
New York, NY 10016-5635
(212) 561–7200

Houghton Mifflin Company
Mount Support Rd.
Lebanon, NH 03766
(603) 448–3838

Ideal Learning Inc.
5005 Royal Ln.
Irving, TX 75063
(214) 929–4201

Jostens Learning Corp.
7878 N.16th St., Suite 100
Phoenix, AZ 85020-4402
(800) 442–4339
(602) 678–7272

Skills Bank
15 Governor's Ct.
Baltimore, MD 21207-2791
(800) 847–5455

Unisys Corporation
Box 500, MS B330
Blue Bell, PA 19424
(215) 542–4583

Wasatch Education Systems
5250 S. 300 W., Suite 350
Salt Lake City, UT 84100-0007
(801) 261–1001

Educational Technology Periodicals
Magazines and Journals

AI Applications
P.O. Box 3066
Moscow, ID 83843

AI Expert
600 Harrison Street
San Francisco, CA 94107

Byte
One Phoenix Mill Lane
Peterborough, NH 03458

CD-ROM Today
GP Publications Inc.
23-00 Rte. 208
Fair Lawn, NJ 07410

*Computer Assisted English Language
Learning Journal*
International Society for Technology in
Education (ISTE)
1787 Agate St.
Eugene, OR 97403-1923

Computer Design
10 Tara Blvd.
Nashua, NH 03062-2801

Computer Science Education
Ablex Publishing Corp.
355 Chestnut St.
Norwood, NJ 07648

Computers and Composition
Michigan Technological University
Department of Humanities
Houghton, MI 49931

Computers and Education
Pergamon Press
600 White Plains Rd.
Tarrytown, NY 10591-5153

Computers in Human Behavior
Pergamon Press
600 White Plains Rd.
Tarrytown, NY 10591

Computers, Reading and Language Arts
Modern Learning Publishers, Inc.
1308 E. 38th St.
Oakland, CA 94602

The Computing Teacher
(See *Learning and Leading with
Technology*)

Curriculum Product News
Educational Media Inc.
992 High Ridge Rd.
Stamford, CT 06905

Digital Video Magazine
80 Elm St.
Peterborough, NH 03458

ED Journal
Applied Business teleCommunications
P. O. Box 5106
San Ramon, CA 94583

EDN
8773 South Ridgeline Blvd.
Highlands Ranch, CO 80126-2329

ED-TECH Review
Association for the Advancement of
Computing in Education (AACE)
Box 2966
Charlottesville, VA 22902

Education at a Distance
Applied Business teleCommunications
P. O. Box 5106
San Ramon, CA 94583

Educational Media International
120 Pentonville Rd.
London, N1 9JN, UK

Educational Technology
720 Palisade Ave.
Englewood Cliffs, NJ 07632

*Educational Technology Research
and Development*
Association for Educational
Communications and Technology
(AECT)
1025 Vermont Ave., N.W., Suite 820
Washington, DC 20005

Electronic Design
1100 Superior Ave.
Cleveland, OH 44114-2543

Electronic Learning
Scholastic, Inc.
730 Broadway
New York, NY 10003

*Hands-On! Technical Education
Research Center*
2067 Massachusetts Ave.
Cambridge, MA 02140

*HyperNEXUS: Journal of Hypermedia
and Multimedia Studies*
ISTE
1787 Agate St.
Eugene, OR 97403-1923

IAT Briefings
Institute for Academic Technology
Box 12017
Research Triangle Park, NC 27709

InfoWorld
InfoWorld Publishing Co.
155 Bovet Rd., Suite 800
San Mateo, CA 94402

Instruction Delivery Systems
Communicative Technology Corporation
50 Culpepper St.
Warrenton, VA 22186

Interact
Interational Interactive Communications
Society
2120 Steiner St.
San Francisco, CA 94115

Interactive Learning Environments
Ablex Publishing Corp.
355 Chestnut St.
Norwood, NJ 07648

Internet World
20 Ketchum St.
Westport, CT 06880

*Journal of Artificial Intelligence
in Education*
AACE
Box 2966
Charlottesville, VA 22902

Journal of Communication
8140 Burnet Rd.
P.O. Box 9589
Austin, TX 78766

Journal of Computer Science Education
ISTE
1787 Agate St.
Eugene, OR 97403-1923

*Journal of Computers in Mathematics
and Science Teaching*
AACE
Box 2966
Charlottesville VA 22902

*Journal of Computing in
Childhood Education*
AACE
Box 2966
Charlottesville, VA 22902

*Journal of Computing in Teacher
Education*
ISTE
1787 Agate St.
Eugene, OR 97403-1923

Journal of Educational Computing Research
Baywood Publishing Company, Inc.
26 Austin Ave.
Box 337
Amityville, NY 11701

*Journal of Educational Multimedia
and Hypermedia*
AACE
Box 2966
Charlottesville, VA 22902

Journal of Educational Technology Systems
Baywood Publishing Company, Inc.
26 Austin Ave.
Box 337
Amityville, NY 11701

*Journal of Interactive Instruction
Development*
Society for Applied Learning
Technology (SALT)
50 Culpepper St.
Warrenton, VA 22186

*Journal of Research on Computing
in Education*
ISTE
1787 Agate St.
Eugene, OR 97403-1923

*Journal of Science Education
and Technology*
Plenum Publishing Corp.
233 Spring St.
New York, NY 10013-1578

*Journal of Technology and Teacher
Education*
AACE
Box 2966
Charlottesville, VA 22902

Learning
P. O. Box 2580
Boulder, CO 80322

Learning and Leading with Technology
(Formerly The Computing Teacher)
ISTE
1787 Agate St.
Eugene, OR 97403-1923

Logo Exchange
ISTE
1787 Agate St.
Eugene, OR 97403-1923

Machine Design
1100 Superior Ave.
Cleveland, OH 44114-2543

Machine-Mediated Learning
Taylor & Francis Inc.
242 Cherry St.
Philadelphia, PA 19106-1906

MacWEEK
Ziff-Davis Publishing Co.
One Park Ave.
New York, NY 10016

MacUser
P. O. Box 56972
Boulder, CO 80321-6972

MacWorld
MacWorld Communications Inc.
501 2nd St.
San Francisco, CA 94107

Media and Methods
American Society of Educators
1429 Walnut St.
Philadelphia, PA 19102

Microcomputers in Education
Two Sequan Rd.
Watch Hill, Rl 02891

Microsoft Works in Education
ISTE
1787 Agate St.
Eugene, OR 97403-1923

*MIPS-the Magazine of Intelligent
Personal Systems*
400 Amherst St.
Nashua, NH 03063

MultiMedia Schools
2809 Brandywine St. NW
Washington, DC 20008

New Directions for Teaching and Learning
350 Sansome St.
San Francisco, CA 94104-1310

NewMedia
HyperMedia Communications Inc.
901 Mariner's Island Blvd., Suite 365
San Mateo, CA 94404

PC Computing
Ziff-Davis Publishing Company
One Park Ave.
New York, NY 10016

PC Magazine
Ziff-Davis Publishing Company
One Park Ave.
New York, NY 10016

PC Week
Ziff-Davis Publishing Company
One Park Ave.
New York, NY 10016

PC World
Box 78270
San Francisco, CA 94107-9991

Pen Computing
P.O. Box 640
Folsom, CA 95763-0640

Personal Computing
Hayden Publishing Co., Inc.
10 Mulholland Dr.
Hasbrouck Heights, NJ 07604

Personal Engineering and Instrumentation
News
Box 430
Rye, NH 03870

Personal Publishing
Hitchcock Publishing Company
191 S. Gary Ave.
Carol Stream, IL 60188

Presentation Products
Pacific Magazine Group
513 Wilshire Blvd., Suite 344
Santa Monica, CA 90401

School Library Journal
P.O. Box 1878
Marion, OH 43305

Science Teacher
1742 Connecticut Ave., NW
Washington DC, 20009

Scientific Computing & Automation
301 Gibraltar Dr.
Morris Plains, NJ 07950-0650

SIGTC Connections
ISTE
1787 Agate St.
Eugene, OR 97403-1923

Syllabus
P.O. Box 2716
Sunnyvale, CA 94087-0716

*T.H.E. Journal-Technological Horizons
in Education*
150 El Camino Real
Tustin, CA 92680

Teaching and Computers
Scholastic, Inc.
Box 2040
Mahopac, NY 10541-9963

Technology and Learning
Peter Li Inc.
330 Progress Rd.
Dayton, OH 45499

Tech Trends
AECT
1025 Vermont Ave. N.W., Suite 820
Washington, DC 20005

Telecommunications in Education
(T. l. E.) News
ISTE
1787 Agate St.
Eugene, OR 97403-1923

Via Satellite
Phillips Business Information, Inc.
7811 Montrose Rd.
Potomac, MD 20854

Virtual Reality Special Report
600 Harrison St.
San Francisco, CA 94107

Virtual Reality World
11 Ferry Lane West
Westport, CT 06880

Windows Magazine
CMP Publications
600 Community Dr.
Manhasset, NY 11030

Windows User
Wandsworth Publishing Inc.
831 Federal Rd.
Brookfield, CT 06804

Wired
544 Second St.
San Francisco, CA 94107-1427

The Writing Notebook, Creative Word
Processing in the Classroom
Box 1268
Eugene, OR 97440-1268

Newsletters

Apple Education News
Apple Computer Inc.
20525 Mariani Ave.
Cupertino, CA 95014

Education Technology News
Business Publishers Inc.
951 Pershing Dr.
Silver Spring, MD 20910-4464

The Heller Report: Internet Strategies for
Education Markets
Nelson B. Heller & Associates
1910 First St., Suite 303
Highland Park, IL 60035-3146

Inside the Internet: Rocket Science for the
Rest of Us
The Cobb Group
9420 Bunsen Parkway, Suite 300
Louisville, KY 40220

ISTE Update
ISTE
1787 Agate St.
Eugene, OR 97403-1923

LCSI Logo Link
Logo Computer Systems Inc.
Box 162
Highgate Springs, VT 05460

Logo Update
The Logo Foundation
250 West 57th St., Suite 2603
New York, NY 10107-2603

Online Searcher
14 Haddon Rd.
Scarsdale, NY 10583

Telecommunications in Education News
ISTE
1787 Agate St.
Eugene, OR 97403-1923

Terrapin Times
Terrapin Software Inc.
400 Riverside St.
Portland, ME 04103

Professional Organizations
General Organizations with Educational Technology SIGs or Interest Groups

American Educational Research
Association (AERA)
1230 17th St. N.W.
Washington, DC 20036

Association for Computers in
Mathematics and Science Teaching
P. O. Box 4455
Austin, TX 78765

Computers in Reading and Language
Arts (CRLA)
P. O. Box 13039
Oakland, CA 94661

National Council of Teachers of
Mathematics (NCTM)
1906 Association Dr.
Reston, VA 22091

National Science Teachers
Association (NSTA)
1742 Connecticut Ave. N.W.
Washington, DC 20009

Organizations with an Educational Technology Focus

American Association for
Artificial Intelligence
445 Burgess Drive
Menlo Park, CA 94025-3496

Association for the Advancement of
Computing in Education
Box 2966
Charlottesville, VA 22902-2966

Association for Computing
Machinery (ACM)
1133 Avenue of the Americas
New York, NY 10036

Association for Educational
Communications and Technology
(AECT)
1025 Vermont Ave. N.W., Suite 820
Washington, DC 20005

Computer Using Educators Inc. (CUE)
Box 2087
Menlo Park, CA 94026

Educational Products Information
Exchange (EPIE) Institute
P. O. Box 839
Water Mill, NY 11976

Educational Technology Center
Harvard Graduate School of Education
Nichols House
Appian Way
Cambridge, MA 02138

ERIC Clearinghouse on Information
and Technology
Center for Science and Technology
4th Floor, Room 194
Syracuse University
Syracuse, NY 13244-4100

International Neural Network Society
1250 24th St. N.W.
Suite 300
Washington, DC 20037

International Society for Technology in
Education (ISTE)
1787 Agate St.
Eugene, OR 97403-1923

International Technology Education
Association
1914 Association Dr.
Reston, VA 22091

International Teleconferencing Association
1650 Tysons Blvd.
Suite 200
McLean, VA 22102

The Logo Foundation
250 W. 57th St., Suite 2603
New York, NY 10107-2603

Microcomputer Software and Information
for Teachers (MicroSIFT)
Northwest Regional Educational Laboratory
100 S. W. Main, Suite 500
Portland, OR 97204

Public Broadcasting Service/Adult
Learning Satellite Service
1320 Braddock Pl.
Alexandria, VA 22314

Society for Applied Learning
Technology (SALT)
50 Culpepper St.
Warrenton, VA 22186

Technology Education Research Center
(TERC)
2067 Massachusetts Ave.
Cambridge, MA 02140

United States Distance Learning
Association
P.O. Box 5129
San Ramon, CA 94583

U. S. Department of Education/
Technology Resources Center
555 New Jersey Ave., N.W.
Washington, DC 20208

Networks and Online Services

Accu-Weather
619 College Ave.
State College, PA 16801
(814) 237–0309

America Online
8619 Westwood Center Dr.
Vienna, VA
(800) 227–6364

AppleLink
Apple Computer, Inc.
20525 Mariani Ave.
Cupertino, CA 95014

ARPANET
Stanford Research Institute
Menlo Park, CA 94025
(800) 235–3155

AT&T Long-Distance Learning Network
295 N. Maple Ave., Rm. 6234S3
Basking Ridge, NJ 07201
(201) 221–8544

BRS (Bibliographic Retrieval Service)
1200 Rte. 7
Latham, NY 12110
(800) 468–0908

CompuServe
5000 Arlington Center Blvd.
Columbus, OH 43220
(800) 848–8990
(614) 457–8600

Delphi
3 Blackstone St.
Cambridge, MA 02139
(800) 695–4005

Dialog Information Services
3460 Hillview Ave.
Palo Alto, CA 94304

Dialog Information Services
Classroom Instructional Program
Administrator
1901 N. Moore St., Suite 500
Arlington, VA 22209
(800) 334–2564

Dow Jones News/Retrieval Service
Box 300
Princeton, NJ 08543-0300
(800) 522–3567
(609) 520–8349

EDUCOM (Bitnet)
Box 364
Princeton, NJ 08540
(609) 520–3377

ERIC
Educational Resources Center
National Institute of Education
Washington, DC 20208

Florida Information Resource
Network (FIRN)
Florida Education Center
325 W. Gaines St.
Tallahassee, FL 32399

FrEdMail Foundation
Box 243
Bonita, CA 91908-0243
(619) 475–4852

GEnie (The General Electric
Information Service)
410 N. Washington St.
Rockville, MD 20850
(800) 638–9636

GTE Education Services Inc.
8505 N. Freeport Pkwy., Suite 600
Irving, TX 75063-9931

Mead Data Central
9443 Springboro Pike
Box 933
Dayton, OH 45401
(800) 227–4908

NASA Spacelink
(202) 895–0028

National Geographic Society KidsNet
Educational Services
Box 96892
Washington, DC 20090

NSFNET
National Science Foundation
1800 G St. N.W.
Washington, DC 20550
(202) 357–9717

OCLC (Online Computer Library Center)
6565 Frantz Rd.
Dublin, OH 43017-0702
(800) 848–5878

Prodigy Services Company
445 Hamilton Ave.
White Plains, NY 10601
(800) 776–0836

Scholastic Network
Jefferson City, MO
(800) 246–2986

SpecialNet
Colorado Department of Education
201 E. Colfax
Denver, CO 80203
(303) 866–6722

Technology Organizations and Vendors for Students with Special Needs

Ability Research
P.O. Box 1791
Minnetonka, MN 55345
(612) 939–0121

AbleNet, Inc.
1081 Tenth Ave. S.E.
Minneapolis, MN 55414
(612) 379–0956
(800) 322–0956

AbleTech Connection
P.O. Box 898
Westerville, OH 43081
(614) 899–9989

Academic Software, Inc.
331 West 2nd St.
Lexington, KY 40507
(606) 233–2332
(800) VIA–ADLS

Access Unlimited
3535 Briarpark Drive, Suite 102
Houston, TX 77042-5235
(713) 781–7441
(800) 848–0311

Adaptive Devices Group
1278 N. Farris Ave.
Fresno, CA 93728
(800) 766–4234

Aicom Corp.
1590 Oakland Rd., Suite B
San Jose CA 95131
(408) 453–8251

AI Squared
P.O. Box 669
Manchester Center, VT 05255
(802) 362–3612

American Printing House for the Blind, Inc.
P.O. Box 6085
Louisville, KY 40206
(502) 895–2405

American Thermoform, Inc.
2311 Travers Ave.
City of Commerce, CA 90040
(213) 723–9021
(800) 331–3676

APT, Inc.
6377 Clark Ave., Suite 111
Dublin, CA 94568
(510) 803–8850
(800) 448–1184

Artic Technologies
55 Park St., Suite 2
Troy, MI 48083
(313) 588–7370

Articulate Systems, Inc.
600 W. Cummings Park, Suite 4500
Woburn, MA 01801
(617) 935–5656
(800) 443–7077

AT&T Accessible Communication Products
5 Woodlawn Rd.
Parsippany, NJ 07054
(800) 233–1222, TT

Berkeley Systems, Inc.
2095 Rose St.
Berkeley, CA 94709
(510) 540–0709 TT
(510) 540–5535

Blazie Engineering, Inc.
105 E. Jarrettsville Rd.
Forrest Hill, MD 21050
(410) 893–9333

BEST
63 Forest St.
Chestnut Hill, MA 02167
(617) 277–0179

Biolink Computer Research and
Development, Ltd.
4770 Glenwood Ave.
North Vancouver, BC V7R 4G8, Canada
(604) 984–4099

Blissymbolics Communication International
250 Ferrand Drive, Suite 200
Don Mills, ON M3C 3P2, Canada
(416) 242–9114

Brown & Co., Inc.
14 Midoaks Street
Monroe, NY 10950
(914) 782–3056

Center for Applied Special
Technology (CAST)
39 Cross St.
Peabody, MA 01960
(617) 531–8555

Closing The Gap, Inc.
P.O. Box 68
Henderson, MN 56044
(612) 248–3294

Consultants for Communication
Technology
508 Bellevue Terrace
Pittsburgh, PA 15202
(412) 761–6062

Costar Corp.
100 Field Point Rd.
Greenwich, CT 06830-6406
(800) 426–7827
(203) 661–9700

Creative Switch Industries
P.O. Box 5256 A-2
Des Moines, IA 50316
(515) 287–5748

Crestwood Company
6625 N. Sydney Place
Milwaukee, WI 53209-3259
(414) 352–5678

Digital Equipment Corporation
P.O. Box CS2008
Nashua, NH 03601-2008
(800) 344–4825
(508) 467–5111

Don Johnston, Inc.
P.O. Box 639
1000 N. Rand Rd., Bldg. #115
Wauconda, IL 60084
(800) 999–4660
(847) 526–2682

Dragon Systems
320 Nevada St.
Newton, MA 02160
(800) 825–5897
(617) 965–5200

Edmark Corporation
6727 185th Ave. NE
P.O. Box 3218
Redmond, WA 98073
(800) 426–0856
(206) 556–8400
(206) 556–8402, TT

Enabling Technologies Co.
3102 SE Jay Street
Stuart, FL 34997
(407) 283–4817

ErgoFlex Systems
4917 Chippewa Drive
Larkspur, CO 80118
(800) 788–2810

Exceptional Computing, Inc.
450 NW 58th St.
Gainesville, FL 32607
(904) 331–8847

Extensions for Independence
757 Emory St. 514
Imperial Beach, CA 92032
(619) 423–7709

Foundation for Technology Access
2173 E. Francisco Blvd., Suite L
San Rafael, CA 94901
(415) 455–4575

Fred Sammons, Inc.
P.O. Box 32
Brookfield, IL 60513-0032
(800) 323–5547

GW Micro
310 Racquet Dr.
Fort Wayne, IN 46825
(219) 483–3625

Handykey Corporation
141 Mount Sinai Ave.
Mount Sinai, NY 11766
(800) 638–2352

Hexagon Products
P.O. Box 1295
Park Ridge, IL 60068
(847) 692–3355

HumanWare, Inc.
6245 King Rd.
Loomis, CA 95650

IBM Special Needs Systems
1000 Yamato Rd.
Internal Zip 5432
Boca Raton, FL 33431
(800) 426–4833 TT
(800) 426–4832 voice

Information Strategies Incorporated
1200 E. Campbell, #108
Richardson, TX 75081
(214) 234–0176

Innocomp
26210 Emery Rd., Suite 302
Warrensville Heights, OH 44128
(800) 382–VOCA
(216) 464–3636

Innovative Designs
2464 El Camino, Suite 205
Santa Clara, CA 95051
(408) 985–9255

Institute on Applied Technology
The Children's Hospital
300 Longwood Ave.
Boston, MA 02115
(617) 735–8391

IntelliTools
5221 Central Ave., Suite 205
Richmond, CA 94804
(800) 899–6687
(216) 464–3636

Kensington Microware Ltd.
2855 Campus Dr.
San Mateo, CA 94403
(800) 535–4242

KLAI Enterprises
P.O. Box 43
Garden City, MI 48136
(313) 425–1165

KRI Communications, Inc.
129 Sheldon St.
El Segundo, CA 90245
(800) 833–4968

KY Enterprises/Custom Computer
Solutions
3039 E. 2nd St.
Long Beach, CA 90803
(310) 433–5244

Kurzweil Applied Intelligence
411 Waverley Oaks Rd.
Waltham, MA 02154
(800) 634–8723

Logitech, Inc.
6505 Kaiser Dr.
Fremont, CA 94555
(510) 795–8500

Luminaud, Inc.
8688 Tyler Blvd.
Mentor, OH 44060
(216) 255–9082

Madenta Communications
941 1A-20th Ave.
Edmonton, AB T6N 1E5, Canada
(403) 450–8926
(800) 661–8406

Mayer-Johnson Co.
P.O. Box 1579
Solana Beach, CA 92075
(619) 481–2489

McIntyre Computer Systems Division
22809
Birmingham, MI 48025
(313) 645–5090

Mouse Systems, Inc.
47505 Seabridge Dr.
Fremont, CA 94538
(510) 656–1117

Pointer Systems, Inc.
One Mill Street
Box 826
Burlington, VT 05401
(800) 537–1562
(802) 658–3714

Prentke Romich Company
1022 Heyl Rd.
Wooster, OH 44691
(800) 262–1984
(216) 262–1984

North Coast Medical, Inc.
187 Stauffer Blvd.
San Jose, CA 95125-1042
(800) 821–9319

OMS Development
1921 Highland Ave.
Wilmette, IL 60091
(847) 251–5787

Penny and Giles Computer Products
163 Pleasant St., Suite 4
Attleboro, MA 02703
(508) 226–3008

Reasonable Solutions
1221 Disk Dr.
Medford, OR 97501
(800) 876–3475

Seeing Technologies Inc.
7047 Brooklyn Blvd.
Minneapolis, MN 55429
(800) 462–3738

Synergy
68 Hale Rd.
East Walpole, MA 02032
(508) 668–7424

TASH, Inc.
Unit 1-91 Station Street
Ajax, ON L1S 3H2, Canada
(905) 686–4129
(800) 463–5685

TeleSensory
455 N. Bernardo Ave.
P.O. Box 7455
Mountain View, CA 94039-7455
(800) 227–8418
(415) 960–0920

Toys for Special Children, Inc.
385 Warburton Ave.
Hastings-on-Hudson, NY 10706
(800) 832–8697
(914) 478–0960

Trace Research and Development Center
Room S-151
Waisman Center
1500 Highland Ave.

University of Wisconsin
Madison, WI 53705
(608) 263–5408 TT
(608) 262–6966 voice

UCLA Intervention Program for Young
Handicapped Children (UCLA/LAUSD)
1000 Veteran Ave., Room 23-10,
Los Angeles, CA 90024
(310) 825–4821

Richard Wanderman
P.O. Box 1075
Litchfield, CT 06759
(203) 567–4307

Words+, Inc.
40015 Sierra Hwy, Bldg. 13145
Palmdale, CA 93550
(800) 869–8521

Zygo Industries, Inc.
P.O. Box 1008
Portland, OR 97207
(800) 234–6006
(503) 684–6006

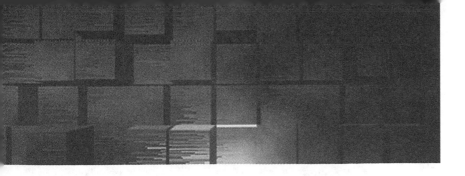

Glossary

algorithm—A step-by-step solution to a problem (e.g., a programming problem)

anchored instruction—Constructivist term for learning environments that focus on meaningful, real-life problems and activities

applications software—Computer programs written to support tasks that are useful to a computer user (e.g., word processing) in contrast with *systems software*

Archie—A system located on the Internet and designed for finding files through lists of archived files from various locations

artificial intelligence (AI)—Computer programs that try to emulate the decision-making capabilities of the human mind

ASCII (American Standard Code for Information Interchange)—A standardized, commonly accepted format for representing data (e.g., characters and numbers) so that programs and files created with one program and stored in ASCII can be used by another program

assembler—A low-level programming language that uses mnemonic commands that are one step up in complexity from "machine language" (binary numbers)

authoring system—A program designed to help nonprogrammers write computer-based instructional programs (e.g., CAI or multimedia); can be either a high-level programming language or a series of nonprogramming prompts

automaticity—A level of skill that allows a person to respond immediately with the correct answer to a problem

backup copy—A copy of a disk that is made to guard against loss of files if an original disk is lost or destroyed

bar code—A set of lines that represent a number (A UPC (Universal Product Code) is an example of a bar code.)

bar code reader—A device that reads and interprets bar codes

BASIC (Beginners All-purpose Symbolic Instruction Code)—A high-level programming language designed for beginning programmers; popular with microcomputers

baud rate—The speed at which data are transmitted across communication lines between computers measured in bits per second (bps); 1 baud = 1 bps

BBS (See *bulletin boards*.)

binary—A condition of two possible states (e.g., on or off, 1 or 0). For computers, *binary* refers to the coding system that uses 1s and 0s

bit—A binary digit, either a 1 or a 0; several bits together (usually eight) make up a byte of computer storage that can hold a letter or a character

BitNet—A network mostly among higher education organizations; replaced by the Internet

boot/boot up—To start up a computer system; a "cold boot" is turning on the device from a power switch; a "warm boot" is restarting from the keyboard without shutting off power.

bps (bits per second)—The speed at which data are transmitted across communication lines between computers

bug—An error in a computer program caused either by faulty logic or program language syntax (named by early programmer and systems designer Grace Hopper, who found a moth in the machinery of a broken computer)

bulletin boards (BBS)—A computer system set up to allow notices to be posted and viewed by anyone who has access to the network

button—A place on a computer screen, usually within a hypermedia program, that causes some action when the user clicks on it using a mouse; also called a *hot spot*

byte—A group of binary digits (bits) that represent a character or number in a computer system; designates a unit of computer storage (See also *K*, *M*, and *G*.)

C—A structured programming language originally designed for use on the Unix operating system and widely used on today's microcomputers

CAI (computer-assisted instruction)—Software designed to help teach information and/or skills related to a topic; also known as *courseware*

card—One frame or screen produced in a hypermedia program such as HyperStudio (several cards together make a stack.)

CAT (computer-assisted testing)—Using a computer system to administer and score assessment measures

CAV (constant angular velocity)—A videodisc format that makes available up to 54,000 images or 30 minutes of full-motion video per side; images can be selected and viewed randomly (See also *CLV.*)

CD-I (compact disk interactive)—A type of CD-ROM used on special players with built-in computer and TV capabilities

CD-ROM (compact disk-read only memory)—Removable computer storage medium that can store images and/or up to 250,000 pages of text

central processing unit (CPU)—The circuitry in a computer that processes commands composed of the controller, the arithmetic/logic unit (ALU), and internal memory

chip—A piece of silicon inside a computer on which electronic circuits have been placed by depositing small paths of a metal such as aluminum; an integrated circuit

clip art—One or more pieces of professionally prepared art work, stored as files and designed to be inserted into a document such as a newsletter

clone—A computer designed to operate like another brand of computer, but usually not made by the same company as the original

CLV (constant linear velocity)—A videodisc format that can hold up to 60 minutes of full-motion video; designed for playing straight through, users cannot access individual frames as CAV discs allow (See also CAV.)

CMI (computer-managed instruction)—Computer software systems designed to keep track of student performance data, either as part of CAI programs or by themselves

COBOL (COmmon Business Oriented Language)—A high-level language designed specifically for business applications

code—Lines of commands or instructions to a computer written in a certain language (e.g., BASIC or C); the act of writing such instructions

cold boot (See *boot.*)

command—One instruction to a computer written in a computer language

compiler—A computer program that converts all statements of a source program into machine language before executing any part of the program (in contrast with *interpreter*)

computer—Usually equivalent to *computer system* (see below); sometimes refers to the CPU part of the system

computer-assisted instruction (See *CAI.*)

computer-assisted testing (See *CAT.*)

computer language—A communication syntax

computer literacy—Term coined by Arthur Luehrmann in the 1960s to mean a set of basic abilities everyone should have with computer systems; now has variable meanings

computer-managed instruction (See *CMI.*)

computer system—A set of devices designed to work together to accomplish input, processing, and output functions in order to accomplish tasks desired by a user

conferencing—Communication between people in two or more places made possible by computer systems connected by communication lines

constructivism—Teaching/learning model based on cognitive learning theory; holds that learners should generate their own knowledge through experience-based activities rather than being taught it by teachers (See also *directed instruction.*)

control unit (controller)—The part of the computer system housed in the CPU that processes program instructions

courseware—Instructional software; computer software used to enhance or deliver instruction

CRT (cathode ray tube)—A TV-like screen on which information from a computer system may be displayed; a monitor; a primary output device for a microcomputer system

cursor—A line, block, or underline displayed on the computer screen that shows where information may be inserted

cut and paste—The act of copying text from one location in a document, deleting it, then inserting it in another location

data—Elements of information (e.g, numbers and words)

database—A collection of information systematized by computer software to allow storage and easy retrieval through keyword searching; the program designed to accomplish these tasks

database management system (DBMS)—Computer software designed to facilitate use and updating of a collection of information in a database

data processing—Organizing and manipulating data for a specific purpose (e.g., to keep track of income and expenses, to maintain student records)

debug—To review a computer program and remove the errors or "bugs"

desktop—The screen that appears first upon starting up a Macintosh computer or an MS-DOS computer with Windows; a graphic user interface (GUI) designed to make it easier for nontechnical people to use a computer

desktop publishing—Term coined in 1984 by the president of the Aldus Corporation to refer to the activity of using software to produce documents with elaborate control of the form and appearance of individual pages

digitized sound—Audible noises (e.g., music, voices) that are transferred to a computer storage medium by first coding them as numbers

directed instruction—A teaching and learning model based on behavioral and cognitive theories; students receive information from teachers and do teacher-directed activities (See also *constructivism.*)

disc—A CD-ROM or videodisc; refers to video storage media as opposed to text storage

disk—A computer storage medium on which data and programs are placed through a magnetic process

disk drive—A device in a computer system used to store data on floppy disks or microdisks and retrieve data

diskette (See *disk.*)

distance learning—Using some electronic means (e.g., modems, satellite transmissions) to make possible teaching and learning at separate sites

DOS (disk operating system)—The systems software that allows a computer to use applications programs such as word processing software; the operating system on IBM-type computers as opposed to those manufactured by Apple

dot-matrix printer—Output device that produces paper copy by placing patterns of dots on the paper to form letters.

download—To bring information to a computer from a network or from a computer to a disk

drill and practice—An instructional software function that presents items for students to work (usually one at a time) and gives feedback on correctness; designed to help users remember isolated facts or concepts and recall them quickly

electronic bulletin boards (See *bulletin boards.*)

electronic gradebook—Software designed to maintain and calculate student grades

electronic mail (e-mail)—Messages (e.g., letters or notes) sent via telecommunications from one person to one or more other people

Ethernet—Type of local area network (LAN) in which several data transmissions among network users can be sent at any given time (in contrast with token-passing networks, which allow only one transmission at a time)

Events of Instruction, Gagne's—The nine kinds of activities identified by learning theorist Robert Gagne as being involved in teaching and learning

expert systems—A form of artificial intelligence (AI) which attempts to computerize human expertise

export—To save all or part of a document in a format other than that in which the program created it (e.g., as a text file) so it can be used by another program

external storage—Devices outside the computer's internal circuitry to store information and/or data (in contrast with *internal memory*)

field—The smallest unit of information in a database

file—The product created by a database program; any collection of data stored on a computer medium

file server—In a local area network (LAN), the computer that houses the software and "serves" it to the attached workstations

File Transfer Protocol (See *FTP.*)

flat file database—Database program that creates single files (in contrast with *relational database*)

floppy disk (See *disk.*)

flowchart—A planning method that combines rectangles, diamonds, and other figures joined by arrows to show the steps of a problem solution in graphic form; used by programmers to show a program's logic and operation before it is coded into a computer language

font—A type style used in a document (e.g., Courier, Palatino, or Helvetica)

footer—A line in a document that can be set to repeat automatically at the bottom of each page; usually indicates a title and/or pagination (See also *header.*)

format—To prepare a disk to receive files on a computer; to initialize a disk; to design the appearance of a document (e.g., the font, type styles, type size)

FORTRAN (FORmula TRANslater)—One of the earliest high-level computer languages; designed for mathematical and science applications

FTP (File Transfer Protocol)—On the Internet, a way of transferring files from one computer to another using common settings and transmission procedures

full justification (also simply *justification*)—In word processing or desktop publishing, a type of text alignment in which text is both flush right on the right margin and flush left on the left margin (in contrast with right or left justified or centered)

fuzzy logic—A logic system in artificial intelligence (AI) designed to parallel the decision-making processes of humans by accounting for subjectivity and uncertainty

G (gigabyte)—Roughly 1 billion bytes of computer storage

GIGO (garbage in/garbage out)—Popular term meaning that if the data that go into a computer are faulty or badly organized, the result will also be inaccurate

gopher—On the Internet, a menu-based system designed to search for and retrieve files

gradebook (See *electronic gradebook.*)

grammar checker—The part of the word processing software designed to check text for compliance with grammar and usage rules

graphics tablet—Type of input device on which pictures are hand-drawn and transferred to the computer as files

graphic user interface (GUI, often pronounced "gooey")—Software that displays options to the user in graphic formats consisting of menus and icons, rather than text formats (e.g., the Macintosh and Windows desktops)

hacker—Computer user who demonstrates an unusual, obsessive interest in using the computer; a computer user who engages in unauthorized use of computer system

hard copy—A paper printout of a computer file from a printer, plotter, or other output device

hard disk (hard disk drive)—In a computer system, a secondary storage device, usually housed inside the computer, but can be external; holds from 20 megabytes to 1 or more gigabytes of information

hardware—The devices or equipment in a computer system (in contrast with *software* or *computer programs*)

header—A line in a document that can be set to repeat automatically at the top of each page; usually indicates a title and/or pagination (See also *footer*.)

high-level language—Computer programming language (e.g., *Fortran*, *Basic*, *Cobol*) designed with syntax and vocabulary like a human language so that less technical people can use it to write programs

hot spot (See *button*.)

HyperCard—Authoring software designed for Macintosh systems to create products called *stacks* which consist of a series of frames or *cards*

hypermedia—Software that connects elements of a computer system (e.g., text, movies, pictures, and other graphics) through hypertext links

hypertext—Text elements such as keywords that can be cross-referenced with other occurrences of the same word or with related concepts

icon—On a computer screen, a picture that acts as a symbol for an action or item

ILS (See *integrated learning system*.)

information highway—A popular term with various meanings associated with the worldwide linkage of information; sometimes synonymous with the Internet

import—To bring into a document a picture or all or part of another document that has been stored in another format (See also *export*.)

information service—A set of communications (e.g., e-mail) and storage/retrieval options made available by a company such as America Online or Prodigy

initialize (See *format*.)

ink-jet printer—Type of output device that produces hard copy by directing a controlled spray of ink onto a page to form characters

instructional game—Type of software function designed to increase motivation by adding game rules to a learning activity

input device—Any device in a computer system (e.g., keyboard, mouse) designed to get instructions or data from the user to the processing part of the system

I/O (input/output)—The process of getting instructions and/or data into and out of a computer system; devices that perform both functions (e.g., disk drives)

integrated circuit (See *chip*.)

integrated learning system (ILS)—A network that combines instructional and management software and usually offers a variety of instructional resources on several topics

integrated software packages—Software products that have several applications in a single package (e.g., word processing, database, spreadsheet, and drawing functions) (e.g., Microsoft Works and ClarisWorks)

intelligent CAI—Computer-assisted instruction with software logic based on artificial intelligence (AI) principles

interactive video—Videodiscs that allow the user to control the order and speed at which items from the disc are displayed on a screen (In Level I, control is through a bar code reader or remote control; in Level III, control is through a menu on a computer screen.)

interface—Cables, adaptors, or circuits that connect components of a computer system; the on-screen method a person employs to use a computer (See also *graphical user interface*.)

internal memory—Integrated circuits inside a computer system designed to hold information or programs; ROM and RAM (in contrast with *external storage*)

Internet—A worldwide network that connects many smaller networks with a common set of procedures (protocols) for sending and receiving information

interpreter—A computer program that converts program code into machine language and executes one statement of a source program before converting and executing the next (in contrast with *compiler*)

ISDN (Integrated Services Digital Network)—A digital telecommunications system in which all types of data (e.g., video, graphics, text) may be sent over the same lines at very high speeds

joystick—Input device, used primarily with games, that moves on-screen figures or a cursor

JPEG (Joint Photographic Experts Group)—A file format for storing and sending graphic images on a network

justification (See *full justification*.)

K (kilobyte)—A unit of computer memory or disk capacity that is roughly equivalent to 1,000 bytes

keyboard—Any of a variety of input devices that have keys imprinted with letters, numbers, and other symbols in order to allow a user to enter information

keyboarding—The act of using a computer keyboard to enter information into a computer system; sometimes used to mean efficient, ten-finger typing as opposed to hunt-and-peck typing

LAN (See *local area network.*)

laptop computer—Small, standalone, portable personal computer system

laserdisc (See *videodisc.*)

laser printer—An output device that produces hard copy by using a controlled laser beam to put characters on a page

LCD (liquid crystal display or diode) panel—A device consisting of light-sensitive material encased between two clear pieces of glass or plastic designed to be placed on an overhead projector; projects images from the computer screen to a large surface

light pen—Input device that allows a user to select items from a screen by sensing light on various points in the display

listserv—On the Internet, a program that stores and maintains mailing lists and allows a message to be sent to everyone on the list

local area network (LAN)—A series of computers connected through cabling or wireless methods to share programs through a central file server computer

Logo—A high-level programming language originally designed as an artificial intelligence (AI) language but later popularized by Seymour Papert as an environment to allow children to learn problem-solving behaviors and skills

Logowriter—A word processing program that incorporates logic and drawing capabilities of the Logo language

low-level language—Programming language such as assembler designed for use by technical personnel on a specific type of computer (See also *high-level language.*)

M (megabyte)—A unit of computer memory or disk capacity that is roughly equivalent to 1 million bytes

machine language—A computer language consisting of commands written in 1s and 0s; designed for a specific type of computer

mainframe—Type of computer system that has several peripheral devices connected to a CPU housed in a separate device; has more memory and capacity than a microcomputer

mark sense scanner—Input device that reads data from specially coded sheets marked with pencil

MBL (microcomputer-based laboratory)—A type of instructional software tool consisting of hardware devices

(probes) and software (probeware) to allow scientific data to be gathered and processed by a computer

MECC (See *Minnesota Education Computing Corporation.*)

memory—Circuitry inside a computer or media such as disks or CD-ROMs that allow programs or information to be stored for use by a computer

menu—List of on-screen options available on a specific topic or area of a program or GUI

metropolitan area network (MAN)—A network whose components are distributed over an area larger that a LAN but smaller than a WAN

microcomputer—Small, standalone computer system designed for use by one person at a time

MIDI (musical instrument digital interface)—A device and software that allows a computer to control music-producing devices (e.g., sequencers, synthesizers)

minicomputer—A type of computer system in a range between microcomputers and mainframes (In practice, minicomputers and mainframes are becoming indistinguishable.)

Minnesota Education Computing Corporation (MECC)—Originally established as the Minnesota Education Computing Consortium, one of the first organizations to develop and distribute instructional software for microcomputers

modem—A device that changes (MOdulates) digital computer signals into analog frequencies that can be sent over a telephone line to another computer and changes back (DEModulates) incoming signals into ones the computer can use

monitor—An output device that produces a visual display of what the computer produces (See also *CRT.*)

Mosaic—One of the first programs designed to allow Internet resources to be displayed graphically rather than just in text

mouse—Input device that a computer user moves around on the table beside the computer in order to control a pointer on a screen, and presses down (clicks) in order to select options from the screen

MPEG (Motion Picture Experts Group)—A file format for storing and sending video sequences on a network

MS-DOS (Microsoft Disk Operating System)—A type of systems software used on IBM and IBM-compatible computers

multimedia—A computer system or computer system product that incorporates text, sound, pictures/graphics, and/or video

network (See *local area network, metropolitan area network,* and *wide area network.*)

neural network—A type of AI program designed to work like a human brain and nervous system

node—One station or site on a computer network

online—Being connected to a computer system in operation

operating system—A type of software that controls system operation and allows the computer to recognize and process instructions from applications software such as word processing programs

optical character recognition (OCR) (See *mark sense scanner.*)

optical disc—Storage medium designed to be read by laser beam (e.g., CD-ROM, videodisc)

optical mark reader (See *mark sense scanner.*)

optical scanner—Input device that converts text and graphics into computer files

output device—Any device in a computer system (e.g., monitor, printer) that displays the products of the processing part of the system

pagination—Automatic page numbering done by a word processing or desktop publishing program

password—A word or number designed to limit access to a system to authorized users only

PDA (See *personal digital assistant.*)

peripherals—Any hardware devices outside the CPU

personal computer (PC) (See *microcomputer.*)

personal digital assistant (PDA)—A small, handheld computer that allows a user to write in freehand on a screen with a stylus and translates the writing into a computer file

photo CD—A compact disc format designed by the Eastman Kodak Company to store and display photographs

piracy (See *software piracy.*)

pitch—The number of characters printed per inch, usually 10-pitch or 12-pitch

pixel—The smallest unit of light that can be displayed on a computer screen

PLATO (Programmed Logic for Automatic Teaching Operations)—One of the earliest computer systems (mainframe-based) designed for instructional use; developed by the University of Illinois and Control Data Corporation

plotter—An output device designed to make a paper copy of a drawing or image from a computer screen

point size—A unit designating the height of a typeface character; 72 points = 1 inch

presentation software—Programs designed to allow people to display pictures and text to support their lectures or talks

primary storage (See *internal memory.*)

primitive—A simple command in the Logo programming language that does one operation (e.g., FORWARD)

printer—Output device that produces a paper copy of text and graphics from a computer screen

print graphics—Software designed to produce graphics on paper (e.g., cards, banners)

probeware (See MBL.)

problem-solving software—Instructional software function that either teaches specific steps for solving certain problems (e.g., math word problems) or helps the student learn general problem-solving behaviors for a class of problems

program (See *software.*)

prompt—On a computer screen, an indicator that the system is ready to accept input; can be any symbol from a C:\> to a ?

proportional spacing—Displaying text so that each character takes up a different amount of space depending on its width

public domain software—Uncopyrighted programs available for copying and use by the public at no cost

QuickTime—Program designed by Apple Computer Company to allow short movies or video sequences to be displayed on a computer screen

RAM (random access memory)—Type of internal computer memory that is erased when the power is turned off

record—In a database file, several related fields (e.g., all the information on one person)

relational database—Database program that can link several different files through common keyword fields

ROM (read-only memory)—Type of internal computer memory designed to hold programs permanently, even when the power is turned off

sans serif—Typefaces that have no small curves (serifs or "hands and feet") at the ends of the lines that make them up; usually used for short titles rather than the main text of a document

search and replace (also *find and replace*)—A function in a program such as a word processing package that looks for all instances of a sequence of characters and/or spaces and replaces them with another desired sequence

secondary storage (See *external storage.*)

shareware—Uncopyrighted software that anyone may use, but each user is asked to pay a voluntary fee to the designer

simulation—Type of software that models a real or imaginary system in order to teach the principles on which the system is based

snail mail—Regular postal service mail, as opposed to e-mail

software—Programs written in a computer language (in contrast with hardware)

software piracy—Illegally copying and using copyrighted software without buying it

speech synthesizer—An output device that produces spoken words through a computer program

spell checker—Part of a word processing or desktop publishing package that looks for misspelled words and offers correct spellings

spreadsheet—Software designed to store data (usually, but not always, numeric) by row–column positions known as *cells*; can also do calculations on the data

sprite—Object in some versions of the Logo programming language that the user can define by color, shape, and other characteristics and then use to do animations on the screen

structured query language (SQL)—A type of high-level language used to locate desired information from a relational database

supercomputer—The largest of the mainframe computer systems with the greatest storage space, speed, and power

synthesizer—Any of a series of output devices designed to produce sound, music, or speech through a computer program

SYSOP—The SYStem OPerator in a network; a person responsible for maintaining the software and activities on the network and assisting its users

systems software—Programs designed to manage the basic operations of a computer system (e.g., recognizing input devices, storing applications program commands)

telecommunications—Communications over a distance made possible by a computer and modem or a distance learning system such as broadcast TV

telecomputing—Term coined by Kearsley to refer specifically to communications involving computers and modems

teleconferencing—People at two or more sites holding a meeting through computers and telephone lines

test generator—Software designed to help teachers prepare and/or administer tests

thesaurus—In word processing and desktop publishing, an optional feature that offers synonyms or antonyms for given words

TICCIT (Time-shared, Interactive, Computer-Controlled Information Television)—An early instructional computing system developed by Brigham Young University and the Mitre Corporation that combined television with computers

touch screen—Type of input device designed to allow users to make selections by touching the monitor

Trojan horse—Type of virus that gets into a computer as part of a legitimate program

turtle—In the Logo programming language, the triangle-shaped object that can be programmed to move and/or draw on the screen

tutorial—Type of instructional software that offers a complete sequence of instruction on a given topic

Usenet group—On the Internet, one of a series of news groups that offer bulletin boards of information on a specific topic

variable—In programming languages, a name that stands for a place in computer memory that can hold one value

Veronica—On the Internet, a program that searches for files across gopher servers

videodisc (also *laserdisc*)—Storage medium designed for storing pictures, short video sequences, and movies

virtual reality—A computer-generated environment designed to provide a life-like simulation of actual settings; usually uses a data glove and/or headgear that covers the eyes

virus—A program written with the purpose of doing harm or mischief to programs, data, and/or hardware components of a computer system

voice recognition—The capability provided by a computer and program to respond predictably to speech commands

warm boot (See *boot.*)

wide area network (WAN)—An interconnected group of computers linked by modems and other technologies

window—A box in a graphic user interface that appears when one opens a disk or folder to display its contents

word processing—An applications software activity that uses the computer for typing and preparing documents

word wraparound—In word processing, the feature in which text automatically goes to the next line without the user pressing Return or Enter

worksheet—Another name besides *spreadsheet* for the product of a spreadsheet program

World Wide Web (WWW)—On the Internet, a system that connects sites through hypertext links

worm—A type of virus that eats its way through (i.e., destroys) data and programs on a computer system (See also *virus.*)

Index

Name Index

Atkinson, Richard, 59–60
Ausubel, David, 59–61

Becker, Henry Jay, 43, 106–107
Bereiter, Carl, 56, 65
Bitzer, Donald, 18
Bloom, Benjamin, 60, 74, 88
Brainerd, Paul, 157
Briggs, Leslie, 59, 62–64
Brown, Ann, 65
Brown, John Seely, 65, 69–70, 72
Bruner, Jerome, 65–66, 68
Bunderson, Victor, 18
Bush, Vannevar, 196

Campione, John, 65
Charp, Sylvia, 239
Clark, Richard E., 28, 73
Clinton, Janeen, 309
Cronbach, Lee, 59, 63

Deming, W. Edwards, 223
Dewey, John, 65

Fletcher, Dexter, 18

Gagne, Robert, 59–63, 74–75, 88–89
Glaser, Robert, 59

Harris, Judi, 229

Kasparov, Garry, 242

Luehrmann, Arthur, 19

Mager, Robert, 63
McCorduck, Paula, 28
Merrill, David, 59, 63, 75
Molnar, Andrew, 43

Nathan, Joe, 157
Norris, William C., 86

Papert, Seymour, 20, 31–32, 68–69, 71,
    112–113, 115–116
Perkins, D. N., 65, 70–73
Piaget, Jean, 65, 67–69

Reigeluth, Charles, 59, 63, 72, 94

Saettler, Paul, 5–6, 62–63
Scardamalia, Marlene, 65
Scriven, Michael, 59, 63
Skinner, B. F., 59–60
Spiro, Rand, 65, 70–71
Suppes, Patrick, 18, 20

Taylor, Robert P., 82
Tennyson, Robert, 58–59
Thorndike, Edward, 59

Vygotsky, Lev Semenovich, 65–66, 70–71

Wagner, Roger, 201
Weizenbaum, Joseph, 86

Subject Index

Accommodation, 67. *See also* Piaget, Jean
ACOT. *See* Apple Classrooms of Tomorrow
Adaptive devices, 44
    equity in providing, 44
AECT. *See* Association for Educational
    Communications and Technology
Americans with Disabilities Act (ADA). *See*
    Exceptional student education
Anchored instruction, 71–72, **349**. See also
    Cognition and Technology Group at
    Vanderbilt (CTGV)
Animation software, 172, 205, 301
Apple Classrooms of Tomorrow (ACOT), 214
Archie, 228, 349. *See also* Internet
ARPANet. *See* Internet
Artificial intelligence (AI), 178, 239–242, **349**
    applications in education, 241
    benefits, 242
    components, 239–240
    expert systems, 241, **351**

fuzzy logic, 240, **351**
    limitations, 242
    neural networks, 240–241, **354**
Art education, technology in, 203, 256,
    297–298, 300–302
    example lesson plans, 302–303
    Internet resources for, 301–302
Assessment, 72–76, 88, 204, 220, 250, 275
    authentic, 72
    changes brought about by distance
        learning, 220
    hypermedia role in, 204
    in science, 275
    preparing for, 88
    using PDAs in, 250
Assimilation, 67. *See also* Piaget, Jean
Association for Educational Communications
    and Technology (AECT), 6
Asynchronous transfer mode. *See* Distance
    learning

Atlases, on-disc, 169–170, 187, 207. *See also*
    Word atlases
Audio-based systems, 216. *See also* CD-ROM;
    Distance learning; Sound
Audioconferencing, 216
Audiographics, 216, 223
Augmentation technologies. *See* Exceptional
    student education
Authentic experiences, 71. *See also* Cognition
    and Technology Group at Vanderbilt
    (CTGV)
Authentic assessment, *See* Assessment
Authoring system, 19, **349**. *See also* Hypermedia
Automaticity, 74, 88, **349**

Bar-code generators, 161–162, 182–183, 280,
    290, 292–293, 304
    example lesson plans, 162, 292
    used with videodisc, 182–183, 280, 290,
        292–293, 304

357

About the Authors

M.D. Roblyer has been a technology-using teacher and a contributor to the field of educational technology for nearly 25 years. She began her exploration of technology's benefits for teaching in 1971 as a graduate student at one of the country's first successful instructional computer training sites, Pennsylvania State University, where she helped author tutorial literacy lessons in Coursewriter II on an IBM 1500 dedicated instructional mainframe. While obtaining a Ph.D. in Instructional Systems at Florida State University, she worked on several major courseware development and training projects with Control Data Corporation's PLATO system. After working as Instructional Technology Coordinator for the Florida Educational Computing Project (the predecessor of what is now the state's Office of Educational Technology), she became a private consultant, working for companies like Random House, Harcourt Brace Jovanovich, and the Apple Computer Company. In 1981 to 1982, she designed one of the early microcomputer software series, *Grammar Problems for Practice*, in conjunction with the Milliken Publishing Company.

She has written extensively and served as contributing editor for various research and popular publications on educational computing and educational technology. Her book with Castine and King, *Assessing the Impact of Computer-based Instruction: A Review of Research* (Haworth, 1988) is widely considered the most comprehensive review and meta-analysis of the effects of computer technology on learning ever written. She was one of the editors and authors for Macmillan Publishing Company's 1992 *Encyclopedia of Computers*.

Since 1984, she has been a Professor of Instructional Technology in Florida A&M University's College of Education. In addition to teaching graduate and undergraduate technology applications and research methods courses, she created and has served as editor of Florida's statewide educational technology journal, the *Florida Technology in Education Quarterly,* now in its eighth year of publication. She has also been principal investigator for a number of the university's funded technology research, development, and training projects such as Florida Educators Online, the Florida-England Connection Project, Teaching with Technology at the Elementary School Level—Technology Training for Teachers, and A Preservice Model for Integrating Technology into Teacher Education.

She is married to William R. Wiencke and the mother of a 7-year-old daughter, Paige.

Jack Edwards has been using educational technology in the classroom since 1988 when he was hired to teach gifted students at the Webster School in St. Augustine, Florida. In that same year, the Webster School was selected to be one of the Florida Department of Education's five Model Technology Schools. In 1990, Jack was one of 28 teachers from throughout Florida selected to spend the summer at the Florida Institute of Technology participating in the Florida Science Videodisc Project.

Mr. Edwards has trained thousands of Florida teachers over the past seven years. His training experience includes three years as a teacher-on-special-assignment with the

University of Central Florida's Instructional Technology Resource Center. During that stint he traveled throughout North Florida consulting with school districts and teachers on strategies for technology integration.

Mr. Edwards also served as the lead faculty member for instructional technology with the University of North Florida's First Coast Urban Academy from 1993 to 1995. This academy focused on initiating systemic change in seven inner-city elementary schools.

Mr. Edwards has worked with the University of North Florida, Florida A&M University, and the Miami Museum of Science to develop training materials for the Florida Department of Education's Bureau of Educational Technology. His credits include training manuals for interactive videodiscs and telecommunications.

Mr. Edwards is president of TechKnow Consulting in St. Augustine and was selected as a "Consultant that Works" in the International Society for Technology in Education's Fall 1993 edition of the *IRM Quarterly*. He is also president of the Florida Association for Computers in Education (FACE), a rapidly growing organization that annually cosponsors the Florida Instructional Technology Conference. In 1992, he was the first runner up for the Florida Instructional Technology Teacher of the Year. He lives in St. Augustine with his son, Jordan Burke.

Mary Anne Havriluk has worked in the field of education since beginning her career in 1972 as an instructor at a North Carolina junior college. After moving to Florida State University in the mid-1970s, she wrote and managed a number of federal and state grants that developed technology-based curricula for postsecondary vocational programs. She also has extensive experience in communications, working as a writer, editor, speech writer, publications coordinator, editorial manager, press secretary, and communications director.

She began her involvement with telecommunications technology in 1991 as the first director of Florida's distance learning program, working to develop a coherent distance learning network from the widely diverse telecommunications activities of Florida's public school districts, community colleges, and universities.

Ms. Havriluk is vice president for Network and Program Development for Distance Learning Associates, a New York-based firm that markets distance learning programs for educational developers. Her responsibilities include developing and delivering staff and inmate programs to correctional institutions, designing and implementing telecommunications networks for distance education, and securing video-based and computer-based programs needed by clients in the United States and other countries. She is currently on the board of directors of the U.S. Distance Learning Association.

Ms. Havriluk holds a Bachelor of Science degree in English/Education from East Carolina University and a Master of Science degree in Educational Administration from Florida State University. She resides in Tallahassee with her husband and youngest daughter.